TECHNICAL COMMUNICATION

About the Authors

Mary M. Lay is a professor of scientific and technical communication in the Department of Rhetoric and Director of the Center for Advanced Feminist Studies at the University of Minnesota. She is past president of the Association of Teachers of Technical Writing (ATTW) and is coeditor of the ATTW journal *Technical Communication Quarterly* and of *Collaborative Writing in Industry: Investigations in Theory and Practice* (Baywood, 1991), which won the 1991 NCTE Award for Excellence in Technical and Scientific Writing for Best Collection. She is author of "Interpersonal Conflict in Collaborative Writing: What We Can Learn from Gender Studies" (*Journal of Business and Technical Communication* 3.2, 1989), which won the 1990 NCTE Award for Excellence in Technical and Scientific Writing for Best Article on Philosophy or Theory of Technical Writing.

Billie J. Wahlstrom is professor and head of the Department of Rhetoric at the University of Minnesota. Her primary areas of research are the impact of technologies on communication practices and the role of gender in communication. Her work has appeared in a variety of publications, including *Computers and Composition, Journal of Business and Technical Communication,* and *Collegiate Microcomputer.* She is the coeditor of *Technical Communication Quarterly* and chairs the College Composition and Communication Committee on Technical Communication.

Stephen Doheny-Farina is an associate professor of technical communications at Clarkson University. His work has appeared in journals such as the *Journal of Business and Technical Communication, Technical Communication,* and *The Technical Writing Teacher.* One of his books, *Effective Documentation: What We Have Learned from Research* (MIT Press, 1989) was named Best Collection of Essays in the NCTE Awards for Technical and Scientific Communication for 1989. His most recent book *Rhetoric, Innovation, Technology* (MIT Press, 1992) is a study of the role of technical communicators in the transfer of technologies from research and development to the marketplace.

Ann Hill Duin is an associate professor of scientific and technical communication in the Department of Rhetoric at the University of Minnesota. In 1989 she won the EDUCOM award for her project titled "Collaborative Writing and Telecommunications'' in which technical communication students used courseware and a computer network to collaborate on technical documents. Her publications related to document design and collaboration have appeared in *The Technical Writing Teacher, Technical Communication Quarterly, Technical Communication,* and the *Journal of Business and Technical Communication.*

Sherry Burgus Little is an associate professor at San Diego State University where she has directed the Technical and Scientific Writing Program since its inception in 1982 and the Composition Program since 1989. Burgus Little has published in the *Journal of Business and Technical Communication,* the *Journal of Technical Writing and Communication, Computers and Composition, The Technical Writing Teacher, IEEE Transactions on Professional Communication,* and other journals and proceedings. She currently serves on the editorial board of the journal *Writing on the Edge,* is review editor of *Issues in Writing,* and is working on a book on the rhetoric of ethics in technical communication.

Carolyn D. Rude is a professor and Director of Technical Communication at Texas Tech University. Rude is editor of the ATTW anthology, *Teaching Technical Editing,* and author of *Technical Editing* (Wadsworth, 1991) as well as author of articles in the *Journal of Business and Technical Communication, Technical Communication,* and others. She is coeditor of the ATTW *Bulletin,* and she served as ATTW executive secretary-treasurer from 1986 to 1990.

Cynthia L. Selfe is a professor of composition and communication and head of the Humanities Department of Michigan Technological University. She is current coeditor of the *CCC Bibliography on Composition and Rhetoric.* She has served on the executive committee of the Conference on College Composition and Communication, and the Committee on Emerging Technologies for the Modern Language Association. Selfe currently serves as coeditor of *Computers and Composition.* Selfe is co-founder and current coeditor of Computers and Composition Press, which sponsors the publication of books on computers and their uses in English classroom.

Jack Selzer is an associate professor and teaches graduate and undergraduate courses in technical communication at Penn State University. A founder of the Penn State Conference on Rhetoric and Composition, a long-time director of composition programs at Penn State, and a recent president of the ATTW, he has published many essays on scientific and technical communication in *College Composition and Communication, Technical Communication, Written Communication, Writing in Non-academic Settings,* and other journals and books.

TECHNICAL COMMUNICATION

Mary M. Lay
University of Minnesota

Billie J. Wahlstrom
University of Minnesota

Stephen Doheny-Farina
University of Minnesota

Ann Hill Duin
San Diego State University

Sherry Burgus Little
Michigan Technological University

Carolyn D. Rude
Penn State University

Cynthia L. Selfe
Texas Tech University

Jack Selzer
Clarkson University

Chicago ● Bogotá ● Boston ● Buenos Aires ● Caracas
London ● Madrid ● Mexico City ● Sydney ● Toronto

IRWIN

© RICHARD D. IRWIN, INC., 1995

Senior sponsoring editor: Craig Beytien
Developmental editor: Lara Feinberg/Jennifer R. McBride
Marketing manager: Kurt Messersmith
Project editor: Lynne Basler
Production supervisor: Laurie Kersch
Designer: Laurie J. Entringer
Cover designer: Jamie O'Neal
Interior designer: Lesiak/Crampton Design: Cynthia Crampton
Cover illustrator: Brian Yen
Interior illustrator: Brian Yen
Art studio: ElectraGraphics, Inc.
Art coordinator: Heather D. Burbridge
Compositor: Bi-Comp, Inc.
Typeface: 10/12 Garamond
Printer: R. R. Donnelley & Sons Company

Library of Congress Cataloging-in-Publication Data

Technical communication / Mary M. Lay ... [et al.].
    p.   cm.
  Includes bibliographical references (p.   ) and index.
  ISBN 0-256-11985-6
    1. Communication of technical information.   I. Lay, Mary M.
  T10.5.T413   1995
    601.4—dc20   94-29205

Printed in the United States of America
  3 4 5 6 7 8 9 0   DO   1 0 9 8 7 6

Preface

The first question you will probably have about this text is "Why so many authors?" The answer to that question represents the nature of technical communication today.

Technical communication—the specialized communication that helps readers, viewers, or listeners respond to the challenges of a technological world—has become more complex, more socially and legally responsible, and more important in everyday life. Each of the eight authors of this textbook is an expert in some aspect of technical communication. That expertise comes from many years in the classroom and many years in industrial, research, and scholarly settings. We believe that by combining our voices, our experiences, and our knowledge we can bring you the latest communication strategies that will enable you to function in this complex technological world.

Some of you reading this book may become professional technical communicators, but almost all of you will become technical professionals who communicate. As a professional, you will need to know how to conduct research about technology, how to help your readers, viewers, and listeners—your audiences—use your information to solve problems. You will need to display technical data visually or present it orally, design and package effective documents, and use technology, particularly computers, to accomplish these tasks.

In addition, you must know what forms and formats are traditionally used on the job, how to choose an appropriate style for your communication, and how to analyze your audience's needs and interests. You will even have to know how to cast that audience into a receptive role as they approach your communication. In order to be effective in your job, you will need to know how to establish your credibility as an author, how to write as part of a collaborative team, how to help your audience make decisions, how to avoid endangering your audience, and how to meet many more communication challenges.

In this text, then, you receive the advice of eight teachers, scholars, and practitioners, and we have worked hard to make that advice reflect a growing body of knowledge about technology and communication.

Organization   *Technical Communication* has four parts. The chapters in Part One, "Understanding Technical Communication," show you how and where communica-
tion is situated in the workplace and help you understand the uses, creators,
and audiences of such communications.

- Chapter 1 focuses on the workplace, assessing how it is affected
  by increasing use of technology, by growing regard for how society affects and is affected by technical communication, and by increasing awareness of cultural diversity. This chapter also provides
  you with some of the history and special characteristics of technical communication.
- Chapter 2 teaches you how to plan your technical communications to solve workplace problems.
- Chapter 3 teaches you how to address your audiences effectively
  by thinking rhetorically and determining your audiences' needs, attitudes, and knowledge.
- Chapter 4 gets you started on the writing process—planning, organizing, drafting, and revising.
- Chapter 5 focuses on the persuasive nature of technical communication and how to establish your credibility, how to appeal to
  your audience's values, and how to provide good reasons for your
  argument.

The chapters in Part Two, "Acquiring the Tools of Technical Communication," give you the strategies for writing effective technical communication
in whatever situation you are called upon to do so.

- Chapter 6 introduces you to the nature of collaborative writing—
  producing documents as part of a writing team—and provides
  practical advice for maintaining effective interpersonal relationships in that team while you are initiating, executing, and presenting your document.
- Chapters 7 and 8 teach you research strategies for collecting and
  generating information, whether you use the library, conduct a survey, or navigate the Internet.
- Chapter 9 tells you how to evaluate, organize, summarize, and document the information you collect or create.
- Chapter 10 focuses on what stylistic choices to make in your technical communications and what to check, change, or correct as
  you revise, edit, and proofread your documents.
- Chapter 11 covers document design and packaging—everything
  from type styles to binding.
- Chapter 12 demonstrates how to display data visually in graphs, tables, charts, and drawings.

Part Three, "Creating Effective Documents," focuses on the typical forms and formats used to organize technical documents.

- Chapters 13 and 14 cover definitions, descriptions, instructions, specifications, and procedures—all traditional systems for organizing technical information.
- Chapters 15 and 16 focus on the longest and often most challenging type of technical communication—the report. These chapters will show you how to recognize and create the traditional types and parts of reports and how reports help readers make decisions, based on technical, managerial, and social criteria.
- Chapter 17 demonstrates how effective proposals, which set forth solutions to problems, can win not only approval but also funding for your ideas.

The final part of this text, "Developing and Maintaining a Professional Edge," concentrates on some common but challenging communication situations.

- Chapter 18 introduces you to the most personal and frequent type of workplace communication—correspondence.
- Chapter 19 takes those correspondence skills one step further, showing you how to apply what you've learned so far to the creation of a successful job search.
- Chapter 20 shows you how to apply much of what you have learned about written documents and visual displays to oral presentations.

Several features in this text make it useful and interesting for you.   **Special Features**

■ SOCIAL CONSTRUCTION

Many of the examples, assignments, and exercises in the text are socially situated. That is, they address not only technical but also social issues. We don't believe that either communication or technology is an isolated phenomenon; instead, we believe that they occur in a rich setting that is shaped by economics, ethics, legal considerations, and social and cultural forces. Therefore, the examples, assignments, and exercises we offer here assume that the technology is created and used by people with various values, interests, and needs—and that technical communicators can and should help audiences understand and use technology, make decisions about technology, and solve problems with technology. When technical communicators fail at their jobs, there are consequences to those failures; communicators can alienate, frustrate, or endanger their audiences.

■ COLLABORATIVE WRITING

One of our goals is to help you see that the need to communicate about technology springs from unique and describable workplace problems, values, and goals. In fact, in this text, we assume that you never write alone—whether you collaborate within a writing team or listen and respond to the needs of people in your organization, your communications are socially constructed.

■ CASE STUDIES

This text also contains a set of documents, upon which many exercises are based. These documents, collected in Appendix A, address both technical and social issues: the AIDS epidemic, liability and legal battles, environmental concerns, and public relations concerns. You'll read how community and medical leaders have responded, through technical communication, to health care workers and the general public's need to know about AIDS and HIV. You'll read about bioremediation, a process used in the *Exxon Valdez* oil spill. You'll see how one university responded to community concern about the number of birds trapped and killed in experimental agricultural fields. And, you'll see how poorly written technical documents have injured or killed users and involved corporations in legal battles.

■ TECHNOLOGY

Throughout this text, we'll teach you how to use technology to communicate. You'll see not only how to find information within computer databases and in the electronic environment of the Internet and how to use the computer to draft and revise your documents, but also how to use the computer software to display data, organize oral presentations, and even collaborate within a writing team.

■ PERSUASION

This text emphasizes the persuasive nature of technical communication. Thus, you'll not only learn how to convey technical information in a clear and concise manner, but you'll also learn how to persuade your audience that you have found solutions for problems, helped an audience make decisions, or deserve to be hired or funded.

■ DIVERSITY

Within this text, we also remind you about the culturally diverse and international nature of technical communication. We point out when to think about your audiences' level of literacy, how to meet the needs of color-blind or hearing impaired audiences, and when to consider an international audience's response to your words and symbols.

■ LEARNING AIDS

Finally, each chapter begins with a quotation from a technical communication scholar, teacher, or practitioner. These statements should help you think about and discuss some of the main principles within each chapter. Throughout the text, we have highlighted important or new words and defined them in the margin, so that you can easily review them. And, we have created Writing Strategy checklists and worksheets for you to use in creating documents now and in the future.

The Instructor's Resource Guide

The Instructor's Resource Guide to this textbook offers a range of supplemental activities and information to the new and experienced teacher. We provide three sample syllabi for the course and directions on what themes and issues can be stressed throughout a 10-week or 15-week term. We offer specific suggestions on how to integrate the technical, scientific, social, and ethical problems within the Appendix A cases. With each chapter, we give an overview, a set of goals, definitions of important terms and concepts, teaching strategies, and supplementary activities and assignments. When appropriate, we provide overhead transparency masters and handouts. Finally, we have reprinted several landmark or recent articles that inform current approaches to technical communication teaching.

Acknowledgments

We want to thank some special people who contributed to either our collective or individual efforts. Linda Jorn, University of Minnesota, shared with us the AIDS documents that appear in this book, and Linda Van Buskirk, Cornell University, alerted us to and helped us gather the *Exxon Valdez* documents. Laurie Gardner, University of Minnesota, helped gather information, document research sources, and produce the Instructor's Resource Guide. Gayle Berry, Clarkson University, advised us on the library research process. Lise Hansen, University of Minnesota, helped coordinate many of our research, design, and planning activities. Patricia Goubil-Gambrell, Texas Tech University, alerted us to the progress report and proposal that appear in Chapters 15 and 17. Andrew Stephenson, at Penn State, offered information on scientific writing and gave permission to use and revise his work, and Gay Gragson enhanced our conversations about audience.

We thank the entire Irwin staff for their help in completing this challenging project. In particular, we thank Craig Beytien for getting all of us together in the first place, Lara Feinberg for helping us maintain a clear vision and a positive attitude, and Lynne Basler for bringing the entire project to completion.

The teachers and scholars who reviewed the text in its various stages provided valuable suggestions and reactions, and at times they even generously offered some of the words, ideas, and opinions that appear in this text:

Jo Allen, East Carolina University

Dennis Barbour, Purdue University Calumet

Steven Bernhardt, New Mexico State University

Virginia Book, University of Nebraska/Lincoln

Sam Geonetta, University of Cincinnati

Hillary Hart, University of Texas/Austin

David Helgeson, British Columbia Institute of Technology

Dan Jones, University of Central Florida

David Kaufer, Carnegie Mellon University

Charles Kostelnick, Iowa State University

Marilee Long, Colorado State University

Carolyn Miller, North Carolina State University

Carolyn Plumb, University of Washington

Kathy Underwood, University of Washington

We are thankful to these people for their assistance. Any errors that remain are ours alone.

Finally, we thank our students, our colleagues, our friends, and our families who continue to give us insight, patience, support, and love.

Mary M. Lay
Billie J. Wahlstrom
Ann Hill Duin
Sherry Burgus Little
Cindy Selfe
Jack Selzer
Carolyn Rude
Steve Doheny-Farina

Brief Contents

xiii

Contents

Understanding the Nature
of Technical Communication

Technical Communication in the Workplace

The History of Technical Communication ▌ The Definition of Technical Communication ▌ Technical Communication as a Socially Situated Activity ▌ Special Characteristics of Technical Communication ▌ The Changing Nature of the Workplace

The important thing is to get started. I'll often put a piece of paper in the typewriter and decide the paper doesn't want to be written on, so I'll throw it away and take another piece of paper.

Russell Baker, *Growing Up* (New York: Plume, 1983).

Introduction

communication skills include reading, speaking, listening, writing, and designing visual displays

professional technical communicators mainly write documents, give oral presentations, and create visual displays of technical information

technical professionals who communicate mainly create, design, analyze, repair, and test technical devices and systems, but must communicate about these products

This is a book about communicating in a technological world. In particular, this book teaches you to create specialized kinds of technical communication for business, industrial, governmental, and educational settings. The skills and insights you gain here will make it easier for you to succeed at work, and they will enable you to be a more knowledgeable citizen.

These may seem like large promises to make for a textbook in technical communication, and in one sense they are. However, good **communication skills** help you deal more easily with a variety of audiences, increasing your chances of being understood. They enable you to understand your world and deal effectively with change. Perhaps most important, good communication skills give you the power to shape your environment, to facilitate and direct change.

Some of you reading this book might become **professional technical communicators,** but almost all of you will become **technical professionals who communicate.** If you are planning to become an engineer, you will find that engineers are often responsible for writing feasibility studies for such technological changes as new buildings, electrical systems, and computer networks. If you are majoring in computer science, you will have to write specifications for hardware and software. If your goal is to manage the technical products division of an industrial corporation, you will find that through written documents, lectures, and videotapes you must teach technicians how to install products and teach customers how to use and adapt those products.

Whatever your technical profession, a significant part of your time, even more than you spend now in school, will be spent communicating. In a survey of 1,400 members of the American Institute of Chemists, the American Consulting Engineers Council, the International City Management Association, the Modern Language Association, the American Psychological Association, the Professional Services Management Association, and the Society for Technical Communication, respondents said they spent 44 percent of their professional time in some kind of writing activity (this includes brainstorming, note-taking, organizational planning, drafting, revising, and editing) (Lunsford and Ede). If you intend to be one of the more successful practitioners of your profession, you can count on spending even more time than the average practitioner engaged in communication activities.

If, instead, you become a professional technical communicator, you will join a growing number employed in business and industry. In 1990, the US Bureau of Labor Statistics classified approximately 70,000 workers as technical writers, practitioners in a field that has only been recognized officially as a career since the 1950s. Because the age we live in depends highly on technology, your work is particularly important, and you will find that more and more fields are coming to recognize the importance of having professional technical communicators on their staffs.

As you work your way through this book, you'll be learning about **technical communication,** regardless of your eventual career goal. Therefore, we'll refer to you and to the authors of the examples we show you as "technical communicators."

In this chapter, we first introduce you to the practice of technical communication as it existed centuries ago. We then define technical communication and distinguish between it and other kinds of communication and make clear the relationship between technical communication and technical writing. Finally, we explain how technical communication is a social activity and list other special characteristics.

technical communication
applied communication that makes technical information accessible to many readers, viewers, and listeners

People have always needed to describe phenomena in the natural environment so that specialists and nonspecialists could understand them. And for nearly two thousand years, teachers and philosophers have addressed the problems associated with communicating complicated information accurately and effectively. This book will teach you both the theory and practice of technical communication—the "why" as well as the "how" of effective communication about technical issues and concerns.

The practice of technical communication has a very long history. Although modern technical communication began during World War II and grew out of a need for manuals and instructions to explain complicated equipment to novice users, technical communication has its roots in the past centuries. Some of the theories behind modern technical communication come from the most famous of communication philosophers, **Aristotle.** In his *Rhetoric,* Aristotle discussed various ways to create effective messages (Example 1–1). His emphasis on **persuasion** and **argument** influences

The History of Technical Communication

Aristotle
wrote *Rhetoric* around 330 B.C. to describe the art of creating the most persuasive message on any subject

persuasion and argument
persuasion motivates a reader or listener, while argument is the arrangement of persuasive points

---

EXAMPLE 1–1   Aristotle from *Rhetoric*

Rhetoric may be defined as the faculty of observing in any given case the available means of persuasion. This is not a function of any other art. Every other art can instruct or persuade about its own particular subject matter; for instance, medicine about what is healthy and unhealthy, geometry about the properties of magnitudes, arithmetic about numbers, and the same is true of the other arts and sciences. But rhetoric we look upon as the power of observing the means of persuasion on almost any subject presented to us; and that is why we say that, in its technical character it is not concerned with any special or definite class of subjects.

Source: Aristotle, *Rhetoric, Book I. The Rhetorical Tradition: Readings from Classical Times to the Present.* Ed. Patricia Bizzell and Bruce Herzberg (Boston: Bedford Books, 1990), p. 153.

what we have to say about creating technical communication. Aristotle was concerned with creating effective messages with whatever means of persuasion were avilable. Today, technical communicators have a range of means—video, print, slides, audio, multimedia—that Aristotle could not imagine, but their primary concern is still the creation of effective messages.

Sextus Julius Frontinus
Roman author whose writing about aqueducts is an early example of technical communication

Sextus Julius Frontinus is generally recognized as having written one of the first pieces of technical communication in approximately 97 A.D. Frontinus wrote a manual for building and maintaining aqueducts while working as the water commissioner in Rome (Example 1-2). Then, as now, government relied on the work of technical communicators to provide the information necessary to maintain a civilization's infrastructure.

Technical communication has changed over the years from the time of Frontinus, but in every age it has aided people in making sense of technology, and it has often helped them apply scientific principles to the development of technology. For example, during the **English Renaissance, 1475–1640,** technical manuals helped people do everything from preparing the ground for planting to curing diseases. In Example 1-3, you can see how Reginald Scot in 1574 described how to cultivate a hop garden. In Scot's time, instructions were just as important as they are today; then, as now, crops could fail. Today, as well, the malfunction of a small part like an O ring can lead to the explosion of a space shuttle and the death of its crew. Contemporary technical communicators create the manuals that help us run everything from nuclear submarines to our home computers.

English Renaissance
1475-1640 A.D., a time when technical writing taught people everything from surgery to planting

The Definition of Technical Communication

Technical communication is a specialized form of communication. It differs from other forms of communication you may already have studied in terms of its **audience** (its listeners, viewers, or readers), purpose, style, format,

---

EXAMPLE 1–2   *Frontinus from Aqueducts of Rome, II*

[Aqueduct] Claudia, flowing more abundantly than the others, is especially exposed to depredation. In the records it is credited with only 2,855 quinariae, although I found at the intake 4,607 quinariae—1,752 quinariae more than recorded. Our gauging, however, is confirmed by the fact that at the seventh mile-stone from the City, at the settling reservoir, where the gauging is without question, we find 3,312 quinariae— 457 more than are recorded, although, before reaching the reservoir, not only are deliveries made, to satisfy private grants, but also, as we detected, a great deal is taken secretly, and therefore 1,295 quinariae less are found than there really ought to be.

Source: Frontinus, *The Stratagems and the Aqueducts of Rome,* vol. II. Ed. Mary B. McElwain. Transl. Charles E. Bennett (Cambridge: Harvard UP, 1930), 70–72.

EXAMPLE 1–3   A Perfite Platforme of a Hoppe Garden

22   *A perfite platforme*

This figure
or roume is
more than
halfe filled
with Poales,
and not filled
up to the top.

one ende of the roume
woulde be full before
the other, whereas
nowe they shall lye e=
uen and sharpe a=
boue, lyke an Haye=
stacke, or the ridge of
an house, and suffi=
ciently defende them=
selues from the wea=
ther.
    If you thinke that
you haue not Poales
ynowe to fylle the
roume, pull downe
the Wyths or bandes
lower, & your roume
will be lesse.

¶ *Of tying of Hoppes to the*
Poales.

When your Hoppes are growne about
one or twoo foote high, bynde vp (with a
Rushe or a Grasse) such as declyne from
the Poales, wynding them as often about the
same Poales as you can, and directing them al=
wayes according to the course of the Sunne, but
doe it not in the morning when the dewe remay=
neth vpon them, if your leysure may serue to doe
it at any other time of the day.
                                                                    If

of a Hoppegarden   23

If you laye softe græne Rushes abroade in
the dewe and the Sunne, within twoo or thre
dayes, they will be lythie, tough, and handsome
for this purpose of tying, which may not be fore=

slowed, for it is most certaine that the Hoppe that
lyeth long vpon the grounde before he be tyed to
the Poale, prospereth nothing so wel as it, which
sooner attayneth therevnto.

¶ *Of hylling and hylles.*

Nowe you must begyn to make your hils,
and for the better doing therof, you must
prepare a toole of Iron fashioned some=
what like to a Coopers Addes, but not so much
bowing, and therfore lykest to the netherpart of
a Showell, the powle wherof must be made with
a round hole to receyue a helue, lyke to the helue
of a Mattocke, and in the powle also a narle
hole must be made, to fasten it to the helue.
                                                                    This

This toole is
here aboue
better pro=
portioned,
than that on
the other side
following.

Source: Reginald Scot, *A Perfite Platforme of a Hoppe Garden*, 1574 (Amsterdam: Theatrvm Orbis Terrarvm, Ltd., 1973). The English Experience 620. Short Title Catalog 21865. As discussed in Elizabeth Tebeaux and M. Jimmie Killingsworth, "Expanding and Redirecting Historical Research in Technical Writing: In Search of Our Past," *Technical Communication Quarterly* 1.2 (1992), 5–32.

situation, and content. Technical communication is applied communication, communication designed to accomplish specific tasks or to help the audience solve specific problems. That task may be to inform users about an update in computer software, instruct line workers how to produce a new product, or to warn customers about unsafe ways to operate a machine. Or, you might help readers solve such problems as repairing a copying machine, selecting a home computer, or choosing the most effective method to clean up an oil spill. To help you understand the special nature of technical communication, we first distinguish between technical and nontechnical communication and then between scientific communication, science com-munication, and technical communication.

audience
readers, listeners, and
viewers

TECHNICAL AND NONTECHNICAL COMMUNICATION

Technical communication differs from other kinds of communication with which you might be familiar, such as the creative or expository writing you might have done in other courses. Technical communication shares with all communication its emphasis on conveying information so that it can be understood. It differs from other kinds of communication in its emphasis on clarity, accuracy, conciseness, readability, and legibility.

Technical communication is usually put to some use rather than enjoyed for its own sake. For example, a reader might follow an instruction manual in learning about a new piece of computer software, a technical report in choosing a model of microwave, or a governmental report in deciding how

---

EXAMPLE 1–4   Technical and Nontechnical Communication

Nontechnical Communication—Poetry

> Revolution is the Pod
> Systems rattle from
> When the Winds of Will are
>   stirred
> Excellent is Bloom
>
> But except its Russet Base
> Every Summer be
> The Entomber of itself,
> So of Liberty—
>
> Left inactive on the Stalk
> As its Purple fled
> Revolution shakes it for
> Test if it be dead.
>
> Emily Dickinson, 1866

Source: Emily Dickinson, *The Complete Poems of Emily Dickinson*, Ed. Thomas H. Johnson. Copyright 1929 by Martha Dickinson Bianchi; Copyright © renewed 1957 by Mary L. Hampson. By permission of Little, Brown and Company.

Technical Communication—Directions

Flower bulbs will do well in any good, well-drained soil. Plant them, with top up, at the depth appropriate to their type as indicated in the chart below.

Tulips	15 cm
Daffodils and hyacinths	10 cm
Crocuses	5 cm

to vote on the creation of a new incinerator. Examine the two passages in Example 1–4. What differences do you see between them in the type of audience, **purpose,** approach, vocabulary, sentence structure, and organization?

Technical communication also differs from other forms of communication in the ways it is practiced. For example, technical communication is far more often a **collaborative activity** than are other kinds of communication. Several people—writers, editors, technical experts, graphic artists—may work together to create one technical document.

Finally, technical communication is most often produced within an organization that has identifiable goals, customs, and values, and each technical document that an organization creates potentially reinforces or changes its corporate image and practices. For example, technical communicators at the Bank of Canada not only research and analyze specific financial questions, but they also help formulate Canadian monetary policy by making recommendations (Smart).

TECHNICAL COMMUNICATION, SCIENCE COMMUNICATION, AND SCIENTIFIC COMMUNICATION

You may hear the terms *technical communication, science communication,* and *scientific communication* used interchangeably, but these terms indicate distinct activities. Although technical communication, science communication, and scientific communication have much in common, they have important distinctions in audience and use. Frequently, the aim of **scientific communication** is to present an informed lay audience with information about the physical world. Example 1–5 shows scientific communication written for an intelligent lay audience—a description of an aspect of chaos theory dealing with fractals. Benoit Mandelbrot, who coined the term *fractal*—which can be both a noun and an adjective—to describe a particular geometry, was trying to calculate the fractional dimensions of real objects, such as shorelines, that have irregular shapes. This example describes in rather general terms an application of scientific principles to a real-world problem, the occurrence of errors in an electronic transmission.

Scholarly papers written by physicists and mathematicians for their peers are examples of **science communication.** Mandelbrot's "On Fractal Geometry and a Few of the Mathematical Questions It Has Raised," which he presented to the International Congress of Mathematicians in 1983, is an example of science communication designed for expert, peer audiences. Science communication is the discourse that occurs within the discipline, and it is used by scientists and mathematicians to present their research hypotheses and findings to their peers.

Technical communication, on the other hand, helps a reader make use of technology. In Example 1–6, the technical communicator teaches the

purpose
why a communication is needed and what it is supposed to accomplish

collaboration
many people with different skills working together on a task or communication

scientific communication
information about the physical world directed toward an informed lay audience

Mandelbrot's fractals
invented to describe a particular geometry

science communication
scholarly papers written for scientific professionals

EXAMPLE 1–5   A Sample of Scientific Communication

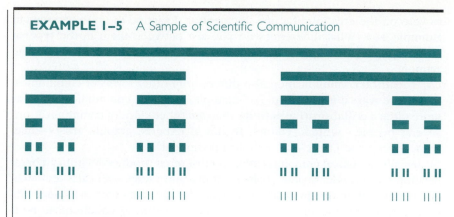

The Cantor Dust

Begin with a line; remove the middle third; then remove the middle third of the remaining segments; and so on. The Cantor set is the dust of points that remains. They are infinitely many, but their total length is 0.

The paradoxical qualities of such constructions disturbed nineteenth-century mathematics, but Mandelbrot saw the Cantor set as a model for the occurrence of errors in an electronic transmission line. Engineers saw periods of error-free transmission, mixed with periods when errors would come in bursts. Looked at more closely, the bursts, too, contained error-free periods within them. And so on—it was an example of fractal time. At every time scale, from hours to seconds, Mandelbrot discovered that the relationship of errors to clean transmission remained constant. Such dusts, he contended, are indispensable in modeling intermittency.

Source: Illustration: Benoit B. Mandelbrot, *The Fractal Geometry of Nature* (New York: W. H. Freeman and Co., 1982). By permission of the author. Text: James Gleick, *Chaos: Making a New Science.* (New York: Viking, 1987), 93.

EXAMPLE 1–6   A Sample of Technical Communication

Instructions on Printing a File Using a Macintosh Computer

To print a file that you have edited, use the Print command from the File menu. When you use this command, you will see a dialog box. Simply click on the Print button to make a single copy of your file.

If you wish to make more than one copy, delete the "1" in the box to the right of "Copies," and type in the desired number.

If you wish to make a copy of just a few pages of your file, in the box to the right of "From," type the first page number of the section to be printed. In the box to the right of "To," type the last page number of the section to be printed. If you wish to print just one page of a file, type that page number in both the "From" and the "To" box.

computer user how to print a document. The reader relies on the technical communicator to give complete and accurate instructions so that the reader can understand, follow, and execute the needed actions. You can see in this example that the technical communicator tried to imagine all the options a reader might require.

TECHNICAL COMMUNICATION AND TECHNICAL WRITING

Technical writing is a subset of technical communication. When communicating about complex scientific or technical information, people use more than written documents. The field of technical communication covers this whole range of communication activities. Although we focus primarily on written communication in this book because it makes up a substantial portion of the communication you will be responsible for at work, technical communication includes oral presentations. Moreover, each written document is *designed*—the words and data are visually arranged on the page. Finally, now that much information is transferred directly from one computer to another and does not appear in print, technical communication includes online, or on-screen, documentation as well.

technical writing
a subset of technical communication

In this book, we generally refer to technical communication when we discuss principles and practices that help you create a communication using any medium for any audience. However, when we discuss specific types of technical communication—for example, reports, interviews, persuasive speeches, or letters—we'll narrow our terms to technical writing for *readers, oral presentations* for *listeners,* and so on.

Defining technical communication is difficult, and some experts in the field resist broad definitions that include practically every type of manual or process description, including cookbooks. Other experts resist narrow definitions that exclude any communication whose main topic is not technology. For our purposes, technical communication conveys complex technical information in easily understandable forms, often to inexperienced users or to lay audiences. Although at times users might be experienced and educated in the subject of the technical document, they cannot share completely the writer's or speaker's education and knowledge of the subject of the document.

Technical communication conveys information and creates understanding—it does not convey meaning. To a large extent, meaning rests in each person. Whether or not what is written is meaningful depends on the needs of the person reading it. As we saw in Example 1-6, the writer helps the reader understand how to print, but if the reader doesn't have a printer or doesn't wish to print a file, then the passage lacks meaning. That doesn't imply that a reader doesn't understand what has been written. Rather, because meaning is individual, it is something that cannot be controlled completely by the writer. Therefore, you will need to concentrate your energy on creating documents that are understandable. You must work to establish a common basis of understanding with your audience based on vocabulary, experience, values, and education, so that your information is accessible.

Technical Communication as a Socially Situated Activity

social contexts
political, economic, cultural, and organizational settings with particular values, goals, and communication styles

Exxon Valdez
tanker that ran aground on March 24, 1989, and released about 11 million gallons of crude oil into Prince William Sound and the Gulf of Alaska

product liability laws
define the responsibility of companies (including their writers) to protect consumers

Technical communication doesn't occur in a vacuum. The work you will do often requires a sophisticated understanding of the **social contexts** within which communication takes place. These contexts involve legal, ethical, and social responsibilities involved in the communication of technical information. You may have heard about some of these examples of technical communication contexts: the reports and memoranda involved in the *Challenger* disaster, the reports and news stories documenting the cleanup efforts of the **Exxon Valdez** Alaskan oil spill, the written policies and news stories growing out of recent research on recombinant DNA, and the studies and reports associated with automobile emissions standards. Increasingly, these cases indicate the need for communicators who not only have clear competencies with language, but who also have a sense of what one could call "public responsibility" and "civic courage" (Giroux).

With growing frequency, technical communicators are realizing the legal and moral implications inherent in their work. **Product liability laws** now consider documentation to be an integral component of a given product. Consequently, if the documentation is faulty and someone is hurt or loses money as a result, then the corporation (and ultimately the person who prepared the documents) is responsible. Similarly, in cases such as the *Challenger* disaster, technical communicators have to ask to what extent they are responsible for the communication failures that permitted the launch of a spacecraft under dangerous conditions.

Our growing understanding of the social contexts within which technical communication takes place has changed how we think about this subject. Two decades ago, you would have learned about technical communication only as a series of formats that could be memorized and mastered—memoranda, feasibility reports, lab reports. Now you also must understand technical communication as it functions within complex organizational and cultural contexts. You need to understand the values and goals of the organization in which you work, for example, and how your communications reflect or can influence those values and goals.

Technical documents are written and read for a wide range of purposes: for informing experts within an organization and persuading nonexperts in cooperating firms; for passing along information to managers and for educating line workers; for describing products to clients and for limiting the liability of a corporate entity should a legal claim be filed; for communicating with regulatory groups; and for representing an organization's work to governmental agencies. Technical communicators work within organizations that have a variety of goals and values, which shape not only what is communicated but also how it is communicated.

The communication you will do on the job will differ significantly from what you may have learned during your college career. Some of the differences will result from having different purposes and different audiences for the communications you create. Other differences will grow from the nature of technical communication itself and from the conditions under which you will be expected to create effective communications.

Special Characteristics of Technical Communication

READER-BASED ORIENTATION

One characteristic of effective technical communication is its **reader-based orientation.** In school you have probably written papers with a keen awareness that your teacher would be reading them. Both what and how you wrote were shaped by that awareness. In other words, what you wrote was reader-based in its orientation. In contrast, writer-based prose, such as journal and creative writing, is characterized by self-expression of feelings, experiences, and emotions, much of which, in fact, is not intended to be read at all.

reader-based orientation creating communications to meet an audience's needs and interests

When you communicate technically, you won't be addressing the teacher, and often you will not even be addressing your boss. You may be addressing an audience of users whom you do not know personally. The reader-based orientation of technical communication helps you to write, speak, and design not from the point of view of what you know but from the point of view of what frequently inexperienced or beginning users need to know. Notice how the second part of Example 1–4 is aimed directly at the readers, telling them what they need to know about planting bulbs.

FOCUS ON SUBJECT

Technical communication is also distinct because of its **focus on subject.** In technical communication, the main subject or topic of the communication determines the kind of information and the form in which it must be conveyed. In creative writing, one can focus on an object, of course, but that focus doesn't necessarily shape the form of the creative piece. For instance, a poet writing about a flower would not have to organize the poem by having a stanza on petals, another on leaves, and a third on color. Instead, the intent of the poem is to evoke an emotional response rather than to describe function or organization.

focus on subject main subject determines information and the form it takes

With this distinction in mind, look again at Example 1-4. You can see that although the poem is about a flower, the poet's focus is the flower's symbolic meaning. By contrast, in instructions below the poem, the writer's intent is to describe a process. Technical communication focuses on subject in this way because its purpose is to instruct or describe rather than to evoke images or emotions in the reader.

CORPORATE REPRESENTATION

organizational culture
corporate sense of identity and value systems

When you are creating documents at work, you are creating them as a representative of the business or corporation that hired you. All corporations have their own **organizational cultures,** their sense of who they are and what is important to them. Corporations have value systems that cover everything from dress codes to ways to create documents. For example, some organizations require that all managers use the same time management system and carry the same kind of planning calendar. Communications within some organizations must reflect corporate philosophy, such as safety, open-door management, outreach to the public, environmental awareness, and such. For example, communications may be printed in the same color and illustrations always placed in the center of the page, as documents may be created by similar teams of writers, editors, product developers, and graphic artists. Because what you create will be read or viewed by people both inside and outside the company, it will need to represent the company in keeping with its corporate culture.

Generally speaking, developing an awareness of yourself as representing the corporation through your documents takes a while. Sometimes technical communicators believe that thinking of themselves as corporate representatives will stifle all of their individuality. As an effective communicator, you will need to remember that your writing has the potential to change the company just as the company has the potential to affect your writing. Your ability to create change within your corporate workplace is one of the things that makes developing your communication skills so important.

COLLABORATION

collaborative groups or teams
people with special skills, knowledge, or experience who work on a common task or communication

As we noted earlier, much communication, particularly writing, in business, industry, government, and education is collaborative. Because people who convey information about technology have special skills, specific education and experience, or special knowledge, the most effective documents are often produced by **collaborative groups** or **teams.** For example, a computer manual might require the input of the software developer, the systems analyst, the technical communicator, the document manager, and the graphics specialist. One person cannot know and write it all, and many people in an organization will review your work. In fact, your organization's name rather than your own may appear on the final product.

Effective collaboration means listening as well as stating your point, understanding others' points of view as well as expressing your own, and working and writing together as well as on your own. Technology, particularly computers, makes collaboration easier in terms of sending and editing documents, but the computer can make it harder to ensure that interpersonal relations build and grow as they can during face-to-face communication.

STYLE AND ORGANIZATION IN TECHNICAL COMMUNICATION

You'll find that because of all the special characteristics of technical communication we have been discussing, technical communication has unique features of **style** and organization. For example, as a technical communicator, you'll probably reveal the topic and your position on that topic at the beginning of a document. Your reader will have a preview of what's to come in the document, and some readers might elect not to read the document but pass it on to others. Much of the writing that you have done in the past was probably organized inductively—you described or listed ideas and evidence and then came to a conclusion. Technical communication uses deductive organization a great deal—you reveal your conclusions and then demonstrate how you reached them.

style
communicator's choices about tone, point of view, level of formality, use of figures of speech, voice, and emphasis

As a technical communicator, you'll also need to know how to communicate directly and concisely so that your audience can easily and safely make use of your ideas. For example, in the past, many people found technical writing very formal and distant, an impression created somewhat by an excessive use of the passive voice. The reader of a technical document encountered passages like the following:

> Explanation was needed for the striking difference in performance between KN-211 subsurface and KN-135 subsurface. There was a need to find the cause of such difference in performance when the starting oil compositions appeared similar.

Most technical communicators now try to write more directly and personally, as in the rewritten example below:

> The striking difference in performance between KN–211 subsurface and KN–135 subsurface clearly needed explanation. What caused such difference in performance when the starting oil compositions appeared similar? (Bragg et al. 59)

These rewritten sentences are much easier to read, rely more on the active voice, and engage the reader more effectively.

Even when describing a product or process or when recording data, you need to persuade your audience that the description or record is complete and appropriate. As a technical communicator, you'll begin and end with conclusions and recommendations, and you will use the discussion segments of your document to convince your reader to accept and act upon your conclusions and recommendations.

VISUAL DISPLAY IN TECHNICAL COMMUNICATION

visual display
design or presentation of
words or data on a page

As a technical communicator, you'll rely on **visual display** of words and data to help your audience understand the subject. Collaborating with a graphic artist or designer and using computer software to generate displays, you can help your audience follow your communication. Example 1–7 forces the reader to sort through information. Example 1–8 uses a deliberate visual display to help the reader understand and act on the information.

A reader relies on visual cues to process information. In creating your own technical documents, you can use visual display to group or chunk information into different units. These chunks are often separated by white space or marked by visual cues such as bullets or numbers. In Example 1–8, different types of material—poetry, prose, illustration, and special works—

EXAMPLE 1–7   A Document Lacking Attention to Visual Display

Subject: Potential Copyrighting Infringement Relative to Course Materials

I am sure that you have received some word that we need to update our copy guidelines for course material. A number of faculty members require that students purchase copies of essays that aren't available in textbooks. Sometimes faculty members just want students to read one essay or example out of a book and don't want to require that students buy the entire book. Ever since the much publicized Kinko's Graphic Corporation court case, we have worried about our own on-campus copying service violating copyright laws. We need to make sure that all faculty members who have students buy duplicated or copied articles do not violate the "fair-use" guidelines. Under those "fair-use" guidelines, you can copy a portion of the copyrighted material without copyright infringement if you are using the copies of the work for a nonprofit educational purpose, or if the time at which you were inspired to use the material and the time when you needed it were so close that you did not have time to request permission to copy it.

However, there are some limitations on these guidelines. For example, if you are copying poetry, you can't copy more than 250 words or two pages. But if you are copying prose, you can't copy a complete article, story, or essay if it is over 2500 words. If it is more than 2500 words, you can copy up to 10% or a maximum of 1000 words. If you are copying illustrations, such as a chart, diagram, drawing, and such, you can copy only one per book or per periodical issue. Finally, if you are copying from a work that combines text and illustrations (like a children's book or poetry), you can't copy more than 10% of the words of the text. If you want to copy more, then you have to ask for permission.

Please learn these guidelines because, as of December 15, our campus copy center will be following them.

EXAMPLE 1–8 Document with Attention to Visual Display

October 20, 1995

TO: All Faculty Members
FROM: Vice President Kofflin
RE: **Copyright and Fair Use Guidelines for Course Material—**
Effective December 15, 1995

On December 15, 1995, the following guidelines will be in effect for the campus copying center. All faculty members must follow these guidelines when requesting that the campus copying center duplicate and sell course material to students. Failure to follow these guidelines could result in litigation between the University, the faculty members, and the copyright holder (for example, Kinko's Graphic Corporation court case).

Fair Use Guidelines: You may copy a portion of a copyrighted work without copyright infringement if:

- You are using your copies of the work for a nonprofit educational purpose.
- The time at which you were inspired to use that copyrighted work and the time at which you will actually use it are so close that a permission request could not be processed (usually two months' time or less).

Limitations: The following limitations on the portion of the copyrighted work must be applied to the guidelines above:

- *Poetry*—No more than 250 words or two pages.
- *Prose*—Complete articles, stories, or essays that are less than 2500 words may be copied. If the work is longer, no more than 10 percent or a maximum of 1000 words.
- *Illustrations*—No more than one chart, diagram, drawing, graph, cartoon, or picture per book or per periodical issue.
- *Special works*—No more than 10 percent of the words in the text in works that combine text with illustration, such as children's books or poetry.

If you have any questions about these guidelines, please contact the campus copy center at 555–1224. Again, these guidelines will be in effect as of December 15, 1995.

---

are chunked in bulleted lists. You can also use visual display to queue information or order chunks hierarchically to convey order and importance. In Example 1–8, *Fair Use Guidelines* and *Limitations* appear as major headings because they appear in boldface and are less indented than the lists that follow them.

You can help your readers filter information by seeing differences and similarities between units or queues. For example, readers see that elements with the same type size, darkness, or placement are similar or in the same category. In Example 1–8, the reader sees that the categories *Fair Use Guidelines* and *Limitations* are of equal importance. Finally, your readers will visually abstract the purpose of the elements in a document based on the elements' relationship to the entire document. In Example 1–8, readers will know that because it comes at the top of the document, following the

RE: (for regarding) cue, *Copyright and Fair Use Guidelines for Course Material* is the title or subject of the entire document (Martin).

Whenever you create a communication, you can help your readers see, understand, and use your works through visual display.

The Changing Nature of the Workplace

As we mentioned earlier, because technical communication involves people—designers, engineers, managers, advertisers, clients, customers, and users—we say it is socially situated. As a consequence, in order to be an effective technical communicator, you must understand the nature of the workplace.

Constant change marks life in the last part of the twentieth century, and nowhere is that change more apparent than in the corporate world. One obvious change is the way information keeps increasing in the workplace. To give you some idea of the magnitude of the increase, consider how much dataflow occurs for one company. In 1981 American Express already had an online communication system that processed 310 million American Express card transactions, 360 million Visa and MasterCard transactions, and more than 350 million American Express Travelers Checques. It completed 56 million insurance premium and claim transactions and executed approximately $10 billion *each day* in international banking transactions. In order to handle these transnational information needs, American Express had to commit resources to hire nine major information processing centers, six worldwide data- and time-sharing network groups, 70 large computer systems, and 229 smaller computer systems for a combined processing capacity of 170 million instructions per second (Schiller 50). By 1993, such information flows were common, and worldwide communication on the Internet alone included 39,000 networks, in 100 countries, with 5,000,000 users worldwide (Marine et al. 1)

Thus, new communication technologies affect the ways in which we communicate and how we interact with each other given this increase in information. Cellular phone service, fax machines, and fax modems, for example, have altered how and where we send and receive information. Computers have given many people the opportunity to design their own newsletters and brochures, to compile and send complex information, and to edit and revise documents fast and efficiently.

Another important change is the increasing number of women and minorities in the workforce. The work environment of the United States is undergoing major changes in the makeup of its workforce. According to *Workforce 2000,* nearly two-thirds of all people entering the workforce between now and the year 2000 will be women.

The workforce is also becoming more varied in terms of race, ethnicity, and cultural background, and businesses have both employees and customers who do not speak English as their first language or who have difficulty reading English. In the near future, immigrants will represent the largest share of the increase in the workforce since World War I. Throughout the rest of the 1990s, approximately 600,000 immigrants will enter the United States and two-thirds of them will enter the labor force. By the year 2000, nonwhites, women, and immigrants will make up more than 83 percent of the additions to the workforce, compared with about 50 percent today (Johnston and Packer).

This diversity in the workplace is augmented by the increasingly global nature of business. Doing business in the global marketplace means dealing with people who have different cultures, languages, and business practices, who live in different time zones, but who all have need for the latest technology to improve their environment and lifestyles and to compete in their own marketplaces. It also means learning to navigate through a sea of legal, ethical, and environmental considerations.

THE GLOBAL CORPORATE WORLD

Many corporations are already multinational and global in nature and have offices in numerous countries, with suppliers and customers from around the world. This change profoundly affects the way these organizations carry out their communication. New trade agreements and free trade zones mean that businesses must become global. This global business environment has had two major effects.

First, corporations are having to take **cultural diversity** into consideration in ways that they have not had to before. For example, much has been made of the differences in work habits among cultures. Additional differences in verbal and nonverbal communication, various views of time and promptness, and various different cultural and religious observances cannot be ignored if companies are to do business on a global scale. May 23rd may not mean much to most American workers, but to Canadian citizens it's Victoria Day, and businesses are closed for the holiday. As a technical communicator, you'll need to be sensitive to cultural diversity, and that means sometimes adopting concepts and processes as appropriate. For example, adoption of some Japanese automobile production processes and standards by US car makers has led to a recent upsurge in US car quality and sales.

Second, businesses are affected by changing standards. Products are less and less likely to be created from parts manufactured on the premises. As a result, national and global standards for manufacturing and for communicating exist both within countries and across them. The **American National Standards Institute** (ANSI) makes available standards that are adopted by business and industry in the United States. Even though businesses are not required to follow these standards, the standards represent what most businesses think desirable. As a technical communicator, you'll need to know

cultural diversity
differences among and between people due to their gender, ethnic background, religious preferences, and such

American National Standards Institute (ANSI)
creates quality and reliability standards

these standards, as they cover such things as the design of safety signs and labels.

International Standards Organization (ISO)
creates international quality and reliability standards, discussed in the five documents of ISO 9000 series

Standards provided by groups such as the **International Standards Organization** (ISO) also affect how you'll create documents and how the company you work for will do business. The ISO 9000 series, for example, makes quality control of supplies and manufacturing processes more realistic. These standards also include guidelines for both the content and visual display of the documents you'll create for your company.

In general, ISO and ANSI standards improve quality control and communication because they allow for the production of generic or uniform processed materials. These materials can be incorporated into manufactured parts or hardware assemblies. Again, these standards cover everything from units of measure, pipe fittings and valves, fasteners and screw threads, to the content of documents. There are, for example, ISO standards for vocabulary and for common symbols (such as the telephone in Figure 1-1). They provide guidelines for how documents are to be formatted (paper type and size, holes, representation of dates and times) and standards for document reproduction (type face, paper type, required signature forms).

Communicating with Other Cultures

culture
socially constructed and transmitted behaviors, beliefs, values, and institutions

The word **culture** refers to socially constructed and transmitted behaviors, beliefs, values, and institutions. Culture affects all aspects of human life, from telling people what time to eat the big meal of the day to what is acceptable dress for a job interview. The rules of a culture are known to its people in an almost unconscious way, so it's hard for someone entering into a new culture to grasp the rules and their significance. For example, in Japan, people arrange themselves around a business table in a hierarchical fashion,

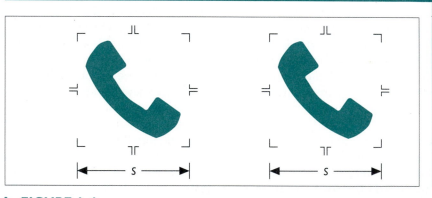

FIGURE 1-1
ISO Symbol Signifying Communication Facilities by Telephone

derived from the way people would be seated at the Japanese tea ceremony, a central ceremony in cultural life.

In turn, communication codes do not easily transfer from one culture into another; for example, in the United States, we associate death with the color black, but in some Asian cultures, mourners wear white instead. Technical communicators who are not aware of cultural rules or who work from their own set of unconscious rules invite insult and misunderstanding.

Diversity and Levels of Literacy

Although we tend to think of corporate employees in terms of the corporate white *man,* such images are very much out of date. In the United States white men have been a minority in the workforce since 1980. By the year 2000, white men will make up less than one-fifth of the new entrants to the labor force (Hage). The labor force that you will be entering will have more women, African-Americans, Asian-Americans, Hispanics, and other groups than ever before. Moreover, these groups are going to bring with them new ideas, values, and different perspectives on how things can be done.

This diversity represents great challenges and opportunities to people in the business world and the way they communicate. Sensitivity to language, a broader repertoire of problem-solving techniques, and an awareness of gender, race, and cultural differences in communication are essential for success as a communicator. The United States is a remarkably diverse and rich environment in which to work. Japan, in contrast, is quite homogeneous, with more than 99 percent of its population being Japanese in origin and therefore almost identical in education and cultural customs. The United States has one of the least homogeneous populations in the world, having brought together people from all over. This diversity may make life richer and more interesting, but it also means that in order to communicate successfully, one has to consider a wider range of variables.

Diversity isn't the only variable that affects communicators. **Literacy** levels in the United States are quite low. The United States Department of Education analyzes to what extent people in the United States possess document literacy: the knowledge and skills that enable a person to locate and use information that exists in such places as indexes and on the stubs of our paychecks. The statistics for 1991, published in the *Digest of Education Statistics,* indicate that although 95.5 percent of the population can perform simple reading tasks such as matching coupons with a grocery list, only 57.2 percent can follow directions using a map, and only 20.2 percent can decipher a bus schedule.

literacy
skills that enable a person to locate and use information

Our declining ability to use practical information, coupled with the increasing number of United States citizens who speak English as a second language or have difficulty understanding English at all, creates special problems as we seek to communicate technical information to the population at large. As we move through this book, we examine a number of ways you will be able to improve your communication skills to reach this increasingly diverse population.

TECHNOLOGICAL IMPACT

One focus of this book is on changing technologies and their impact on communicative processes. Knowledge of these technologies is essential for success in your field. For example, many large and small businesses and industries not only use new communication technologies to plan activities, but also they market these communication technologies, produce user documentation to accompany them, and train workers to use them. As Figure 1–2 shows, people generally spend nearly twice as much of their time using electronic media as using conventional media such as pens, pencils, and typewriters (Halpern).

Corporations are investing heavily in new communication media because they help in decision making, in collaboration, and in access, storage, and transfer of information among distant sites. Because the business world relies increasingly on high technology products and processes, you will be surrounded by these technologies and expected to use them in your work.

Technological changes in communication practice usually affect the business environment first. Thus, you will be exposed to an ongoing array of changes in communication technology that you will have had no experience with at home. You will be expected to incorporate these devices and processes into your communication procedures quickly. And, you will be called upon to explain to others how to use these devices and processes.

Additionally, technical communication relies heavily on new communication technologies. You will be expected to understand the computer and

Method of Communication	Percentage of time
Communicating face to face	37
Using new media	23
Using conventional media	12
Conducting other activities	28

Source: Jeanne W. Halpern, "An Electronic Odyssey," *Writing in Nonacademic Settings.* Ed. Lee Odell and Dixie Goswami (New York: The Guilford Press, 1985), p. 163. Adapted by permission of the Guilford Press.

FIGURE 1–2
How People Communicate in the Workplace

its use for collaborative decision making and collaborative writing; you will be expected to have at least minimal expertise with word processing and document design, including some skill with graphics programs used for creating print documents and programs for creating presentation graphics. You will be expected to make judgments about what medium would best present a particular message. You will work with e-mail, teleconferencing, voice mail, and more. And, you will be expected to use an increasing number of technologies from your desktop, both at work and at home.

SOCIAL IMPACT

Because of the complexity of current life and the rate at which change occurs in it, individuals often feel they cannot make a difference. Many people who work for large corporations feel that they are basically anonymous workers whose contributions to the corporation are not of a personal nature. Yet, as we understand more about the role communicators and communication play in an organization, we realize that people can and do make a difference and that they can initiate and carry out change within organizations. Sometimes those changes are not good ones because writers and other communicators fail to see the connection between their acts and the larger world. The human loss and pain associated with the *Challenger* accident, the Johns Manville asbestos case, or the Dow-Corning silicon breast implant, among others, underline how important communication can be.

What we write—or don't write—as well as how effective our communications are profoundly affect others. Engineers involved with the *Challenger* wrote memoranda explaining their reservations about a cold weather launch, but found themselves overridden by other managers, and the result was an explosion, loss of life, ruined careers, and a horrified nation who watched the explosion replayed on television. As such disastrous lapses in communications show, if we do not write or communicate effectively, lives may be lost and society may be profoundly affected.

This chapter provided a brief introduction to the field of technical communication and how important it is becoming in today's changing work environment. The work environment is affected by rapid technological change, increasing numbers of women and minorities in the workplace, and the global nature of the economy. As a technical communicator, you will help people cope with technological change and make the most of technology in ethically and socially responsible ways.

Although technical communication was only recently recognized as a new profession, people wrote about technology centuries ago. In current times, technical communicators work in a variety of fields from computer science, to natural resources, to biomedical technology. Technical professionals in a variety of fields are expected to create effective and accurate documents as part of their daily work.

SUMMARY

Technical communication varies from other forms of communication in a number of ways. It is distinguished from other forms of writing by being reader-based, focused on the subject, dependent upon technology (both for its subject matter and its means of production), representative of the corporate culture which produces it, collaborative, and marked by its technical style and visual display.

ACTIVITIES AND EXERCISES

1. Example 1–9 is a section of directions on how to use a self-injection system to combat migraines. Analyze this example for ways in which it is reader-based communication and for ways in which it focuses on its subject. Be prepared to discuss your findings in class.

    When you've finished your analysis of Example 1–9, look at Example 1–10. This information represents adverse reactions to a medication documented in clinical trials. What characteristics of technical communication does this passage have? Does it have characteristics of science writing? If so, what? Who is the reader? What is the purpose? Why was the passage written in this fashion as opposed to Example 1–9? Be prepared to discuss your findings in class.

2. Go to your college or university library and find copies of scientific and technical magazines from the 1940s or 1950s. Select an article that addresses scientific or technological concerns. What discovery, health issue, or technological innovation is discussed? What background or education would a reader have to have to understand the article? What social impact did the author(s) of the article predict the topic would have? How did the topic or innovation relate to past discoveries

---

EXAMPLE 1–9   Sample Directions

How to Use the Self-Injecting Unit

- *Before injecting*—Clean the skin area to be injected (top of the thigh or the outside of the upper arm).
- *Injecting*—Pick up the self-injecting unit and pull off the blue cap. This will remove the gray needle guard.
- Pull black safety lock away from the red release button. DO NOT TOUCH THE RED RELEASE BUTTON. The safety lock is now OFF.
- Place the gray tip of the self-injecting unit securely on the skin area to be injected.
- Hold the self-injecting unit steady. Push in the red release button until you hear a click. Hold the unit against the skin at least 5 seconds before removing the unit from the skin. If the unit is removed from the skin too soon, not all the medicine will be released.
- Slowly remove the self-injecting unit. Do not touch the needle.

Source: Cerenex Pharmaceuticals. Division of Glaxo Inc.

EXAMPLE 1–10 Information about Adverse Reactions

Sumatriptin succinate (Imitrex™) Injection
ADVERSE REACTIONS Sumatriptin may cause coronary vasospasm in patients with a history of coronary artery disease, known to be susceptible to coronary artery vasospasm, and, very rarely, without prior history suggestive of coronary artery disease.

There have been rare reports from countries in which Imitrex™ Injection has been marketed of serious and/or life-threatening arrhythmias, including atrial fibrillation, ventricular fibrillation, ventricular tachycardia, myocardial infarction, and marked ischemic ST elevations associated with Imitrex™ injection.

Other untoward clinical events associated with the use of subcutaneous Imitrex™ injection are: pain or redness at the injection site, atypical sensations (such as sensations of warmth, cold, tingling or paresthesia, pressure, burning, numbness, tightness, all of which may be localized or generalized), flushing, chest symptoms (pressure, pain, or tightness), fatigue, dizziness, and drowsiness. All these untoward effects are usually transient, although they may be severe in some patients.

Source: Cerenex Pharmaceuticals. Division of Glaxo Inc.

---

or issues? What particular writing or organizational strategies did the author(s) use to capture the readers' attention, explain complicated scientific or technological ideas, and establish authority to write the article? What do you think the author(s) wanted the readers to think, feel, or do after reading the article?

Record the results of your investigation in a memo to your instructor. Attach a copy of the article you read to your memo.

3. Go to the library and read at least two of the following essays that attempt to define technical communication or technical writing:

Allen, Jo. "The Case Against Defining Technical Writing." *Journal of Business and Technical Communication* 4.2 (September 1990) pp. 68–77.

Britton, W. Earl. "What Is Technical Writing? A Redefinition." *The Teaching of Technical Writing*. Ed. Donald H. Cunningham and Herman A. Estrin. Urbana: NCTE, 1974, pp. 9–14.

Dandridge, Edmund P., Jr. "Notes toward a Definition of Technical Writing." *The Teaching of Technical Writing*. Ed. Donald H. Cunningham and Herman A. Estrin. Urbana: NCTE, 1974, pp. 15–20.

Dobrin, David N. "What's Technical about Technical Writing." *New Essays in Technical and Scientific Communication: Research, Theory, and Practice*. Ed. Paul V. Anderson, R. John Brockmann, and Carolyn R. Miller. Farmingdale: Baywood, 1983, pp. 227–50.

Harris, John S. "On Expanding the Definition of Technical Writing." *Journal of Technical Writing and Communication* 8.2 (1978) pp. 133–38.

Hays, Robert. "What Is Technical Writing?" *Word Study* 36.4 (1961) pp. 1–4.

Hogan, Harriet. "Distinguishing Characteristics of the Technical Writing Course." *Technical and Business Communication in Two-Year Programs.* Ed. W. Keats Sparrow and Nell Ann Pickett. Urbana: NCTE, 1983, pp. 16–21.

Kelley, Patrick M., and Roger E. Masse. "A Definition of Technical Writing." *The Technical Writing Teacher* 4.3 (1977) pp. 94–97.

Limaye, Mohan R. "Redefining Business and Technical Writing by Means of a Six-Factored Communication Model." *Journal of Technical Writing and Communication* 13.4 (1983) pp. 331–40.

Stratton, Charles R. "Technical Writing: What It Is and What It Isn't." *Journal of Technical Writing and Communication* 9.1 (1979) pp. 9–16.

Walter, John A. "Technical Writing: Species or Genus?" *Technical Communication* 24.1 (1977) pp. 6–8.

Zall, Paul M. "The Three-Horned Dilemma: Technical Writing, Business Writing, and Journalism." *The Practical Craft: Readings for Business and Technical Writers.* Ed. W. Keats Sparrow and Donald H. Cunningham. Boston: Houghton Mifflin, 1978, pp. 14–21.

In a memo to your instructor and the students in your class, explain how the authors of the essays you read define technical writing or communication. Then explain whether you agree or disagree with their definitions (some of the articles are more than 15 years old, so changes in technology may affect the definitions). Finally, find a brief sample of technical writing and explain in your memo why you believe this sample is technical writing. Be sure to attach your sample to your memo.

4. To understand how an organization develops and displays a particular culture, think about a job you have had or an organization you belong to. How did you hear about the job or organization? How does the organization describe itself? What characteristics or values did that organization appear to have at first? Did your impression of those characteristics or values change after you got to know the organization better?

   Describe that organization in a brief oral presentation to your class. Be sure to use examples either from your own experience or from public documents that the organization has created, such as job advertisements or public relations pieces.

5. Investigate a recent technological change in your environment. For example, your college or university library might have recently adopted a new computerized card catalog, a local fast-food restaurant might have purchased new cash registers, you may have purchased a home computer, or you may have recently encountered the many choices that businesses such as airline reservation offices can give someone who calls on a touchtone phone.

   In a brief memo to the students in your class, describe the benefits of this technological change as well as the possible losses, complications, or problems. In particular, describe how this change has affected the communication patterns within that particular environment.

6. Test the communication requirements of your future profession by interviewing a professor who is active in consulting or research as well as teaching, a graduate

student who also has some workplace experience, or a professional in your community. Ask what communication media (oral, graphic, online, or written) they most frequently use. Ask whether the person ever collaborates with others in planning, writing, or revising. Ask whether the organization in which the person communicates has a prescribed style or format (formal or informal; letters, phone calls, memos, and reports; and such).

In a table or chart display the information you gather during your interview. Be prepared to describe your graphic display orally to your instructor and the students in your class.

7. Example 1–11 is a piece written much like an essay. Working in pairs during class time, revise the passage so it has the necessary features of technical communication.

8. In Appendix A, Case Documents 2, you will find five documents written by the Exxon Company and their associates to explain the bioremediation process used to clean up the 1989 *Exxon Valdez* oil spill in Prince William Sound. The documents are: (a) "How Tiny Organisms Helped Clean Prince William Sound" (look at the first four paragraphs); (b) the Executive Summary of "A Report Detailing the Effectiveness and Safety of Bioremediation in Prince William Sound";

EXAMPLE 1–11   Adding the Features of Technical Communication

Some people call the photoreceptor the photoconductor—it's all the same. The photoreceptor is covered with photoconductive material that will conduct electricity in the light but not in the dark. The photoreceptor accepts the latent electrostatic image and provides the surface where this image can be developed and then the image is transferred to paper. The whole process begins with the charge—which happens when the photoreceptor is exposed to an electrostatic charge. This charge goes across its entire surface and is uniform. And then the photoreceptor is exposed to the image of a document—this image is projected onto the photoreceptor. The light that comes from the white background areas of that document dissipates the electrostatic charge on the photoreceptor surface. The charge remains on the photoreceptor in the dark areas where no light has been reflected from the original. A sheet of paper is placed in contact with the photoreceptor and the image is transferred. But before that happens, a developing substance is spread over the photoreceptor surface. The particles of this substance adhere to the projected image areas because of electrostatic attraction. Then again during transfer, the toner particles are transferred from the photoreceptor to the paper by imparting an electrical charge of opposite polarity to the paper surface. Finally, the surface of the photoreceptor is cleaned of residual particles after the image is fused to the paper by heat. If it is not fused by heat, it can be fused by pressure. To understand this process, you have to remember that the developing substance used here is often called the toner and that opposite electrical charges attract, while similar ones repel. Now you know all about the photoreceptor in the copying process. Remember that the photoreceptor provides the medium for forming an image that is developed and transferred onto paper.

(c) the introduction to "Bioremediation Technology Development and Application to the Alaskan Spill"; (d) "The Prince William Sound Bioremediation Story"; and (e) "Bioremediation for Shoreline Cleanup Following the 1989 Alaskan Oil Spill" (look at the first section, "The Bioremediation Process.")

In a brief report to your classmates, answer the following questions:

- What are the audience, purpose, and situation for each communication? How are they similar and how are they different?
- What stance do the writers take in representing the corporation—particularly in terms of the social and environmental impact of the spill?
- Do the writers use inductive and deductive organization?
- What characteristics of technical communication do you see in each communication?
- Finally, if you were to rank these communications in terms of clarity, conciseness, effective visual display, focus on subject, reader-based orientation, and such, is one communication more effective than the others?

Works Cited

Bragg, James R., et al. *Bioremediation for Shoreline Cleanup Following the 1989 Alaskan Oil Spill.* Florham Park, NJ: Exxon Research and Engineering Company, 1992.

Giroux, Henry. *Border Crossings: Cultural Workers and the Politics of Education.* New York: Routledge, 1992.

Hage, Dave. "White Male Influence Waning?" *Star Tribune,* February 23, 1988, sec. D: pp. 1–4.

Halpern, Jeanne W. "An Electronic Odyssey." *Writing in Nonacademic Settings.* Ed. Lee Odell and Dixie Goswami. New York: The Guilford Press, 1985, pp. 157–189.

Johnston, William B., and Arnold H. Packer. *Workforce 2000: Work and Workers for the 21st Century.* Indianapolis: Hudson Institute, 1987.

Lunsford, Andrea, and Lisa Ede. *Singular Texts/Plural Authors: Perspectives on Collaborative Writing.* Carbondale, IL: Southern Illinois UP, 1990.

Marine, April; Susan Kirkpatrick; Vivian Neou; and Carol Ward. *Internet: Getting Started.* Englewood Cliffs, NJ: PTR Prentice Hall, 1993.

Martin, Marilyn. "The Visual Hierarchy of Documents." *Proceedings from the International Technical Communication Conference.* Chicago, 1989. VC-32-35.

Smart, Graham. "Genre as Community Invention: A Central Banks' Response to Its Executives' Expectations as Readers." *Writing in the Workplace: New Research Directions.* Ed. Rachel Spilka. Carbondale, IL: Southern Illinois UP, 1993, pp. 124–40.

Effective Technical Communication for Problem Solving

Personal Goals and Personal Habits ▌ Three Ways to Think about Purpose ▌ Multiple Purposes and Multiple Audiences ▌ Planning the Process

One thinks one has learn to write quickly; and one hasn't. And what is odd, I'm not writing with gusto or pleasure: because of the concentration. I am not reeling it off; but sticking it down. . . . Whenever I make a mark I have to think of its relation to a dozen others. And though I could go on easily enough, I am always stopping to consider the whole effect.

Virginia Woolf, "Friday, October 11th [1929]." *A Writer's Diary.*
Ed. Leonard Woolf (London: Hogarth Press, 1958).

Introduction

planning
deciding on a project's structure, scope, purpose, audience, strategies, tactics, and timetable

Now that you understand more about the nature of the modern workplace, you need to see how technical communication can solve problems found there. By definition technical communication is applied communication; that is, it's designed to accomplish specific tasks or help an audience solve specific problems. In this chapter you will learn to engage in the activity that we call **planning,** the first activity you will need to do when you create technical communications to solve problems. Planning involves deciding upon a project's structure, scope, and purpose; the audiences to which it will be directed; the strategies and tactics you will use to achieve various goals; and ways of handling timetables and organizational constraints. Like every communication activity, planning is constrained by social factors and is often also collaborative in nature. In this chapter we focus on two major planning activities: arriving at a clear sense of purpose—the problem you want to solve and the means of solving it—and deciding on the stages and timing necessary to create your problem-solving communication.

Even though it may seem that this planning happens early in the process of creating a document and is then finished, the fact is that planning occurs throughout the process. Deciding early on what problems you need to address and the steps needed to create your document is very important, but this planning continues throughout the rest of your work, involving constant refinements and adjustments as the project develops. Good writing plans are like Critical Path Diagrams in civil engineering, Project Task Flow schedules, or similar planning documents used to coordinate technical tasks in any number of industries. Successful projects always begin with careful planning about goals and resources.

Imagine yourself, for example, in your first job, a few months after your graduation from a program in mechanical engineering. After completing your initial on-the-job training, you've been working with your first work team—a group of mechanical and chemical engineers like you charged with finding a more dependable means of applying a protective coating to new automobiles. Your company is engaged in a research-and-development effort to find a coating that will protect the finish of new automobiles from stones, gravel, water, ice, and other enemies. The company has developed some promising possibilities, but it needs to make sure that those possibilities are practical: workers on the assembly line must be able to spray this new coating easily onto finished automobiles.

That's where you and your small group of co-workers come in. You have been brought together to test various kinds of spray guns to see if the new coatings can be applied effectively by such devices. You've been working on the project for a few weeks now, testing various spray guns with various versions of the new coating, when your supervisor asks you to write a progress report on how things are going. "These things are very routine, so don't worry about it too much," she tells you. "I just need to know how you are coming along on the project so that I can keep my own bosses informed on how things are progressing. A few pages will be just fine. By Friday." Then she vanishes, and leaves you to write the report.

You say to yourself, What do you mean, don't worry? It's my first report! And how can you say it's "very routine"? How can it be routine if I've never done one before and have no idea what you and your bosses want or need? And Friday is only three days away!

What *do* readers want and need? Where *do* writers begin a document like this? How *do* writers make sure the job is done on time? Although many writers might begin by knocking out a few sentences in something of a panic, we suggest that you take a more deliberate approach—that you begin by doing some careful planning before you actually draft any sentences. Establish clear objectives, and plan the production process. No matter how pressed you are for time and no matter how sure you are of what you are doing, you will probably have better success if you begin by collecting your thoughts, at least informally, on what you need to accomplish and how you need to approach the task. Plan well, and you are likely to find the actual task of writing going more smoothly. Having a clear plan with a clear set of objectives will guide your decision making throughout the process of developing your communication or document.

In this chapter we suggest a number of approaches you can use to arrive at a clear set of objectives that will be useful when you communicate on the job. We also offer some ideas to keep in mind as you plan the development of your documents.

Personal Goals and Personal Habits

For many students in college, it is relatively easy to think in terms of **personal goals** and to plan activities in a personal way. You can decide what you want to get out of a particular course or assignment, and you can plan your activities to fit your personal goals: so much time to study, to work, to socialize, and so forth. In a sense, you are the boss of your college workplace because you have a fair degree of personal freedom in establishing the goals of your work and the amount of time that you devote to various tasks. "I want to get at least a B on this report," you might say. Or you might want to leave a course knowing how to perform a certain technical activity. Or you might seek to impress a professor so that you can get a good recommendation. In all three instances, your goals could be described as *personal.*

personal goals
what one wants to accomplish and when, including one's desire for professional reward and achievement

In the workplace, you will also have substantial freedom in determining what you will spend your time doing; after all, you will be hired in large part for your ability to handle responsibility and for your sense of personal initiative. And, certainly, you will have personal and professional aspirations as well. The recently hired mechanical engineer at the beginning of this chapter will probably approach his first report with some highly personal goals—to create a good first impression, for instance, or to establish his credentials. And personal goals, including the desire for professional reward and advancement, will remain goals in almost everything that engineer will write over the course of a professional career.

To a certain extent you will be able to accommodate your writing habits on the job—the space, the uninterrupted time, the technology, and so forth—just as you do as a student. But at work, when you are part of a cooperative enterprise, your goals and habits must be more than purely personal. They must be social as well, tied to the goals and practices of other employees and to the goals and practices of the enterprise itself. Your schedule must accommodate others; you'll have to deal with interruptions and emergencies; or the production capabilities of your workplace might have to be shared. Ironically, the key to achieving your personal goals is to concentrate and plan the social ones.

Three Ways to Think about Purpose

Successful planning can help you to include social goals along with personal ones when you develop documents. Three different approaches are included in the coming paragraphs—thinking rhetorically, thinking of yourself as a problem solver, and thinking of yourself as solving a problem for some people. Try each one out, become familiar with it, and then you will probably find that one or two approaches work especially well for you. Each approach involves purpose. When you think of purpose, imagine both what you want to find out and the intended effect of your discovery on your audience. What problem do you want to solve? What do you want your audience to do, think, or feel after reading or listening to your solution? What action do you want the audience to take after reading or listening to your communication?

THINK RHETORICALLY

think rhetorically
considering the communicator, audience, content or topic, genre, and medium of a communication

genre
the form a communication takes, such as a memo, a letter, or speech

medium
the way in which one communicates, such as through print, speech, videos, and telephone

Learning to **think rhetorically** is the first way you should think about purpose. To think rhetorically means to consider the elements of the rhetorical situation that make up all communications: the *communicator* (you and your collaborators), the *audience* (those who receive the communication), the *content* or the *topic* (what the communication is about), the **genre** (the form that the communication takes), and the **medium** in which you make your communication (print, speech, video, graphics, etc.). One easy way of planning effectively is to make sure that all of those factors are considered in your plans and are expressed in a **statement of purpose.**

The mechanical engineer described earlier, for example, should think of all five factors. The genre in this case is easily defined: the engineer was told to write a progress report. And the topic seems easy to discern as well: the report will concern progress on a certain technical task—what's been done so far and what remains to be done. But it may not be so simple as that. The content may be easy to describe if all is going well, but if any difficulties have arisen, the readers will need to be reassured somehow or warned that no solution appears workable. And the engineer will have to decide if it is enough simply to report progress or whether he should have other information or recommendations to offer to readers. And who are those readers, anyway? Is he really writing to his supervisor or to his supervisor's supervisors, or both? Do any future readers need to be considered? Will co-workers also read the progress report?

Finally, the engineer will need to think about himself as a communicator and his aim in writing this document. He will want to report progress, of course (assuming that there's some progress to report). But the writer may also wish to introduce himself and his fellow workers, to reassure the readers that any difficulties are being dealt with satisfactorily, to brag a little bit, or perhaps even indicate somehow that the project should go in a certain promising direction or that it should continue to get support because it is likely to have a good outcome.

It's hard to be too specific when you are talking about a hypothetical example, of course. The point is that by identifying and thinking about the elements of the rhetorical situation, you'll have a good sense of purpose. After pondering these elements and before going on to the actual writing, the engineer we've been discussing might summarize his purpose this way (maybe even in writing):

> I need to write a progress report that will remind my supervisor—and especially my supervisor's supervisors—of what our team has been working on. I also want to explain the good work we've been doing and the interesting results that we've been getting. I think I'll send copies to all my co-workers so that they know about the report and so that they'll have a sense of the high regard I have for them. I want to make a good first impression, too, to indicate that I write well and that I can deliver my work on time.

Notice that in this formulation the writer has attended to all the elements of the rhetorical situation: writer (and the writer's aims), audiences, topic, medium, and genre. By writing this statement of purpose down, the writer has a good benchmark to use to evaluate decisions that he will make in his developing text.

Developing statements of purpose like these is generally quite easy. Their ease and informality lend them to the composition of routine memos and correspondence and oral presentations, for instance. Even in routine circumstances or when time is short, get in the habit of taking a few minutes to think rhetorically. Example 2–1 is a memo that grew out of a need to convey an important, but brief, piece of information about a product line—a kneeling chair for clerical workers who use computers within an organization. The information about the chair line was important because it involved safety issues, production delays, and possible client satisfaction. Working with this memo, come up with an informal statement of purpose that the

statement of purpose
expresses all the factors of planning: communicator, audience, content, genre, and medium

EXAMPLE 2–1   Purpose—Thinking Rhetorically

The Computer Company
For Your Computing Needs

DATE: 28 December 1994
TO: Dan Musgrove, Shipping
          John Loomis, Purchasing
          Carol Adamsson, Division Manager
FROM: Carla Etchison, Technical Sales and Customer Relations
RE: Kneeling Pads for Computer Chair #332

CC: Lemonia Williams, Production Manager
          Sandra Blevins, Product Engineer
          Ethel Walmera, Research and Development

As Carol requested (memo dated 5 February 1994), I have checked into the 16 cases of defective kneeling pads on Computer Chair #332. All these chairs were traced to the same production batch #332-49-7711 (dated 21 December 1993). The pads on these chairs were manufactured from a single shipment (Invoice #769-1) received from our supplier, Amos Stuffing, on 12 December 1994. The supplier at Amos has acknowledged that some pads in this shipment represented a slightly different grade of high-density foam. According to their estimate, approximately 14 percent of the shipment may have been involved.

According to Amos, the composition of this foam will degrade with heavy use, especially in situations involving high pressure demands (>523 psi) over minimal surface ares. Amos has offered to reship the affected portion of the shipment immediately, free of charge to our firm.

As Carol has requested, I have sent cards to all of our client retailers who may have purchased chairs with similarly defective pads and offered immediate replacement, without charge.

From our similar experience three years ago, we can anticipate that approximately 1,500 chairs were affected and that approximately 900 claims will be made within the next month. Lemonia Williams has outlined a production plan for reconditioning the affected chairs to specification, and Dan anticipates a two-day delay in reshipping during the next 30-day period.

Thank you all for your cooperation on this project.

memo's author, Carla Etchison, might have used to aid her in drafting the memo.

THINK OF YOURSELF AS A PROBLEM SOLVER

ideal and actual situations
to think as a problem solver, one imagines purpose as a bridge between ideal situations and actual situations

A second way to determine purpose is to think of it as the bridge between an **ideal situation** and an **actual** one. In this case, what you discover and often what you recommend to your audience can help turn the actual situation into the ideal one.

The following examples will help clarify this way to think about purpose:

- Your company is on the brink of developing an effective coating for protecting the finishes of automobiles *(ideal situation),* but it cannot market the product until it can be sure that the coating can be sprayed on easily during the production process *(actual situation);* consequently, you must test the feasibility of a number of spray guns that could be used to apply the product *(purpose).*

- Your company and its customers would like a reliable and inexpensive adhesive that can be applied under water *(ideal situation),* but so far no such product exists *(actual situation);* therefore, your team is trying to develop such an adhesive or at least to contribute to its creation *(purpose).*

- Domestic animals such as cows sometimes develop a disease called malignant catarrhal fever, so it would be a good idea to have a treatment for the disease handy at your farm *(ideal situation),* but you currently have no idea how to treat the condition *(actual situation);* so you need to compose a literature review to summarize effective treatments *(purpose).*

- People who fight forest fires should be able to count on absolutely reliable equipment, including chain saws *(ideal situation);* however, in real life the chain saws often break down during use *(actual situation).* Consequently, you need to write a document that will equip firefighters to make routine repairs on chainsaws when they are out in the field *(purpose).*

- Your microbiology professor would like to have absolutely brilliant lab experiences to offer students *(ideal situation),* but in recent years her lab materials have been getting stale *(actual situation);* so she assigns you to research some better possibilities that she might be able to incorporate into her courses *(purpose).*

As both Example 2-2 and the preceding examples indicate, you can come up with a good sense of purpose if you can supply three different sorts of statements.

First, there should be a statement of a goal or desired state that you and your organization consider important. The goal can be a large, general goal ("We want to reduce our costs"), but the more specific you can be, the better ("We want the parents of children placed in body casts to treat congenital hip dislocations to know how to care for the children and their casts at home; that way the children will heal better"). Then, after naming a connecting word—"but" or "however" usually do very nicely—offer a statement about things as they are. This statement articulates a clash between the ideal state and the actual one: there's something that keeps the ideal statement from being true ("Right now the children's parents have no idea how to care for a toddler in a body cast"). Finally, in a third statement, name some activity that you could do to bring the actual state in line with the ideal one ("Let's write a brochure that we can give to parents to explain how to care for the kids and casts at home"). If you can come up with a sentence that includes all three statements, you probably have a good sense of your aim.

EXAMPLE 2–2   Purpose—Thinking of Yourself as a Problem Solver

First statement   Describe a goal, desired state, or value that you or your organization considers important.
  Example: "A laptop computer would make it easy to collect data on site."

Add a connecting signal word: BUT or HOWEVER

Second statement   Describe a condition that prevents the first statement from being achieved or realized at the current time. The statement could describe the status quo or define a competing goal or value or desired state, but it should reveal a clash with the first statement.
  Example: "But we can't purchase a laptop computer until we know more about their capacities, features, and costs."

Third statement   Describe some action—a study, a research program, etc.—that could resolve the difficulty.
  Example: "So I'll do some checking into what kinds of laptops are available and what they cost."

One disadvantage of this way of thinking about problems, however, is that it can be a little impersonal, with a focus on organizational or institutional problems, rather than human problems. Because communications or documents are directed to people, you may need some mechanism for humanizing problems. To put it another way, this method for arriving at purpose is very satisfactory for defining precisely what an enterprise is up to, but by using it you run the risk of forgetting the needs of the audience. For this reason, we recommend that you also consider the approach to purpose outlined in the next section.

THINK OF YOURSELF AS SOLVING A PROBLEM FOR *SOME PEOPLE*

Within organizations, writing tasks generally follow from the problems and needs of real people. In fact, as a technical communicator you might think of defining problems not as technical or institutional problems but as people problems. No problem or need exists unless it is a problem or need for someone, and this understanding provides us with a third way to think about purpose.

The purposes for which writers compose documents, or for which managers assign writing projects, are tied to the needs of the people who will read a document. Consequently, you might think about problem solving in a very human way in order to have a truly robust sense of your goals. As in the previous section, again think of yourself as a problem solver. But now, instead of thinking of yourself as engaged in solving a *technical* problem, see yourself as working on a **human problem.** See yourself as solving a problem *for someone.*

Often it is easy enough to define an audience's problem because at work you often write at someone else's request and that person will define his or her needs for you. But other times it isn't so easy to uncover your audience's needs. If your audience is available, you can ask the person to describe exactly what he or she needs and why; indeed, you can (and should) be assertive about talking to your prospective reader about what problems or needs you should be addressing. But sometimes it isn't easy to approach your audience. Consider once more, for example, the progress report on the coating spray gun project. Remember how vague the boss was about defining the purpose of the report? Remember how vague she was about who would be reading it? Should that report be simply a routine record of what has been done, or should it be designed to assist the readers in directing future work in especially promising directions? Is this the place for the writer to request additional resources that might be handy in assisting the work? What can a technical communicator do in those circumstances, when it is difficult to learn an audience's needs and when a specific sense of purpose is absolutely essential?

Throughout the first chapters of this book, you will learn several ways of thinking about your audience that can improve the quality of the document you eventually create. Advice about audience will especially aid you in developing the kind of information your audience will find useful in your communications. But in thinking about purpose in planning, you must consider audience in certain ways. In the act of planning, take some time to think about your goals in the audience-centered way, as outlined in the first half of Writing Strategies 2–1. Identify your audience and your audience's problem, identify what you will do and what kind of communication you will produce to solve the audience's problem, and name explicitly the action(s) that your audience will take as a result of receiving your communication or document. Keep this writing strategies worksheet in mind as you read more about thinking of yourself as a problem solver for some people.

human problems
thinking about purpose as solving a problem for someone involves identifying human problems rather than technical problems

Identifying Your Audience's Problem

In thinking of yourself as a problem solver, you learned one way of identifying problems within an organization: state an ideal or desirable situation, then state an actual situation so as to imply a clash between the two. That clash between the ideal and the real can be defined as the organization's problem — and your purpose in communicating is to discover and propose a solution to that problem. But an organization's problem is not the same as an *individual*'s problem, although the two may be closely related. In order to think effectively about solving a problem for an audience, you need to consider what task(s) that person or those people need to complete. The first step in understanding

---

WRITING STRATEGIES 2–1   Planning Your Document

I. Questions about Purpose

   A. Who is my audience and what are my audience's specific problems?

   B. What will I do and what communication will I produce in order to solve my audience's problems?

   C. What will my audience do as a result of receiving my communication?

II. Questions about the Process

   A. What resources will I need for this document—people, books and articles, existing reports, etc.?

   B. What visual presentations might I need to support my document, and who will prepare them?

   C. When should I schedule the activities noted in questions II A and B?

   D. When should I schedule my writing and production activities so that I can be sure to meet my deadline?

---

an individual's problem may be to identify the organizational task that the person is engaged in. In addition, you should also identify the kind of problem the person has, such as the following:

- a procedural problem (a problem of "know how");
- an informational problem (a problem of "know that"); or
- a decision-making problem (a problem of "what to do").

procedural problems
problems of "know how," or the need for information about how to do something

People have a **procedural problem** if they need information on how to do something. A parent may not know how to care for a toddler placed in a body cast to cure a dislocated hip; a field worker may not know how to use a new piece of equipment; a lab worker may not know how to care properly for lab animals; an office worker may not know how to use a new piece of software; a customer may not know how to install a new smoke detector. Obviously there are many people who need "know how" of various kinds. Technical writers produce instructions and procedures manuals to solve those kinds of problems.

informational problems
problems of "know that," or the need for detailed information on specific topics

Informational problems are just as common as procedural problems. Just how expensive will it be to maintain this piece of equipment? What do we know right now about malignant catarrhal fever? What will take place at next week's meeting? Will a certain spray gun be able to handle the new coating we are developing? Is the research group making progress on a specific genetic engineering problem? When questions like these are asked, the answers generate informational reports and brochures, empirical research reports and reviews of the literature, research articles and news reports, progress reports and meeting minutes, and so on.

A **decision-making problem** occurs when a person doesn't know what to do. Whom should we contract to design a safer intersection? Should we buy this product or that one? What manufacturing process will be most efficient in this case? Which employee should we hire? To whom should we award our grant money? Which experiments should we use in our lab work next semester? Questions such as these frequently involve the word *should* (or a synonym). They generate proposals, resumes and application letters, recommendation reports, feasibility studies, and so forth.

decision-making problems
problems of "what to do," or the need for information that identifies what action the audience should take

Of course, an audience often will have several questions or problems, even several kinds of questions, and a given document will need to address all of them to be successful. Sometimes the questions may be equally important ("I wonder what kinds of software are available to handle this situation, and should we purchase one?"). Other times, some questions might take precedence (a recommendation report primarily designed to aid decision making might also suggest how the decision might be implemented). Therefore, an important part of arriving at a strong sense of purpose is to identify the problem that your audience has.

Planning How to Solve Your Audience's Problem

Once you determine your audience's problem, consider what you must do to address it and what kind of document you will need to write to solve it. Often you will have to perform some kind of test or study or use some other information-gathering activity to find the answers you need. In this sense, naming what you must do will remind you of the technical solutions to problems described in the previous section.

But "what you must do" might involve as well (or instead) other activities designed to generate information that you will learn later in this book—interviewing people or doing a literature search, for instance. The mechanical engineer charged with writing that progress report, for example, must collect information on the goals of his group's work and must determine exactly what's been done already and what remains to be done. He must consult with others about what information might go into the report and look at other sample progress reports (because this is his first one) so that he'll know what his report should look like.

The kind of communication you must produce usually follows easily enough from the kind of problem at hand. Decision-making problems generate proposals or recommendations, and so forth. Informational problems generate oral presentations, literature reviews, empirical research reports, and the other items named above. And know-how problems generate instructions and procedures manuals. As we discuss these kinds of documents in the third part of this book, remember that they all are designed to help an audience solve a problem.

Planning What Action You Want Your Audience to Take

Before you are satisfied that you have a complete sense of how to solve a problem for some people, ask one last question about purpose: What *precisely* do you want to happen as a result of your document? Do you want the members of your audience to change their minds about something in a specific way? Do you want your audience to be able to do something? Do you want your audience to see things in a certain way? Do you want some other kinds of outcomes, even some personal ones ("Even though I don't want to say so explicitly, I want my boss to assign me to a new project")? If so, this is the time to make a note of your specific desires.

Take a look at Examples 2–3 and 2–4 to see how you can plan what action you want your audience to take.

This method of arriving at purpose—considering how to solve a problem for some people—reinforces the other two methods described in this chapter. Thinking about your audience's problem and what you can do to solve it has much in common with problem solving in general and with thinking rhetorically. However, this third method puts a greater emphasis on your audience and your audience's needs, while the previous methods put the emphasis on the organization's problem and your technical task. In addition, recall once more that we are trying to give you several options for arriving

---

EXAMPLE 2–3   Following Writing Strategies 2–1 for Planning
a Typical Document: Sample One

I.   *Questions about Purpose*

   A. Who is my audience and what are my audience's specific problems?
      My supervisor and her supervisors need to know how we are coming along on testing various kinds of spray guns that might be useful for applying our new sealant for car finishes. The "big bosses" are concerned that the coating can be easily sprayed on during the production cycle, so they assigned our group to test various spray guns; now they want to know how things are coming along. Are any of the spray guns working? Which ones have we eliminated, and which ones might work? Which ones should be tested more (and more intensively) in the future? My immediate supervisor has a pretty good idea of how the project is going because I see her frequently; but she needs a good summary to keep her bosses happy, and she needs them to feel confident that this project is turning out well. Her supervisors need to keep an eye on things so that they can respond to difficulties and change directions if necessary. And my co-workers are an audience too, if I copy them; they want to be sure that their work is being seen in a positive light.

   B. What will I do and what communication will I produce in order to solve my audience's problems?

EXAMPLE 2–3 *(concluded)*

First of all, we better produce some progress so that I'll have some progress to report! That's not too tough because the work has been going pretty smoothly. But I want us to finish the next phase of our work before I turn in the report so that I can report that we're ahead of schedule. That means some hard work in the next few days; I hope I can convince the other members of the group to grind hard this week so that I'll be able to write a progress report that can make us all look good.

C. What will my audience do as a result of receiving my communication?

That's hard to say. I hope that they will feel reassured that the work is coming along. I also want them to know about the excellent preliminary results that we've had with that one kind of spray gun; it's my opinion that they should push hard for more testing of that one, even if it means that some of the testing of other ones is put off (I don't want to push hard on this, though). I also want my readers to feel good about those of us working on this project. We're relatively new, and we want to be noticed—and kept in mind for other interesting assignments.

II. *Questions about the Process*

A. What resources will I need for this document—people, books and articles, existing reports, etc.?

Besides making a careful inventory of the tests we have run so far on four spray guns, I'll have to review the charge that we got at the beginning of the project so that I can define what we're doing in my readers' own terms. And since I haven't written a progress report here, I want to get a sense of how long the report should be and what it should generally look like. I also want Fran to take a look at my rough draft, if possible—I'll need to tell her ahead of time.

B. What visual presentations might I need to support this document, and who will prepare them?

Don't need any visuals, except for a table that shows how each spray gun is doing against the criteria for success that we've set up. At least I can't think of any others right now . . .

C. When should I schedule the activities noted in questions II A and B?

See answer to next question!

D. When should I schedule my writing and production activities so that I can be sure to meet my deadline?

Holy cow, I've only got until Friday to finish—just four days. If I want Fran to take a look at a draft, I'll need to have that by late Thursday (if I can convince Fran to look at it at home Thurs. nite!). The boss says it shouldn't be too long, so if I collect my information by the end of Wednesday, I'll have Wednesday nite and Thursday morn to write (though I also have to help with the testing too on Wednesday and Thursday). Better cancel my plans to see that movie on Thursday. Better yet, change the movie date to Friday nite and make it a celebration!

EXAMPLE 2–4 Following Writing Strategies 2–1 for Planning
a Typical Document: Sample Two

I. Questions about Purpose

A. Who is my audience and what are my audience's specific problems?
Dr. Jane Horvath, my chemistry prof, has been having trouble with some of the laboratory experiments in her Chem 23 course. Some of the experiments take too long to complete during the lab period. Others seem unclear or poorly conceived for carrying out her course goals.

B. What will I do and what communication will I produce in order to solve my audience's problems?
Dr. Horvath assigned me (and some other students) the task of developing a new experiment as a replacement for a poor one. I'll need to do some background reading, decide on a good experiment, design it, pick the reactants and do some tinkering in the lab to discover the best mix of chemicals, and test the experiment myself and with some others. Then I'll need to write up the experiment and attach a cover letter explaining what the experiment is, what it's designed to do, and how I developed it.

C. What will my audience do as a result of receiving my communication?
Dr. Horvath will agree that my experiment is well conceived, reward me for my efforts, and then incorporate my experiment into next year's Chem 23 lab manual.

II. Questions about the Process

A. What resources will I need for this document—people, books and articles, existing reports, etc.?
I'll need to interview Dr. Horvath about shortcomings in some current experiments and determine what she'd like to accomplish with the experiments that she hopes to improve on. Then I'll head to the library to do some reading; maybe Dr. Horvath can suggest things for me to look at and other ways of getting ideas. I'll also need some lab time to try out the experiment I design. And I'll have to recruit (and schedule) some student guinea pigs to test the experiment out on so I'll know how to fine-tune it.

B. What visual presentations might I need to support my document, and who will prepare them?
I'll have to produce my own visuals of the lab set-up; if my experiment is chosen by Dr. Horvath, she'll have a professional prepare a better version using my drawing as a rough draft.

C. When should I schedule the activities noted in questions II A and B?
This project is due on November 1. Considering everything else I'm doing, I need to leave at least a week for writing up my experiment (including getting suggestions on my rough draft, revising it, and typing it up to look sharp). So I should know generally what I want to try out by October 1, so I'll have time to have it drafted and tested and then work out any bugs. Before that I'll have to get an appointment to speak to Dr. Horvath, do my reading, and

EXAMPLE 2–4   (concluded)

> then try out my ideas in the lab. Yikes! Today is September 5—so I better get moving right away!
>
> D. When should I schedule my writing and production activities so that I can be sure to meet my deadline?
>
> I'll need to reserve time in the last 10 days of October to write, revise, test, and produce my document, so get that on my schedule now! Juggle my other responsibilities around that date!

at a good sense of purpose and to move you beyond merely personal goals to social ones as well. These methods are mutually reinforcing. Use one, and you can count on having a respectable sense of purpose. Use more than one, and you can count on having an excellent sense of your aims.

Multiple Purposes and Multiple Audiences

As we've seen already, purpose is rarely a simple thing. You will probably have several purposes for the documents that you produce; you'll have several changes that you hope will happen as a result of your reports, correspondence, or presentations. In addition, your purposes will be influenced by other people:

- The person who assigns you the task of producing a communication certainly has something to say about its purpose.
- Your co-workers, sales and marketing representatives, consultants, clients, and others may have their own agendas for the communications you create.
- Each of your readers will have special needs and idiosyncrasies that you cannot even know, let alone plan for.

These facts make arriving at a purpose difficult. But they also make purpose more important. With so many people contributing perspectives on purpose and so many elements involved in shaping it, you must do everything you can to clarify your sense of a document's aim early and continually throughout a project. Thinking about purpose can provide you with an excellent direction as you begin a writing or speaking task and as you see it through to completion.

Thinking about purpose is also complicated by the fact that audience is not a simple concept either. Many documents are read by different kinds of readers or by readers with different kinds of needs. If purpose is directed at least in part by audience, and if audiences are **multiple and heterogeneous,** then it can be especially difficult to have a clear sense of purpose.

multiple and heterogeneous audiences
audiences that include a number of readers, listeners, or viewers with different goals, needs, values, literacy levels, and cultural background

Writing Strategies 2–2 provides some specific suggestions for clarifying purpose when you have multiple audiences.

To help you see how to answer the questions in Writing Strategies 2–2, we have examined one of the case documents in Appendix A, the Minnesota AIDS Project brochure (Case Documents 1). See how these writers would have analyzed their multiple audience for this brochure in Example 2–5.

One final note about purpose. Sometimes it will not be possible to accomplish all the purposes you hope to achieve through a given document. Or, you may turn up conflicting aims—one audience might require one thing from a document and someone else might be hoping for something different. For example, you might intend to write a document that will build better relationships with all your co-workers—but also you might be faced with making some recommendations that specific co-workers will resent or take personally.

If you find yourself with multiple purposes and/or conflicting purposes, try to keep your perspective. You cannot achieve everything in such a case, so try to establish a hierarchy:

- What do you want or need to accomplish most?
- What is second most important to you?
- What can't you possibly accomplish in this one document (and what can you do to accomplish that in a separate document)?
- What special or extraordinary strategies can you adopt that might permit you to achieve your conflicting goals after all?

WRITING STRATEGIES 2–2   Clarifying Purpose with Multiple Audiences

Try the following strategies for clarifying the purpose of a writing task from several perspectives:

- In one sentence, identify the most important purpose(s) for the writing task you are undertaking. Write as if you were describing this project to a familiar and trusted colleague.
- In one paragraph, identify the purpose(s) you think the audience(s) might have for the documents you are producing. Write from the perspective of each audience.
- In a list, identify briefly the purposes other important people or groups might have for this document. Write from the perspective of this person or group.

Fill in the blanks in the following sentence:

- My purpose for this _____ (report, document, etc.) is

_____

EXAMPLE 2–5   Clarifying Purpose with Multiple Audiences:
The Minnesota AIDS Project (MAP) Brochure

In one sentence, identify the purpose(s) you think most important for the writing task you are undertaking. Write as if you were describing this project to a familiar and trusted colleague.
My purposes for this brochure are to inform interested people about the organization, to give basic definitions of AIDS and HIV, to offer a way of contacting MAP offices and using the AIDSLine, and to give some simple precautions to prevent the spread of the virus.

In one paragraph, identify the purpose(s) you think that the audience(s) might have for the documents you are producing. Write from the perspective of each audience.
The most important thing to accomplish in the brochure is to inform Minnesota citizens about the existence and efforts of MAP. The audience would be curious about what MAP does and who works with MAP. Minnesota citizens are known for their volunteer spirit and might be interested in working for or donating to MAP. The audience might keep the brochure around if it contained simple definitions of AIDS and HIV. They might remember some brief, clear instructions on how to avoid transmission of the virus. Addresses and phone numbers would make it easy for them to call or write. Overall, we want to impress the audience enough so they will think MAP is doing a worthy job and deserves support.

In a list, identify briefly the purposes that other important people or groups might have for this document. Write from the perspective of this person or group.
Other purposes:

1. Counselors working with high school and junior high school students just learning about AIDS could use the brochure for basic information.
2. Clearinghouses who publish information about volunteer organizations could send out or display the brochure to answer inquiries easily.
3. Educators could use the brochure as a starting basis for lectures and discussions.
4. MAP volunteers would offer the brochure to those wanting the services of MAP.
5. People living with HIV and AIDS would find out how to take advantage of our client services.

The challenge of writing (and a large part of its fun) is that audiences and their problems are so various and dynamic that writing seldom becomes routine. In fact, if you attend carefully to purpose in your writing and to the dynamics of audience and aim, your writing will probably never get dull no matter how frequently you compose the same sorts of documents.

Planning the Process

Once you have a strong sense of purpose, you are well on the way to finishing your planning. But before you proceed to information gathering, take a few minutes to plan the process of communicating too.

PERSONAL PLANNING

Effective communicators are able to change strategies and tactics when the rhetorical situation changes—when a different audience is addressed, when a different subject is taken up, or when different goals are important. Similarly, effective communicators are just as flexible and resourceful in the *act* of composing. They can succeed when a deadline is far away *and* when a document is needed very quickly. They can compose alone when they have to, and they can cooperate with others when that is called for. They can use sophisticated word-procssing technology *and* handle a paper and pencil. They know how to profit from peer reviews when reviews are available, and they know how to live without advice when they must.

One key to such resourcefulness in creating documents is good planning. When you plan your efforts, you can be clear about your goals. But to be a good communicator, you also will have to be good about planning the act of composing itself. Good writers know how to schedule carefully the various composing activities required by every document.

Look again at Writing Strategies 2-1 as you study our recommendations. As you plan your process, jot down the resources you will need for a particular project. Determine if you will need to speak to certain individuals to clarify your goals or to collect information. Decide if you will have to spend some time reading. Identify what you need from others and whether you will have to produce visual aids for the finished project.

After you have compiled a list of the resources you'll need, put into your daily or weekly schedule specific times when you will do your reading, your interviews, your actual composition, your revision (see Example 2-6). If you are counting on others to help along the way, schedule them into your own daily plans in a formal way. Make firm appointments to talk to a colleague or customer in an distraction-free environment, carve out time for others to review your drafts and offer suggestions, find the time necessary to permit others to collaborate with you, arrange for those in your organization who do artwork or graphics to contribute their parts to your effort, and so on. Set deadlines for certain parts of the document or your drafts, and stick to them. Finally, although you need to stick to your deadlines, remain flexible in your composing habits—don't write every document in exactly the same steps—and protect jealously the time you need to write well.

One way to get a good sense of the time and planning required for effective documents is to keep a log of how you spend your time composing and how much time you spend. During or after your writing, note in a log what you did and how long it took. Every time you consult or collaborate

EXAMPLE 2–6   Page of a Weekly Planner Showing Times Set Aside to Gather Resources for a Short Report on Emerging Plastics

	Week at a Glance								
	8 9 10 11 12 1 2 3 4 5								
Monday	Breakfast with Mike re: statistics			Library ⟶					
Tuesday	Read plastics articles from library ⟶ Meet Susan from R&D re: new plastics								
Wednesday	Draft plastics memo ⟶ Make sketches					Give draft to Jane to read			
Thursday	Revise memo ⟶ Put sketches on PageMaker								
Friday	Final draft, proofreading, printing ⟶ Turn in to boss!!								
Saturday	SLEEP IN !!								

with someone, note that. Every time you stop to read something in the process of gathering information, note it. Every time you sit down to compose, or revise, or type, write down how long it takes. From such a log you can begin to get a reliable and realistic sense of the time you are spending on your documents and how the various activities that you employ might be fit into your schedule in the future. That way, if you are asked to prepare a document for your supervisor, you will have a realistic notion of the time it will take you to do a good job, and in that way you will be able to keep promises you make to have materials ready at a certain time.

As you think more about planning, read the following testimony of a very successful transportation engineer who runs a consulting firm in the Chicago area.

When I first began working, I thought I was being paid to do engineering work. I used to think all of the specialized knowledge that I got and all the techniques that I employed were what I produced at work. The designs and the calculations were what I was paid to do, I thought. I looked at writing as relatively unimportant—it was just copying down the important stuff into words. If there wasn't time to do it well, that was tough.

I don't think that way now. I see myself more as a writer. Our clients don't care all that much about our specialized knowledge and our tests and our sophisticated models and calculations. All they care about is that report that we give to them in the end. That and the quality of our proposals. Those reports and proposals are our products, not the engineering techniques. We sell ''paper.''

When this engineer was asked what difference that thinking has on his work, he said the following:

I used to fit writing in whenever I could. Like I say, if I got rushed and it didn't turn out too well, I didn't mind. But now that I see paper as our product, I make time for reports and proposals, schedule them into the day, and I make the people who work here do the same thing. And I even make our clients pay for it: When they see our Project Task Flows, they see time in there for writing. That way I feel like I am scheduling time for writing right into the project, the writing is a lot less painful, and I know I'm billing clients for writing time too. They get billed for writing time as well as engineering time, just like lawyers charge. I like it that way, and you know what? Our clients like it that way too. One reason we're in demand is that we deliver ''good paper.''

COLLABORATIVE PLANNING

collaborative planning
coordinating and scheduling the activities of a collaborative group or team

Finally, before we leave the subject of planning, we need to remind you that, within organizations, planning is often a social, public, and collaborative activity. Because many people with differing levels of expertise are often involved in a communication project, all their activities must be coodinated and scheduled. For example, your manager might plan not only the time line for product development but also the interaction of those involved in that development. Such careful planning, which encourages frequent communication and feedback, can help a group avoid conflict and misunderstanding while creating a communication.

A different version of the log that you saw in Example 2–6 could also be used to plan a complex communication project, involving many people at different stages in a project. Example 2–7 shows how such a complex project might be planned.

In addition to pencil and paper devices (such as those we've shown you here) to help you plan your projects, there are many software products that can assist you with planning. Many kinds of inexpensive planning software exist that allow you to create, update, and print out plans. This planning software often provides you with a template for entering the steps that you need to accomplish in putting together your project. Once entered, the software will arrange your steps in a variety of formats to suit your needs.

EXAMPLE 2–7   Sample of a Monthly Planner Showing Schedule for Coordinating Activities for New Software Documentation for a Personal Computer

	December 1994			
	Week 1	2	3	4
Monday	Manager meets with writing group to explain project time line and assign project	Writing project begins		
Tuesday	Lead writer meets product development staff to discuss software use and audience		Writing group meets to check consistency of various sections of the document	Editor receives complete document for review
Wednesday	Lead writer meets with writing group to divide documentation into sections	Graphics staff presents recommendations for document design		Product development staff checks document for technical accuracy
Thursday	Lead writer meets with graphics staff to discuss visual displays in documentation	Writing group begins to share drafts		
Friday		Writing group, manager, and lead writer meet with product development staff to discuss any last-minute changes in the software.		Final draft sent to graphics staff

For example, you can have your steps arranged in Gannt chart form which allows you to see which tasks overlap, or you can choose a critical step model that helps you identify which critical steps must be completed on time because other steps rely on them. Such software devices are particularly helpful when you must share your plans with others in a group so that all of you know what is due when. They are also personally helpful when you are faced with several multitask projects at a single time and have a hard time keeping all the various steps clearly in mind.

SUMMARY

In this chapter we have stressed the problem-solving nature and the planning process involved in technical communication. The personal habits and goals you have encountered so far in your educational experience are usually transferable into the workplace. However, practicing the three ways of thinking about purpose—thinking rhetorically, thinking of yourself as a problem solver, and thinking of yourself as solving a problem for some people—will help you develop effective documents in your profession.

When you think rhetorically about the purpose of a document, you consider yourself as a communicator, you analyze your audience, you explore your topic or issue, and you choose the genre that is best for your particular communication or document. When you think about yourself as a problem solver, you can balance the ideal situation with the actual situation and then propose a way of coming closer to that ideal situation, usually a technical one. And, when you think of yourself as solving a problem for some people, you consider the human situation. Your audience might have a procedural, information, or decision-making problem that your document can respond to or solve. You need to ask what action you want your audience to take. Although these three ways of thinking about purpose are linked and mutually informative, they are complicated by multiple audiences and purposes.

Finally, in this chapter you began to see how weekly planning sheets and writing logs can help in planning a simple or complex document.

ACTIVITIES AND EXERCISES

1. Suppose that you are a new graduate from your university with a major in forestry. You feel that the company that has hired you ought to be doing more in the way of reforesting the federal lands that it has rights to log. You feel that the company should sharply increase its investment in reforestation even though there are few immediate gains to be reaped from such an investment, and you feel that the company would over the long run gain the trust and confidence of the public and the government by initiating reforestation. What kind of document could you write in this situation? Using the three methods described in this chapter, come up with plans for the document.

2. Here are four statements of purpose. How adequate do you think they are?
   a. KVC Corporation has recently announced that "unless something unexpected happens," production at its Carr Forks Mine near Mineralton, Utah, will have to be cut by more than half because several major ore veins have been lost in a complex fault zone. As chief consultant hired to make something unexpected happen, my job is to locate the lost ore veins. I first must initiate an intensive study of the local geologic structure. Using the information I find about the individual fault zones across which the ore bodies were lost, I should be able to locate the positions of any lost ore zones. This method has proven to be highly effective in the past, and it should permit the mine once again to operate at an economically feasible level.

b. In the past few years, many experts and other bird lovers have observed an unusually large number of white-chested tufted titmice dying of starvation, sometimes close to regular stock bird feeders. Because our company, Bird N. Hand, Inc., is the largest producer of bird seed designed for these titmouse birds, we've been working to develop a better bird feeder so that these birds can more successfully get at the seed. I'm the chief engineer, so to speak, on this project; I was picked because I have experience designing feeders for related species. I need to write a report that will permit our product designers to manufacture a successful bird feeder that will keep these birds from starving to death.

c. My manager in our Systems Department, Truman Furgerson, has assigned me to evaluate the quality of two new software packages designed to assist with inventory problem like ours: Computer Products Corporation's InvenSort 3, and Roberts Computer Software's Sort 6.3. Both of these packages are highly compatible with the computers that we currently use in our company. So I will have to write a report that will permit Mr. Furgerson and other managers to determine whether one of these new packages is better than the one we now use.

d. The purpose of our brochure is to inform citizens of Minnesota about the services offered by the Minnesota AIDS Project (MAP) and to give basic information about AIDS and HIV so that people can protect themselves. The brochure might also encourage people to donate time and money to MAP. In short, we want the brochure to inform citizens about AIDS and HIV, persuade them to take steps to protect themselves, and consider MAP to be a worthy cause. The brochure must also make it easy for citizens to contact MAP for more information. We intend to use the brochure in many ways: in direct mailings to citizens, in response to telephone requests for information, to pass out at events. Thus, it is possible that there could be secondary uses: Could high school counselors and teachers use the brochure to help them educate teenagers? Could newspapers refer to the brochure in the process of writing their stories? Could people who have contracted the disease use it to take advantage of our client services?

3. Exercise *2d* above refers to a document you can find in Appendix A, Case Documents 1. Look at one or two of the other documents from Case Documents 2, 3, or 4 collected at the back of this book. Can you recover in writing the kinds of purpose statements that the authors of those documents might have employed?

4. Name the purposes of two communications (oral or written) that you have composed in the past two months or that you are currently in the process of composing. These could be school assignments, communications required at work, or something that you might be producing for an organization that you belong to. If you can, try to use the questions on the Writing Strategies 2–1 worksheet to guide you.

5. Using Example 2–6 as a guide and following the suggestions in the section on planning your time and activities, plan for a project you must do this term in school or at work. Be prepared to share your plan with another student for feedback and suggestions.

6. Planning is particularly useful as part of designing documents that will be translated into other languages. Many communications fail because planning of this sort is ignored. Imagine that you are designing online help that will be translated ultimately into Japanese and German. Thinking rhetorically as a problem solver and as a solver of problems for particular people, make a list of the kinds of questions you will want to consider as you create your document.

Audience and Technical Communication

Introduction ▌ The Meanings of Audience ▌ Determining Who Will
Read Your Document ▌ Determining Your Audiences' Needs,
Attitudes, and Knowledge ▌ Determining the Conditions under Which
the Audience Receives Your Message ▌ Audience Analysis: Sample
Working Papers ▌ Creating an Audience or a Reader

I am writing for myself and strangers. This is the only way I can do it.
Gertrude Stein, *The Making of Americans, Selected Writing of Gertrude Stein,*
ed. Carl Van Vechten (New York: Vintage Books, 1972).

As authors, the physicists faced the challenge of persuading the biologists and
chemists, audiences that were unfamiliar to them and quite possibly indifferent to
their arguments, that a variation of Monte Carlo was better than molecular dynamics.
To do this, they employed a number of strategies to learn about their audiences,
including one seldom addressed by current scholarship in rhetoric: They entered
into collaborative relationships with actual members of their audiences. The aims
of these collaborative interactions included obtaining feedback on their paper and
learning, firsthand, the needs and concerns of these unfamiliar readers. The physicists
were addressing individuals in different social contexts who quite possibly possessed
different values, goals, and approaches.
Ann M. Blakeslee, "Readers and Authors: Fictionalized Constructs
or Dynamic Collaborations?," *Technical Communication Quarterly*
2.12 (1993), pp. 223–35.

Introduction

In the last chapter, we advised you to spend a great deal of time and effort planning. We do so because on all but routine assignments, successful technical communicators spend about three-fourths of their time not writing sentences and paragraphs but *planning* what they want to go into those paragraphs and sentences, *developing* their information and arguments, and later *revising* their drafts into professional work. The documents you write and the oral presentations you make will turn out better if you give proper attention to planning, inventing, arranging, and revising your communications. As in the case of praiseworthy engineering projects, a good communication *product* depends on a good production *process.* Part of that process is thinking about your audience.

In this chapter, we want to suggest some general ways of thinking about audience that you will find productive and stimulating. We have already discussed in Chapter 2 how audience analysis can help you plan your documents, and the various chapters that follow this one will have more to say about audience because audience is one of the central concerns in technical communication.

Thinking about audience is not something that you do at one point in planning a communication and never again; rather, considering audience is an ongoing activity for good technical communicators. They think about audience as they conceive of their task; as they develop and organize content; as they draft individual paragraphs, sentences, tables, graphs, and illustrations; as they test the effectiveness of their document; and so on. In this chapter, we concentrate on how you can think about audience in the process of **invention,** in the act of *developing your content.* Our goal in this chaper is to get you to think about audience in ways that will influence what you decide to put into your documents.

invention
the act of developing the content of a communication

The Meanings of Audience

What exactly is "audience," anyway? The answer to that question is, of course, easy to supply in many cases; audience refers to the person or persons who actually hear an oral communication or who actually read a written one. Audience denotes the real consumers of communications. The concept of a **real audience** is especially concrete for you when you are speaking because a listening audience is sitting right in front of you and is hard to ignore. Audiences who are lost or confused by an oral presentation can make their discomfort known by body language ("If this thing doesn't end soon, I'm going to scream," their bodies say), by their closed eyes and steady breathing, or by their verbal responses ("Excuse me, Michael, but I'm really confused by what you just said").

Audiences for written communications can be just about as real (and just as responsive) as the ones you see when you give a presentation. If you are writing to people who know you well or who work closely with you,

real audience
the audience who actually hears, views, or reads a communication

your audience is real and immediate indeed. For example, turn to Appendix A, Case Documents 3, where you'll see some letters that were exchanged between representatives of a citizens' group and a university dean about trapping and killing birds in experimental fields. Here are the first and last sentences of one of those letters:

> The Coalition of Citizens Concerned with Animal Rights (CCCAR) was alerted by a concerned person two weeks ago that the University is trapping and killing large numbers of birds in the experimental crop fields on its campus . . .
> For these reasons, we insist that the University immediately stop the killing of these birds.

The letter writer, a member of CCCAR, addresses a real and immediate audience: Dean Karen Wartala. The letter writer announces a topic of concern and calls for a specific action from that audience.

When we use the term *audience* in this book, then, we are most often referring to that real, immediate, and concrete presence. But you may also write to address an **imagined audience:** a construct in our minds that helps us to compose our documents. In this sense, an audience is an imaginative concept in the mind of the writer, rather than a flesh-and-blood presence.

imagined audience
the audience as constructed in the mind of the writer or speaker, an imagined, writer-centered concept

You probably are aware of this writer-centered sense of audience yourself from your own writing. You have an awareness of individuals whose presence in your mind helps guide your decision making during writing. This imaginary, writer-centered concept of audience can be especially important when your sense of your "real audience" is rather vague, when you are writing to readers known to you barely or not at all. In these cases, audience can be a pretty abstract notion, as abstract as the proverbial "dear sir or madam" at the top of a form letter. For example, if you look in Appendix A, Case Documents 2, you will see a public relations release from the Exxon Company entitled, "How Tiny Organisms Helped Clean Prince William Sound." Here are the first few lines of this document:

> Alaska's Prince William Sound today is essentially clean. There are several reasons for the rapid cleanup of the area, but among the least known and most important factors in the recovery process were microscopic petroleum-eating organisms that were assisted by a process known as bioremediation.
> Prince William Sound is home to naturally-occurring single-celled creatures that continuously feed on hydrocarbons from natural petroleum seepages and evergreen droppings. . .

Here the writers had to imagine a great many people who might read their document, people who knew little or nothing about bioremediation and the environment of Prince William Sound, people who might be hostile to the company and upset by the *Exxon Valdez* spill, and people who needed to be reassured that cleanup had proceeded rapidly.

The goal of this chapter then is to help you make the audience in your mind a very concrete and useful concept for you, whether you know the real people who will ultimately encounter your work or not. As a writer you should think of audience not only as a group of real, flesh-and-blood people who encounter your work, but also as a *concept in your mind that will help you create effective documents.* Real audiences exist, and you can think about them in productive ways. But even if your real audience doesn't have a concrete presence for you, you can make audience a **creative concept;** you can make audience an element in your mind as you communicate to help you develop more effective documents and presentations. Whether the term audience refers to a real presence or to something in the mind of the writer might not make much difference in the long run. In either case, whether your audience is very real to you as a living presence that animates your imagination, or whether it is something vaguer, there are certain questions about audience that you can ask that will invigorate your communications. That's the premise of this chapter and will become much clearer to you in the next few pages.

Finally, before the chapter ends, you'll also be introduced to a third way of thinking about audience. You'll learn that audience can also be a **textual presence**—an element on the page of the documents you write, a "character" in your prose that you can actually create in your documents in order to achieve certain effects. That is, toward the end of the chapter you'll meet the "audience in the text"—a character that you can create in your work to help you meet your rhetorical aims.

creative concept of audience
an element in the writer's or speaker's mind to help develop more effective communications

textual presence
a "character" in the text that the writer or speaker creates to achieve certain effects; a role that the reader or listener is asked to take on

Determining Who Will Read Your Document

A good question to ask yourself early in the composing process is who will actually encounter and use your documents. It's important to know who your real audience is, and many times it is simple enough to identify that audience. If you are speaking before a group of people or writing a personal letter to someone you know, your audience is easy to identify. However, it's not always so easy to figure out who your audience is. To get an inventory of everyone who might read your documents, you should think a bit about whether your audience is simple and homogeneous, or complex and heterogeneous; and you should think of how your documents will be used.

SIMPLE, HOMOGENEOUS AUDIENCES

When you write a letter to a personal friend, you know exactly who your audience is, and you know that no one else is likely to see your letter. Sometimes in technical writing you will communicate with a single individual when you write routine memos to a colleague or collaborator, and you will occasionally write personal or semipersonal letters to clients and customers.

These single and well known audiences qualify as relatively simple to identify and often simple to address.

Other times your real audience might consist of many individuals, but you can conceive of them as a simple audience because their needs are all roughly the same. For instance, if you give an oral presentation to a group of colleagues with similar backgrounds and needs who are gathered together to hear you for a similar purpose, you can think of them as a **simple, homogeneous audience.** Or if you write an article for a trade journal, you can address everyone in your audience in rather similar terms because your readers can be considered as very homogeneous: they have the same needs and responsibilities, and they read with the same goals. Although many individuals read the journal, they can all be addressed as if you were writing to a single one of them.

simple, homogeneous audience
readers, viewers, or listeners with similar backgrounds and needs who use a communication for a similar purpose

MULTIPLE AUDIENCES

In other situations, however, as we have already seen in Chapter 2, it is not always so easy to identify all the audiences of a communication. Take a look, for example, at Example 3-1, which reproduces a deceptively simple memo from a university writing coordinator, Robert Topping, to his supervisor, Marianne Angell. Can you imagine who else might read this document or receive a copy?

This memo seems to be a simple report on the developments described in the memo, having to do with efforts to change the testing procedures used in the writing program at the university. The audience seems to be easy to identify—it's Dr. Angell, named as the recipient of the memo. But with a bit of thought, you can easily imagine that others will also be likely to receive and/or read the memo, that the memo is in fact a very complex attempt to address several readers at once.

Besides Ms. Angell, Mr. Topping is also addressing several other people in his own department. First, there are obviously members of the Writing Committee, who want to know how efforts at reform are coming along. At least one important member, Sarah Maxwell, is addressed directly; but the other committee members are also very likely to receive copies of the memo. Second, Brenda Chait will need a copy of the memo in order to arrange the next meeting on the subject.

Then there are people outside his department that Mr. Topping is addressing both directly and indirectly. You can tell that he is planning to send copies to them because of the way he elaborately outlines the current situation and because the final paragraph prepares them to receive a phone call from Ms. Chait. These other readers include the following:

multiple audiences
several readers, viewers, or listeners, in the present or future, inside or outside an organization, who receive and use a communication

- Mr. White, who apparently opposes the proposed plan but who must be acknowledged with respect;

EXAMPLE 3–1   A Memo to a Complex Audience

Megalithic State University
Interoffice Correspondence

DATE: October 17, 1994
TO: Marianne Angell, Head, Department of English
FROM: Robert Topping, Director of Writing
SUBJ.:   Recent meeting on placement of undergraduates

As you know, Marianne, last month the Writing Committee, after lengthy debate, recommended that the department should seek to improve the method used to place our students into the required first-year writing courses here at Megalithic State. I just wanted to bring you up to date on recent developments.

The committee (and hence the department) wishes to get students placed more reliably in the appropriate course. Most students complete their writing-course requirement by enrolling in English 101, Rhetoric and Composition, but about 15 percent take English 101H, the "honors" version of the course; and another 15 percent must first complete English 100, Basic Writing, before they go on the English 101. Students enroll in the three courses on the basis of the First-Year English Placement Test administered by the Academic Testing Program during the summer orientation program. The trouble is, many English faculty (and the committee as a whole) feel that a great many placements are incorrect: faculty report that many students placed into English 101 and 100 belong in the other course, and they feel that many students now placed into English 101 actually belong in English 101H. These incorrect placements result from the fact that the Placement Test is an indirect test—a multiple-choice, machine-graded exam—instead of a direct test that assesses students' actual writing samples. The committee, particularly Professor Sarah Maxwell, also feels that our indirect test is unprofessional: most experts in testing agree that indirect testing is far less accurate than direct tests, particularly for nontraditional and minority students.

Consequently, the committee asked me to investigate alternatives to our current testing method. I shared our concerns informally in meetings with Stanley White, coordinator of the Academic Testing Program, and several of his colleagues; with Gene Poritzky, who chairs the University Committee on Testing; and with Carol Thigpen, Dean of Undergraduate Studies. Dean Thigpen understands our perspective on this issue and hopes that a direct writing test can be incorporated into the summer testing program. Mr. White and his colleagues, on the other hand, are worried about changing the testing program. He wonders how time for an expanded direct test can be carved out of a busy orientation program, how these tests can be scored quickly and reliably enough to return results to students in time for registration, and who will pay to score a direct test.

I am personally convinced that all of Mr. White's objections can be answered. In my memo to him of September 21, I tried to address his concerns. Dean Thigpen feels we all should get together to discuss the testing plan and has asked me to schedule a meeting for early November. November 8 or 9 might be good possibilities. Could you ask Brenda Chait as Administrative Assistant to contact everyone on this memo and arrange for a mutually convenient meeting time? And would you be interested in attending the meeting yourself?

- Dean Thigpen, who apparently doesn't know much about the testing program (hence all the detail in paragraph 2 in a letter ostensibly to someone who knows that information already) but who still supports the reform (and who does not wish to alienate Mr. White); and

■ Members of the University Testing Committee, including Mr. Poritzky, who have to agree to any changes in testing policies.

Finally, there might also be a host of future readers who will encounter the document:

■ Some people might be called upon to help implement the change once it is agreed to, and read the memo as background.
■ Other people who might need to have background on the matter some years in the future, people like Mr. Topping's successor.

You can probably think of other possible future readers yourself. You can see then that even a short memo like this one can have a very complex readership. In fact, the nominal audience for a document—in this case, Ms. Angell—may not even be the most important audience for a communication at all.

To get a good inventory of all your readers, then, you should think carefully about all those people who might encounter your message now and in the future, both inside and outside your organization. Take note of people that you'll send copies to yourself, as well as people who might be shown the document by the people who get it, and do your best to address all of those people in your communication. When you join an organization, you usually pick up quickly a sense of how things get done, a sense of who likes to be kept informed on what. Incorporate that sense into your inventory of audiences and write with them in mind. In addition, think about who might read your communication in the future. Some of your readers will discard your communication immediately, but others will file it for future use. What will those future uses be? And how can you facilitate future users in how you communicate now?

SINGLE AUDIENCES WITH MULTIPLE NEEDS

Sometimes a single or homogeneous audience may not be as simple as it looks if that audience has **multiple needs.** Take, for instance, the user manual that came with your computer, CD player, or bicycle. You may be the real audience for that manual, but it will be a *very different you* who uses it each time. When you consider buying the computer, for instance, you might take a look at the manual to learn about its features and capabilities. In that sense, the manual addresses you as a potential buyer. Once you do buy the machine, you'll want to install it, and the manual must address you as an installer. Then you'll want to use it, so the manual must address you as a novice user. Later, as you use the computer and become more confident, you'll return to the manual, now as an expert, to pick up the nuances or to perform repairs or to troubleshoot.

Or, to give a second example, consider the various audiences addressed by a booklet describing the benefits enjoyed by employees of a certain

audiences with multiple needs
readers, viewers, or listeners who may seem similar but who have different needs for a communication

company. Because it is read by potential employees who are considering joining the company, it must address readers as a kind of recruiting device. Once the new employee signs on, that employee might read the booklet to learn exactly what benefits he or she possesses (the employee might need to know if he or she should purchase additional health or life insurance, for instance). Later, after an illness, the booklet might be consulted because the employee wants to learn how he or she might submit a claim for reimbursement for a medical expense.

Often individual readers wear several different hats as readers. When you take inventory of your audiences, remember to account for all those different hats, all those different uses that your document might be put to. From the perspective of the writer, individuals are complex and multiple to the extent that they have multiple needs. One individual or not, each reader in some cases might as well be many different kinds of audiences.

SOME ADDITIONAL ADVICE

Simple audiences. Multiple audiences. Future audiences. Possible audiences. Simple audiences with multiple needs. It can get pretty difficult, as you can see, to identify with any certainty the real audiences of anything you write. In fact, identifying the real audiences is often impossible because you just can't control who will actually read your work.

But identifying who actually will read your work is less important than identifying who *might* read it—and preparing for as many possible readers as you can, within practical limits. As a writer, what you need to do is to anticipate as many readers as possible so that you can design a document that is as useful as possible. In this third part of this book, you'll learn how to adapt documents to your many readers—how to address heterogeneous, multiple readers in one document. But, for now, take inventory of your audiences by thinking about who will receive your communication directly from you, who will receive it indirectly (from someone else), and who might read it in the future. Then think about any multiple needs that your communication might have to satisfy. If you do all that, you'll very likely have a good sense of your real audience, and you'll have begun the process of inventing information to address those needs.

Determining Your Audiences' Needs, Attitudes, and Knowledge

Once you've identified your readers, you can begin thinking about how to adjust your communications to them. You should ask three basic questions about all of the audiences that you turn up in your inventory:

1. What are your audiences' needs?
2. What are your audiences' attitudes toward you and toward your message?

3. What do your audiences know, and not know, already (Young, Becker, and Pike)?

Asking questions like these will help you to develop appropriate content for your communications.

WHAT ARE YOUR AUDIENCES' NEEDS?

We discussed to some extent in Chapter 2 the importance of assessing your audiences' needs before you formally prepare your communication. You learned to think of yourself as a problem solver and to think of your audience as someone with a problem that you need to solve. That is, you learned to address your readers' needs—to think of your readers' needs for answers to procedural problems ("How do I do this?") or informational problems ("Can you give me the information I lack?") or decision-making problems ("What should I do?"), or some combination of those problems. You learned to think of your purpose as an attempt to alleviate problems in specific ways.

One other consideration will help you to develop appropriate content for your communications: think of your **audiences' role(s).** Within those roles, what will your audience do as a result of reading or hearing your communication? Will your audience transmit the message elsewhere? Will your audience make a decision or take some action? Will your audience advise a decision maker or action taker? Or will your audience implement a decision or action? We discuss below four such roles: *transmitters, decision makers, advisors,* and *implementers.*

audience role
the role of transmitter, decision maker, advisor, or implementer that an audience takes after reading, viewing, or listening to a communication

Transmitters

Transmitters are people who receive a message and then simply direct the communication to the appropriate decision maker or action taker. For example, people who write grant proposals often submit them to people who simply receive the applications and pass them on to the advisors who actually rate them. Similarly, often job application materials are submitted to someone who simply receives the applications and passes them along to other people actually in charge of hiring.

Of course, not every message is directed to a transmitter; quite often the person addressed in a document is not a transmitter but the person who is actually involved in the action of the communication. But transmitters exist often enough that you should consider their needs, typically by making it easy for them to direct your communication to the right audience. Transmitters especially appreciate summaries, subject lines (in memos), title pages (in reports), introductions, and other references that permit them to route communications to the right people without having to digest the entire communication.

transmitters
audience members who direct a communication to a decision maker or action taker

Decision Makers and Action Takers

Decision makers and action takers, by contrast, may often be considered your primary audience. For them, you should provide the information and arguments they need to make decisions or to do the activities described in your communications. Much of the information in reports, proposals, instructions, and other technical documents is included with these readers in mind. Action takers and decision makers often hold executive positions within organizations and have many things competing for their attention; they appreciate completeness, care, and thoroughness.

Advisors

If a decision maker lacks the time or expertise necessary to take action as a result of a communication, he or she may rely on **advisors.** Advisors are often given the role of reading communications carefully and offering advice on them to decision makers. They appreciate the details of a presentation. If you anticipate that your report or proposal or other communication will also be read by advisors, you will need to anticipate their needs as well as your executive's needs; you need to be aware of their needs as you draft your work.

Implementers

Sometimes your audiences may be in the role of **implementers,** too. Once a decision is made, an executive may take action himself or herself. Or he or she might instead turn a communication over to someone else for implementation. If you anticipate that this will be the case, you will need to design your communications with implementers in mind. You'll have to provide the details on implementation that such readers will require.

Of course, it is somewhat artificial to speak of readers' roles as divisible in these ways. In real life roles often shift and overlap, and one person may carry several roles. Nevertheless, thinking about your audiences' roles—thinking of audiences as transmitters, action takers, decision makers, advisors, and implementers—might suggest things that you ought to include in your communications.

WHAT ARE YOUR AUDIENCES' ATTITUDES AND MOTIVATIONS?

In addition to thinking about your readers' roles, you will need to consider their attitudes and motivations—their attitudes toward your message, their motivation to read or listen to your message, their attitudes in the act of reading, and their attitudes toward you.

Attitudes toward Your Message

It would be convenient if we could behave as if readers were dispassionate people who take in messages without getting emotionally involved in them. But the reality is that people are emotional as well as reasonable, that they often have **attitudes** about things that you must take into consideration as you write or speak. These attitudes are related to your readers' motivations for using a document as well as their prejudices and assumptions about the document's content.

audience attitudes
motivations or emotions that determine how an audience may respond to a communication

Look again in Appendix A, Case Documents 2, at "How Tiny Organisms Helped Clean Prince William Sound," the public relations release concerning the cleanup of Prince William Sound in Alaska after an accident involving the oil tanker, *Exxon Valdez,* led to a major oil spill. This document includes sentences such as, "Alaska's Prince William Sound today is essentially clean," and "Repeated tests have shown that in Prince William Sound bioremediation was both effective and safe."

What are your attitudes toward those sentences? What might be the attitudes toward them of committed environmentalists? What would be the attitude of typical executives (if there is such a thing) in the oil industry? What would those audiences think of the use of the terms "accident" and "spill" in the sentences above? Depending on the reader's prejudices and assumptions, the phrase "major spill" rather than a "major environmental disaster" would be met with very different attitudes.

Obviously not all audiences share the same attitudes or share yours. Therefore, even in less controversial circumstances, you must think carefully about your audiences' attitudes toward what you have to say or write. The information and arguments contained in your documents and speeches need to be considered against your audiences' attitudes.

If possible, you need to anticipate those various attitudes and act accordingly. Often you will have to include material in your communication that will show your awareness of differing attitudes, and you will want to include material to counter those attitudes without insulting your audience. As a general rule, the more your audiences' attitudes differ from your own, the more you will have to do in a communication to address and counter their attitudes. As we will see in more detail in the next chapter, it's one thing to state a position on an issue; it's another thing to include everything necessary to change your audiences' views on that issue. As a start, it makes sense to do what you can to determine your audiences' attitudes before you write or speak, and then to create your communication accordingly.

Motivation to Read or Listen to Your Message

A particularly important attitude toward your communication is your audiences' motivation. How motivated is your audience to read your communication or to hear your presentation? With many things competing for their attention, you should not assume that your audience sees your communication as just as important as you do. You must assess your audiences' level of interest and act accordingly. Again, not everyone who hears you or reads your document shares the same motivation. Some may be eager and others bored; some may be interested in the whole presentation, others in only a few parts.

If your audience is unlikely to be interested in your presentation or document, you may wish to begin by **dramatizing** what you have to say in order to win your reader's interest. An especially effective tactic is to stress in your introduction the importance of the problem that you have been working on so that your audience will see its importance. You might also want to use more visual presentations—illustrations, graphs, and tables (in written communications), and slides, photographs, and overhead projections (in oral presentations). You would ask the audience to envision the current situation and contrast it to a future situation, both with and without the changes you are recommending.

If your audience seems unmotivated to attend carefully to your message, it also usually makes sense to be brief. No one minds attending to something they consider unimportant for a short time (and if something is important they don't mind paying attention for a long time!); but audiences do become impatient if they are asked to spend lots of time with what they perceive to be unimportant.

When dealing with motivation as with other attitudes in your audience, it is often useful to acknowledge that you are aware of the variety of motivations in your audience. You might want to acknowledge that variation by saying something like, "I know that some people may be tempted to find this presentation a bit dry, but if you will bear with me, I promise a good payoff in the end." If you try using a technique like this, be sure that your statement isn't an apology for your work but rather an acknowledgment of difference. Or, you might try some humor or wit.

Attitudes toward You

Your audiences' attitudes toward you are as important as your audiences' attitudes toward your message. Does your audience know you at all, either by reputation or from previous work? Are you considered a colleague, a superior, or a subordinate? How your audience regards you will affect the tone and presentation of your message. In many cases, particularly as you address audiences outside your organization or readers elsewhere in your organization, your audience will know little about you. When that is the case, you can build trust in your reader by following the advice on what we'll call *ethos* in Chapter 5 of this book. For example, you might mention your common concerns or your experience with your subject. In general,

dramatizing
a tactic to use with unmotivated audiences to involve them in the communication, including emphasis and visual display

what you'll have to do is consider yourself part of your presentation and make sure that you present yourself as you wish to be seen.

Attitudes in the Act of Reading

The preceding sentences might give the impression that an audience's attitudes are relatively firm before a communication is encountered and that attitudes change only after an audience has completely taken in a communication. But that is not the case. Attitudes are more difficult to chart than that, and responses to communications are continuous, not reserved until the end.

Communicators and their audiences are highly active, and there is a dynamic interaction between speakers and listeners, writers and readers. This active, dynamic communication is a process of intellectual negotiation that occurs in a lively (and, unfortunately, somewhat uncontrollable) social setting.

If you think about your own reading processes, you will remember that they are anything but passive. You have no obligation to do much "receiving" of a writer's message that is simply presented to you. You expect writers to attend to your needs, not simply to express themselves without considering their audience. As you read, you actively make notations, you mentally take issue with some points, and you take note of others as you interrupt your reading. In other words, you actively create meanings relevant to your own needs and experiences, rather than simply decoding the writer's intended meanings.

Moreover, you don't necessarily accept the writer's invitation to start at the beginning of a message and proceed directly to the end, but instead you might preview the reading first (by inspecting pictures and abstracts and headings, maybe even the final paragraph). Sometimes you may read every word of something in order (personal correspondence often gets consumed that way). But more often you read selectively: some sections get your full attention and may even be reread; other sections you read straight through without much reaction; other parts you might skip completely; sometimes you quit reading altogether after taking in a few sentences.

Even documents that readers are quite interested in are read in ways that suggest that readers are very active indeed. For example, you read some things at one sitting with full attention, but other times you break the reading up and take it in at several sittings and read other things in between.

In keeping with this view of reading, then, we advise you to see your audiences as reading very actively and responding as they read. Anticipate possible responses and adjust your presentation accordingly. Understand that the way your audience responds to one part of your communication will affect their responses to other parts. See your reader as a dynamic presence who must be led through your documents like a respected colleague.

WHAT DOES YOUR AUDIENCE ALREADY KNOW—AND NOT KNOW?

You can also turn up information for your communications by asking yourself what your audience already knows—and doesn't know. Here, as you plan your document or presentation, you have to consider your audience's experience, organizational role, expertise, and education. In addition to thinking about the main ideas that you have to present, you will need to consider the audience's background knowledge. If your audience lacks the background information to understand your presentation, you will have to supply that information as you go along. Because people remember by tying new information to knowledge they already have, you should plan to use comparisons and analogies that your reader will appreciate.

In general, then, size up your **audiences' knowledge** before you decide what to include in a communication. Here are some things to consider that might help you to do so, particularly if you do not know your audiences well.

audience knowledge
what readers, viewers, and listeners know already about the topic of a communication

What Are Your Audiences' Organizational Roles?

Your audiences' positions or **organizational roles** may give you useful clues to their needs and to what they already know. The tasks particular employees handle give them specialized knowledge that you can take account of as you decide on the content of your documents and presentations. Managers, for example, know different things than technicians do. You will want to consider the job titles (environmental engineer, nursing supervisor, assistant manager, group leader, systems analyst) that describe your audiences and consider what those roles imply about your audiences' knowledge.

audience organizational role
the task reader, viewer, or listener is assigned; includes job and professional titles

What Expertise and Education Does Your Audience Have?

People are the product of their experiences, and so taking account of your audiences' previous work experience and education may help you to determine how much they know. If you know your **audiences' background,** you will be able to assess with some reliability whether you are including enough or too much information for them and whether your vocabulary needs adjustment. You will need to ask whether your audience reads or speaks English well, or at all.

When you think about your audiences' education, think less about how much formal education they have had than about what they have studied. After all, a reader's educational level really has rather little to do with his or her specialized knowledge. Far more important is whether your audience has had any experience with the ideas and concepts you plan to present. If

audience background
what expertise and education readers, viewers, or listeners bring to a communication

your document involves specialized knowledge—information about technology or scientific processes—it will be especially important for you to consider your audiences' expertise and education.

Is Your Audience Familiar with the Topics and Terminology?

Related to the question of expertise is the matter of **audience specialty.** If your audience knows your specialty well, you can count on a shared knowledge of concepts, issues, controversies, and techniques, and you may not have to provide much in the way of background information. On the other hand, if your audience comes from outside your specialized area, or if you are dealing with an issue with which your readers are unfamiliar, be careful to provide explanations and accessible vocabulary throughout your communication.

The famous British scientist John Maynard Smith, perhaps the preeminent evolutionary biologist of our day, once told one of the authors of this textbook that when he writes he often has a very specific image of his audience. "I usually try to keep in mind an actual person, someone who happens to be a relative of my wife, an intelligent but rather ignorant British civil servant who is bent on improving his mind." You may not be able to generate a formal portrait of your audience like this. But the more you *do* to ascertain what your audiences know and don't know, the more successful you and your communications will be.

> **audience specialty**
> what topics, terminology, concepts, issues, controversies, and techniques readers, viewers, or listeners understand

---

Once you have inventoried your audiences and considered their needs, attitudes, and knowledge, consider the conditions under which readers will encounter your document or listen to your oral presentation. Most people behave as if audiences will encounter their communications under ideal circumstances. They anticipate communicating with a reader who is never interrupted, who is passionately interested, and who has unlimited time.

Your own experience will tell you that **reading conditions** are often anything but ideal. Sometimes you are hurried and can read only superficially. Sometimes you read only parts of documents and skip parts irrelevant to your needs. Often you read in snippets, getting interrupted now and again as you try to navigate a text. And sometimes you read in less than ideal circumstances. For example, maybe you are trying to read a manual under the car as you do a repair, maybe you have to use a document on a construction site in the wind or rain, and maybe you must use a document in an emergency circumstance. As a writer, then, you need to anticipate the conditions under which your document will be used.

Try to picture when, where, and how your documents will be consulted. As you work with this aspect of audience, you should do your best to plan for a document that will work wherever and however it is needed. Early

Determining the Conditions under Which the Audience Receives Your Message

> **reading conditions**
> the circumstances in which a reader encounters, consults, and uses a communication

recognition of what is needed on your part is essential if the final product is to be successful.

We illustrate and summarize the advice we've just given, particularly about audiences' needs, attitudes, and knowledge, by offering two concrete examples for you to consider.

SAMPLE WORKING PAPER ONE

Listen in (so to speak) as Ming Tchen thinks about the needs, attitudes, and knowledge of the audience of a manual he is writing for the director of his summer internship (Example 3–2). A prospective industrial engineer from

---

EXAMPLE 3–2   Showing Thoughts about an Audience's Needs, Attitudes, and Knowledge

Ming Tchen's Informal Chart

My audience is simple and homogeneous—I'm writing my successor in this internship. My audience isn't Paul, even though Paul has given me this assignment; my reader is my successor and I'm writing for that person's approval (as well as for Paul's!). Paul is just my transmitting audience. I guess I can imagine that I'm writing to the person I was three or four months ago. My goal is to prepare him or her to take my internship next summer without wasting a lot of time getting organized and oriented.

Needs: To know about IE Consulting—the company's history and specialties, key employees, and company policies. What the internship involves and what you can expect to do (dwell on some examples?). A profile of Paul, the internship supervisor? Miscellaneous stuff too: parking, lunch, reimbursement policy, pay, etc. In general, teach everything I learned in the first two weeks. For Paul, my transmitting audience, I'll have to compose a cover note to attach to my communication.

Attitudes: My successor's attitude toward what I have to say should be ideal—he or she will be eager to read what I am offering and very curious. Highly motivated reader. As for attitudes toward me, I'll identify myself as this year's intern, mentioning that I too am an IE student at Tech. That way, the reader should be friendly to me.

Knowledge: Again, not a big problem—the person should know everything that I knew in May, so I can assume lots of IE knowledge (no problem with terms and concepts). IE 436 prepares people perfectly for this internship, so I'll build on that in detail and use comparisons to what was covered in that course. My successor will have no real knowledge of IE Consulting, so be careful there—details, details!

This shouldn't be too tough to write after all!

Kansas City who has just completed his junior year of studies, Ming has been working all summer as an intern for IE Consulting, Inc., a firm that specializes in industrial engineering work. Ming's supervisor (Paul Grecco), recalling all the time it took to get Ming acquainted with the job and up to speed on everything, has given him a final task: to write a document that Ming's successor can use to prepare for the same internship next summer. To be sure his information will be complete, Ming considers his successor's needs; to do so, he remembers his own needs from earlier that summer.

Because he doesn't know his reader personally, he thinks of his audience as the "self" he was four months earlier. That reminds him to tell his successor what he or she should know about IE Consulting and about his supervisor; what he or she should do to prepare for the internship; the details about the conduct of the internship; and so forth. Having thought about his reader's needs, he considers that reader's attitudes. On consideration, Ming thinks that his reader will have ideal attitudes, from his point of view. He assumes that reader will be eager to learn all he or she can from the document Ming will prepare, and so is likely to read the document with great interest and motivation. He also feels that his reader is likely to have positive attitudes about Ming himself: Ming can count on his reader's respect because he comes from the same university and majors in the same field, and has had the benefit of just the kind of experience that his reader is seeking.

As for his reader's knowledge, Ming feels that he is again in a good position. Because his reader will have a similar educational background, Ming can count on him or her to know the technical processes and vocabulary that Ming plans to use; and Ming will also be able to build on his audience's familiarity with Industrial Engineering 436, a course that is a prerequisite to the internship and that prepares students for just the kinds of experiences they will encounter at IE Consulting. At the same time, Ming's reader will also need some background information on processes and procedures unique to IE Consulting.

SAMPLE WORKING PAPER TWO

Now, listen in as Mandy O'Brien plans an oral presentation for a meeting that she has scheduled for the next afternoon (Example 3–3). Mandy directs the employee training program at her large manufacturing company, which has a major facility in Pennsylvania and smaller ones elsewhere in the northeastern United States. Employed by her company for two years now, Mandy has been noticing for several months some weaknesses in the kind of continuing training that is being offered by the company, especially at its facilities outside the main office.

In addition, she would like to suggest some improvements of various kinds in some of the training programs traditionally offered by the company.

EXAMPLE 3-3   Showing Thoughts about an Audience's Needs, Attitudes, and Knowledge

Mandy O'Brien's Informal Chart

Tomorrow's presentation. Audience: Emilia, Sam, Harry Moore

Needs: They don't know it, but they need improvements in our company's training programs. I've spoken to Sam about this informally, and he tends to agree with me that the current programs are getting old-fashioned. And I feel that the ones being put on by Harry Moore at our other plants aren't very good at all. They are poorly coordinated with our other facilities and not offered very professionally at all. In fact, Harry M shouldn't even be offering training programs—it's my job, not his. But some of these programs predate both of us so I've never complained. Still, we can be doing much better both here and at our other locations. I need to establish that early (especially for Emilia, who knows next to nothing about all this) with all that data I've collected and by indicating that I know of better alternatives. Show that our programs are broken and need to be fixed. And get myself assigned to the task of formally proposing a solution—that's my specific goal.

Attitudes: I'm not sure, but I think Emilia will be receptive to my ideas—she's given me responsibilities and she hired me because I could make our training better. She also stands to lose nothing, and she'll be very interested when I show her I might be able to deliver better training for less money. Same with Sam—we've already talked about improvements that we'd like to see at the main office, and I think he'll like some of the ideas I've been developing since we talked last. Harry will be a problem maybe. Will he see it as a personal criticism if I come on too strong about a "problem" with the training at his plants? Got to be tactful there, especially since he doesn't like to get ideas from people like me (I hear he doesn't like getting advice from younger women) and since Emilia will be there. Will she be protective of Harry's views, since he's an assistant VP? Oh, well—expect the best, especially with Sam and Emilia there. Harry still sees me as a rookie, too, so I'm going to be a real pro in my presentation—don't let it get too informal; try to keep the agenda on what I've called the meeting to discuss. Sam is friendly and respects me. Emilia probably still thinks of me as inexperienced—I want her to leave this meeting very happy that I was hired; I want her to think of me as a pro.

Knowledge: Emilia has only a vague and general idea of our programs, so I'll have to make sure that my introduction goes over old ground for her. Don't bore the other two, though—Harry and Sam know most of the details of the training programs. But their expertise isn't in training, so I will need to explain my new ideas in at least some detail. Go easy on the jargon! Don't be pretentious. But show that I know my stuff, especially for Harry's sake. Maybe it would be wise to have some of my plans in short written (but tentative!) form to pass out if the occasion presents itself.

So she has called a meeting of the company vice president (Emilia Black), the assistant vice president in charge of production at the main office (Sam Priest, who is also her boss), and the assistant vice president in charge of the other plants (Harry Moore); all of them have only a vague sense of what the meeting is for. Anticipating some reluctance from one or two members of her audience (all of whom outrank her), Mandy has called for and is planning for an informal oral presentation rather than making a written proposal; her goal is to leave the meeting with a mandate to devise a written plan that can be formally considered at a later date. She feels that it would be unwise for her to draft a formal plan herself without creating some kind of demand for it—hence the meeting.

Mandy has recorded informally her thoughts about her audiences for the presentation. Note that although the problems she is addressing are really her own responsibility, she tries hard to conceive of them in terms of her audiences' needs.

Then she thinks carefully about her audiences' attitudes. Her boss, she feels, is likely to think favorably about her planned changes, because she has spoken to Sam about them in the past; in fact, some of the ideas she has for improvements were suggested by her boss. The vice president is also likely to be cooperative, Mandy feels, because she has no real personal stake in the matter and is always interested in improvements. But the other assistant vice president is unlikely to be receptive to her ideas, Mandy feels, because he has initiated his own employee training programs for other plants—some of them precisely the programs that Mandy wants to reform and coordinate. With Emilia at the meeting, he may be especially protective of his training programs, especially since, strictly speaking, he should have been operating those programs through Mandy's office.

Mandy also thinks about her audiences' attitudes toward her. She has a satisfying professional and personal relationship with Sam, but Emilia, the company's vice president, doesn't know her well and probably views her as somewhat inexperienced yet. As for the assistant vice president in charge of other plants, Mandy again anticipates trouble. Having come up from the ranks and having worked for many years at other company facilities, he tends to be protective of his territory and to view suggestions as interference. He also sometimes finds it difficult to take advice from younger women.

So Mandy is thinking of information to include in her presentation that will show a special awareness of his protectiveness toward his territory and has resolved to take a rather formal approach to the tone of her presentation so as to be as professional as possible. As for her audiences' knowledge, Mandy is the expert in this case—she's the only one who knows a lot about employee training programs. So she plans to provide careful explanations of the proposals that she wants to kick around; in fact, the more she thinks about it, the more she thinks it might be wise to have some written explanations to hand out to support her presentation. She'll also have to provide some background information for Emilia.

Whether you use notes like the ones illustrated in Examples 3-2 and 3-3 or whether you use other more or less formal means of thinking about your audiences, the important thing is that you do such thinking. Doing this can make the difference between a document or presentation that speaks

---

WRITING STRATEGIES 3–1   Questions to Ask about Audience in Preparing Documents and Presentations

Who will read this document or hear this presentation?

- Is your audience simple and homogeneous? or multiple and heterogeneous?
- Do any of your audiences have multiple needs?
- How will the above affect the contents of your communication?

What are your audiences' needs?

- What problem of your audiences are you trying to solve?
- What are your audiences' job classifications?
- What are your audiences' roles: transmitters? action takers? advisors? implementers? a combination?
- How will the above affect the contents of your communication?

What are your audiences' attitudes toward you and toward what you have to say?

- What do your audiences think of you as a person and co-worker?
- What will your audiences' responses be to what you have to say or write?
- How motivated are your audiences to attend to you and your communication?
- How will the above affect the contents of your communication?

What does your audiences know—and not know?

- Do you need to supply any background information, explanations of terms, or other information to your audiences?
- What do your audiences know already that you can build on?
- What are your audiences' professional background, educational background, and field(s) of expertise?
- Do your audiences have experience with the ideas and concepts you are presenting?
- How will all that affect the contents of your communication?

Under what conditions are your audiences reading?

- Is there anything unusual or less than ideal about the time and place that your audience will read your communication?
- How does that affect the contents and design of your communication?

generally to everyone, or specifically to the needs, attitudes, and knowledge of your intended audiences. The four questions we have raised about audience so far in this chapter are summarized in Writing Strategies 3-1.

Creating an Audience or a Reader

So far in this chapter on audience, we've encouraged you to think about your real audiences—the actual individuals who will encounter your documents and presentations. But at the beginning of the chapter we told you that we would also be talking about the audience not *of* the communication but *in* the communication—the audience in the text.

The concept of the audience in the text is rather difficult but important to grasp. We will try to explain this concept with a few examples. Consider the short poem in Example 3-4 that you may well have encountered sometime already. It is called "Spring and Fall: To a Young Child" and was written around 1880 by Gerard Manley Hopkins, a Jesuit priest who died in Dublin in 1889. The speaker in the poem is addressing a young child, "Margaret," who is crying because she feels a sense of loss when she sees the "Goldengrove unleaving"—the beautiful autumn leaves falling off the trees in front of her and turning into late November "leafmeal" (bits of leaves).

Who is the audience of this poem? Despite its title and the presence of Margaret in the poem, we know that "Spring and Fall" wasn't really written for a young child after all. Children cannot understand poetry this complicated, and the theme of the poem—that we must remember that we will all turn to leafmeal one day—isn't exactly something young children are

---

EXAMPLE 3-4   The Audience in the Text

Margaret, are you grieving
Over Goldengrove unleaving?
Leaves, like the things of man, you
With your fresh thoughts care for, can you?
Ah! As the heart grows older
It will come to such sights colder
By and by, nor spare a sigh
Though worlds of wanwood leafmeal lie;
And yet you will weep and know why.
Now no matter, child, the name:
Sorrow's springs are the same.
Nor mouth had, no nor mind, expressed
What heart heard of, ghost guessed:
It is the blight man was born for,
It is Margaret you mourn for.

audience in the text
a fictional character created by the author to aid in making a rhetorical point; an implied reader

ready for. The real audience for this poem is people "grown older" whom Hopkins wants to remind of what is most important in his eyes, namely the certainty of death. As for Margaret, she is an **audience in the text**— a fictional character created by the author in order to aid him in making his rhetorical point.

Think of it this way: Hopkins creates in his poem a sort of miniature drama, complete with characters (Margaret and the speaker) and action (the leaves falling, the tears, the speaker's sermon). As real readers, we eavesdrop on and observe this created drama in order for the author to achieve his rhetorical purpose—to remind us about our ultimate end.

Now what does all this have to do with nonfiction writing, in particular technical writing? All writing, believe it or not, can be seen as a sort of verbal drama akin to "Spring and Fall." It has fictional "speakers" just as this poem has a speaker; it has fictional "listeners" just as this poem has its Margaret; and it has verbal action. Sometimes the fictional listeners, the "audiences in the text," if you will, are quite overt, as overt as young Margaret. For example, the "letters to the editor" in your local newspaper are really directed not to the editor (the "dear editor" is simply a created character within the text) but to readers in the community. An "open letter" to someone that is printed in a newspaper is also really written to people in the community. *The Wall Street Journal,* for instance once published an open letter from the economist Milton Friedman to President Reagan's drug czar, William Bennett, that began, "Dear Bill," and that contained other ostensible references to Bennett; yet the letter was really addressed to the public (readers of the *Journal*), not to Mr. Bennett. If Friedman wanted to write to Bennett, he would have sent it through the mail, not published it in *The Wall Street Journal.*

Another example that also comes to mind is Martin Luther King, Jr.'s famous "Letter from Birmingham Jail." The letter begins "To My Fellow Clergymen" and then seems to address those clergypeople as "you" throughout the course of the letter. But in fact the letter wasn't really addressed to the clergy (otherwise King would have sent it to them in the mail rather than publishing it in many other places); rather, the clergypeople were created and put into the letter by King in order to help him to win assent from the general public (King's real audience) for his civil rights efforts.

So far we have focused on dramatic fictional characters within the text in order to begin to explain what we mean by an *audience in the text.* But we can argue that every document is addressed to a fictional reader that is present, or implied, within the text. Occasionally, as our examples indicate, that character is indeed an overt presence in the text. But more often— especially in technical writing—the reader in the text is only implied.

A reader in the text is implied by the language and conventions employed by the writer and by background knowledge assumed by the writer. Just as every document contains an implied author, a narrator (or *persona,* as we'll call the implied author in the next chapter) who may or may not have a lot in common with the real author, so too documents contain implied readers (created by writers) that may or may not have lots in common with the real audience.

Let's look at the two documents in Example 3–5 for another illustration of creating an audience or reader in text (you will also see these documents in Chapter 5 when we discuss persuasion). The Apple Computer manual creates an implied reader who even has a name, "you." That "you" is a "concerned" type of person, the kind of employee who worries about the company's equipment as if it were his or her own. That "you" is also assumed to be a worker. And that "you" is an informal and approachable worker, the kind of person who likes to drink coffee at the desk and perhaps likes the socializing that goes with coffee drinking at work. In addition, that "you" has a sense of humor. On the other hand, the "you" in the Motorola manual is constructed not as the informal worker of the Apple document. Rather this "you" is rather more formal—the kind of person to whom the writer assumes he or she must say "welcome" and "thank you" to before giving

---

EXAMPLE 3–5   Two Examples of Creating an Audience or Reader

Apple Computer

Many people are concerned at first about making a mistake that will damage their computers or their work.

But you don't need to worry. If you set up your Macintosh according to the instructions, avoid bumping it while it's turned on, and don't spill coffee on the keyboard, you won't hurt your equipment.

Source: Apple Computer, Inc., *Getting Started with Your Macintosh*, "Welcome," ix.

Motorola

Thank you for selecting Motorola and welcome to cellular telephone service. Your Digital Personal Communicator™ represents the state-of-the-art in personal cellular telephones today. The listing below shows just a few of the exceptional features that the telephone contains. . .

To cover all of these features properly, we will take you through a logical step-by-step learning procedure that explains everything you need to know to operate your new telephone.

Source: Motorola, Inc., "Introduction," *Digital Personal Communicator User Manual*, p. 4.

directions. In addition, the "you" is constructed as "logical" and thus presumably unlikely to panic when faced with instructions on how to use an unfamiliar product.

Thus, you see that writers create implied, fictional readers in their texts; they create fictional roles for their implied readers by means of language choices and conventions. And then authors implicitly invite their real readers to assume the roles implied for them. Sometimes the authors of computer manuals, for example, create a reader in the text who is uncomfortable with the machine or needs to be cajoled into using it. The implied reader is given all sorts of commands and warnings and directives ("do this, don't do that"); the implied reader seems anxious and error-prone (there is lots of emphasis on errors and the reader's fears); and the machine seems to call all the shots. Other times the writers of computer manuals create a different kind of reader in the text—a responsible person patient enough to sit still for a tutorial, an adaptable and intelligent and unintimidated person who makes decisions and creates things.

CREATING READERS IN TECHNICAL DOCUMENTS

To look more carefully at how the created reader works in technical documents, compare the introductions of two scientific documents reproduced in Examples 3-6 and 3-7. These two introductions were written for exactly the *same real audience*—researchers in evolutionary biology, most of them university biology professors and their graduate students. We know that because the documents were published in scholarly journals read by those kinds of readers.

But note that the two documents construct very *different fictional audiences* within the text. Both of them imply audiences of professional biologists—that's clear from the scientific jargon ("natural selection," "Darwinian fitness," the authority "Lewontin," "adaptation," "constraint," and so on) used in both introductions without the explanation that would be required if those terms were offered to readers outside evolutionary biology. The articles also situate their readers as biologists by invoking some of the shared assumptions, beliefs, and common interests of the field, for example, that "it is assumed that evolution has occurred by natural selection," that "there is nothing particularly new in this logic," and that "we do not impose our biological biases."

But beyond that, very different implied audiences are created. The first implies an audience of scientists who are *only* scientists. By employing passive and impersonal constructions ("It is assumed," "are to be interpreted," "has been a growing attempt"), by treating his adversaries with respect and impartiality, by offering standard qualifications, and by keeping his language neutral and dispassionate, the author implies that his readers are

idealized, almost stereotypical models of their kind—objective, impersonal, fairminded, careful, reserved, and reasonable pursuers of factual truths.

In addition, the first introduction creates scientists who are somehow subtly inferior to the scientist doing the talking. It is as if the author-director first puts implied readers into white lab coats and then conducts them to a lecture hall where they are asked to sit and take notes from a master scientist in front of them. The decisive, personal "I" that is used to forecast the essay's contents establishes the author as an authority and the implied audience as a relative novice in need of direction. Like a student, the implied reader is directed to take notes—but not to take issue—with what is to follow from this presentation from a confident teacher.

The second introduction, by contrast (Example 3-7), although it too conceives of its implied readers as professional scientists, creates its readers as *more than* scientists. By beginning with St. Mark's Cathedral and King's College Chapel, Voltaire and Dr. Pangloss, Biblical tradition and art history, metaphors and allusions, the author creates readers who are not merely scientists or specialists; these scientists are *worldly and well educated and cosmopolitan*. By using the active voice and avoiding neutral phraseology, the author makes his readers less stereotypically scientific. In addition, by using "we" and avoiding heavy, pedantic forecasting, the author creates readers who are equals, not subordinates—readers who are invited to take issue with (not simply take note of) what is presented. If the first introduction cues readers into the role of students taking notes, the second removes readers from school and addresses them as colleagues, not subordinates.

Both of these introductions are well written; both are effective. They illustrate that there is a range of possibilities in scientific and technical writing for how you want to construct your implied reader. Your job as a writer is, in a sense, to create a drama, to cast readers into certain roles. As the producer and director of this drama, you can't help but create roles and action, so it makes sense to think carefully about the roles you are creating. In Chapter 5, you will learn how to create effective narrators or effective implied speakers that we will call *personas*. Here we want you to consider what kind of roles you are implying for your readers. What specific signals place implied audiences into certain roles? How will your real audiences respond to the implied audience that you create? Are you creating an implied reader who is an ally and trusted partner, or are you unwittingly creating readers who are adversaries?

SUMMARY

In this chapter you have learned how to think about audiences—the real consumers of your communications. You have learned that audiences are often real, immediate, and concrete presences to whom you address your communications. But you have also seen that audiences can be constructed in your mind to help you compose your documents. Finally, you have learned that audiences can be a textual presence that you can create to achieve certain effects.

At times you may address audiences who are simple and homogenous, with the same traits and similar identities. These audiences may be easy to imagine and understand. However, you will often encounter multiple audiences who will make use of your communication. Meeting all their needs is more challenging. Finally, you may be challenged by single audiences with multiple needs, who use your communication differently at different times.

To understand and address your audiences effectively, you need to determine your audiences' needs, attitudes, and knowledge. Audiences may need to solve procedural, informational, or decision-making problems, as we discussed in Chapter 2. However, your audiences' organizational roles must also be part of your consideration in creating a communication. If the members of your audience are transmitters, you'll need to make it easy for them to direct your communication to the proper audience. If they are decision makers, they will use your information and arguments to make decisions and to do activities. If they are advisors, they will use your communication to offer advice to decision makers. Finally, if they are implementers, the information you convey will help them take or direct specific action.

You will need to access your audiences' attitudes, assumptions, and prejudices toward your message and to anticipate, address, reinforce, or counter these attitudes within your communication. Your audiences' motivation or their level of interest can be enhanced by dramatization, visual presentations, or even humor. You must also keep in mind that their attitudes toward you, their knowledge of and respect for you, will either encourage or discourage them from listening to you or reading your words. Finally, if you assume that your audience is actively, constantly engaged in intellectual negotiation as they encounter your words, you can better decide how to appeal to them.

Your audiences' experiences in school and within their organizational roles give them specialized knowledge that you need to access in designing your communication. This specialized knowledge makes your audience familiar with topics and terminology, concepts, issues, controversies, and techniques within a given field. And, the conditions under which your audience encounters your communications—when, where, and how your documents will be consumed and consulted—make a difference in how you design them. Remember that in our sample working papers, Ming's thinking about a manual for a summer internship and Mandy's preparing for an oral presentation for a meeting on a training problem involved all these aspects. The Writing Strategies worksheet in this chapter offers questions that you might get in the habit of using as you think about or construct your audience.

Finally, audiences are created in texts, as we illustrated with the Hopkins poem. They are characters created by authors to help make a point. Real audiences observe a sort of miniature drama involving audiences implied in the text, and authors invite their audiences to assume these roles, as we saw in the two introductions to the scientific articles.

In this chapter we have tried to challenge you to think not only about real, concrete, and immediate audiences but also about audiences you might construct in your mind or in your text. These are difficult concepts, but as you read through the rest of the chapters in this book, you will see how these concepts enable you to design effective technical communications in the workplace.

---

1. Look again at the document, "How Tiny Organisms Helped Clean Prince William Sound" in Appendix A, Case Documents 2. The document, produced by Exxon and intended for publication in newspapers and similar publications, emerged from the aftermath of the *Exxon Valdez* oil spill in 1989. After the spill, Exxon used the bioengineering technique described in the article to help clean up the mess and reported on its success in this news release. Read the article not as a student but (as much as possible) as a member of the general public. And in your role as a member of the general public, try to consider the following:

   ■ What is your reaction to the first paragraph? To the first section? Do you react to the first sentence cynically? Favorably? Neutrally? Do you feel open-minded and interested? Does the article interest you or not? Note: there are no incorrect answers to these questions.

   ■ What do you think the purpose of this article is? Does your sense of its purpose change at all as you read?

   ■ How does your reaction to the first section affect your reading of the rest of the document? Do your attitudes remain consistent throughout the article, or do they change? What specific words and sentences arouse your reactions?

   ■ Do you lose interest at any point? Do you find yourself skipping or skimming certain parts? Did you stop reading altogether before finishing? Did you read the article in the order in which it was written, or did you preview it first or skip around at all?

   ■ What was your general reaction after reading the article?

   ■ What did the authors assume about their readers of the article? How can you tell? How accurate are their assumptions?

   Compare your answers to these questions to those of your classmates.

2. Read the document, "Bioremediation for Shoreline Cleanup Following the 1989 Alaskan Oil Spill" that appears in Appendix A, Case Documents 2. How would you describe the audience(s) of this document? How do you know? What devices do the authors use to engage or create their audiences?

   Write your findings in a brief report to your instructor.

3. Imagine that you are Dr. Karen Wartala, Dean of the College of Agriculture at the University of the Midwest. On August 2, you receive the letter from the Coalition of Citizens Concerned with Animal Rights that is reprinted Appendix A, Case Documents 3. Remembering to do so not as a student but as Dean Wartala, read the letter as she would. Afterwards, answer these questions (there are no incorrect answers):

   ■ Did you react to the letterhead? Did the fact that the letter is coming from the Coalition of Citizens Concerned with Animal Rights make you put your guard up at all, or did you still read the letter "neutrally"? Would it make a difference if Dean Wartala was a member of the coalition herself?

- Did you react to the salutation of the letter, "Dear M. Wartala"? Would a salutation like "Dear Dean Wartala" or "Dear Dr. Wartala" or "Dear Ms. Wartala" have elicited a different response?

- How did you react to the first sentence of the letter? To the first paragraph? To the first three bulleted accusations? Is your response a neutral processing of the information, or are you reacting emotionally to it?

- How do your reactions to the first half of the letter shape your responses to the rest of it?

- Are you likely to call the author of the letter at the numbers listed or not? Or will you respond by letter? Or ignore this communication?

- How do you react to the fact that copies of the letter were sent to the news media before you have had a chance to respond? Do you resent that Mr. Younger wrote the letter (or copied it to others) before you have had a chance to respond to the accusations?

Compare your answers to these questions to those of your classmates.

4. Fill in the answers to the questions in the Writing Strategies 3–1 worksheet for a document (or two or three) that you wrote recently—for a class, for a job or internship, as part of your work in a social or extracurricular group, or whatever. Or fill in the answers for a document that you plan to write. Be prepared to turn your work in to your instructor.

5. By yourself or along with your classmates, examine carefully any of the documents reproduced in Appendix A and consider these questions: Who is the implied audience of the document? What textual features create this implied reader? How effective is the writer's choice of this implied reader? (For example, take again the letter to Dean Wartala that is printed Appendix A, Case Documents 3. What kind of person does the writer assume Dean Wartala to be. Conscientious? Morally concerned or unconcerned? Easily intimidated, or not?).

WORKS CITED   Young, Richard; Alton Becker; and Kenneth Pike. *Rhetoric: Discovery and Change.* New York: Harcourt, Brace, and World, 1970.

The Writing Process in the Workplace

True ease in writing comes from art, not chance,
As those move easiest who have learned to dance.

> Alexander Pope, "An Essay on Criticism," in
> *The Norton Anthology of English Literature,* 4th ed., vol. 1
> (New York: Norton, 1979).

Invention should be reconceived, I have suggested, as a social act: one in which individuals interact with society and culture in a distinctive way to create something. With this view, invention may be seen as an act encompassing symbol-using activities such as speaking and writing, often involving more than one person, and extending over time through a series of social transactions and texts. The generation of what one comes to know and say is brought to completion by others who receive and execute the action.

> Karen Burke LeFevre, *Invention as a Social Act*
> (Carbondale: Southern Illinois UP, 1987).

Introduction

writing process
the activities involved in the writing task, including exploring, generating, and representing information; planning documents and arriving at goals; structuring and organizing; drafting; revising; and editing.

This chapter traces, in broad strokes, processes and activities that technical writers typically employ when they write in workplace settings. Although many of our suggestions will also be helpful in preparing oral presentations, we focus now on the **writing process.** We begin by outlining the activities that comprise a writing task and offer some thoughts about how these activities are shaped by the workplace setting we described in Chapter 1. We then offer some strategies for getting started on writing projects, drafting documents, and revising and testing documents.

You will find that writing gets increasingly complicated as you learn more about it. You know how to write and probably write often and with success—notes to yourself, correspondence to friends and family, and papers for school. However, the business contexts within which you will work often require you to have a more sophisticated understanding of writing than you may have at the present. Some of your writing in the workplace will consist of routine exchanges (announcements and minutes of meetings, orders for supplies, requests for information, and such), but your writing will also help your co-workers and supervisors make major decisions and solve important problems, perhaps within the global marketplace we describe in Chapter 1.

In Example 4–1, we remind you of the many audiences and purposes of technical writing. In some cases technical documents help sell products or services to clients or educate workers, but in other cases technical documents help organizations negotiate with community groups and with governmental agencies. Keeping in mind audience and purpose, part of your planning process, you will now learn more about how to reflect that planning in your writing.

The Social Nature of the Writing Process

social process
the writing process in the workplace involves many people, such as supervisors, co-workers, and clients as well as company goals, values, and needs

Within the workplace, you will probably produce a wide range of drafts for major writing projects. You produce these drafts because the writing process in the workplace is a **social process.** By social, we mean many people are involved in helping you plan, draft, and revise. At times, you may sit at your desk or in front of your computer alone and write, but the opinions, requirements, and needs of your supervisors, co-workers, and clients will influence your writing.

Depending on the organization within which you work, you might well submit your plans and drafts to other people for reading and reaction. You may seek additional information from other people that may figure into the revision of documents. You may test your documents on multiple users who are representative of the various audiences your communication will eventually reach. And, you may design your document using procedures prescribed by your organization's publications group (the professional writers, editors, and graphics staff).

EXAMPLE 4–1   Range of Purposes and Audiences of Technical Documents

Example 1—Describing Products or Services to Clients:

The Minnesota AIDS Project (MAP) has long been a leader in the fight against AIDS. We are a statewide, private nonprofit agency combining the strength of dedicated volunteers and trained professional staff in the fight. Dedicated to serving those most affected by HIV/AIDS, MAP provides direct service and practical support to partners, families, and caregivers.

Source: *The Minnesota AIDS Project Brochure* (see Appendix A, Case Documents 1, for complete document)

Example 2—Educating Workers in the Field:

Keep resuscitation devices handy. Although saliva has never been implicated in the spread of HIV, try to minimize the need for direct mouth-to-mouth resuscitation. Keep mouthpieces, resuscitation bags, or other ventilation devices available in areas in which the need for resuscitation is predictable.

Source: *Taking Care: A Manual for AIDS Caregivers* (see Appendix A, Case Documents 1, for more of the document)

Example 3—Informing Experts of New Information:

Who Should Use AZT and How Much? AZT is officially recommended for HIV-infected individuals who have a CD4 count of less than 500, with or without symptoms. Many studies have confirmed added benefits for individuals who begin AZT before symptoms develop in addition to a decreased incidence of side effects in these individuals. The current dosages recommended are either 500 mg a day or 600 mg a day. Many physicians recommend 500 mg for individuals who are asymptomatic and 600 mg for individuals experiencing HIV-related symptoms.

Source: *AZT: Step Fact Sheet* (see Appendix A, Case Documents 1, for more of the document)

Example 4—Negotiating with Regulatory or Citizens Concern Groups:

The CCCAR and the University agreed that the crops research is important, that it must be continued, that birds do damage to the research plots, and that this damage must be controlled . . . The University is willing to work with CCCAR during the remainder of the current season to transport the trapped birds to a release site provided that an acceptable site can be found that meets all community and state regulations.

Source: Letter to Coalition of Citizens Concerned with Animal Rights (CCCAR) from Karen Wartala (see Appendix A, Case Documents 3, for complete letter)

EXAMPLE 4-1   *(continued)*

Example 5—Representing an Organization's Work to a Governmental Agency:

The joint ADEC/USEPA/Exxon biodegradation monitoring team successfully organized and implemented a comprehensive program for assessing the utility of fertilizer amendments for enhancing the biodegradation of surface and subsurface oil, and for characterizing the associated ecological risks . . . The Final Report presents all the monitoring data, and reinforces the earlier conclusion that biodegradation is an important mechanism in the removal of oil from shorelines in Prince William Sound.

Source: Executive Summary prepared by the Alaska Department of Environmental Conservation, the US Environmental Protection Agency, and Exxon to the US Coast Guard (see Appendix A, Case Documents 2, for the complete document)

Moreover, you will be writing within specific circumstances. For example, let's say that you are working with your manager to plan a report on purchasing new pollution control devices for your company's factories. You'll need to read governmental guidelines for such devices, assess local community reactions by reading newspaper articles and interviewing individuals, read previous studies done by your organization and others on the topic, check with the financial department on your budget for the project, and read scientific studies on the effectiveness of the pollution devices you'll recommend.

You'll probably have a co-worker or editor give you an opinion on your first draft, perhaps someone who can help you decide how much technical detail should go into such a report. You may have your graphics staff assist you in producing technical drawings of the pollution devices. And, finally, you will need to ask your manager about the company's position on and history with pollution control devices—perhaps your company has been praised for being a leader in such areas before or criticized for lagging behind. All these people and their concerns and needs make your writing social.

In addition, you may find that the company you work for prescribes a standard document production process involving a specified number of stages and checkpoints that you must go through in creating documents. This process allows companies to control the quality of documents, to coordinate the timing of any writing that is tied to specific product releases, and to ensure a consistent style and organization of all writing projects. Generally, these document production processes have been negotiated over time, often in cooperation with the company's publication department, and are subject to change through the negotiation of a number of people and interests. Whether you are an engineer involved in product manufacture and design, a member of the sales force that markets products and communicates with clients, or a manager and administrator who conceives of company direction, you will have to learn your own company's document process.

If you work on a collaborative writing team to produce your document, your writing process will be affected by the individuals who comprise the team and the leadership and negotiations taking place within the team. And, finally, your own writing process will reflect the ways you have developed so far in your life. The ways in which you plan for writing, begin writing, revise writing, and involve others in reading your writing come from the people who guided your experience and education so far. All these aspects too make your writing social.

Remember that as we discuss the writing process in the workplace, you'll need to select the strategies that best fit your social circumstances for any particular project. Writing Strategies 4-1 lists a number of questions to ask about the writing process of any project. Add these factors to those you already know about audience and purpose, from Chapters 2 and 3.

WRITING STRATEGIES 4–1   Some Questions to Ask about the Writing Process

Social and Cultural Concerns

What is appropriate? What is polite?

What is typically done? What has been done in the past?

What is responsible writing in terms of our organizational obligations?

What is responsible writing in terms of my own professional obligations?

How does my organization typically present itself to the public?

Concerns about Language

What tone or style will be best for conveying this information?

Does my audience have difficulty with English? Will my work be translated?

Should I address my readers? Second person, or more formally?

Can I refer to myself in my document? Or do I speak for the organization as a whole?

Who can help me define technical terms?

How can I engage all my readers? Is a glossary needed for nonspecialists?

Concerns about Organization and Structure

What information is necessary? What is extraneous?

What sections are required or standard in my company? Is an abstract needed? An index? A glossary?

What organizational frameworks are useful or typical in this kind of document? This section?

WRITING STRATEGIES 4–1   *(continued)*

Concerns about Genre

What is typical of this genre (type of document)? Is it a memo? A year-end report?

Have similar documents been written before in this organization?

Do we have a prescribed format for this information?

Would visual presentations (tables, charts, graphs) be typical or helpful? Where? When?

Concerns about Managing the Writing Process

What are the deadlines on the project? Are they firm?

What are the various parts of the production process? Who is in charge?

How many writers and technical experts will be involved? What is their level of skill?

Does an editor plan to review my work?

What checkpoints or review stages will be required of my work? Who is involved in these stages? My manager, editor, or co-workers?

An Overview of Writing Process Activities

As you can see, getting an accurate and complete sense of all the factors influencing writing processes is challenging, but we can start with *pieces* of this picture. Below we offer you an overall sense of the activities effective communicators generally engage in as they write—exploring, generating, and representing information for documents; planning documents and arriving at goals; structuring and organizing; drafting; revising; and editing. We'll devote entire chapters to some of these activities later in the book.

In the last half of this chapter, we give some writing strategies, particularly for those activities that help you get started. We'll reserve discussion of other writing strategies for specific chapters on the tools of technical communication and on particular types of documents.

EXPLORING, GENERATING, AND REPRESENTING INFORMATION

Some exploratory, representation, and developmental activities can give you a sense of the writing task you face and help you plan your writing process. Although you can use these activities in the early stages of a writing task, they may also help you refine and improve your document at any point in the writing process.

brainstorming
a quick, informal writing session in which ideas and information are generated without excessive evaluation or criticism

As you explore your writing task, you may wish to **brainstorm;** to try some quick, informal writing sessions; or to present your tasks visually in a flowchart or diagram. These activities help you to flesh out your writing goals and begin to gather the information you will need. These activities help you to get a good initial sense of your audiences, the potential structure of your document, your focus, your timelines, and other important factors.

As you explore writing tasks and begin to generate your content, you get a sense of the social factors associated with that task: the impetus for your project, the people involved, any ethical and legal issues, the nature of conventions involved in a particular document, and more.

Later in this book, you'll find out how collaboration can be a productive means of exploring and generating ideas, how to gather potential information from primary research (e.g., surveys, interviews, and observations) and from library sources, and how to come up with the best reasons for the solutions you want to propose.

PLANNING

Although some people see planning (i.e., getting a sense of your goals and planning the writing process itself) as an activity that takes place *before* the real business of drafting and redrafting begin, as we noted earlier, it is better to think of planning as occurring *throughout* a writing task. As we discussed in Chapters 2 and 3, in planning you consider the audience(s) and purpose(s) of your document. You also plan the development, structure, ways of handling timetables and organizational constraints, sources to tap for needed information, and so forth. You'll think about the timing of a writing project and the people involved in it, including start and stop dates, budget, collaborative opportunities, interview schedules, and deadlines.

Given all of these factors, you will find yourself making and re-making plans at many levels. These plans are dynamic representations, continually shifting, constantly re-forming as you think of new possibilities and respond to new ideas and your own developing text. Indeed, when you write, you make plans (at least in part) so that you can make writing decisions. Writing plans also are shaped by a broad range of social, historical, political, and economic factors. These factors include your position within a company, the ways in which documents have been constructed in the past, the relative importance of a given project within the context of an organization's larger efforts or within the context of existing cultural priorities outside an organization, and the cost of doing business in any given sector of the economy.

STRUCTURING AND ORGANIZING

Often, your plans may include ideas about organization and structure. For example, as you explore how a document might develop, you may think about its overall **organizational structure,** including the number and types of sections, graphics, appendices, tables, and headings. You may think as well about subsections, paragraphs, and sentences.

Structuring and organizing take place both as you prepare to draft and as you proceed throughout the writing task. Some writers like to make detailed, rather formal outlines of a document before they begin. Others

organizational structure the sentences, paragraphs, sections, graphics, headings, and such, and their order in a communication

prefer to make outlines after their writing team produces a second or third draft of a document as a way of checking on the overall organization of a project. You might use concepts of structuring and organization for revision, thinking about the topic in each paragraph, for example, as a strategy for working closely and productively with a final draft.

DRAFTING

drafting
physically producing words and graphic displays in a document

Drafting (physically producing words and graphics) may be what you picture when you think of writing—although it may actually represent a relatively small portion of everything involved in a writing task. Drafting is really then a framework within which all other writing activities seem to happen as you draft.

The physical behaviors that you exhibit while drafting are also unique to you. You might compose in fits and starts, in planned sessions of concentrated activity; you might swear by a one-draft method, or by multiple versions; you might prefer to draft by hand or on a computer. In any case, your behaviors will differ according to various constraints associated with your writing tasks, including:

- The length of a document, its organization, and its ultimate audience(s) and purpose.
- Your familiarity with the material and your educational background.
- The strategies accepted within an organization.
- The extent of collaborative activity.
- The nature of the writing task.
- The use of writing technologies such as computers or dictaphones.
- The availability of information.

REVISING

revising
a critical examination, reading, or rereading; a re-visioning of a communication

Although it is impossible to separate activities of drafting from those of revision (so often these activities happen almost simultaneously), we use the term revision here to refer to a critical examination, reading, or rereading of a text by a writer and the changes that grow out of these considerations. As we have seen with all the composing activities, **revising** (or *re-visioning*) takes place at many different levels, at many different times.

Moving among various kinds of revising efforts, you might rethink written plans and notes early in the process of a writing task; recast sentences, paragraphs, and sections throughout a drafting sequence; and restructure entire documents as a piece emerges from a collaborative team effort. You might also revise in any number of ways: making notes with a pencil in the margin of a document, creating entirely new texts on a computer, or cutting and pasting sections of your draft with the drafts of other writers. You also may respond in your revision to the testing of your document by representative users.

EDITING AND PROOFREADING

For you, as an individual writer, **editing** and **proofreading** involve another kind of critical attention to writing, often the kind of attention that focuses on the surface level of text and graphics. The surface level includes: proper spelling, grammar, and usage; complete headings; punctuation; use of white space; and such. Editing and proofreading look at how this surface level reflects the document's purpose, communicates to intended readers, and satisfies your conception and goals.

Next, we'll look at some strategies you might use in several of these writing process activities—particularly those that will help you get started.

editing
a substantive look at how well a communication meets stylistic, grammatical, spelling, punctuation, usage, and consistency conventions

proofreading
a careful reading for surface errors, such as missing words, spacing problems, misspellings, and typos

Some Writing Process Strategies to Try

There is no easy way to go about the actual drafting of a writing project. You just have to start putting words down on a page. Even the most complicated writing task begins with this activity that seems so simple and so difficult at the same time. There are, of course, some time-tested strategies writers have found helpful in the drafting process. You can be sure, however, that no single strategy will work for all writers. Thus, you will have to experiment to find which strategies work best for you or for the writing groups of which you are a member. And you'll have to be flexible and resourceful: The more strategies you have in your repertoire as a technical communicator, the more effective you can be as a problem solver.

One of the keys to success in generating early drafts is to keep yourself from being too critical about your words and ideas at the beginning of a writing project. When you are just starting your project, you need to be willing to accept imperfections in your expressions and ideas, instead of stopping to search for the very best possible phrase. Of course, you must eventually devote time within a writing project to taking long, critical, detailed looks at individual words and expressions, but generally this kind of activity comes when you have produced at least one draft of a project. Stopping to evaluate and worry about individual words or ideas before thoughts ever get down on paper, or on a computer screen, can make it difficult to generate even a first draft. The following activities should help you as you begin your writing process.

WRITING PLANS

You may find it useful to write brief planning notes for creating a draft before attempting to write an entire draft. **Writing plans** help you identify the specific goals you have for a task and the timeline necessary for completing the writing project. To create such a plan, you may want to use questions such as the following as prompts:

- What is this document's primary purpose? What major sections will be needed in order to achieve this purpose? What do I want to accomplish in each of these sections?
- What things do I want to accomplish in the beginning or introduction of this document?
- What things do I want to accomplish in the body of this document? Do I need to create several subparts or sections within the body of the document? If so, what are they and what do I have to accomplish in each section? How will I go about creating/structuring each of these sections?
- What do I want to accomplish in the end portion of the document or the conclusion? How will I go about accomplishing these goals? Are there subparts of the end of this document? If so, what are those parts and what do I want to accomplish in each part? How am I going to go about accomplishing these goals?

You learned a lot about purpose in Chapter 2. In a writing plan you ask yourself about the purpose of each part of your document.

Example 4–2 offers a sample set of informal notes for a writing plan of a quick-start manual for users of a new word-processing program. You can see how the writer used the questions to plan the Write Soft document.

START DRAFT OR ZERO DRAFT

You might find it useful to think of your first draft simply as a quick list of notes and ideas that you put down on paper without too much regard for order or value at this point. Of course, you will have your audience and purpose in mind, but a **start or zero draft** can be done in one short sitting of 15 to 20 minutes and still be quite useful. The start draft represents a ''brain dump,'' a quick listing of everything you know or think about a subject. These drafts often include elements of your writing plan, brainstormed lists, references to other documents that need to be checked, ideas for organization of a document, and notes for later interview or group sessions—anything that will be useful in making some sense of a writing task.

Start drafts are especially helpful when you just don't seem to know where or how to begin. To create a start draft, sit down at your usual writing

EXAMPLE 4–2   Writing Plan for Write Soft Document

1. State manual's purpose and scope. Remind audience that the document is intended for quick reference, not power users. Get statements from project description—see Al.
2. Familiarize users w/features of program and basic vocabulary—Take from the tech manual in R&D and from program specs. Arrange first section by feature—within each feature, highlight things that make product different from key competitors—try MS Word, Word Perfect, etc.
3. Include tutorial to get readers going on short doument. Get ideas from User Testing folks. Use real writing task and work through it. Consult Jason in clerical about typical tasks for substantial document—organize tutorial around a case scenario involving one of these tasks.
4. Write short troubleshooting guide covering major problems people encounter—use Jim's last one as template for organization. Save this until last and get ideas from testing team too.
5. Include glossary and index. Consult Sharon about indexing capabilities of document and production program.

space and just begin writing. Keep writing—whatever comes to mind about your topic—for 15-20 minutes. You can write down words, phrases, or complete sentences; do whatever is best for you. You might have learned this technique, often called freewriting, in earlier writing courses and found it useful in getting started.

When you are done with your start draft, examine it for material that seems promising. Once you have completed this draft, rearrange any promising material it contains into some order that will help you work on later drafts. Various important points from your start draft may suggest essential sections of a document to you or may lead you to a more carefully articulated statement of your focus or purpose. You can reduce your start draft to an outline and then check for completeness and consistency. As you work with the start-draft, add or delete material as needed.

Example 4-3 shows a start draft for the introduction to an engineering report that evaluated the feasibility of a branch of a national bank moving into an existing building. This writer has done much of the research before beginning the start draft, but the activity revealed some gaps in that research. Example 4-4 shows what happened when the writer of the engineering report in Example 4-3 rearranged the start draft. In this second version, the writer reduced thoughts to phrases and lists and added subheads to help group the thoughts. Now the writer has a solid outline for the introduction to the engineering report.

EXAMPLE 4–3   Start Draft for an Engineering Report

Our purpose was to evaluate the existing building—Central Bank. We wanted to see if they can move in and not look or build elsewhere. Everything looks like a "go." We decided that Central Bank can move in and get the things they want after some reconstruction. We considered their budget and schedule, but they will have to pay overtime to get into the building when they want to. We covered electrical, plumbing, and heating as well as air-conditioning. The improvements that need to be made can be done—and not be too costly, at least not as much as it would cost to build or improve elsewhere. The building (220 Columbia St.) is close to one we worked on before at 112 Columbia so we used some of the information from that report. We did that for United Plumbing—their needs were close enough to be useful. We looked at 114 Columbia too but the electrical systems in that building had been improved three years ago—and more complicated than Central Bank wants—not any good information there. Because we looked at heating and air-conditioning, we had to look at ventilation. Central Bank needed better and more centrally located elevators than United Plumbing. We made sure that handicapped clients of the bank would have access to everything, even hand-dryers in the restrooms—a major electrical challenge installing a handicapped-access elevator from the parking garage in the rear. Broughton, Horn, Karis, and Grant looked at this building 7 years ago—not many changes since then, so we used their report too. But our survey needed to be very detailed—both the systems that exist now and what improvements were necessary. We talked to the VIPs at Central Bank and then the people who actually maintain the building. Better to check with Central Bank to see if they will use the same maintenance company. We came up with a construction schedule and included the overtime in it. Made sure we look at fire protection and codes, of course. We got existing blueprints and plans from the architect—building isn't that old and hasn't been modified greatly. Made sure we really knew exactly what this client wanted—the type of facilities and arrangement.

GROUP BRAINSTORMING, OUTLINING, AND DRAFTING

If you are part of a writing team, perhaps a group of engineers, writers, editors, and graphics experts who must produce one cohesive and effective document, you can engage in **group writing process activities.** For example, your group can brainstorm for 5 to 10 minutes on any of the issues that drive the writing project: purpose, audience, information, organization, and so on. In a brainstorming session, all ideas are recorded—none are rejected. Once your group has come up with everything it can think of related to the topic, gather your ideas into categories and create a rough outline for the document. Let's say that your group is producing the engineering study for Central Bank (Examples 4-3 and 4-4), and during your brainstorming session, you describe your audience as supervisors who will check the report for an impression of your work, co-workers who may depend on your technical details in this report for future projects, and Central Bank (your clients) who

group writing process activities
brainstorming, outlining, and drafting activities of a collaborative writing group or team

EXAMPLE 4–4 Outline Based on Start Draft of Example 4–3

Purpose:
To evaluate and recommend whether Central Bank, Albuquerque Branch, can occupy an existing building at 220 Columbia Street.

Scope of Report:
Heating, ventilating, and air-conditioning
Electrical
Plumbing
Fire protection
Elevators
Provisions for handicapped

Sources of Information:
Previous study of building (7 years ago) by Broughton, Horn, Karis, and Grant.
Study of similar building located at 112 Columbia Street for United Plumbing, Inc.
Blueprints and plans from architect.
Discussions with building personnel.
Interviews with banking officials.
Detailed survey of facility and systems.

Basis for Recommendations:
Arrangement and type of facilities needed.
Central Bank's budget.
Moving-in schedule.
Construction schedule.
Overtime needed.
Code considerations.

---

will use the report to make a decision. Your group can create an outline for the report that places technical details in appendices, is most persuasive and conclusive in the introduction and in the final recommendations, and is most complete and thorough in the body of the report. This structure should meet your client's needs but also satisfy or aid your supervisors and co-workers.

If you work on a computer, changes during brainstorming, outlining, and drafting activities are easier. If you have a networked computer system, your group may want to try meeting alternately in a face-to-face setting and in an online setting. Face-to-face meetings may be difficult to arrange, but often you need to see each other to produce the personal atmosphere your group may need to work over the long term.

As your group outlines your draft, try the following strategies:

- Give the document a working title to help focus your thinking.
- State the purpose of the document.

- Identify and list the major parts or sections the document will include. Put Roman numerals next to these and experiment with their order until you are satisfied for the time being.
- Identify and list the main kinds of information that should be included in each section. Experiment with the order for this information within each section until you are satisfied for the time being. Put capital letters next to these, starting with "A" in each section.
- Identify and list all the subpoints you can think of under each main piece of information. Experiment with the order of these subpoints until you are satisfied for the time being. Put small numbers, starting with "1," next to each point.

In the early stages, an outline may be incomplete and lack a fully articulated structure, and that's fine. It's often counterproductive to get so worried about adhering to a rigid organizational structure that you lose track of content. Also, as a project progressees and your draft emerges, your organization may change dramatically.

PURPOSE, AUDIENCE, INFORMATION DRAFT

purpose, audience, information draft
a draft in which goals (purpose), outcomes (audience), and information (structure) are addressed

A **purpose, audience, information draft** can be useful to you as an individual writer or to your writing team. Begin this draft by reviewing project specifications and other individuals' notes about purpose, audience, and information about the writing task you are addressing. If possible, work on a computer using a word-processing file. Create a file that has three headings: Goals (Purpose), Outcomes (Audience), and Information (Structure). Under these headings, begin drafting. If you are working with a group of writers, you might each compose a draft of one of these sections. When you are done, you can cut and paste the individual efforts into one group document and discuss the results in a face-to-face group meeting.

- **Goals:** Keeping the audience and the information content of the document in mind, write one or two sentences that identify the document's primary purpose. At some later point, you will probably want to include these purpose statements in the document's introduction. Use them to set the tone for the document you want to create.
- **Outcomes:** Keeping the document's purpose and information in mind, make a bulleted list of the outcomes that should result from the document. List what readers should be able to do after reading the document that they could not do before reading it; what they should know about reading the document that they did not know before reading it; what problems readers should be able to

solve after reading the document that they could not solve before reading it; and so on. At a later point, these outcome statements may also be included in the document's introduction and recommendations, and you can certainly also use them to test the document's effectiveness. They should, of course, be reexamined and revised at several points during the writing process.

- **Information:** Keeping in mind the document's purpose and the audience(s) who will use it, make a quick outline of the information you have to convey in the body of the document. Think in terms of the organizational structures suggested by the information itself. For example, your engineering report might be best organized by each facility you had to study: Heating, Ventilating, and Airconditioning; Electrical; Plumbing; Fire Protection; Elevators; and Provisions for Handicapped. Or you might organize the report according to the time table needed to meet the moving-in day that Central Bank specified.

DIAGRAM DRAFTS

You might find that representing information visually and working manually with these physical representations may be a productive way for you to begin drafting your documents. So, instead of writing sentences to represent information, you can draw a diagram of the draft that you want to compose. A **diagram draft** enables you to visualize the components of your document and aids in the planning process.

diagram drafts
a visual display of a draft, using designs such as boxes and squares, to represent the sections or categories of the draft

To use this strategy, you create designs (boxes, squares, circles, and so on) for each of your draft's sections or categories. Then you create designs of a different shape for as many pieces of information that you can think of. You label each of these designs with a key word or two to represent the information or the content of the draft. Then you experiment with how these parts fit together.

For example, if you decide to organize the engineering report for Central Bank by each facility you had to study, you could represent Heating, Ventilating, and Air-conditioning by a square; Electrical by a circle; Plumbing by a triangle; and so on (Example 4–5). Then you would list your pieces of information, assigning each a different shape or simply placing each in a separate box; for example, you might list hot water system, plumbing stacks, general office lighting, refrigeration machines, primary air systems, sump pumps, sprinkler systems, and so on. Then you could draw connecting lines (for example, you would connect general office lighting to Electrical, sump pumps to Plumbing, plumbing stacks to Plumbing, and so on). If you prefer, you can list these pieces of information and your key word categories on separate index cards and shift them around to find the best matches. Or, you can use color to code your information and key word categories.

EXAMPLE 4–5

○ Electrical
    ○ general office lighting

☐ Heating, Ventilating, Air-conditioning
    ☐ refrigeration machines
    ☐ primary air systems
    ☐ sprinkler systems

△ Plumbing
    △ hot water sytem
    △ plumbing stacks
    △ sump pumps

As you work with the categories and specific information in a diagram draft, ask the following questions about how they relate to one another:

- Is there some system to these categories or pieces of information? Can they be arranged from more general to less general? Simple to complex? Are some subordinate or superordinate? If so, arrange the subordinate categories or pieces of information inside or underneath the superordinate categories or pieces of information to show how they relate. Arrange the superordinate elements in order of importance or occurrence depending on their relationships. For example, in your engineering report, you might find that Heating, Ventilation, and Air-conditioning should be subordinate to Electrical (Example 4–6).

- Are some categories or pieces of information related by association? If so, place a major category in the center of your page. Then brainstorm as many pieces of information as possible that are related to it. Draw lines between and among related pieces of information. Examine these connections. If a cluster of pieces suggests itself as important because there are many lines leading to it, try placing this cluster at the center of your page and continuing your diagram. For example, you might discover that most of the problems with the building are related to the plumbing (Example 4–7).

- Can you classify pieces of information in a system of some sort? Try applying various principles of division to the pieces of information on the page. Classify them by importance, by kind, by magnitude, by success rate, by topic. Do they lend themselves to comparison because many of them are alike? If so, cluster these items by likeness. Do the pieces of information lend themselves to con-

EXAMPLE 4–6

○ Electrical
    ○ general office lighting
    ▢ Heating, Ventilating, Air-conditioning
        ▢ refrigeration machines
        ▢ primary air systems
        ▢ sprinkler systems

△ Plumbing
    △ hot water sytem
    △ plumbing stacks
    △ sump pumps

trast? If so, cluster them by dissimilarity. Using this technique, you might have discovered that when considering overtime and scheduling in your engineering report to Central Bank, the time needed to change the Plumbing systems was much greater than that needed to change the Electrial facilities, a matter of "magnitude." Or, you might decide that discussions with the building personnel and interviews with the banking officials should be discussed in the same section of the document, a matter of "kind."

■ Can the pieces of information be represented in a flowchart that shows movement from one instance/decision/event to another? If so, start with the first step in the process or narrative and work through to the end. Represent decision points in this chart with diamonds or some other design that indicates branching (Example

EXAMPLE 4–7

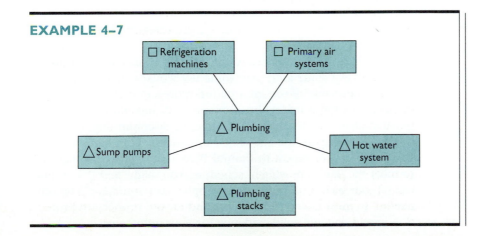

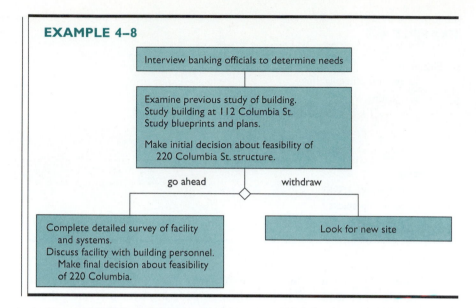

EXAMPLE 4–8

Interview banking officials to determine needs

Examine previous study of building.
Study building at 112 Columbia St.
Study blueprints and plans.

Make initial decision about feasibility of
  220 Columbia St. structure.

go ahead   withdraw

Complete detailed survey of facility
  and systems.
Discuss facility with building personnel.
Make final decision about feasibility
of 220 Columbia.

Look for new site

4-8). Do the pieces of information have a cause and effect relationship? Draw lines to indicate these relationships; then list evidence points beneath each relationship. Do these pieces of information represent problem/solution relationships?

Then identify the problems with one design (bullets, squares, and such) and the solutions with another. Connect problems with their solutions. Do the pieces of information illustrate a before-and-after relationship? If so, clarify the connections between the pieces of information that precede and those that follow. For example, when organizing your engineering report, you might have realized that recommending that the contractors start with a major overhaul of the sewage system in the Central Bank building would solve the other problems listed under the category Plumbing (Example 4-9).

■ Do the categories and pieces of information represent an evaluative context of some kind? Are some decisions? Criteria? Outcomes? Does some of the information represent evidence? Place the evidence items under the appropriate claims. Brainstorm additional evidence. Brainstorm alternative, or disconfirming, evidence as well to anticipate the audience's arguments. For example, if you argue in your report to Central Bank that overtime is required to meet the Bank's moving-in schedule, you might find that a number of your pieces of information support that argument. The time needed to meet code requirements and ensure proper fire protection would become solid evidence for your argument.

EXAMPLE 4-9

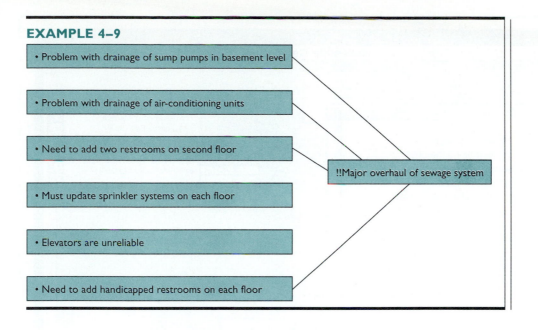

- Problem with drainage of sump pumps in basement level
- Problem with drainage of air-conditioning units
- Need to add two restrooms on second floor
- Must update sprinkler systems on each floor
- Elevators are unreliable
- Need to add handicapped restrooms on each floor

!!Major overhaul of sewage system

If you have available a computer-supported drawing program or a hyper-text writing environment (e.g., *StorySpace*™ or *Writing Environment*™) to create these diagrams and elements, you can move your categories and pieces of information around in several different ways. We have included a sample screen from a diagram document in Figure 4-1.

INTRODUCTION DRAFT

You might find it helpful and even necessary to draft the introduction of a document in quite a bit of detail before going on to draft the rest of the document. Often writers spend a great deal of time within an **introduction draft** laying out the following items:

- *Purpose statement* or *problem statement:* a concise statement that articulates the purpose of the document or the problem that the document is designed to address.
- *Goals statement* or *outcome statement:* a list of the document's major goals or the outcomes that the document is designed to produce.
- ***Information map:*** a concise statement that maps out the structure of the document that follows. This statement reveals the organizational structure of the document's presentation for readers.

Example 4-10 represents an introduction draft.

introduction draft
a draft of the beginning of a document that includes a purpose or problem statement, a goals or outcome statement, and an information map

information map
a concise statement that maps out the structure of a communication; a preview of the organizational structure

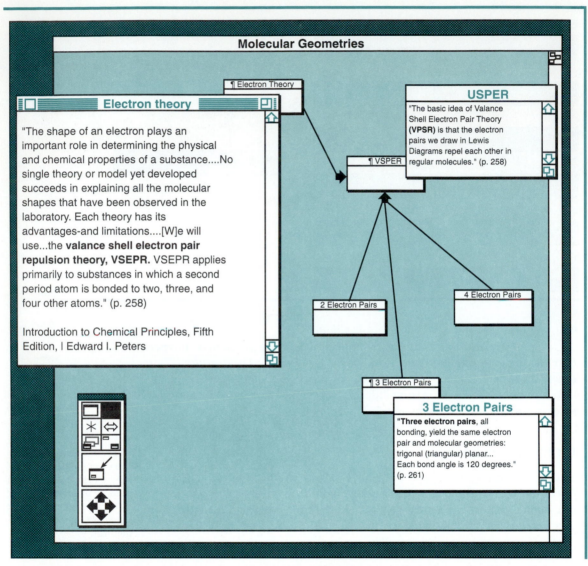

Molecular Geometries

Electron theory

"The shape of an electron plays an important role in determining the physical and chemical properties of a substance....No single theory or model yet developed succeeds in explaining all the molecular shapes that have been observed in the laboratory. Each theory has its advantages-and limitations....[W]e will use...the **valance shell electron pair repulsion theory, VSEPR.** VSEPR applies primarily to substances in which a second period atom is bonded to two, three, and four other atoms." (p. 258)

Introduction to Chemical Principles, Fifth Edition, I Edward I. Peters

¶ Electron Theory

USPER

"The basic idea of Valance Shell Electron Pair Theory **(VPSR)** is that the electron pairs we draw in Lewis Diagrams repel each other in regular molecules." (p. 258)

¶ VSPER

4 Electron Pairs

2 Electron Pairs

¶ 3 Electron Pairs

3 Electron Pairs

"**Three electron pairs,** all bonding, yield the same electron pair and molecular geometries: trigonal (triangular) planar... Each bond angle is 120 degrees." (p. 261)

Source: Text from Edward I. Peters, *Introduction to Chemical Principles.* 5th ed. (Philadelphia: Saunders Golden Sunburst Series, Saunders College Publishing, 1990). Diagram created by authors.

FIGURE 4–1
Working with a Diagram Draft in *StorySpace*™

EXAMPLE 4–10   Introduction Draft—"Buying Your First Home" Brochure

First-time homebuyers are seldom confident homebuyers—and it shows. More importantly, a lack of knowledge about the procedures involved in financing and purchasing a home may result in homebuyers who pay more than they need to. With today's economic pressures, few families can afford to overspend this way.

This brochure is designed to provide first-time homebuyers with the information they need in order to be more confident with lending institutions, realty companies, and moving companies. We cannot tell you everything within these pages, and we can't promise to make your experience problem free, but we can give you a clear outline of the steps involved in obtaining financing for a new home, planning for your move, and making it through moving day.

By the time you have studied the information in this pamphlet carefully, you should be able to accomplish the following steps toward buying your home and moving into it:

- Step 1:   shop for mortgage financing
- Step 2:   apply for and obtain a mortgage loan
- Step 3:   plan for your move
- Step 4:   sail through moving day

We have organized the information into four easy-to-follow steps, so let's get started. Read carefully, jot down questions, and get ready to MOVE!

OUTLINE REVISION

So far, we have discussed several writing process activities that help with getting started on your drafts, but if you're like most people, you may feel more comfortable when you have a draft to work with than when you are facing a blank page or an empty computer screen. Because the processes involved in revising or rewriting are also complex, all writers wonder, one time or another, where to begin revising and when to stop, whom to include as readers in a revising effort and whom to avoid, and how complete or how skeletal a draft should be during each stage of developing a document. These concerns are magnified when revision efforts must be coordinated among several writers. Furthermore, within an organization, additional constraints of deadlines, standardized procedures for revising and editing, management supervision, and production requirements may complicate revision processes.

Although many writers use outlines as a way of working toward a *first* draft, we also suggest trying outlines as tools for revising an *existing* draft, even very late in a writing project. Sometimes, for example, a **revision outline** can be useful as a final check on the organization of a document. Here's how it works. Using a marker or highlighter, highlight the purpose statement in your existing draft. If the document is short, you can highlight

revision outline
an outline used for a final check on the purpose, organization, coverage, and logical relationship of sections in a communication

the main idea of each *paragraph* by identifying one phrase or sentence. If the document is longer, you can highlight the main idea of each *major section* by selecting a particular sentence in the introduction to each section. Next, using the highlighted items, you should construct an outline of major sections and the points included in each section. (If you have access to a computer, you may want to accomplish the process of selecting, cutting, and pasting electronically.)

When you have an outline of your existing draft, examine it carefully. Ask yourself the following questions:

- What are the major sections of the outline? What are the minor sections or subparts under each major section? Make these major and minor sections your outline headings. Are these appropriately identified in the outline or should some of the items identified as major become minor points, and vice versa?
- How does the document, as represented in this outline, meet the purposes(s) identified for this task?
- How does this order (of headings and the sections, points, and elements under them) help to reveal or present the document's information to the audience?
- Where does the outline look thin? Which sections are over-crowded with information?
- Which sections need further information? Which needed sections are missing? Which extraneous sections are included?
- Are the sections related logically? If so, how? What is the principle of organization and what are the methods of organization that provide the framework for this document, this outline? Do they suggest a useful conception of genre, or conventional form, for this task?
- Is this system of logical relations and organization articulated clearly for the reader? Where would a reader, trying to move from section to section, get lost?
- Are the elements at each level representing information that is equivalent in importance or status? Does further work need to be done with organizational structures to clarify what's most important?

group revision
a collaborative group or team effort to examine, re-read, or revise a communication

GROUP REVISION

In the early stages of writing a document, you may want to ask group members to help you articulate a document's purpose and goals, to suggest alternative organizational structures, to identify weak sections, or to create a draft outline. As a document develops, you may want to seek help on refining individual sections, on focusing and organizing individual paragraphs, on identifying places where examples and graphics are needed, or on finding inconsistencies in tone and vocabulary. In the later stages of a project, you may want to get the group's help in attending to stylistic conventions, to layout and design features, or to sentence-level revisions.

Although groups can provide a great deal of help with revision, they can also produce so much information that it is difficult to manage productively. When you are planning a **group revision,** the following suggestions will help you make your group revision productive:

- Hand out the text in plenty of time for everyone to read it carefully before the revision session. Make sure that all group members have a copy of the document's purpose statement and the project timeline.

- Ask group members to focus on particular revision tasks that will prove helpful at particular stages of a project. Do not allow the group to expend effort needlessly by concentrating on minutiae before attending to larger structural, audience, or purpose concerns that plague a document—this is like putting siding on a house that needs major structural attention to its joists and load-bearing walls. You are bound to waste time and money if you try to cover up major structural or conceptual problems with surface-level cosmetics.

- Ask group members to write in response to drafts or to make notations on drafts. Group members may want to adopt a set of conventions or strategies for responding to drafts: using a "G" to denote places where explanatory graphics are needed; using a highlighter to identify the key sentence or phrase in every paragraph that should serve as the focal point (the highlighting will differ among group members and serve as a productive focal point for revision discussions); and using a "?" to indicate areas that prompt confusion or questions.

COMPUTER-SUPPORTED REVISION

Technical communicators functioning in business and industry often use computer-supported tools, **groupware,** for collaboration and revision efforts. Groupware works on a computer network that writers use as a medium for exchanging text and graphics; for trading commentary on or information for drafts; for "real time" or synchronous discussions about writing projects; or for online meetings. But computer-supported collaboration and revision need not involve the use of networks. These activities can be supported by computers even if no networks are available.

The following strategies will help you think in creative ways about revision activities that can be supported by computers:

groupware
computer-supported tools for collaborative writing teams; particularly useful in revision efforts

- Send a disk with the draft around to several writing team members. Have readers use the "all caps" feature of a word-processing program to distinguish their comments from those of the author(s).
- Circulate a draft on disk among members of a writing team. Have each reader rename an electronic copy of the document with a different file name that indicates the responder's identity. Or maintain only one disk copy of a document and have each group member comment using a different type font.
- Using a computer network, circulate drafts electronically so that people do not have to coordinate busy schedules to meet face-to-face for every draft. Once an electronic draft has circulated among all the members of a writing team, send the fully annotated commentary to all group members so that everyone can see the group's comments. If possible, use a computer-supported response program like *PREP*™ or *Prose*™ to allow various readers to flag copy electronically. We have included a sample copy of a *PREP*™ commentary in Example 4–11. Some computer-supported response programs also allow authors to cut and paste suggested revisions electronically.

EXAMPLE 4–11   Computer-Supported Group Commentary as a Revision Strategy—Using *PREP*™

Year-End School Board Report on Instructional Computing Effort

Draft Content	**Reader Number 1**	**Reader Number 2**
Much time and attention have been paid recently to the potential of computers for addressing educational inequities in connection with poor and disadvantaged populations and women. This report examines the claims that have been made and the evidence that confirms or disconfirms each of them.	I don't really like the use of "much" here—it sets a tone that's overly academic—even for the school board	

EXAMPLE 4–11   *(continued)*

One of the first claims to be made about computers in educational settings when personal computers first came on the market in the early years of the past decade had to do with democratization. Indeed, researchers claimed that personal computers would provide disadvantaged school-age populations with access to information, to publications, and to learning approaches that have been denied them in traditional classrooms.

Even the briefest glance at the current uses of technology in our school systems belies these early claims. Computers, far from providing disadvantaged populations with access to new information, are actually being used to exacerbate the gap between those who have and those who have not. In many inner city schools, for example, computers are being used to . . .

I wonder if we need to include some citations to actual research studies here to lend credibility. A good bibliography might help—maybe we can get some teachers to annotate one.

Democratization really might not go over well with this crowd—politically, I think we're dealing with a pretty conservative group in this district and we might not want to be so overt in our approach.

John, get stats here from DOE documents in the government archives in Jeff City—16 of the 22 Board members are business people.

- Use a synchronous network discussion to hold an online discussion of a document or section. **Synchronous discussion programs** like *Daedalus' Interchange*™ allow groups to print transcripts of their discussions and thus have a record of group decision making. Example 4-12 shows a sample *Interchange*™ session.

synchronous discussion programs
groupware that allows collaborative writing teams to hold online discussions of communications

EXAMPLE 4–12   Online Synchronous Discussion as a Group Revision Strategy Using *Daedalus' Interchange*™

Revision Session: 27 July 1991

In this session, brainstorm budget considerations are offered for the report section on finances. As it now stands, the draft of this section is pretty sketchy. We've done enough site visitation at this point to be able to get down to specifics on hardware and software. You might also want to suggest school district resources that can be employed in support of individual facilities.

Eileen:

We need to know more about the size of the project we are working on! Has the Board provided any clarification about whether they are willing to contract for six facilities or twelve?

Cornelius:

One budget consideration would be what size each site chooses to establish—they do have some choice within the specs that we provide, so how can we indicate this in a budget for the Board?

Mac:

Software choices will certainly impact on budget—what kind of things do the teachers want a lab to be able to do, what student populations will it accommodate? This section has to show some variance.

Eileen:

Remember they also wanted us to consider what machines they already have at the Milton site that we can use there: 14 old Mac SEs. I don't think any of them can really be used. Class sizes there max in about 24, so even if we could, we'd better figure on 10 new workstations, one file server, one instructor station, one back up. Cabling for all 24 stations. Positions: A director, student assistants.

Mac:

Printers, too—and what about CD ROM players, scanners, etc.?

Cornelius:

Would it help to provide budget details for one of the proposed facilities and then separate out elements that are optional or tailored to specific sites? Certainly this might help in the case of the four magnet schools, but I wonder if all of this should go in the appendices rather than the Finances section of the report?

Mac:

Before you determine that those old SEs can't be used, Eileen, you need to determine what kinds of things those instructors will want the lab to be able to do. SEs can be upgraded and can work very nicely as long as you aren't asking them to do things beyond their capacity. With 4 megs of memory SEs do OK for word processing.

Eileen:

Software purchases—How much of a budget should we set aside for software? Let's see. . . . We need

word processing—25–30 station packages.

EXAMPLE 4–12   *(continued)*

> online reference—"   "   "   "
> telecommunication—"   "   "   "
> ?
>
> **Cornelius:**
>   A ten-pack at educational prices for Word 5.0 is $410. Excel 10 pack runs $700; do you need spreadsheet and data capability? But is all this really going in the Finances section—think, folks!

electronic mail
called e-mail; computer-generated mail systems that allow communicators to mail, receive, and share documents

■ Use **electronic mail** or e-mail to gather information to be used in redrafting a document. Try surveying potential readers or expert consultants via e-mail. Consider sending excerpted sections of a document for commentary and critique.

■ Use electronic archives to find **boilerplate** material that can used in your drafting and revision efforts.

boilerplate
an existing material that can be used as a model for new documents or can be placed within a new document

SUMMARY

In this chapter, we wanted to help you see the writing process as a whole and to provide you with some practical strategies to use when you face a technical writing or rewriting task. You should now also understand more about why the technical writing task is somewhat different from other writing you have done in school. The social nature of workplace writing creates this difference. That is, the many managers, technical experts, writers, editors, and graphics experts who participate in the writing process; the deadlines and timetables that dictate when you must begin and complete your writing; the company procedures you must follow to ensure consistency with company goal and image in your writing; and the styles and genres that are common in technical writing all demand that you listen to, work with, and seek advice from others as you write. You write within a context that is formed by your organization, the people in that organization, and the products, goals, values, and challenges that organization faces.

Although we will go more into detail on writing process activities—exploring, generating, and representing information; planning; structuring and organizing; drafting; revising; and editing and proofreading—later on in the book, you should now see these as activities that occur and reoccur as you produce a document.

Also, we have introduced you to a number of writing strategies, particularly those you can use as you start and as you rethink or revise a draft. Such strategies as writing plans; start drafts; group brainstorming, outlining, and drafting; purpose, audience, and information drafts; diagram drafts; and introduction drafts are designed to help you get a good start on your writing. Outline revision, group revision, and computer-supported revision help you revise or rethink your drafts. Remember that although we have focused on writing in this chapter, you can use many of these strategies to create an oral or visual presentation. And, these strategies are designed to help you not only when you are the sole author of a document but also when you are a member of a writing team.

We suspect that you will need to turn back to this chapter often as you read this book. Every time you encounter a writing task, dip into your toolbox of strategies.

ACTIVITIES AND EXERCISES

1. Review the first section of this chapter, in which we describe the social nature of the writing process. Recall a document that you produced, or a presentation that you made, as part of your summer job or as a member of some other organization to which you belong (a religious group, a service or social organization, an extracurricular group). How did the group that you belonged to affect your composing? For example, with whom did you discuss your plans or drafts? Why did you make the revisions you did? Why did the document or presentation take the form it did? Share your thinking with the instructor and class.

2. Read carefully the public policy paper from the Minnesota AIDS Project on Tuberculosis and AIDS (Appendix A, Case Documents 1). Imagine the writing processes for such a document. How did the writers define (or create) their audience? What purpose does the paper serve? Can you identify categories and clusters of information? Write a short memo to your instructor and classmates in which you share your observations.

3. Recall the last two documents of real substance you wrote and try to remember *how* you wrote them. Did you pursue each task in the same way? Did you work alone or with others? How did you gather your information? How did you organize it all? Did you write everything at one time, or did the writing get spread out over a longer period of time? How, and how much, did you revise? Did you revise alone or get suggestions from others? How did you produce the document— longhand? by typewriter? on computer? What else can you recall about the activities you went through to produce the documents? Which activities worked well, and which not so well?

4. Read carefully the case involving the chemical company DuPont, in which a person who accidentally drank some pesticide died (Appendix A, Case Documents 4). Choose one section of the case, and, using a marker or highlighter, highlight the main idea of each paragraph in the section by identifying a main phrase or sentence. Next, using these items, construct an outline of the major points in the section. If this case was designed to be read by the general reader instead of lawyers and law students, what additional information might you want to add to your outline?

5. The following activities ask you to assume the role of an employee of Consumer Trends, a small urban-based company that interprets consumer trends for its customers by obtaining and analyzing various sources of consumer statistics. Many of the company's customers, as you might imagine, are in the manufacturing industry. These groups want to know what customers are likely to want and how much they'll spend on various products. Such information is exceedingly valuable in making predictions about a product line. Now the company is trying to branch out, to find other customers that need consumer information on a consistent basis, regardless of economic trends. One of their new projects involves contracting with a local independent school district to provide some instructional units in Social Sciences classes.

Within the company, started by Joan McIllich (a recent college graduate with a liberal arts degree and a minor in statistics), there are six to nine employees, depending on the number of analyst writers currently employed. Ms. McIllich is the president and chief executive officer (CEO); Joseph Janalleo is the company's financial officer and chief statistician. The analyst writers, with backgrounds in various areas, must take consumer statistics published by various sources (e.g., local, state, and national governments; consumer advocacy groups; business and industry; think tanks) and interpret them to identify what value such data might provide for various customers.

Read Examples 4–11 and 4–12, and then assume the role of one individual on the team of analyst writers. Example 4.13 is a memo from Joan McIllich to the Analysis Section. Example 4–14 are the Consumer Price Index tables that were attached to the memo.

These activities illustrate how you can clarify and start on writing tasks by focusing on the purpose, the audience, and the nature of information you must deal with.

A. Take five minutes alone to brainstorm, in writing, about the following questions in connection with each of the two documents (the summary and the memo mentioned in Example 4–8) your analysis section has been charged with writing.

When you are done, share your ideas with your writing team and discuss the responses. Focus both on those responses most group members agree on and those that you do *not* agree on. See if you can make some productive sense of the points on which you disagree; use them to think about the range of perspectives involved in thinking about purpose and audience:

- What is the purpose(s) of—and the audiences for—each document (summary and memo)? What problems are meant to be solved? Are there several purposes? If so, how are they related? Think about each document (and the information it contains) both in terms of its purposes and audiences within Consumer Trends, Inc., and its purposes and audiences outside the company when the information is presented to the DISD people.

- For each document, what is involved in addressing the purpose(s) and the audience(s)? Who? When? Where? Why?

B. As a team, assign each member of your group at least one of the following writing tasks. Work for 10 minutes on these tasks. When you are done, share your ideas with your writing team and discuss the responses. Focus both on those responses most group members agree on and those that you do *not* agree on. See if you can make some productive sense of the points on which you disagree; use them to think about the range of perspectives involved in thinking about purpose and audience:

- In one sentence, identify the purpose(s) and audiene(s) you think most important for both the summary and the memo your section is charged with writing. Write as if you were describing this project to a familiar and trusted colleague. Write about each document (and the information it contains) both in terms of its purposes and audiences

EXAMPLE 4–13   A Memo from Joan McIllich

Consumer Trends, Inc.
23 Paloma Blvd., Suite 63
Tele. (901) 487-2338
Fax #901/487-3347

To: Analysis Section
From: Joan McIllich
Subject: New Customer, DISD Project
Date: 23 November 1993
cc: Joe Janalleo

    I have been continuing negotiations with the Deerwood Independent School District (DISD) about contracting for a portion of the district's Consumer Education program in Social Science classes. Some details still need to be worked out in these negotiations, and the deal is far from done. To begin with, we are one of the first private educational contractors the DISD has hired. My interpretation is that they remain unconvinced that we can accomplish the job better and at less expense and hassle than they can. Added to this, we have no track record in public school education and cannot direct them to previous customers for recommendations.

    I have asked the Superintendent, Dr. Jan Laasila, to let us present a sample lesson at McAuliffe High School in a senior level Social Sciences class on December 15 (9:15 A.M., Rm. 45). They have selected the data for this one—dry stuff, straight from the Consumer Price Index (CPI). I think they are trying to gauge our ability to present data in a manner that is both informative and interesting. I will present the lesson itself; in attendance will be Superintendent Laasila, the Social Sciences Coordinator for the DISD (Dr. Bezell Wallingford), the classroom teacher (Tom Drummer), and twenty-seven high school seniors.

    Please come up with some general suggestions in connection with this project for our staff meeting on Tuesday when I will present a run-through of the actual presentation. To prepare me more specifically for this meeting, Joe and I need the following documents from your section by Monday at 9:00 A.M.:

- For Joe, a brief **summary** about the consumer trends suggested by the data—think HIGH SCHOOL STUDENTS!!
- For me, a **memo** detailing your suggestions about the presentation itself.

Thanks, as always.

Attachments: CPI tables

within Consumer Trends, Inc., and its purposes and audiences outside the company when the information is presented to the DISD people.
- In one paragraph, identify the purpose(s) and audience(s) you think most important for both the summary and the memo your section is charged with writing. Write as if were describing this project from

EXAMPLE 4–14   Consumer Price Index Tables

Consumer Price Index for All Urban Consumers (CPI-U): US city average, by expenditure category and commodity and service group

(1982–84 = 100, unless otherwise noted)

Group	Relative Importance, December 1991	Unadjusted Indexes Apr. 1992	Unadjusted Indexes May 1992	Unadjusted Percent Change to May 1992 from— May 1991	Unadjusted Percent Change to May 1992 from— Apr. 1992	Seasonally Adjusted Percent Change from— Feb. to Mar.	Seasonally Adjusted Percent Change from— Mar. to Apr.	Seasonally Adjusted Percent Change from— Apr. to May
Expenditure category								
All items	100.000	139.5	139.7	3.0	0.1	0.5	0.2	0.1
All items (1967 = 100)	—	417.9	418.6	—	—	—	—	—
Food and beverages	17.627	138.8	138.3	.7	−.4	.5	.0	−.3
Food	16.007	138.1	137.4	.4	−.5	.5	−.1	−.4
Food at home	9.921	137.4	136.2	−.5	−.9	.7	−.2	−.7
Cereals and bakery products[1]	1.426	150.6	150.7	3.7	.1	.3	.6	.1
Meats, poultry, fish, and eggs	3.030	130.3	130.0	−2.0	−.2	.0	.2	.2
Dairy products[1]	1.229	127.4	127.0	2.1	−.3	−.2	−.3	−.3
Fruits and vegetables	1.854	162.0	155.1	−4.8	−4.3	3.8	−1.1	−4.2
Other food at home	2.382	128.6	128.9	1.3	.2	.0	−.3	.5
Sugar and sweets[1]	.344	133.0	132.9	2.9	−.1	.4	.1	−.1
Fats and oils[1]	.260	129.6	130.4	−1.7	.6	−1.1	−.2	.6
Nonalcoholic beverages	.739	114.4	114.5	−.3	.1	−.3	−.6	.6
Other prepared food	1.039	139.5	140.0	2.6	.4	.4	−.1	.4
Food away from home[1]	6.085	140.2	140.4	2.1	.1	.1	.1	.1
Alcoholic beverages	1.621	147.2	147.4	3.3	.1	.5	.4	.1
Housing	41.544	136.5	136.7	2.9	.1	.4	.1	.1
Shelter	27.894	150.2	150.2	3.4	.0	.3	.1	.1
Renters' costs[1]	8.003	160.1	159.5	3.4	−.4	−.3	.3	.4
Rent, residential	5.835	146.2	146.3	2.5	.1	.5	−.1	.1
Other renters' costs	2.168	183.7	180.9	6.0	−1.5	−2.2	1.1	1.0
Homeowners' costs[2]	19.683	154.2	154.4	3.5	.1	.6	.1	.1
Owners' equivalent rent[1]	19.303	154.4	154.6	3.5	.1	.6	.1	.0
Household insurance[1,2]	.380	141.1	141.4	2.5	.2	.1	.1	.2
Maintenance and repairs[1]	.208	128.0	128.1	.9	.1	.1	−.3	.1
Maintenance and repair services[1]	.125	132.2	131.9	.4	−.2	.0	.2	−.2
Maintenance and repair commodities[1]	0.83	122.4	123.0	1.7	.5	.0	−.9	.5
Fuel and other utilities	7.327	115.8	116.8	2.3	.9	.4	.4	.2
Fuels	4.057	105.1	106.5	1.0	1.3	.4	.6	.1
Fuel oil and other household fuel commodities	.419	89.9	89.8	−1.2	−.1	.2	−.1	.8
Gas (piped) and electricity (energy services)	3.638	111.3	113.0	1.3	1.5	.4	.7	.0
Other utilities and public services[1]	3.270	142.2	142.4	3.7	.1	.4	.4	.1
Household furnishings and operation[1]	6.323	118.0	117.9	1.4	−.1	.3	.3	−.1
Housefurnishings[1]	3.699	109.7	109.2	1.0	−.5	.6	.3	−.5
Housekeeping supplies[1]	1.154	129.0	129.5	.4	.4	−.5	.3	.4
Housekeeping services[1]	1.469	130.5	131.0	3.1	.4	.2	.2	.4
Apparel and upkeep	6.097	133.3	133.1	2.9	−.2	.6	−.7	.4
Apparel commodities	5.535	131.1	130.9	2.8	−.2	.6	−.8	.4
Men's and boys' apparel	1.451	127.8	127.5	1.0	−.2	−.1	−1.1	−.6

EXAMPLE 4–14   *(continued)*

Group	Relative Importance, December 1991	Unadjusted Indexes		Unadjusted Percent Change to May 1992 from—		Seasonally Adjusted Percent Change from—		
		Apr. 1992	May 1992	May 1991	Apr. 1992	Feb. to Mar.	Mar. to Apr.	Apr. to May
Women's and girls' apparel	2.517	133.1	132.6	3.5	−.4	1.1	−.8	.9
Infants' and toddlers' apparel[1]	.218	131.3	130.3	.6	−.8	.6	3.3	−.8
Footwear	.800	125.6	126.0	3.5	.3	.5	−.5	.6
Other apparel commodities[1]	.550	141.5	142.8	4.5	.9	.7	−1.7	.9
Apparel services[1]	.562	146.7	146.8	3.2	.1	.1	.1	.1
Transportation	17.013	125.2	126.3	2.4	.9	.7	.5	.3
Private transportation	15.523	122.9	124.3	2.3	1.1	.6	.4	.6
New vehicles	5.032	129.1	129.2	2.6	.1	.7	.4	.3
New cars	4.055	128.2	128.4	2.4	.2	.4	.5	.2
Used cars	1.135	117.9	120.5	3.0	2.2	.0	1.9	1.7
Motor fuel	3.304	95.0	99.4	−.8	4.6	.9	.2	1.1
Gasoline	—	94.8	99.4	−.8	4.9	.8	.2	1.2
Maintenance and repairs[1]	1.520	140.5	140.8	4.5	.2	.4	.1	.2
Other private transportation	4.533	152.4	152.5	3.2	.1	.5	.4	.2
Other private transportation commodities[1]	.678	104.8	104.8	1.2	.0	.3	−.4	.0
Other private transportation services	3.855	163.2	163.2	3.6	.0	.5	.6	.2
Public transportation[1]	1.490	154.7	151.6	3.8	−2.0	1.9	.8	−2.0
Expenditure category								
Medical care	6.689	188.1	188.7	7.7	0.3	0.5	0.5	0.5
Medical care commodities	1.256	187.9	187.6	7.0	−.2	.6	.5	−.1
Medical care services	5.433	188.1	188.9	7.9	.4	.5	.5	.6
Professional medical services	3.213	174.1	174.7	6.3	.3	.3	.6	.5
Entertainment[1]	4.357	142.0	142.0	3.0	.0	.4	.6	.0
Entertainment commodites[1]	2.026	131.4	131.2	2.4	−.2	.4	.5	−.2
Entertainment services[1]	2.330	155.2	155.3	3.5	.1	.4	.6	.1
Other goods and services	6.674	180.3	181.3	7.2	.6	.5	.7	.7
Tobacco and smoking products	1.665	214.5	219.3	9.9	2.2	.6	.8	2.4
Personal care[1]	1.187	138.5	138.0	2.3	−.4	.3	.4	−.4
Toilet goods and personal care appliances[1]	.632	137.0	136.1	2.3	−.7	.2	.7	−.7
Personal care services[1]	.555	139.8	139.8	2.3	.0	.3	.1	.0
Personal and educational expenses	3.822	193.9	194.0	7.7	.1	.6	.7	.4
School books and supplies	.243	188.7	188.4	6.1	−.2	.5	.5	.4
Personal and educational services	3.579	194.5	194.7	7.8	.1	.6	.7	.4
Commodity and service group								
All items	100.000	139.5	139.7	3.0	.1	.5	.2	.1
Commodities	44.487	128.8	129.1	1.8	.2	.5	.1	.2
Food and beverages	17.627	138.8	138.3	.7	−.4	.5	.0	−.3
Commodities less food and beverages	26.860	122.5	123.4	2.5	.7	.6	.2	.4
Nondurables less food and beverages[1]	16.224	125.6	126.9	2.7	1.0	1.1	.5	1.0
Apparel commodities	5.535	131.1	130.9	2.8	−.2	.6	−.8	.4
Nondurables less food, beverages, and apparel[1]	10.689	125.7	127.9	2.7	1.8	.3	.7	1.8
Durables	10.636	118.2	118.4	2.2	.2	.5	.5	.2
Services	55.513	150.8	150.9	4.1	.1	.5	.3	.1
Rent of shelter[1,2]	27.273	156.3	156.2	3.4	−.1	.4	−.1	−.1
Household services less rent of shelter[2]	8.915	128.2	129.1	2.5	.7	.1	.6	.2

EXAMPLE 4–14   *(continued)*

Group	Relative Importance, December 1991	Unadjusted Indexes		Unadjusted Percent Change to May 1992 from—		Seasonally Adjusted Percent Change from—		
		Apr. 1992	May 1992	May 1991	Apr. 1992	Feb. to Mar.	Mar. to Apr.	Apr. to May
Transportation services	6.864	155.7	155.1	3.9	−.4	.7	.6	−.3
Medical care services	5.433	188.1	188.9	7.9	.4	.5	.5	.6
Other services	7.027	166.6	166.7	5.6	.1	.5	.5	.2
Special indexes								
All items less food	83.993	139.7	140.1	3.5	.3	.5	.3	.2
All items less shelter	72.106	136.6	136.9	2.9	.2	.6	.2	.1
All items less homeowners' costs[2]	80.317	141.1	141.3	2.9	.1	.4	.3	.1
All items less medical care	93.311	136.7	136.9	2.7	.1	.4	.2	.1
Commodities less food	28.480	123.5	124.4	2.6	.7	.6	.2	.4
Nondurables less food[1]	17.844	126.8	128.0	2.7	.9	1.0	.5	.9
Nondurables less food and apparel[1]	12.310	127.0	128.9	2.7	1.5	.3	.7	1.5
Nondurables[1]	33.851	132.4	132.8	1.7	.3	.8	.2	.3
Services less rent of shelter[2]	28.241	156.0	156.3	4.6	.2	.5	.6	.3
Services less medical care services	50.080	147.2	147.3	3.7	.1	.3	.4	.1
Energy	7.361	99.5	102.4	.3	2.9	.6	.4	.6
All items less energy	92.639	144.9	144.9	3.3	.0	.5	.2	.1
All items less food and energy	76.633	146.6	146.7	3.8	.1	.5	.3	.2
Commodities less food and energy commodities	24.757	132.4	132.6	3.0	.2	.5	.2	.4
Energy commodities	3.723	94.6	98.6	−.7	4.2	.8	.1	1.1
Services less energy services	51.876	154.8	154.8	4.2	.0	.5	.3	.1
Purchasing power of the consumer dollar:								
1982–84 = $1.00[1]	—	$.717	$.716	−2.8	−.1	−.6	−.1	−.1
1967 = $1.00[1]	—	.239	.239	—	—	—	—	—

[1] Not seasonally adjusted.

[2] Indexes on a December 1982 = 100 base.

— Data not available.

Note: Index applies to a month as a whole, not to any specific date.

Source: *Consumer Price Index Detailed Report.* May 1992. US Department of Labor, Bureau of Labor Statistics.

Joan McIllich's perspective. Think about each document (and the information it contains) both in terms of its purposes and audiences within Consumer Trends, Inc., and its purposes and audiences outside the company when the information is presented to the DISD people.

■ In one paragraph, identify the purpose(s) and audience(s) you think most important for the information contained in the presentation that McIllich is going to give. Write as if you were describing this project from Superintendent Laasila's perspective.

■ This time, describing the project from the high school student's perspective, identify in one paragraph the purpose(s) and audience(s) you think most important for the information contained in McIllich's presentation.

C. As a writing team, complete the following activities to help you focus on purpose and audience:

■ Choose one member of your writing group to play each of the DISD audiences you have identified. As a writing team, role play an interview with this individual. Ask questions such as the following:

What kind of information do you need? In what form? Why? When? What information should be included? Left out?
Why would you *want* to listen to McIllich? Why *not?*

D. Now that your writing team has thought about some of the purposes for the information you are going to put in each document, take another look—a closer and more analytical look—at the CPI data tables (Example 4-9) you have available for this task. As a team, brainstorm about the most promising ways of representing these data for the DISD audiences.

E. As an individual, draft one of the two documents with which your section has been charged. Next, schedule a team meeting at which members will critique these drafts. Well before the team meeting, distribute each member's draft documents. Do this in plenty of time for all group members to read the drafts carefully before the revision session. Make sure that all group members have a copy of each draft document and that *they have made written comments on the draft that are limited to the following questions:*

■ How does the information in this document seem suited to the purpose(s) of the document, both within Consumer Trends and in terms of the DISD context?

■ How does this information in this document seem suited to the audience(s) of the document, both within Consumer Trends and in terms of the DISD context?

■ How does the information in this document seem suited to the structure of the CPI data?

In a team meeting, compare each member's comments for the two documents. Focus both on points of agreement and disagreement. See if you can make some productive sense of the points on which you disagree; use them to think about the range of perspectives involved in thinking about purpose and audience and structuring information.

The Persuasive Nature of Technical Communication

A Continuing Case to Demonstrate Persuasive Strategies ▌ *Ethos:* Creating a Reliable, Trustworthy Persona ▌ *Pathos:* Appealing to Your Audience's Values ▌ *Logos:* Persuading with "Good Reasons" ▌ Supporting Good Reasons ▌ Which Good Reasons to Use?

Understanding practical rhetoric as a matter of *conduct* rather than production, as a matter of arguing in a prudent way toward the good of the community rather than constructing texts, should provide . . . a locus for questioning, for criticism, for distinguishing good practice from bad. That locus is not the individual or any particular set of private interests but the human community that is created through conduct. . . .

> Carolyn R. Miller, "What's Practical about Technical Writing?" in *Technical Writing: Theory and Practice,* ed. Bertie E. Fearing and W. Keats Sparrow (New York: Modern Language Association, 1989), pp. 14–24.

Introduction

When you think back on all you have learned about technical communication so far, you probably have realized that technical communication is quite often persuasive in nature: you persuade your audience that your solution to their problem will be effective; you persuade the public that bioremediation is the best technique for cleaning up certain oil spills; you persuade a university to question its practice of trapping and killing birds in its experimental crop fields; you persuade health care workers to take precautions against HIV infection; you persuade a future employer that you can fulfill job expectations. In fact, Aristotle, in his *Rhetoric,* encouraged people to take a broadly argumentative perspective on *all* communication. For that reason, he defined **rhetoric** in general as "the faculty of observing in any given case the available means of persuasion" (Bizzell and Herzberg, 153).

Once, the argumentative perspective that people routinely assume in courtrooms, editorial pages, and other public forums seemed inappropriate for technical communication because those arguments were often meant to change listeners' and readers' beliefs. Technical communication seemed more objective, and the facts (often scientific facts in technical documents) seemed to speak for themselves, in particular without the emotional associations of argument and persuasion.

These attitudes toward technical communication did indeed prevail until two decades ago or so. But since the 1970s, scientists, engineers, and technical communicators themselves have come to recognize that technical communicators do indeed argue rather than describe or demonstrate, do indeed try to win adherence for their claims by debate and social interaction. Even when writers or speakers are describing a product or a technique, summarizing events, or giving instructions, they are trying to persuade their audience that their version of things is correct. And, if you remember how we described purpose in Chapter 2, you realize that technical communicators must persuade their audiences that the solutions they recommend are effective, legal, ethical, financially feasible, and safe. Only if these audiences are persuaded will they use the technical document to help make decisions and to take action.

In fact, we now see that the sciences and the applied sciences or technology are not coldly rational disciplines dedicated to dispassionate, transparent descriptions of reality. They are instead very human disciplines, dedicated to developing arguments that create knowledge and solve problems in a social context.

Moreover, technical work is a very human enterprise—emotional as well as rational, messy and unpredictable, as competitive in nature as it is cooperative, and as inevitably colored by circumstances. Scientists and engineers seek to find and share the truth, of course. But those "truths" are established through informal argument and reasoning more often than through formal, syllogistic reasoning. Therefore, **technical communication**

rhetoric
the faculty of observing in any given case the available means of persuasion (as defined by Aristotle)

technical communication
applied communication that makes technical information accessible to many readers, viewers, and listeners; a set of arguments laid before a jury of readers, listeners, and viewers

is less an impersonal and rational demonstration than it is a set of arguments laid before a jury of readers or listeners or, sometimes, viewers.

Given that situation, you will have to convince your audiences. You will have to develop effective arguments that will be productive in technical and industrial settings. Because rhetoric, as Aristotle noted, is "the faculty of observing in any given case the available means of persuasion," we can use his account of persuasion as a place to start.

For Aristotle the available means of persuasion were what he called

- *ethos* (or the trustworthiness, credibility, and reliability of the speaker);
- *pathos* (or the appeal to the readers' or listeners' most basic, most deeply held values and beliefs);
- *logos* (or the appeal of the evidence and the reasoning process).

In reality it is sometimes difficult to separate out *ethos, pathos,* and *logos* in particular communications; it is difficult to tell where one kind of appeal begins and the other ends. But for the sake of discussion, we will separate them here. Learning these means of persuasion will allow you to develop more successful communications that are overtly argumentative—things like resumes, proposals, sales presentations, and recommendation reports—as well as documents that are less overtly persuasive, such as informational reports and instructions.

In order to illustrate the concepts of *ethos, pathos,* and *logos,* we refer to a continuing case as well as to the case documents collected in Appendix A, with which you are now somewhat familiar. To make the concepts more concrete, let's imagine that we are building an argument together in a specific situation with which you will probably be able to identify. Imagine that you are an engineering student at a state university and a member of your local student chapter of the American Society of Civil Engineers (ASCE), a group of prospective civil engineers like you.

Recently you and the other members of the chapter have read and heard a lot about the growing internationalism in engineering and the increasing internationalism of our economy. The articles and speakers that you've encountered have been emphasizing the advantages of bi- and multilingualism as well as the benefits to engineers of knowing about other cultures. Just last week you heard a respected engineer visiting the campus bemoan the fact that civil engineers interested in roads spend so little time studying the techniques of Italian engineers, even though Italian highways are beautifully engineered. The speaker concluded that the single best way for an engineering student to distinguish himself or herself from other new engineers would be to gain fluency in another language. You and some of the other members of ASCE are impressed—and a little frustrated. You know all too

**A Continuing Case
to Demonstrate
Persuasive Strategies**

well how difficult it is to study foreign languages within the constraints of an already demanding engineering curriculum. Faced with so many requirements already, you wonder how you can even think about taking courses in another language.

You and your friends wonder whether space in your engineering curriculum might be carved out somehow to permit students at your university to take more courses in other languages. Indeed, you wonder whether it wouldn't be wise for your university to *require* engineering students to show competency in a second language (at present only the liberal arts students on your campus have such a requirement).

You decide to ask your campus administrators to explore the possibility of encouraging or requiring engineering students to study a second language. You're not really recommending any change in the curriculum right now; all you want is some serious exploration. After all, you aren't even sure yourself if it's a good idea—if it's feasible to require a second language or worth the time and trouble to learn one, especially if it's at the expense of some other aspect of your education. But you think the issue has merit; you do want a serious study of the issue.

In short, you need to write a proposal that engineering administrators will use to decide whether a full fledged feasibility study is worth the bother. Your goal is to ensure that a serious study is indeed authorized. You don't want to be brushed off with a polite ''thanks for the suggestion; we'll think about it.'' And you don't want a superficial study that can be easily ignored.

Now, as we proceed through this chapter, let's try to develop *ethos, pathos,* and *logos* for an argument in this and the circumstances surrounding the case documents in Appendix A.

Ethos: Creating a Reliable, Trustworthy Persona

ethos
the trustworthiness, credibility, or reliability of the communicator; the ethical appeal

persona
the created narrator within a communication; relates to the *ethos* of the communicator

You have probably noticed that in the course of reading or listening to someone, you often get a very real sense of the kind of person who is doing the writing or speaking. Even if you have never read or listened to the writer or speaker before, you still often get a sense of his or her character and personality just by reading or listening to the performance. And you respond to the message to some extent according to how you respond to the person who delivers it.

Ethos refers to the persuasive value associated with the ''person'' that is created in the text of a document or presented through a speech. Sometimes people use the term **persona** to denote this person who narrates a document (so to speak), in order to underscore that this narrator of the document is not necessarily the same person as the author. You have encountered a great number of personas in your reading. Sometimes it is obvious that the narrator is not the author, in the way that Huckleberry Finn is a persona to be distinguished from his creator, Mark Twain. Other times it may be much less obvious that the role of the narrator or persona is a creation.

But in fact every document, every speech, is delivered by a created character, a persona, who may or may not have much in common with the real author of the piece. For instance, what kind of person seems to narrate Examples 5-1 and 5-2? (These are the original and a revised version of a case document reprinted in Appendix A, Case Documents 3.)

Both of these letters contain the same information, but their *ethos*— their ethical appeal—is very different. One of the writers comes across as a concerned but not confrontational person; the Ed Younger in Example 5-1 seems eager to "discuss this matter" and to "alert" the reader to a problem that both he and his reader might deal with together. Respectful and concerned, this first Ed Younger also creates the impression that he is a careful and reasonable person: note the cool accounting of factual observations and how the bulleted lists suggest that this persona lets facts speak

---

EXAMPLE 5-1   The Original Letter to Dean Wartala

August 2, 1992

Dr. Karen Wartala
Dean, College of Agriculture
University of the Midwest

Dear Dean Wartala:

The Coalition of Citizens Concerned with Animal Rights (CCCAR) was alerted by a concerned citizen two weeks ago that the University is trapping and killing large numbers of birds in the experimental crop fields on its campus. In our surveillance and investigation we have learned that:

- Birds are lured to the traps by bait placed in them, and the number of birds in these large traps has been observed to be at least 50 at times.
- The reported method of killing the birds was suffocation, by placing them, 400–500 at a time, in a bag.
- Approximately 500 birds are killed every day on this campus; 10,000 were killed in a three-month period.

We object to this for the following reasons:

- It is a waste of animal life.
- It is ineffective. Trapping and killing birds will not permanently reduce the bird population in the area. Even *temporarily* it will have no more than a minimal effect.
- The trapping and suffocation of these birds is cruel in the extreme.

For these reasons we insist that the University immediately stop the killing of these birds. If you wish to discuss this matter, please contact me at 555–5537 or 555–5856.

Sincerely,

Ed Younger

Ed Younger
Vice President, CCCAR

EXAMPLE 5–2   Revised Version of Letter in Example 5–1

August 2, 1992

Karen Wartala
College of Agriculture
University of the Midwest

Dear Ms. Wartala:

The Coalition of Citizens Concerned with Animal Rights (CCCAR) learned recently that your University is trapping, torturing, and killing large numbers of birds in the experimental crop fields on your campus. These outrageous actions must cease immediately.

In the course of our investigation of your organization, we have learned a number of shocking facts. Birds are being lured into traps when your employees place bait into them; we have observed that the number of birds in these large traps has often exceeded 50 at a time. In addition, your employees kill the birds by suffocating them; the employees murder the birds by placing them 400–500 at a time, in a bag, sealing the bag, and watching them die a slow painful death. Finally, your employees murder approximately 500 birds every day on your campus; we know that over 10,000 birds were slaughtered in a three-month period.

We object to your reign of terror for several reasons. You are wasting animal life that we consider to be precious indeed. In addition, your actions are ineffective. Trapping and murdering birds will not permanently reduce the bird population in the area—that's a simple fact. Even *temporarily* the killings will have no more than a minimal effect—birds always quickly repopulate decimated areas. Finally, the trapping and suffocation of these birds is cruel in the extreme. While your employees trap the birds and joke about it, the birds flap about, desperately trying to get some air. They die slowly and painfully.

For these reasons we insist that you immediately stop the killing of these birds. Otherwise we will be forced to take direct action.

Sincerely,

Ed Younger

Ed Younger
Vice President, CCCAR

over feelings. Not that the first Ed Younger lacks feelings; his persona has strong feelings indeed about animal rights and the abuses of animals that his organization has observed, but he prefers understatement, prefers to let the facts speak for him in creating his argument, prefers to create a self that seems relatively objective, unbiased, and respectful.

On the other hand, the second Ed Younger, in Example 5-2, seems to be more confrontational than cooperative. Note how he constructs for himself a position of moral superiority by addressing Dr. Wartala and her colleagues as ''you'' and himself and his organization as ''we.'' Note how he uses morally charged language to suggest that his position is passionately felt—the Ed

Younger who writes the last sentence even seems rather threatening! The language he chooses says something about his personality; notice how he refers to his reader and how he manages somehow to come off as an overconfident, less reliable, more biased investigator.

You may feel (and for good reason) that the second Ed Younger is a less effective persona than the first one, that the first Ed Younger seems more credible, more persuasive than the second. However, our point is not that one persona is more effective than the other, but that such a thing as a persona exists. Although their content is essentially the same, both letters seem to be the product of specific personalities with particular characteristics. Those personalities are created by the language and organizational choices that the writer makes. And the nature of those personalities is responsible in a significant way for the persuasiveness of these documents.

People react to the persona created in a message as well as to the contents of the message. Your job as a writer or speaker, then, because you are aware that people will be responding to *you* as well as to your message, is to *create the kind of persona that will be most persuasive in a given case.* In most circumstances, that means creating a persona that is reliable, trustworthy, fair, and in every other way the kind of person that an audience is likely to find credible.

In the following sections in this chapter, we introduce you to some of the principles for creating an effective persona in technical communication:

- an appropriate voice;
- a credible and responsible persona;
- an appropriate attitude toward persuasion; and
- attention to the *ethos* of your company or organization.

CHOOSING AN APPROPRIATE VOICE

As you know from the previous discussion and from your own experience, documents can seem stern, friendly, condescending, gracious, or such. Our response is largely because the persona of those documents seems stern or friendly or condescending or gracious. Thus, you will want to make sure that the **voice** that comes across in your documents is one that you intend.

voice
the nature of the persona; can be stern, friendly, condescending, formal, gracious, and so on

For example, take a look at the four versions of the first paragraphs in an article on botany and ecology (Examples 5-3, 5-4, 5-5, and 5-6). The paragraphs discuss the flowering patterns of plants that are pollinated by ants, birds, and bees, not by the wind. The author is introducing his study of how flowering patterns attract or fail to attract the ants, birds, and bees—and how all that affects a plant.

First, note carefully the differences in the voice of each version. Can you see that one or two of them might be appropriate for a publication in a professional journal? That another might be appropriate for a classroom

EXAMPLE 5–3   Version One of an Article on Botany and Ecology

The flowering pattern of many animal-pollinated plants may be divided into three phases: a) an initial phase in which a low but increasing number of flowers is produced per day; b) a short peak phase in which most of the total flower production occurs; and c) a final phase in which a low and decreasing number of flowers is produced per day. In recent years studies by a number of evolutionary ecologists have focused much attention on the effects of this mass-flowering pattern on pollinator attraction and movement (e.g., Janzen, 1978, 1981; Gentry, 1984; Stiles, 1985; Augsperger, 1986). In these studies it is demonstrated that many species of floral visitors preferentially forage on individual plants that produce large numbers of flowers per day. Presumably these individuals are both conspicuous and highly rewarding to pollinators (Ravens, 1982; Stiles, 1983; Carpenter, 1986). In contrast, the low levels of daily flower production that characterize the initial and final phases of the flowering pattern may be insufficient for pollinators to be attracted from alternative floral resources. Consequently, this mass-flowering pattern poses a dilemma.

This study seeks to answer two questions: During what phase of the flowering pattern does outcrossing occur, and what factors promote outcrossing? Specifically, this study documents the flowering pattern of *Catalpa speciosa* (Warder ex Barney) Engelm (Bignoniaceae) and examines the effect of different floral abundances on the proportion of flowers that are self- and crosspollinated on an individual plant.

Source: Adapted from A. G. Stephenson, "When Does Outcrossing Occur in a Mass-Flowering Plant?" *Evolution* 36 (1982), pp. 762–67. Reprinted with permission.

presentation or other oral presentation? That another might be best in a popular article discussing the research? And can you note the various differences in each version, differences that create the different voices? For example, what are the differences in word choice, punctuation, and sentences that distinguish these versions? These differences imply differences in the personas for each document.

As these examples indicate, there is really no single, "correct" voice that you can adopt for all documents. Although it is likely that you will develop a more-or-less characteristic voice that you use most often and that people will recognize and become accustomed to (after all, much of your writing will occur in recurring rhetorical situations), it is equally true that effective writers can change their voices when the situation calls for it. Look at Example 5–3 and Example 5–5. Both assume the "insider's" voice of the professional scientist, but the personality of one scientist seems somehow different from the other.

Just as you can change your voice orally when the situation calls for it—just as you can speak in a friendly way, in an excited way, in a stern

EXAMPLE 5–4   Version Two of an Article on Botany and Ecology

Many plants that are pollinated by animals—examples would be anything pollinated by ants or the proverbial birds and bees—develop their flowers in three fairly discrete phases. First, there's an initial phase; during this period the plant produces a small but increasing number of flowers each day. Next comes a short peak phase in which most of the total number of flowers are produced. And finally comes a final phase when the plant produces a small and decreasing number of flowers each day. Recently many evolutionary ecologists, such as Dan Janzen, Al Gentry, Gary Stiles, and Carol Augsperger (right down the hall), have been concentrating their efforts on studying the effects of this mass-flowering pattern—on how pollinators like bees and ants are attracted to plants and how and why they move from plant to plant. These people are showing that many kinds of floral visitors (for example, the ants) prefer to eat individual plants that produce really large numbers of flowers per day, presumably because these individual plants are easy to spot by the pollinators (and are very tasty to them) and because it takes less effort for the pollinators to do all their pollinating when flowers are so close together.

In contrast, when daily flower production is low, during the first and last phases of the flowering pattern, the plant through its flowers might not be able to attract enough pollinators from other flowers. Consequently, this mass-flowering pattern poses a dilemma: When flowers are few and far between, the bees or ants fly or walk around and visit many different trees, since flowers are scarce; that's good in that crosspollination occurs (and crosspollination makes for a more vigorous next generation), but it's bad in that the birds and ants spend a lot of their time and energy visiting various trees and therefore don't make as many stops as usual. On the other hand, when flowers are many, the pollinators can save their energy and work more efficiently on one plant; that's certainly more efficient, but it cuts down on the crosspollination because the pollinators are sticking to one location. So what's a plant to do?

In our study, Terry Haas and I tried to answer two questions: During what phase of the flowering pattern does outcrossing occur? And what factors promote outcrossing? Specifically, we've examined the effect of different floral abundances on the proportion of flowers that are self- and crosspollinated on an individual plant. What's more, we've documented the flowering pattern of the catalpa tree (*Catalpa speciosa*, Bignoniaceae).

way in different circumstances, just as you have been able to develop many oral voices in order to coexist happily with friends, rivals, co-workers, and loved ones in a variety of situations—so too you should be able to modulate your voice in prose, depending on some of the following factors:

- Your relationship with your audience (your voice can be less formal when you know someone well or when you are communicating with people in the same circumstances, as opposed to when you are communicating with a relative stranger or with someone elsewhere in a hierarchy);

EXAMPLE 5–5   Version Three of an Article on Botany and Ecology

A great many animal-pollinated plants develop their flowers in three phases:

   a. during an initial phase, a low but increasing number of flowers is produced per day;
   b. during a short peak phase, most of the total flower production occurs; and
   c. during a final phase, a low and decreasing number of flowers is produced.

Recently a number of evolutionary ecologists have been studying how this mass-flowering pattern affects the activities of various pollinators (e.g., Janzen, 1978, 1981; Gentry, 1984; Stiles, 1985; Augsperger, 1986). According to their studies, many species of floral visitors prefer to forage on individual plants that produce large numbers of flowers per day, presumably because these individuals are both conspicuous and highly rewarding to pollinators (Ravens, 1982; Stiles, 1983; Carpenter, 1986). In contrast, when daily flower production is low (during the initial and final phases of the flowering pattern), pollinators may not be sufficiently attracted from alternative floral resources. Consequently, this mass-flowering pattern poses a dilemma.

   We sought to answer two questions: During what phase of the flowering pattern does outcrossing occur? And what factors promote outcrossing? Specifically, this study documents the flowering pattern of *Catalpa speciosa* (Warder ex Barney) Engelm (Bignoniaceae); we examined the effect of different floral abundances on the proportion of flowers that are self- and crosspollinated on an individual plant.

■ Your audience's personality (different people respond sympathetically to different voices);

■ Your message (some messages are more routine that others; some messages contain more welcome news than others);

■ Your aim (sometimes you will desire an outcome that implies a certain voice);

■ Your genre and medium—the kind of document and the medium in which it will appear (some documents call for a certain characteristic voice: think of the sober voice of your newspaper's front page, for instance, and compare it with the tone of the opinion or sports or comics sections inside; or compare the tone of a published academic essay with the voice you hear in a letter of application).

As a final example of the importance of voice, imagine for a moment our continuing case—the short proposal that you might write to the administration of your university suggesting a look at the feasibility of foreign language instruction for engineering students. Imagine the kind of voice that you would want to adopt for such a proposal. You'd probably want to sound rather formal and sober and respectful—but emphatic as well. Imagine (by

EXAMPLE 5–6 Version Four of an Article on Botany and Ecology

Ever wonder why the birds and bees seem to be attracted to some flowering plants and not to others? Well, some university ecologists have been wondering, too—and trying to find out.

Ecologists know that a great many plants that are pollinated by animals (not by the wind) develop their flowers in three phases. First, a plant will produce a small but increasing number of flowers each day. Next comes a brief (but glorious-to-behold) peak phase during which the plant produces most of its flowers. Finally, the plant will produce a small and decreasing number of flowers per day. In recent years many ecologists—most notably Daniel Janzen of the University of Pennsylvania, Albert Gentry of the Missouri Botanical Gardens in St. Louis, Gary Stiles of the University of Costa Rica, and Carol Augsperger of the University of Illinois—have been studying the effects of this mass-flowering pattern on pollinators; they influence their movements. They and others have established that many species of floral visitors prefer to forage on individual plants that produce large numbers of flowers per day, presumably because these individuals are both conspicuous and highly rewarding to pollinators. To put it another way, birds and bees are attracted to large displays of flowers because such displays are easier to spot and because once the displays are located, the birds and bees can easily feed there and not have to search long and hard for more flowers. In contrast, when a plant produces just a few flowers per day, during the first and final phases of the flowering cycle, few pollinators may visit the plant; the birds and bees may be attracted to alternative flowers or fail to notice the few flowers produced.

This mass-flowering pattern poses a dilemma to birds and bees and ants—and to the ecologists who study them. On the one hand, when flowers are relatively scarce, the bees and ants fly or walk to many different plants. From an ecological perspective that can be beneficial, because plants often benefit from crosspollination: crosspollination tends to make for a hardier next generation, while self-pollinated plants tend to be somewhat less vigorous. On the other hand, when flowers are out in full, the bees and ants tend to stick to one location; instead of visiting many different plants, they concentrate their efforts. From an ecological perspective that too can be beneficial: though such behavior leads to self-pollination and hence to relatively less vigorous offspring, it is highly efficient; many more flowers can be pollinated when the pollinators don't have to spend their time and energy locating new flowers. So what is a plant to do?

University professor Andrew Stephenson and his colleague Terry Haas are studying how plants handle this dilemma. In particular, they want to know the answers to two questions. First, they want to determine at what phase of the flowering pattern "outcrossing" occurs. (Outcrossing is the term used to denote when a plant "mates" with a neighbor as opposed to itself; most plants can either crosspollinate or self-pollinate.) Second, they wonder what factors promote outcrossing? To find answers to these questions, they have been spending several summers looking carefully at the flowering patterns of the catalpa tree; they have been examining the effects that different floral abundances have on the proportion of flowers that are self- and crosspollinated on an individual plant.

way of contrast) how a presentation to fellow students on the same issue might go. There it might be best to be rather informal or colloquial, perhaps to even show a sense of humor or irreverence. Remember that the voice you choose to use in a situation will shape your audience's perception of your persona and ultimately affect the credibility and success of your document or presentation.

CREATING TRUST AND CREDIBILITY

Let's concentrate for a moment longer on the problem of *ethos* in the proposal that we are creating on foreign language instruction for engineering students. Even more important than the problem of voice in this instance is the problem of credibility: How can you get the experienced educators whom you are addressing, educators who by definition consider themselves your superior in experience and knowledge, to take you seriously? The answer lies in developing your facility with *ethos*. *Ethos* is always closely associated with **credibility**—with the creation of a persona that seems knowledgeable and trustworthy. Technical communicators always need to establish credibility, and we suggest three ways in which you can do that.

credibility
associated with *ethos,* the creation of a persona who seems knowledgeable and trustworthy

1. Showing that you have done your homework.
2. Establishing your special expertise.
3. Creating a sense of fair play.

Show that You Have Done Your Homework

An audience appreciates arguers who come to a case after careful study, and they distrust arguers who seem to be jumping into an argument before thinking things through completely. To give the audience the sense that you have done your homework, you might depend on two strategies:

1. Cite the work of experts, either informally in the course of your argument or in the form of notes, references, and works cited. By referring to published scholarship on the subject at hand informally in the text of your document ("Professor Marcia Tawney of Southwestern State University, who has studied foreign language instruction among engineering students since 1986, has noted that . . .") or through more formal means ("see Tawney, 1986, 1988"), you can indicate to your audience that your persona deserves attention. You conduct and refer to interviewers with chairs of other engineering departments at your school or with the registrar or provost of your institution to establish your knowledge of curricular and budget constraints. Look at Examples 5-3—5-6 to see how experts are cited in various ways, formally and informally, as a means of establishing the writer's credibility.

2. Indicate that your ideas are the product of some sort of systematic study or methodology. By indicating that you have done some systematic study

("After surveying the 50 top engineering schools identified by Dean Antonin Cellucci, we concluded that a number of fine universities are moving to find ways to permit their prospective engineers to study foreign languages"), or that you have performed some sort of experiment ("Our tests show . . ."), or that you have completed considerable library work ("Our search of the published literature on this matter indicates that . . ."), you establish the credibility of your persona.

Establish Your Special Expertise

Sometimes your credentials are established by virtue of the homework that you have done. But other times you can mention expertise that you have already established by virtue of previous experiences.

Sometimes such expertise is assumed as a matter of course. If you are given an assignment at work, it is usually assumed that you have the expertise necessary to do it, and so it would be pretentious to mention your credentials. But other times establishing your credentials (note how the word **credentials** is associated with *credence* and *credibility*) is essential to your persuasive efforts. Sponsors of many proposals, for instance, typically require proposers to include information about previous work as a routine part of the proposal. If you have no special background on a given topic, that need not be a fatal flaw in your argument, especially if you follow some of the other advice in this chapter. But if you do have special expertise, be sure to let your audience know about it, if only subtly and indirectly.

> **credentials**
> experiences that give a communicator expertise

Create a Sense of Fair Play

Personas are usually most persuasive if audiences perceive them as fair-minded. You can give your audience assurance that you are fairminded in several ways:

- By attending responsibly to alternative ideas—for example, always summarize an opponent's argument in a way that the opponent could agree with.
- By showing respect for other points of view.
- By indicating somehow that your personal biases are under control.

The last point doesn't mean that you should never reveal that you have a personal stake in the question at hand. In fact, if you do *not* indicate that you are personally involved or engaged somehow, your audience may accuse you implicitly of covering up your personal involvement, or may feel that you are too remote from a situation to be trusted, or may suspect that you somehow stand to gain personally if your position is adopted.

There is nothing wrong with being personally implicated in an argument, as long as you indicate that your involvement is not coloring your judgment improperly. For example, in our hypothetical proposal you might want to confess that your concern about foreign language instruction for engineers derives from self-interest: you want to get an edge on other new engineers who are with you on the job market. But you would also add that your concern is not purely self-interested; there are other good reasons for adopting your proposal.

Before reading on, you might take a moment or two to look again at Examples 5-1 and 5-2 as well as the four versions of the botany article in Examples 5-3—5-6. Note how the writers of those documents establish, or fail to establish, their expertise and their fairness.

ASSESSING AND CHANGING YOUR ATTITUDES TOWARD ARGUMENT

The preceding discussion may remind you that a sense of fairness is often missing in our approaches to argument. In our culture many people grow up thinking that argument and persuasion mean winning and losing in a sort of all-out verbal combat. In courtrooms we see prosecutors and defendants arguing for one kind of clear-cut victory or another—guilty or not guilty—in front of a judge and jury. In our legislatures we see political parties arguing for victory on some piece of legislation. During election campaigns we see candidates engage in debates and then be declared the "winners" or "losers" of those debates.

But now we'd like to discuss other, more productive ways of thinking about argument. The twentieth-century rhetorician Kenneth Burke, having observed the consequences of seeking all-out victory in rhetorical exchanges—consequences that have played out ultimately on battlefields—has suggested a different theory of argument. He proposes that people seek not victory, advantage, manipulation, or coercion. Instead, Burke proposes that writers and speakers conceive of themselves as *cooperators,* not competitors. To do so, writers and speakers must abandon the role of aggressive adversaries and instead become mutually supportive partners in the process of building bridges between people.

agonistic
competitive, rather than co-operative, interaction

Burke and others like him would have us conceive of language and rhetoric not in **agonistic** terms, not in terms that suggest battle and strife, but in terms of finding or inducing cooperation. Burke says an arguer is like "one voice in a dialogue." Try thinking of argument not as a matter of vanquishing an opponent but as a productive exchange, part of a continuing conversation that is designed to generate new insights.

As a general rule, construct a persona in your persuasive technical documents that your audience will see as a cooperator, not a competitor; as a partner, not a rival. Show that you care about both your ideas and your audience. There may be times when you'll need to lay down the law, so to speak, in the workplace. But, most often in our culture, audiences will appreciate it if you cast yourself as a respectful partner, as one whose arguments are put forth in the spirit of caring, mutual support, and negotiation, in the interest of finding the *best* way, not "my way." Here are a few suggestions for how you can create a persona with whom your audience will want to cooperate.

- Maintain an attitude of courtship, not coercion. It's important to articulate your position vigorously, but you don't want to "defend" your "position" so vigorously that your "opponent" is vanquished or silenced. Try to argue in such a way as to invite a response; seek dialogue and consensus, not silent acquiescence.

For example, the original letter to Dean Wartala (Example 5-1) could have included the following statement to seek consensus:

> The Coalition of Citizens Concerned with Animal Rights wishes to propose a solution that would protect both the birds within the vicinity and the seeds and crops of the experimental fields.

- Show that you understand and genuinely respect your audience's position, even if you think the position is ultimately wrong. Often that admonition amounts to remembering to argue against an opponent's position, not against the opponent. And often it means representing your opponent's position in terms that your opponent would accept. Look hard for common ground that you already share with your audience. See yourself as a mediator. Consider that neither you nor the other person has arrived at a best solution, and carry on in the hope that dialogue will lead to an even better course of action than the one you now recommend.

For example, the original letter to Dean Wartala (Example 5-1) might have contained a statement such as the following:

> The Coalition of Citizens Concerned with Animal Rights believes that the University crop experiments are important to the state. However, equally important to the state environment are the many birds that flourish here. It is in the mutual concern for state environmental welfare that we propose . . .

- Expect and assume the best of your audience, and deliver your own best as well. Avoid threats, even if your opponent uses them. In fact, it is often when someone else resorts to reprehensible tactics that your persona will be most appreciated if you respond by taking the high road. Avoid sarcasm and remain respectful.

For example, the original letter to Dean Wartala (Example 5-1) might have contained a statement such as the following:

> Rather than pitting the welfare of the birds against the interests of the University and the experimental fields, Coalition of Citizens Concerned with Animal Rights suggests a means of controlling, rather than killing, the birds. . . .

- Cultivate a sense of humor and playfulness, a legitimate tool to be used by skillful technical communicators. However, like any other technique, humor should not be used thoughtlessly, and you will need to take culture, race, and gender into consideration when you make use of it. Humor does not mean necessarily telling a joke or making light of someone who is truly upset about something. But the fact is that nothing creates a sense of goodwill quite so much as good humor. Your persona will be seen as open to new possibilities and open to cooperation if you occasionally show a sense of humor. And a sense of humor can sometimes be especially welcome when the stakes are high, the sides have been chosen, and tempers are flaring.

Perhaps a statement such as the following in the original letter to Dean Wartala (Example 5-1) would have engaged the reader's interest and lightened the mood:

> It seems appropriate to propose this conversation to the largest university in the state most known for its populist spirit. Let's talk! We'll even bring the coffee.

In order to see this advice at work, consider once more your proposal to have administrators at your school authorize a study of the foreign language issue. In this instance you might work hard to show why foreign languages have traditionally not been associated with engineering education. You might indicate that you know that there are good reasons why such instruction is not currently possible (it could get expensive!), and acknowledge that you are not the first person to think of such an idea. You might name what everyone agrees to be the aims of engineering education at your institution and indicate your agreement with those aims (foreign language instruction, after all, is probably in keeping with such aims).

You might indicate that you appreciate the time and effort that would be required by a study like the one you are recommending. You might put forward your proposal for a formal and serious study of the issue as an effort toward negotiating something mutually satisfactory over the long run. You might in some way show a sense of humor that indicates that your consider yourself an equal, and that you are a colleague pleasant to work with.

ATTENDING TO THE *ETHOS* OF YOUR COMPANY OR ORGANIZATION

Because writers often cooperate with other writers and because documents often speak not just for their creators but for the organizations that sponsor them, you will need to attend to corporation's *ethos* as well as your own. Apple Computer, for example, has a **corporate *ethos*** that is very different from IBM's, and the people who work in those corporations must often adopt a voice that is in keeping with their surroundings. For example, examine Example 5–7, which contains selections from an Apple Computer manual and a Motorola instruction guide, for differences in corporate *ethos*. As a member of a specific enterprise, you should consider your company's collective *ethos* when you are developing your own.

There are a number of ways you can discover your corporation's *ethos* in addition to looking at some representative documents and doing a bit

corporate *ethos*
the values and beliefs of a corporation that determine what persona the corporation assumes when communicating to listeners, viewers, and readers

---

EXAMPLE 5–7   Two Examples of Corporate *Ethos*

Look over these examples of corporate writing and write a sentence or two to describe how they differ in terms of their *ethos*, the image they seek to project, and the way they see their relationship to the user of the product they produce. Note as well that *ethos* accounts for differences in what the introductory sections of the manuals are called (see the source lines below).

Apple Computer

Many people are concerned at first about making a mistake that will damage their computers or their work.

But you don't need to worry. If you set up your Macintosh according to the instructions, avoid bumping it while it's turned on, and don't spill coffee on the keyboard, you won't hurt your equipment.

Source: Apple Computer, Inc., *Getting Started with Your Macintosh*, "Welcome," p. ix.

Motorola

Thank you for selecting Motorola and welcome to cellular telephone service. Your Digital Personal Communicator™ represents the state-of-the-art in personal cellular telephones today. The listing below shows just a few of the exceptional features that the telephone contains. . .

To cover all of these features properly, we will take you through a logical step-by-step learning procedure that explains everything you need to know to operate your new telephone.

Source: Motorola, Inc., *Digital Personal Communicator User Manual*, "Introduction," p. 4.

of analysis on your own. You can get a sense of the corporation's *ethos* from:

- The company's or organization's style manual (which might specify such things as the use of second or third person in all documentation; or it might encourage or forbid the use of contractions depending on how formal it wants to appear).
- Senior employees who oversee your first writing efforts.
- The changes that professional technical editors make in your documents.

Considering your company's *ethos* doesn't mean you have to give up your own. There will be space for you to create your own persona within the context of your organization. Especially in documents that are intended for audiences within the organization, you will have plenty of freedom to develop your own *ethos,* and, even in documents that leave the organization, you can exert considerable influence over the *ethos* of the final product—especially if your choices are well considered.

Pathos: Appealing to Your Audience's Values

pathos
the appeal to the audiences' most basic, deeply held values, attitudes, and beliefs

You already know that you must take your audience's attitudes into consideration when you select and organize information for your communications. The technique of **pathos** fits right in with that approach by appealing to your audience's most basic, most heartfelt attitudes and values.

Sometimes *pathos* is defined simply as an "emotional appeal," an overt appeal to an audience's basest instincts. Critics of advertising, particularly critics of television ads, contend that advertisers depend on purely emotional arguments when they tie automobiles to a drive for success, or when a certain beer is depicted as part of a happy social life, or when insurance ads are associated with our fear of death and loss of security, or when sexually suggestive scenes are used to market a new movie.

Therefore, on first glance, *pathos* seems an unlikely approach to use with the logic of technology (and of science). Yet, there is nothing irrational in acknowledging human emotions in our arguments. For example, it would be appropriate to argue that "Our city should attend to its solid waste problems because we want our children to live in safe and healthy and attractive communities," even if there is an emotional appeal inherent in that argument.

Instead of thinking of *pathos* as an irrational appeal, think of it as an appeal to the most basic values of your audience. Suppose that a civil engineer writes that "If we make the changes recommended in this report, the intersection at Wayside Avenue and Salem Road will be much safer; injuries from traffic accidents will almost certainly decrease." It is true that this statement is couched in logical and rather matter-of-fact terms. But it is also true that this argument carries emotional overtones, "pathetic" overtones that make it much more powerful than the following argument: "If we make the changes recommended in this report, traffic at the intersection of Wayside Avenue and Salem Road will move much more efficiently." That's because the appeal to safety attends more to our basic human values. Example 5–8 illustrates some instances of *pathos* in workplace documents.

As a technical communicator you should use *pathos* with care. Explicitly emotional arguments are distrusted in technical discourse. But *pathos* can be used effectively. For example, if you are tying to convince the group of college administrators to authorize a study of the feasibility of foreign language instruction for engineers, you'll be on firmer ground if you depend mostly on logical appeals as described in the next section. However, you could mention that knowing a foreign language can help individuals in a multicultural world understand and appreciate one another. And, by indicating (perhaps in an introduction or conclusion or some other emphatic point) that you feel strongly about something, or that something deserves strong and close attention, you may convince your audience to give special weight to your argument.

Moreover, you need to be aware of *pathos* in others' arguments. Many of our current arguments about applied technology center on *pathos.* For example, as a society we often seem paralyzed in our ability to agree on what to do with nuclear waste, with our old growth forests, with our range-lands, with our prison systems, and with our cities' infrastructures. Audiences often resist logical arguments because of their deeply held beliefs. In public discourse then, we often use *pathos* in our own arguments or must overcome, if at all possible, *pathos* in others' arguments.

Logos: Persuading with "Good Reasons"

In persuading people to take a certain action or to accept some assertion as true or likely, writers and speakers typically provide "good" or effective reasons for their arguments. You might think of good reasons as a series of clauses that begin with the word *"because"*—a series **because-clauses** used to support assertions. Argument, then, may be seen as a process of devising because-clauses that your audiences will find convincing, even compelling. A new highway should be put through downtown *because* of improved safety, better traffic flow, and better aesthetics. A certain paint should be used on automobiles for good reasons *because* it resists chipping, for instance, and *because* it is inexpensive and easy to use.

because-clauses
good reasons that support arguments or proposed ideal situations

These because-clauses support the ideal situations that you saw in Chapter 2 when we discussed communicating to solve problems. The *ideal situation* is to have foreign language study within the engineering curriculum, but at the moment it is not required or encouraged *(actual situation),* so you are proposing that foreign languages be included in the curriculum

EXAMPLE 5–8   Examples of the Use of *Pathos*

Directions: Look over these passages taken from the Case Documents in Appendix A. What emotions or values are being appealed to in these examples? If you had to arrange these examples from most use of *pathos* to least use, what would be their order?

Liability Case Documents

On May 7, 1974, Venable purchased another sealed container of Vydate L from a clerk at Midkiff and Carson Hardware Store. The employee who sold the product to Venable did not caution him in any way about the use of the product around humans. Venable placed the chemical in the back of his pickup truck under some old coveralls to keep it from falling over. Later that same day, he put a blue mason jar of iced water and paper cups in the back of his truck for the refreshment of his field hands and drove to where they were pulling tobacco plants.

Upon arriving arriving there, Venable offered the water to the workers and poured himself a cup from the mason jar. A short while later, Mrs. Ziglar, a laborer, walked over to the truck to get a drink of water. She did not, however, pour from the mason jar; instead, she opened the container of Vydate L, which was located on the same side of the truck as the carton of paper cups and drank some of the poison.

Bird Control

Birds are trapped in large cages baited with bread. The trapped birds are supplied with water and food until they are removed from the cages. There are eight traps located in the plot area. Approximately 150–200 birds in total will be trapped daily on the average. The number of birds per trap will vary a great deal. On unusual days as many as 500 birds may be trapped. The average number of birds trapped per year is usually about 5,000 with a range from year to year of 3,000–10,000. Bird populations are never permanently reduced as a result of the program. Our bird control program attempts to reduce bird populations in the plot areas only during the growing season. All songbirds are released. Only pest birds like blackbirds, starlings, and English sparrows are destroyed.

Exxon Valdez

After the spill, the US Environmental Protection Agency (EPA) approached Exxon to discuss ways to clean up the oil. A joint effort between EPA, Exxon, and the Alaska Department of Environmental Conservation was organized to test the effectiveness and safety of bioremediation—a process in which nitrogen and phosphorus fertilizers are fed to naturally occurring, oil-hungry microbes, allowing them to multiply and consume oil faster.

AIDS

Tuberculosis (TB), caused by *Mycobacterium tuberculosis,* is a contagious, airborne disease of the lungs. Sometimes TB affects other parts of the body. It is a major cause of death in many parts of the world today and was the leading cause of death in the early years of this century in the United States. As recently as 1950, Minnesota reported 5,000 new cases. TB has been called a "social disease with medical aspects" because it can be found wherever poverty, poor nutrition, homelessness, and substance abuse occur.

(purpose) because other schools are doing it, because it might make students into more flexible and informed and culturally aware engineers, and because it might give students an edge on the job market. You are often *arguing* for these *ideal situations* by providing good reasons for those ideal situations.

Finding good reasons is always a necessary challenge for writers and speakers. Sometimes good reasons come from thinking about *pathos*. By pondering your audience's most fundamental values and beliefs you can usually find some powerful because-statements. More often, however, good reasons derive from mulling things over reasonably as well as emotionally; that is, they derive from *logos*. *Logos* refers to the logic of what you communicate; in fact, *logos* is the root of our modern word *logic*. Good reasons are thus commonly associated with logical appeals.

logos
the appeal of the evidence or the reasoning process; the logic of what is communicated

In the rest of this section, you will find a set of questions for developing good reasons for your arguments. These questions will help you develop effective arguments, and they will likely equip you to communicate more effectively in various circumstances—even when you must think fast at a meeting or speak extemporaneously. But do not expect every question to be productive in every case. Sometimes a certain question won't get you very far, and often the questions will develop so many good reasons and strategies that you will not be able to use them all. Ultimately, you will have to select the ones most likely to work in a given case.

CAN I ARGUE BY DEFINITION—FROM "THE NATURE OF THE THING"?

Probably the most powerful kind of good reason is an argument from definition. If you can get your readers or listeners to think of a thing in certain terms, you enable your audiences to think of things in the way you want them to. You need not think here in terms of formal definition (such as those we will discuss in Chapter 13). Rather, think in terms of an informal definition, or what is called a **categorical proposition.**

A categorical proposition sounds like something complicated, but it really amounts to the kind of simple definition statement—X is Y, something is something else—that we make all the time. To put it more formally, a categorical proposition consists of a subject (the word to be defined), a linking verb (such as "is" or "was"), and a predicate—something to be said about the word being defined. Here are some examples:

categorical proposition
an informal definition that consists of a subject (word to be defined), a linking verb, and a predicate (something about the word to be defined)

- Science and engineering curricula are usually rather demanding.
- A recycling program is something that this campus badly needs.
- Ms. Protevi was an ideal boss.
- Nursing is a more interesting profession than most people imagine.
- Certain climactic changes now in evidence are truly disturbing.

In each case it is easy to identify the elements of the categorical proposition. For example, in the first sentence above, "science and engineering curricula" is the *subject,* the *linking verb* is "are," and the *predicate* is "usually rather demanding."

We all use more categorical propositions than it might seem at first. This statement, "new farming techniques are revolutionizing agriculture in the developing world," is really a version of the categorical proposition "new farming techniques are some things that are revolutionizing agriculture in the developing world." Likewise, the statement, "That substance contains toxins," can easily be transformed into categorical proposition form: "That is a substance that contains toxins."

Writers and speakers can develop categorical propositions to create because-clauses. If we go back to the case study involving foreign language instruction for engineers, we see that one way to persuade people to consider the idea is to define it in positive terms: "Foreign language study is likely to make engineers more flexible and resourceful employees in the 21st century." You have *defined* foreign language study as something very attractive to your audience, and you have in fact developed a good reason—a because-clause—that you can use in your proposal: "You should study the feasibility of foreign language study for engineers *because* it is a course of study likely to make engineers more flexible and resourceful employees in the 21st century."

One of the best characteristics about arguments from the nature of the thing is that you can develop any number of definitions in support of your arguments:

- Foreign language instruction can be defined as a means of ensuring that engineers can compete better in international markets.
- Foreign language can be defined as a way of helping engineers become more comfortable with the other cultures they are likely to encounter in their work.

Of course, you have to do more than state a definition or categorical proposition. You must also present evidence that your definition is *accurate.* That is especially true when your audience either works from a different definition than you do or is unsure of the concept being defined. An example of the first instance—working from another definition—would be the person to whom foreign language study might be "an unnecessary frill for an engineer," or "more painful than root canal work," or "an awfully expensive part of a curriculum," or "something that liberal arts students need, but not engineers." As a consequence, you should think of your categorical proposition as a substitute for the definition (formal or informal) that your audience may bring to your proposal. You are offering a different predicate to replace the one your audience now has.

In the other case—an audience unsure of the concept being defined—you are providing the definition for your audience, not changing their views of the thing defined. For instance, if we are explaining a term like *categorical proposition* to someone like you, we'll be providing a complete, new definition rather than changing the one you already have.

Once you get a good definition to support your argument, you must defend it before moving on to your next good reason. Defending it will probably involve using some of the other strategies described below: citing examples and statistics, offering testimony, providing comparisons, and so forth. It may take you some time to defend your definition, but investing time in such a defense is worthwhile, for argument from definition is one of the most powerful ways of winning support.

If you can get your audience to accept your definitions of things, you've won more than half the battle. That's why the most heated issues in our culture—abortion, affirmative action, gay rights, pornography, women's rights, gun control, the death penalty, among others—tend to turn on matters of definition. Is abortion a crime or a medical technique? Is pornographic speech protected by the First Amendment, or is it a violation of women's rights? Is the death penalty "just," or "cruel and unusual punishment"? People may not even care about the consequences: If a definition of something seems airtight, it often overwhelms questions of practicality. In other words, if you can prove that foreign language instruction is something that will make you closer to the ideal engineer, or show that foreign language instruction is in keeping with the engineering program's own definition of its mission, then you will have achieved a large part of your persuasive goal, even if some of the consequences of your proposal are rather negative (e.g., it might be costly or inconvenient, or require other difficult curricular adjustments).

CAN I ARGUE FROM CONSEQUENCE?

A second powerful and less complicated source of good reasons comes from considering the possible **consequences** of your position: the good things that will follow from as well as the bad things that will be avoided by your position.

consequences
the positive and negative results of a proposition

If we return to our example of prospective foreign language instruction for engineers, we can see how arguing from consequences might work. You could, for instance, make your audience think positively about the prospect of such a curricular change if you established that good things would follow from the idea:

- Engineers with a second language could make their companies more profitable because they might open up new markets.

- Engineers with a second language would be more likely to profit from (and help import) engineering techniques developed in other nations.
- Engineers who speak more than one language are more sensitive to the needs of other cultures as well as to American subcultures, and therefore are likely to develop better products.

You could probably imagine other good consequences, but don't forget to think as well about what bad things would be eliminated by your idea:

- Engineers with proficiency in a second language would eliminate the insularity associated with American engineering.
- Engineers with a second language would help to eliminate errors in documents that are not translated correctly by others.

Thus, after considering consequences, you'd be left with one or more good reasons, or because-clauses, to support your arguments: Administrators on your campus should study the feasibility of requiring foreign language instruction *because* it would make for more flexible, resourceful, and productive engineers and/or *because* it would permit engineering firms to compete in international markets and/or *because* it would make engineers more culturally sensitive in their work. Here again, you also have to show that these consequences would indeed follow from the idea or course of action that you are arguing for by citing statistics, providing examples, or offering the other kinds of supporting evidence that will be detailed in the next section of this chapter. But by thinking of consequences, you can indeed think of good reasons that will provide the spine of your arguments.

Thinking about consequences will also encourage you to consider the feasibility of any proposals or recommendations that you might be offering. A good idea has to be a practical idea, and you need to remember to indicate that your ideas are truly feasible if they are not obviously so. Anyone proposing to make foreign language possible in engineering education, for example, has to deal with the practicalities: curriculum changes, accrediting agencies, budget, institutional support. Thinking about the consequences of your ideas will bring questions like these readily to mind.

Arguments from consequence are especially common in proposals and recommendations. In the pragmatic, results-oriented, bottom-line culture of industry, results matter most; and most arguments will turn on arguments from consequence, particularly when moral and ethical issues are not seen as especially important to a given case. As you will learn in Chapter 17, proposals typically derive from a problem that needs to be solved, and the consequences that a proposal writer cites therefore typically include that the problem will indeed be solved. The same is true in recommendations: When you are recommending a certain course of action, it is often in order to eliminate a problem or to answer a question—which computer software should the organization purchase, for instance, or what reforms will eliminate excessive costs. In these cases, it makes sense to describe the consequences of the actions that you are recommending.

CAN I ARGUE FROM VALUE?

A special kind of argument from definition that often implies consequences is the argument from value. These arguments require that you support your thesis with one or more because-clauses that include an **evaluative component.** Your argument is, essentially, that your position is the best one. Arguments of value sound like these:

evaluative component
a part of argument from value

- In order to produce good engineers, you need to study the feasibility of foreign language instruction.
- You should give strong consideration to the treatment I'm suggesting because it is the most efficient and safe one available.
- The telephone system that you are currently employing is now outdated and cumbersome.
- Your crops will turn out better if you follow certain principles.

And they generate powerful because-clauses like these:

- You should authorize this study *because* foreign language instruction can make people into good engineers.
- Take up this program of physical therapy *because* it is truly the best one for you.
- Buy our telephone system *because* it is the best one for your office.

Evaluative arguments usually proceed from the presentation of **criteria,** and these criteria come from the definitions of good and bad, poor and not so poor, that prevail in a given case. A "really good chemist," for example, meets certain criteria or specifications; so does an effective treatment program, or an outstanding telephone system, or a good harvest. In arguing by means of evaluation, therefore, you will need to define the criteria that hold for the item under discussion and then demonstrate patiently that the criteria are met. Sometimes the criteria—what makes something good or bad—are relatively obvious; a piece of machinery, for instance, should be cost effective, safe, easy enough to operate or maintain, and capable of performing the task that you need it for. In making a case for it, you might need just one sentence to name the criteria, and you will probably spend most of your time applying the criteria to the piece of machinery to show that it has those characteristics. But in other situations the criteria aren't so clear. What makes a good scientist or a poor telephone system, for instance? In these cases you might need to spend some time defending your choice of criteria and the

criteria
used in evaluative arguments or arguments from value

order in which you rank them before applying them to the topic under evaluation.

Evaluative arguments often are associated with *comparisons*. If you need to decide which piece of equipment to buy, which treatment to follow, or which employee to hire, you usually need to offer some choices; otherwise, there may be no reason to argue about the matter. Once the choices are clear, then all you have to do is apply criteria to the available choices, weigh the results, and declare a winner. Offering choices can keep an argument from getting too abstract.

CAN I COMPARE OR CONTRAST?

comparative terms
expressions of what things are like or unlike a topic in a communication; argument by similitude

Again, evaluative arguments often can generate comparisons, but even if they don't, you may need to think in **comparative terms.** Thinking in comparative terms means asking what things are like or unlike the topic you're discussing. Aristotle called this habit of mind *argument by similitude.*

Thinking in comparative terms can help you generate because-clauses like these:

- You should study the feasibility of foreign language instruction *because* our competitor schools are studying the feasibility.
- You should offer that internship to me *because* it's just the kind of thing I'm suited to (because of the experience I gained last summer) and *because* I have qualifications similar to the people whom you've hired as interns in the past.

In turn, contrasts (in a way, negative comparisons) can be equally effective in an argument, such as:

- You should study the feasibility of foreign language instruction for engineers *because* no one else seems to be offering it these days; it would give us a competitive edge.
- You should hire me for the internship *because* my qualifications are unlike (and superior to) those of the many people whom you usually consider.

If you anticipate that your audience will be confused by or unfamiliar with your topic, think of comparisons that will make the topic more easily understood. Comparisons are an effective way of building common ground and effective premises.

metaphor
a figure of speech that compares an unfamiliar item with something more familiar in order to bring new insight to the topic of a communication

Particular kinds of comparisons are *metaphors* and *analogies.* A **metaphor** is a figure of speech that compares an unfamiliar item with something more familiar in order to bring new insight to the topic under discussion. For example, here are some metaphors used in the Case Documents included in Appendix A:

- AZT is a "decoy" that interferes with a virus's ability to integrate with a cell's DNA.
- Bioremediation refers to a process by which microorganisms "eat" oil.

Similes use "like" or "as" to make comparison (e.g., AZT is like a decoy). Metaphors and similes are especially effective when you are explaining a concept that your audience might find obscure or hard to understand.

Scientists especially are comfortable offering comparisons to explain their work:

- "Well, the nectar that flowers produce can serve as a sort of intoxicant to the ants that consume it," explains an ecologist. "When the ants consume the nectar on a catalpa tree, they tend to wander around disoriented afterward. It's like they are drunk."
- An entomologist studying how trees in New England defend themselves against the gypsy moth might say that "The trees seem to talk to each other, in a sense. When one tree is infested with gypsy moths, it gives off a substance that seems to communicate with other trees—that seems to tell them to get their defenses up, to produce a toxin that the moths don't like too well."

The benefits of metaphors and similes are apparent to all who use and encounter them, and using them can be especially arresting and emphatic, but like other rhetorical techniques they should be used with care.

You need to make sure that your audience is indeed familiar with the concept with which you compare your topic. Also, your audience might find the image of a drunk so negative that the point about the ants and nectar is lost. Finally, you don't want to reduce your topic too much within a metaphor or simile that stresses only one characteristic.

An **analogy** is a *comparison that is extended for explanatory or persuasive purposes.* Analogies are typically developed over several sentences or a paragraph. Here is an analogy provided by Richard Brennan in *The Dictionary of Scientific Literacy* to explain certain key terms in cell biology:

> The relationship of the various elements of a cell may be thought of as follows: The nucleus of a cell is a library containing life's instructions. The chromosomes would be the bookshelves inside the library, the DNA would be individual books on each shelf, genes would be the chapters in each book, and the nucleotide bases making up the strands of DNA would be the words on the pages of the individual books. (p. 121)

Analogies such as Brennan's are especially valuable, but they work best with willing listeners or readers. They are less effective in convincing unwilling readers or listeners in more argumentative situations. Such an audience is likely to want other sorts of reasons, perhaps ones of values and consequence.

simile
a figure of speech that uses "like" or "as" to compare an unfamiliar item with something more familiar

analogy
a figure of speech that extends a comparison for explanatory or persuasive purposes

CAN I CITE AUTHORITIES?

Sometimes you can buttress an argument by citing other people, highly respected by your audience, who hold to the same position that you are promoting. Arguments from authority ("believe this *because* so-and-so believes it") don't make for the strongest because-clauses. Few people are persuaded to believe something on the basis of authority alone, but arguments from authority do work well as supporting arguments—as one because-clause among many. Not only can references to authorities give your audience another good reason to believe what you ask of them, but they can build your credibility as well.

CAN I COUNTER OBJECTIONS TO MY POSITION?

Another good way to find convincing because-clauses is to think about possible objections to your position. If you can imagine how your audience might counter your argument, you can include in your argument precisely those points that will address your audience's particular needs and objections and win your audience's assent. The strategy of mentioning counterarguments first and refuting them before they occur is called the **inoculation theory.**

inoculation theory
a persuasive strategy of mentioning counterarguments and refuting them before they occur to an audience

Inoculation theory means that you inoculate your audience with the opposing view; then when they hear it, they already have your response in mind. Sales people often use this technique effectively. They sometimes anticipate another company's argument and then supply a counterargument. For example, a car salesperson might say, "You'll be able to buy a utility vehicle for considerably less money from my competitor down the street, but when you hear the price, be sure to ask the salesperson about the company's plan to discontinue that line next year because it's not selling. You'll never be able to get parts if something goes wrong." In the case of our foreign language requirement example, you would have to consider the following questions:

- Why would anyone oppose foreign language instruction for engineers?
- What practical objections might someone have?
- What philosophical objections might someone have?
- What keeps engineering educators from pursuing foreign language instruction right now?

By asking such questions as these in your arguments, you will develop effective because-clauses.

WHAT ARE THE SPECIAL LINES OF ARGUMENT THAT PEOPLE IN MY FIELD APPRECIATE?

So far the questions we suggested have been ones that would pay off in any situation, whether the argument is carried out in your discipline or not, whether the argument is in an area of technical discourse or in the arena of personal or public affairs. But before you complete the quest for convincing, good reasons, you need to consider the specific lines of argument that people in your company or field especially respect. In fact, what you learn in those specialized courses in your major is not so much technical information and techniques as it is the specific lines of argument and special appeals that people in your chosen field will find compelling.

Specific lines of argument will differ from company to company and discipline to discipline. Scientists in certain research laboratories, for example, describe their methods by means of general formulas and methodological rules that only members of the particular research community can appreciate. Transportation engineers rely so commonly on ridership surveys in their transit plans that a transit plan without a ridership survey would probably be found unpersuasive. Civil engineers looking at ways to improve highway safety rely on special topics such as sight distance, back plates, traffic signal phasing, channelization, total access control, the 85th percentile speed role, and other specialized concepts that civil engineers who design highways understand and appreciate.

When you join a discipline or a specific company, you learn to speak and argue like a member of that discipline or company; you learn to incorporate into your arguments the special lines of argument that distinguish your community from others. It will be very natural for you to employ special lines of argument along with the more common ones included in this chapter.

Once you have a proposition to defend with because-clauses, supported by good reasons, you must also support or verify those good reasons with evidence. Evidence consists of hard data, examples, narratives, episodes, or statistics that are seen as relevant to the good reasons you are putting forward. To be effective, you will have to put forward not only propositions and good reasons but *also* evidence that those good reasons are true. Your evidence will consist of examples, personal experiences, comparisons, statistics, calculations, quotations, and other kinds of data that an audience will find relevant and compelling.

If your audience is likely to find one of your because-clauses hard to believe, then you should be aggressive about offering support: You should present detailed evidence in a patient and painstaking way. As the one

Supporting Good Reasons

presenting an argument, you have a responsibility not just to *state* a case but to *make* a case with evidence. And so you must stick with a particular point until your audience is forced to agree with your conclusions. Arguments that are unsuccessful fail not because of a shortage of good reasons, but because the audience doesn't agree that there is enough evidence for the truth of the good reason that is being presented.

FAILURES IN PROVIDING EVIDENCE

If you don't provide satisfactory evidence for a because-clause, an audience may feel that there has been a failure in your reasoning process. In fact, in your previous courses in writing and speaking you may have learned about various *fallacies* associated with faulty arguments: faulty definitions, faulty analogies, hasty generalizations, faulty causal arguments (the so-called "*post hoc* fallacy"), *ad hominem* arguments (i.e., name calling), red herrings, and so forth.

Strictly speaking, there is nothing "false" about these so-called logical fallacies. The fallacies most often refer to failures to provide enough convincing evidence to convince your audience. You can avoid such accusations if the evidence you cite is both *relevant* and *sufficient.*

relevance
the appropriateness of the evidence to the case at hand in a persuasive communication

Relevance refers to the appropriateness of the evidence to the case at hand. Some kinds of evidence are seen as more relevant than others in science and industry. Personal experiences or anecdotes are often seen as of limited relevance, while experimental procedures and controlled observations have far more credibility. Compare someone who defends the use of a particular piece of computer software because "I like it" with someone who argues that "According to a journal article published last month, 84% of the users of the software were satisfied or very satisfied with it."

On the other hand, in personal and more intimate circumstances, personal experience is often considered more relevant than other kinds of data. You might be convinced to use a particular over-the-counter remedy not because of the scientific studies cited in a brochure but because your roommate used the product with success.

sufficiency
judgment about the amount of evidence presented in a persuasive communication

Sufficiency refers to the amount of evidence cited. Sometimes a single piece of evidence, a single instance, will carry the day if it is especially compelling. Generally, however, people expect more than one piece of evidence if they are to be convinced of something. For example, you would probably need more than one anecdote about one engineer who succeeded after learning a foreign language to persuade engineering administrators to require foreign language study for all students.

If you anticipate that your audience might not accept your evidence, think carefully about the argument you are presenting. If you cannot cite adequate evidence for your assertions, perhaps those assertions must be modified or qualified in some way. If you remain convinced of your assertions, then think about coming up with additional evidence through additional research in the lab or library. Or, if you anticipate that your audience will think that you have overlooked important information, try to reassure them that you have not.

Another strategy might be to acknowledge explicitly the limitations of your evidence. Acknowledging limitations doesn't shrink the limitations, but it does build your credibility as a communicator and does convince your audience that alternatives have indeed been explored fully and responsibly.

Which Good Reasons to Use?

We have now seen that using the questions described in the last sections can generate a series of because-clauses that you might use to achieve your aims. In fact, if used conscientiously, these questions are likely to generate far more information, far more premises or because-clauses, than you can possibly use. The challenge then becomes how you can decide *which points are likely to be most persuasive*. In choosing which good reasons to use in your arguments, consider your audience's attitudes and values, and the values especially sanctioned by your community.

THINK OF YOUR AUDIENCE'S VALUES, NOT YOUR OWN

When they communicate, people tend to present the lines of thought that have led them to believe as they do. People have lots in common, and it is natural to think that the evidence and patterns of thought that have guided *your* thinking to a certain point will also guide others to the same conclusions.

But people are also different, and what convinces you may not always convince others. When you decide what because-clauses to present to others, try not so much to recapitulate your *own* thinking process as to influence the thinking of *others*. Ask yourself not just why you think as you do but what will convince others to see things your way, what you can say or write that will be convincing to others. Pick the because-clauses that will seem most compelling to your audience, not to you.

To test whether your arguments are likely to appeal to your audience, consult your audience during the composition of your communication, or ask someone to take on the role of your audience and review your draft. Before you commit your arguments to a final draft, do something to get feedback from your audience (or a representative of that audience), and you are almost sure to find the places in your argument that "leak."

THINK OF THE VALUES SANCTIONED BY YOUR COMMUNITY

It might be easy to pick arguments compelling to your audience if you belong to the same community. If you work for the same organization as your audience or if you belong to the same discipline (e.g., nursing, environmental engineering), you have probably learned to value many of the same things as your audience. As you decide what because-clauses to present, therefore, think of the ones sanctioned by your institution or discipline. For example, an argument that "foreign language study may need encouragement because it will make for more broad-minded engineers" might work within a university setting because broad-mindedness is a value prized by the university community. However, an argument based on personal development might be somewhat less successful at work, where personal development is seen as less important than corporate development.

SUMMARY

In this chapter you have learned to think about technical communication as less an impersonal and rational demonstration and more as *a set of arguments laid before a jury of readers, listeners, and viewers.* We have given you ways of thinking about argument that may be somewhat different than the ways you thought about argument before—especially by encouraging you to think of your reader as a partner, not an adversary. And in this chapter we have encouraged you to construct arguments that are particularly responsive to your audiences' views and values; we have encouraged you not simply to state your case, but to make your case by means of an appropriate use of *ethos, pathos,* and *logos.*

To help you achieve these goals, we discussed *ethos* as creating a reliable, trustworthy persona. To create the kind of persona that will be most persuasive in a given case, you need to choose an appropriate voice, based on your relationship with your audience, your audience's personality, your message, your aim, and your genre and medium. You also must create trust and credibility, and you need to show that you have done your homework by citing your sources and indicating how you have conducted your study or gathered your information. Establishing your special expertise and creating a sense of fair play also contribute to a reliable and trustworthy persona. And, finally, engaging in a cooperative dialogue with your audience, and attending to the *ethos* of your company or organization, are essential to creating persuasive technical communications.

While at one time it was unusual to discuss *pathos* in technical communication, now we see that appealing to the emotions, values, and heartfelt attitudes of your audience can be effective.

Logos, the third factor we discussed in persuasion, is a matter of selecting good reasons. These reasons help you persuade your audience that the *ideal situations* that we discussed in Chapter 2 are possible and needed. In this chapter, you learned to create because-clauses that reflect compelling reasons for your audience to accept your solutions and recommendations. And, to help you find good reasons for your argument, we provided you with a set of questions:

- Can I argue by definition?
- Can I argue from consequence?
- Can I argue from value?
- Can I compare and contrast?
- Can I cite authorities?

■ Can I counter objections to my position?
■ What are the special lines of argument that people appreciate in my field?

To support your good reasons, you learned that verifying by relevant and sufficient evidence is essential. And, finally, you saw that to be persuasive you need to think of your audience's values and those sanctioned by your community.

We hope that the chapter will make you not only more successful in constructing your own arguments, but that it will make you a better, more discriminating consumer of the arguments that others present to you as well.

ACTIVITIES AND EXERCISES

1.   Individually or in groups, try to develop as many good reasons as you can for supporting a position on your campus or in your major. Some examples would be a reform in the curriculum or an improvement in the facilities and processes at your school, but you should pick something current and topical. (Perhaps you could discuss a technical or scientific controversy current on your campus.) Once you have devised a series of good reasons for that proposition, try to develop a set of good reasons for *not* taking that position.

2.   Examine any three documents in the appendices of this book. First, determine whether the document could be considered an argument. If it can, discuss the writer's *ethos* in the argument: How does the persona present himself or herself? What things make the persona seem credible (or not)? Then consider the "good reasons" that are given in support of the thesis: What kinds of good reasons are presented? Could any of them be considered as appeals to *pathos?* Does the author provide sufficient support for the propositions offered? In sum, how successful is the argument?

3.   Reexamine Examples 5–3 through 5–6. Working with a partner, make a complete list of the differences between the documents. Pay attention to audience and purpose, but focus on all aspects of voice discussed in this chapter. Be prepared to present your findings to the rest of your class.

4.   Select and review carefully one of the two cases in Appendix A, Case Documents 4. Assume that you must argue the "winning" side of this liability case in a closing statement before the court or judge. Consider how you might use all aspects of *ethos, pathos,* and *logos* in arguing this case.

   Then prepare an outline of your basic argument for your side of the case for an oral presentation to your classmates.

   Finally, write a one- to two-page summary of your closing comments to the court. Assume that all evidence has now been presented, and you are summarizing your case before a decision is reached.

5.   Using the position you selected in Activity 1 above, write a letter to your campus newspaper, your department head, your major advisor, or the most appropriate person to argue your case. If you completed Activity 1 in a group, complete the letter individually and then share results.

6.   Throughout this chapter we offered as an extended example the argument for a foreign language requirement for engineering students. The goal was to write

a proposal that engineering amdinistrators would use to decide whether a full fledged feasibility study is worth the bother.

Although you'll learn more about the format of proposals in Chapter 17, now write an initial draft of that proposal (your instructor may ask you to revise the proposal after you study Chapter 17). You may use and add to any of the reasons or strategies suggested in our extended example.

Bizzell, Patricia, and Bruce Herzberg. *The Rhetorical Tradition: Readings from Classical Times to the Present.* Boston: Bedford Books, 1990.

Brennan, Richard. *The Dictionary of Scientific Literacy.* New York: Wiley, 1992.

Burke, Kenneth. "Rhetoric — Old and New." *Journal of General Education* 5 (1950-51) pp. 203-9.

Miller, Carolyn R. "What's Practical about Technical Writing?" *Technical Writing: Theory and Practice.* Edited by Bertie E. Fearing and W. Keats Sparrow. New York: Modern Language Association, 1989, pp. 14-24.

Acquiring the Tools of Technical Communication

Collaboration

Initiating a Collaborative Writing Project ▮ Executing a Collaborative
Writing Project ▮ Preparing Your Collective Work for Public
Presentation ▮ Employing Collaborative Technologies

Rarely do even Big Ideas emerge any longer from the solitary labors of genius.
Modern science and technology is too complicated for one brain. It requires groups
of astronomers, physicists, and computer programmers to discover new dimensions
of the universe: teams of microbiologists, oncologists, and chemists to unravel the
mysteries of cancer. With ever more frequency, Nobel prizes are awarded to
collections of people. Scientific papers are authored by small platoons of researchers.
      Robert Reich, *Tales of a New America* (New York: Times Book, 1987).

I feel it is important to learn how to work with other people in whatever you do
because it is essential when you are out in the working world.
                                            College senior in a technical writing class

Introduction

As we mentioned in previous chapters, professionals frequently collaborate or work in groups or teams in the production of documents in the workplace. In 1982, for example, Faigley and Miller's survey of 200 college-educated individuals in a range of occupational settings indicated that 73.5 percent contributed to collaborative writing efforts. A broad range of collaborative writing goes on in the workplace, including multi-person authorship of reports, articles, books, briefs, memoranda, and proposals.

Professionals collaborate because groups often produce better solutions than those produced by individuals. Although people collaborate regularly to create proposals, plans, reports, and instructions, they often talk of the difficulties surrounding the collaborative experience. These difficulties may arise because most writing teams begin by attacking tasks without first completing some necessary preparatory steps. Teams function best when they first build the interpersonal environment necessary for their creative problem solving, consider possible approaches toward collaborating, decide on the ways they might handle conflict, access computer technologies to facilitate their work, and identify the strengths of their diverse members. Teams that generate a common understanding concerning their collaborative relationships and the process they will follow to complete their tasks are those that succeed.

This chapter helps you prepare for the collaborative writing activities you will encounter in your classes and in the workplace. Deciding how and when to collaborate is an important part of any writing plan, as you learned in Chapter 4. However, you may have little or a lot of control over your collaborative writing assignments. At times you will be assigned to a team and the schedule and process will be mapped out for you by a supervisor; at other times you might have the freedom to design your collaborative process within your writing team. And, perhaps in a future managerial position, you may be assigning and directing the collaborative efforts of others. Thus, this chapter helps you be successful whatever your situation by providing you with a common language for initiating, executing, and presenting a collaborative project. To begin, let's define what it means to collaborate on writing tasks.

collaborative writing
team or group writing on a project

Collaborative writing is often defined according to how people work together on a project. In one survey of 1,400 professionals across the United States seven basic forms of collaborative writing emerged:

1. Team plans and outlines. Each member drafts a part. Team compiles the parts and revises the whole.
2. Team plans and outlines. One member writes the entire draft. Team revises.
3. One member plans and writes the draft. Team revises.
4. One person plans and writes draft. This draft is submitted to one

or more persons who revise the draft without consulting the writers of the first draft.

5. Team plans and writes draft. This draft is submitted to one or more persons who revise the draft without consulting the writers of the first draft.

6. One member assigns writing tasks. Each member carries out individual tasks. One member compiles the parts and revises the whole.

7. One person dictates. Another person transcribes and revises (Ede and Lunsford 63).

Certainly, there are additional variations to these seven forms.

Because collaboration is embedded in the structure of many organizations, writers know they are working and writing with others, but they rarely can explain the exact process they follow when collaborating. Likewise, you have collaborated in the past, but you probably remember more about the people involved and the working relationships than you do about the exact process you followed.

At the heart of collaborative writing is the fact that you produce something with the direct or indirect help of others, that what you produce takes on a new form based on this input and interaction, and that your identity as an individual writer is likely to take second billing to the needs of the team. However, as you work together to solve a problem, create something, or discover something, your collective work results in the best possible response, system, or discovery. Real challenges demand collaboration, and as society and technical communication needs become more complex, we witness even greater challenges.

In order to understand challenges and generate solutions, to function as a team, groups need to develop on both relationship and task levels. When researchers from Bell Communications Research interviewed 50 pairs of collaborators from social psychology, management, and computer science, they found that successful collaborators followed three stages: **initiation, execution,** and **public presentation.** What follows describes a process for initiating your collaborative project, executing it, and presenting it to your audience. We'll look at the relationship level and the task level during each of these stages, and we'll discuss some common structures of collaborative writing groups after our description of the initiation stage.

initiation, execution, and public presentation the three stages of successful collaboration; each stage has a relationship and a task level

---

At each stage of the collaborative process, and especially at the initiation stage, your writing team needs to develop at both the **relationship level** and the **task level.** Look at the collaboration model in Figure 6-1. The

Initiating a Collaborative Writing Project

Initiation Stage	Execution Stage	Public Presentation Stage
Relationship Level		
Find team members	Maintain trust	Establish final division of credit
Get to know each other	Establish ways to resolve conflict	Acknowledge and appreciate each person's work
Share background assumptions	Clarify work styles	
Task Level		
Clarify goal(s)	Share information	Reach closure on your work
Choose approach and schedule meetings	Establish procedures to assess progress	Package your work for public presentation
Generate ideas and plan	Do the work (plan, draft, revise, edit)	

Source: The three stages and two levels in this table come from the research of Robert E. Kraut, Jolene Galegher, and Carmen Egido, "Relationships and Tasks in Scientific Collaboration," *Journal of Human-Computer Interaction* 3.1 (1988) pp. 31–58, specifically p. 34, and are used with permission. The majority of these points have been modified by Ann Hill Duin.

FIGURE 6–1
Collaborative Model

relationship level
the interpersonal activities in which collaborative teams get to know each other, share assumptions, maintain trust, resolve conflict, and acknowledge each other's work

task level
the task completion activities in which collaborative teams clarify goals, generate ideas, establish procedures, draft, revise, and edit, and present their work

relationship level comes first; it's on top. If you neglect relationships, your task level work will not be as effective.

RELATIONSHIP LEVEL

When you initiate a collaborative project, you first need to find partners and share background assumptions (refer back to Figure 6–1 as you read the rest of this chapter). The most fundamental requirement is for the group members to make contact with each other and to become acquainted and identify their common interests and the mutual benefits that might come from working together.

Carl Larson and Frank LaFasto, from interviews with numerous teams ranging from presidential commissions to championship football teams, note that all too often people are chosen as collaborative team members for the wrong reasons or that people choose to work with other people for the wrong reasons: "Laura should be on the team because she's interested in the topic." Or, "Al's feelings would be hurt if he were left off." Or, "Jeremy should be included because he reports to Donna." Instead, what should be paramount is selecting or choosing people who are best equipped to achieve the team's objective. You should choose people based on the personal competencies required to achieve excellence while working with others on the project and on the technical skills and abilities necessary to meet your challenge.

Personal competencies include the qualities, skills, and abilities necessary for your team to function and write together. Part of creating a successful collaborative team is deciding what type of personal qualities and interpersonal skills you want members to have. For example, the team leader of the group that developed McDonald's Chicken McNuggets, Bud Sweeney, states that he specifically looks for people who are "intelligent, creative, tenacious, and a bit of a maverick—willing to buck the system and not be bound by traditional thinking and reporting relationships" (Larson and LaFasto 63).

> **personal competencies**
> the interpersonal qualities, skills, and abilities that are necessary in each collaborative team member for the team to function well

Glenn Parker notes that collaborative team members should have the ability to:

- Participate in open discussions which include consensus-building, ambiguity, and disagreement.
- Listen carefully and employ such listening techniques as questioning, paraphrasing, and summarizing other's ideas.
- Assume a leadership role when necessary because effective team members often change roles.
- Share all information freely with other members of the team as well as with members of contingency teams.
- Build relationships with others outside the team, including important resources in other parts of the organization.
- Appreciate different learning and working styles.

Although needed personal competencies vary from team to team, it's extremely important to take the time to identify and promote interpersonal skills or the relationship level of your team.

Technical knowledge enables a collaborative team member to solve a problem, create something, or execute a specific task. Linda Alvarado, CEO of Alvarado Construction, notes her need for a balance between "nuts and bolts" members and creative and conceptual members. In her construction business, she organizes collaborative teams with members who can see the big picture and know when major changes are necessary as well as members

> **technical knowledge**
> the technical qualities and competencies that are necessary in each collaborative team member for the team to solve a problem, create something, or execute a specific task

who are very detail-oriented (Larson and LaFasto 63). In order to create a sucessful team, you must ask if you need people who know specifics about a topic, people who are creative and conceptual, or a blend. You must consider if you need people whose technical skills are similar; if you need people whose skills are radically different, you need to integrate them carefully into the team at the initiation stage.

Because so much of success in collaborative teams involves intangibles such as attitude and energy, remember to take time at the initiation stage to find the right collaborators (if you are choosing your own collaborative team) and to become acquainted and to identify your common interests and needs (regardless of who initiates the collaborative team). You may even find that going out to eat together or playing volleyball, for instance, will help you and your team members bring your assumptions about the project out in the open, share your concerns and your excitement.

TASK LEVEL

Only after you have talked informally at the relationship level should you proceed to the task level. At this step you can begin to generate ideas and plans (look again at Figure 6–1). If team members are located near each other, plan numerous face-to-face meetings; if not, use e-mail, fax machines, voice mail, and conference calls to link yourselves (note possible ways to use technology in the section entitled, ''Employing Collaborative Technologies'' later in this chapter).

At the task level, you should first clarify your goal by asking exactly what it is that your team wants to accomplish. Successful teams have a clear understanding of their goal, a sense of mission. Kiefer and Senge give this example concerning clarifying a goal:

> The Apollo Moon Project provided a superb demonstration of the power of a clear and compelling vision. By committing themselves to ''placing a [person] on the moon by the end of the 1960's,'' the leaders of the project took a stand. The clarity and conviction they generated touched people at all levels of the enterprise. One can imagine how much less spectacular the results might have been if they had adopted an alternative mission statement such as ''to be leaders in space exploration.'' (112)

In addition to clarifying your team's goal, you need to visualize your final product. Will you develop a proposal that gets funded? Will you design a process that your company will follow? Will you write a position report that clarifies a current controversy? Whether you have been assigned a final product, been given the freedom to choose a genre and design within your team, or are directing the collaborative efforts of others, you need to be sure all members of the team agree on what that product will look like and can participate in the production of it.

THE STRUCTURE OF COLLABORATIVE WRITING TEAMS

Once you have clarified and visualized your goal, decide how to approach the collaborative writing task. Collaborative writing is rarely an assembly-line process. Perhaps a division head or manager can jump into a group and say, "You do this; she'll do that; he'll do this; and I'll do that." And if work was always routine and predictable, this might be a suitable model. But because solving problems is rarely so simple and because people's strengths vary, you will want to consider a variety of models or approaches for structuring your collaborative writing team. We consider five models in this chapter. Perhaps your supervisor will suggest one of these models. If you have the freedom to structure your own group, you need to read about each model and talk about them in your group. Your group will need to decide which will best take into account your group's chemistry, commitment, strengths, weaknesses, and challenges.

Divide-and-conquer Model

The **divide-and-conquer model** is the simplest one. In this case each team member takes a segment of the project and writes, revises, and edits only that part. This model assumes that every member of the team is alike, that each person could replace another at any time. Although industry often resorts to this model to expedite the collaborative process, most writing challenges rarely allow for this simple approach. However, in cases in which your problem is relatively simple and your timeline extremely short, you can divide the project or assignment into subtasks, decide on a group leader and have that person assign jobs, or have members choose jobs themselves.

divide-and-conquer model
the collaborative model in which each team member writes, revises, and edits one part of the project; assumes each team member is alike and replaceable

Specialization Model

The **specialization model** operates on the principle that each member gets a task that relates to his or her special talent or expertise. For example, one member may work best as a writer or editor; another may work best as a graphics specialist or a project manager. This approach doesn't allow you to exchange tasks, assuming as it does that the group's strength is based on its collective expertise. You should follow this approach when the project requires the expertise that certain members can offer. Whether or not you choose to follow this model, one of the first things you should do when establishing a collaborative team is identify each person's special talents.

specialization model
the collaborative model in which each team member is assigned a task according to his or her special expertise

Dialogic Model

The **dialogic model** is marked by flexible roles that may shift as a project progresses. Working in this model, a member might first be a project manager and later take on the role of editor or graphic artist. As the project proceeds,

dialogic model
the collaborative model in which team members' responsibilities shift during the project; requires a high level of interaction

needs and roles change. For example, after producing first drafts, the writers can serve as editors for their own and each other's words. This model best accommodates possible gender or cultural differences among team members. Because the collaborative process is not rigidly structured, a hierarchy or power structure seldom emerges. And, as team members assume new roles as the project proceeds, a high level of interaction is required. When you use this model, everyone will need to be heard from frequently, and voices that might receive less attention or might be less welcome in other situations are essential here. As you develop a flowchart of how your project might proceed, consider how your roles might change and how you might take on the strengths of others.

Sequential Model

sequential model
the collaborative model in which one team member writes a draft and then passes it to another for comment or sign off, who in turn passes the draft to the next team member, and so on

In the **sequential model,** one person writes a draft and passes it to another, who comments and/or revises it and forwards it to the next person. This next person again comments and/or revises and forwards the document to the next person. In many industries a required series of "sign-offs" or signatures of approval results in a sequential model of collaboration. This review process allows a corporation to transmit its values and corporate culture quite effectively. The greater the impact of the document, the more likely it is that the sequential review will be extensive and involve drafts that are passed back and forth among people and corporate levels. For example, when Susan Kleimann studied how reports were created at the United States General Accounting Office, she found an elaborate review process (Figure 6-2). If you are completing a task for an outside client or corporation, you may need to have your client(s) sign-off at specific stages of your project. Even if you do not need these approvals, you may still want to follow this model if you are on a tight schedule.

Synthesis Model

synthesis model
the collaborative model in which team members are brought together because of their different views or perspectives

Often team members are brought together because of their different views or perspectives toward a controversial issue or problem. Such teams might follow the **synthesis model.** For example, if a company needs to solve a problem and three subdivisions have three distinct solutions, it's best to bring a subset of people from each division together to collaborate. In this case the members must synthesize their different perspectives; they must meld them into a recommendation acceptable to all. In the synthesis model, members belong to the team because of their different views or solutions. They are likely to work on those aspects of the project that represent their individual perspectives; however, through the process of collaborating, they will begin to resolve some, though probably not all, of their differences. It's also natural that new areas of difference will emerge. In other less controversial situations, you might follow this model if each team member supports

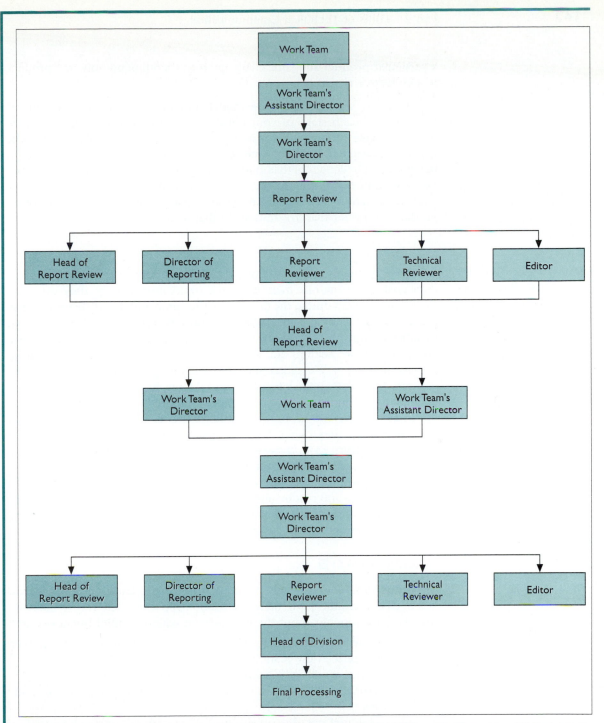

Source: Susan D. Kleimann. "The Complexity of Workplace Review." *Technical Communication* 38.4 (1991) pp. 520–26, especially p. 523. Reprinted with permission from *Technical Communication*, the Society for Technical Communication, Arlington, VA.

FIGURE 6–2
The Approval Process at the United States General Accounting Office

a particular product and your challenge is to recommend only one product in a collective report.

Be sure to study each of these models and identify which one you intend to follow. At each stage of your collaboration, evaluate how the model is working, make refinements to the model, or switch to a different one. Or, if you are assigned to follow one model by a supervisor, be aware of the benefits and challenges of that model. Finally, if you are supervising a collaborative writing team, choose the model that will best help the team meet your organization's goals and ensure that collaborative process will bring satisfaction, rather than frustration, to that team.

Executing a Collaborative Writing Project

In 1985 a 20-member team attempted to climb Mount Everest and identified their goal as getting one member to the top. Unfortunately, they overused their best climbers during the first part of the climb and didn't share responsibilities evenly. They got to 28,200 feet, only 800 feet from the top. Team member George McLeod stated that "I think we should have sat down as a group, which we seldom did, and discussed how we were going to make this happen. . . . We didn't take the summit" (Larson and LaFasto). In contrast, the first women's team to climb Mount Kongur in China identified a different goal and structure. Their goal was more consensus-based and was not to get one person on top, but rather to get as many people as high on the mountain as they could. They were successful in that goal.

What we learn from these experiences is that at the execution stage, a collaborative team needs to continue to develop at both the relationship and task levels. Team members need to sit down and, with each other, clarify their work styles and reach consensus about how they plan to proceed. Part of this process means coordinating how to share information as you plan, draft, revise, and edit your collaborative text.

RELATIONSHIP LEVEL

As you execute your project, you need to maintain trust. In other words, you need to continue to attend to the relationship level. One way to maintain trust is to talk openly about the five areas in which conflict might occur:

1. Struggles for leadership: efforts to control, dominate, exert power over, or lead the team.
2. Unequal workloads: unequal contributions from members.
3. Personality differences: differences due to interpersonal likes and dislikes.
4. Procedural matters: organizational, procedural, or mechanical problems.

5. Ideational matters: problems relating to ideas, goals, and values associated with the team's task (Wall and Nolan).

One study of numerous teams found that when a conflict centered on the task level **(procedural or ideation matters),** 83 percent of the teams chose to discuss the conflict and work out a procedure for resolving the conflict. However, when the conflict focused more on the relationship level (struggles for leadership, unequal workloads, personality differences), only 17 percent of the teams chose to resolve the conflict while the others chose to avoid or ignore the problem (Wall and Nolan). To ignore disputes is not a good strategy; to establish a procedure to resolve them is the best strategy.

We present three procedures for managing conflict depending on its source. Notice that we talk about managing rather than resolving or eliminating conflict. As long as you are working with human beings, you have to expect conflict. It's not always a bad thing, and it can almost always be managed in such a way that your project and the interaction of group members don't suffer. How you manage conflict will depend on whether the source of conflict is a misunderstanding of the problem or solution **(pseudo-conflict),** individual disagreement over which course of action to pursue in a project **(simple conflict),** or defense of a member's ego **(ego conflict).**

Figure 6–3 presents these three sources of conflict and suggestions for managing them. For example, remember that continuing case in Chapter 5? Let's say that you decided to write the proposal that engineering students should take foreign language courses, but the task was too large for you alone. You asked other members of your local chapter of the American Society of Civil Engineers to form a writing team. Perhaps one member of the writing team, Sandy, thought that the proposal should include a recommendation that engineers take more history courses, because history courses might teach engineering students more about other cultures than foreign language courses. Sandy's point is well taken, and she feels very strongly about it, but you might suggest an alternative to putting too much into one proposal. You could assign the history proposal to another group (see suggestion 6 under Simple Conflict in Figure 6–3), delay this recommendation until you see how your foreign language proposal flies (suggestion 8), or postpone vetoing Sandy's proposal until your group can do some research on the history question (suggestion 9).

If Sandy felt so strongly that her suggestion was better than yours, you might encourage her to describe the problem as she sees it and how her solution would work, rather than labeling one suggestion good and the other poor (see suggestion 5 under Ego Conflict in Figure 6–3). If Sandy misunderstood the central issue, that engineering students need to understand *contemporary* cultural differences, the modern Italian highways, for

procedural or ideation matters
procedural matters include organizational, procedural, or mechanical problems; ideation matters include idea, goal, and value problems

pseudo-conflict
a misunderstanding of the problem or solution

simple conflict
disagreement over which course of action to pursue during a project

ego conflict
defense of a collaborative team member's ego

Pseudo-Conflict	Simple Conflict	Ego Conflict
	Source of Conflict	
Misunderstanding individuals' perceptions of the problem.	Individual disagreement over which course of action to pursue.	Defense of ego; individual believes he or she is being attacked personally.
	Suggestions for Managing Conflict	
1. Ask for clarification of perceptions. 2. Establish a supportive rather than a defensive climate. 3. Employ active listening: 　stop 　look 　listen 　question 　paraphrase content 　paraphrase feelings	1. Listen and clarify perceptions. 2. Make sure issues are clear to all group members. 3. Use a problem-solving approach to manage differences. 4. Keep discussion focused on issues. 5. Use facts rather than opinions for evidence. 6. Look for alternatives or compromise positions. 7. Make the conflict a group concern. 8. Determine which conflicts are the most important to resolve. 9. If possible, postpone the decision while additional research can be conducted. The delay helps relieve tensions.	1. Let members express their concerns but do not permit personal attacks. 2. Employ active listening. 3. Call for a cooling-off period. 4. Try to keep the discussion focused on issues (simple conflict). 5. Encourage parties to be descriptive, rather than evaluative or judgmental. 6. Use a problem-solving approach to manage differences of opinion. 7. Speak slowly and calmly. 8. Develop rules or procedures that create a relationship which allows for the personality difference.

Source: Stephen A. Beebe and John T. Masterson, *Communicating in Small Groups: Principles and Practices,* Copyright (c) 1990, 1986, 1982, by HarperCollins College Publishers. Third ed., p. 214.

FIGURE 6–3
Suggestions for Resolving Conflict

example, you can try to clarify your perceptions and have Sandy do the same (suggestion 1 under Pseudo-Conflict in Figure 6–3).

Another way to resolve disputes is to appoint one member to regularly ask each member to talk about any difficulties he or she may be having. This person then brings these issues up at the next team meeting. Another possibility is to ask a person outside of your team to talk to each member about difficulties and then share these with your team. A third approach is to have each person write about difficulties, pass this information around during a team meeting, and let each person write constructive responses to the concerns. After that process is completed, discuss these responses as a team.

As you share background information concerning a conflict, you may discover that the conflict stems from your different work styles or learning styles. Establishing a procedure to discuss rather than avoid conflict will mean taking time to listen and explore the conflict, considering its complexity and the situation that surrounds it. However, such a procedure will benefit your team and the collective decisions you make.

TASK LEVEL

At the task level, you need to share information, prompt and support each other, and complete the collaborative writing task. A number of variables influence collaborators as they execute a writing task:

Control   Talk about the degree of control each member has over the text. How does each member's work contribute to the whole? Let members share how they see their work contributing to the project.

Credit   Begin to discuss how credit will be given for the work. (If you are collaborating at school, ask your instructor this question. Will he or she evaluate each of you individually? Will he or she evaluate only the final team product? If you are collaborating at work, ask who will get credit for the assignment and who is responsible for it.)

Response   Decide how you will continue to meet as a team and respond to each other's work. How will you respond to the modifications made by others?

Flexibility   Decide how flexible you can be with an established format (e.g., how closely you should follow a proposal or report format). Likewise, how flexible should you be with your approach to this collaborative project? If you have been following a specialization model, should you now follow a more dialogic model?

Constraints   Share the very real constraints facing you. What's your deadline? Who's your primary audience? On what criteria will this project be evaluated? (Ede and Lunsford)

At the task level, one form of conflict noted earlier can work to your advantage: pseudo-conflict. Pseudo-conflict is similar to what some call **sub-**

substantive conflict
disagreements about content, purpose, audience, organization, support, and design; enhances decision making

WRITING STRATEGY 6–1   Prompts to Expand Ideas, Consider Alternatives, and Voice Disagreements

Content

What information might we add or delete?

What will our audience expect? Need?

Purpose and Key Points

I can't quite see why you've decided to _____. Could you explain why?

What do we want the audience to do or think?

What is the point we want to get across in this section?

Audience

What problems (conflicts, inconsistencies, gaps) might our audience see?

How will our audience react to this (content, purpose, organization, design)?

Conventions of Organization, Development, and Support

How will we organize (develop, explain) this?

What support (or evidence) could we use?

What examples could we use?

Conventions of Design

What do you like _____ better than _____ as a way to present this information?

How will this design reinforce our point?

Synthesis of Plans About Content and Rhetorical Elements

Why do you think _____ is a good way to explain this point to our audience?

Will changing the example clarify or cloud the point we're making here?

Source: Rebecca Burnett, "Substantive Conflict in a Cooperative Context: A Way to Improve the Collaborative Planning of Workplace Documents." *Technical Communication* 38.4 (1991) pp. 532–39, especially p. 536. Reprinted with permission from *Technical Communication*, the Society for Technical Communication, Arlington, VA.

stantive conflict or considering alternatives and voicing explicit disagreements about both content and rhetorical elements such as purpose, audience, organization, support, and design (Burnett 534). Conflict of this sort typically enhances decision making. In this situation, you prompt members to clarify their perceptions, their positions, and their texts. Here you employ active listening; you stop, look, listen, question, and paraphrase. Writing Strategy 6-1 presents a list of prompts intended to help collaborative writing teams consider issues yet defer consensus. Use these questions to expand your ideas, consider alternative ideas, and make your disagreements explicit. These questions will move your team toward a better collaborative document within a cooperative climate.

Preparing Your Collective Work for Public Presentation

The last stage of the collaborative process involves preparing your document for public presentation. At this stage, you need to establish final division of credit for your collective work, and you need to package your document for public presentation to your audiences.

RELATIONSHIP LEVEL

Before you present your document to these audiences, you should have resolved any disagreements you may have had. Now is not the time to rehash conflicts because your project must assume an outstanding public face. Even though you might be hurrying to complete the project, you need to take special care not to forget inadvertently to acknowledge each person's contribution. It's at this stage that members often feel as if they're the only ones working or that they're working the hardest. Remember to acknowledge each person fully and to respond positively to each other's work.

TASK LEVEL

At this final stage, you obviously need to reach closure on your work—after all, you need to complete your project. You might want to revisit the decisions you made during the execution stage in regard to control over the document. You will need to decide if one or two members should be in charge of the final packaging of the document, or if all of you will meet as an entire team. The team will need to consider what constraints might bog it down at this stage (the computer programs you are using, your audience's expectations for the final packaging of the document, and so on). You also need to establish guidelines covering how flexible you will be with each other as well as with the established format of the document.

Last, remember that a document created collaboratively, like any other piece of writing, can always be revisited and revised up until the final deadline. For example, a team member might come up with a useful way to color code your document. Or, a team member might suggest, just at the end of your collaborative process, that the software your company is producing for university classrooms should be called "courseware," and you need to search your document for this change. A document is never finished, only abandoned. But, you must reach consensus at this point; you must reach closure as you present your document to the public.

collaborative technologies
computer tools and software that support collaboration on the relationship and task level

In today's classroom and workplace, **collaborative technologies** are often available to support your team's work. These tools are designed to support your work at both the relationship level and the task level. Although some schools cannot afford this kind of technology now, many companies can. These discussions of collaborative technologies, therefore, will help you when you enter the workplace and join an organization that uses such equipment.

Collaborative technologies can be used not only to help you produce a complete and polished document; they can also help you communicate with team members, maintain records, receive feedback, and structure the final document. Whatever system you choose or have access to, remember to structure its use to support conversations, sketches, arguments, and discoveries; it should not merely help you write, edit, and design your final text.

computer-supported team
a writing group that uses computer technology to support the collaborative process

Once you decide to use computer technology to support your collaborative process, you are, in effect, a **computer-supported team.** Four factors should determine the technologies you use:

- Your team's goal.
- Whether or not your members are physically separated.
- The deadlines you must meet.
- The type of computer systems available and their compatibility.

What follows are nine approaches to consider as a computer-supported team (for a more complete discussion, see Johansen). If you do not have access to this technology, you can sometimes conduct the activity with pen and paper, flip charts, blackboards, and other tools. Although all of these technologies could be used at any stage of your collaborative process, most are best suited to a particular stage. In the following sections, we describe how each might be used at a particular stage.

TECHNOLOGIES BEST SUITED FOR THE INITIATION STAGE

The technologies we discuss below are most useful as you are discussing your audience and purpose and planning your writing process—and, of course, getting to know the members of your team.

Face-to-Face Meeting Facilitator

face-to-face meeting facilitator
a computer technology that enables group members to participate in meetings and record meeting notes

Throughout your project, but especially at the initiation stage, you need ample face-to-face meetings to get to know each other, identify and clarify your goal, and generate plans. At this point you need frequent, open, and often lively discussions as you explore your options.

One way computer technology can support your collaborative process is through a **face-to-face meeting facilitator.** In this scenario, you ask a person from outside your team to facilitate by recording summary phrases from your meeting into a computer. If this computer is connected to an overhead projection system, the facilitator can project these statements on a screen for the team to see. Essentially these notes are electronic versions of notes you could put on a flip chart or black board. However, by following this process, each group member is free to participate in the discussion rather than take notes. In addition, after the discussion each person has the same recorded version of the meeting to use when generating his or her part of the collaborative document.

Computer Conferencing Software

If your team members do not work in close proximity, if your work schedules do not permit you to meet face to face, or if some members prefer to work at night and others during the day, you need a tool that allows you to collaborate without having to meet face to face. Various forms of computer conferencing software provide such a service.

Although e-mail systems provide an electronic means for one-to-one communication, **computer conferencing software** provides for collaborative members to communicate within one electronic forum. For example, students in technical communication courses at the University of Minnesota meet in a computer laboratory during class, and AppleShare™ software (and its connection to the University's LANmark Ethernet service) allows students to collaborate by giving feedback to and receiving it from each other during class as well as outside of class from remote locations.

computer conferencing software
a computer technology that enables group members to communicate within an electronic forum

Many campuses today have some form of computer conferencing software available. By using such a system, you can send messages to team members; you can use the network as a place to store and share information; you can argue and reach consensus on content, tone, style, and direction of a document; and you can request feedback from people outside of your team as well as from your instructor.

Desktop Videoconferencing

Teams that collaborate via computer conferencing systems often miss the nonverbal dimension to communication present in face-to-face meetings. If you need to collaborate across distances, you might use a **desktop videoconferencing** system that provides simultaneous audio and video. The simultaneous audio and video connections allow you to generate goals and plan ideas as well as get to know each other faster than with traditional computer conferencing systems. However, this system supports only simultaneous or synchronous collaboration unless you have a storage device (such as CD-ROM or a laser disk) capable of handling large amounts of data. Thus, although collaborators can ''meet'' while at different locations, all members must be in front of their computers at the same time.

desktop videoconferencing
a computer technology that enables group members to collaborate over distances with simultaneous audio and video transmission

Desktop videoconferencing systems, although ever more common in the workplace, are slightly less available on campuses. However, look around your media resources center, which is likely to have some system that supports videoconferencing. If you need to get to know each other quickly and make decisions across distances, these systems are a valuable tool. For example, if you are developing a document for a corporate client, the corporation might have such a system available.

TECHNOLOGIES BEST SUITED FOR THE EXECUTION STAGE

The technologies we discuss in this section are most useful as you coordinate your research and drafting.

Project Management Software

Most collaborative teams have obvious and often pressing needs for planning and coordinating their work. Specialized software exists to help you plan what needs to be done, track your progress in reaching your goals, and coordinate activities of individual team members.

project management software
a computer technology that enables group members to keep track of task assignments and schedules

At the execution stage/task level, your team needs to focus on generating the content of your work; however, you also need to keep track of what has been accomplished, what needs to be accomplished, and who is in charge of what. **Project management software** can help. While your team focuses on generating answers to the problems facing you, the project management software keeps a basic record of tasks to be done, task assignments, subtask breakdowns, and schedules. At least weekly, team members review their progress with this system.

Professionals who use such systems or who have been required to use them often complain that the systems fail because they make tasks and deadlines so explicit. Commitments made in the workplace are often ambiguous, and minor deviations from a set schedule are often overlooked; thus, systems that make explicit any minor deviation, perhaps even flagging the person who first upset the schedule, sometimes create resentment.

A decision to use this technology must be made after considering members' work styles. As with any collaborative technology, to be useful, ALL team members have to use it. Because your lifestyles and work styles inevitably vary, structure some flexibility into any project management software you might use.

Group Writing Software

group writing software
a computer technology that enables group members to make document revisions without meeting face to face and to record who made what changes

When you meet face-to-face, how do you write as a group? Perhaps you share printouts and scribble on them, offering ideas in the margins for how to rephrase or rearrange the text. Perhaps you all squeeze around one computer and designate one person to enter and edit text as the rest of you rattle away. **Group writing software** allows team members to make document revisions without meeting face to face, and such systems record and remember who made what changes. Thus, team members can revise without wiping out the original file, and members can compare alternative drafts and suggestions.

Group writing software works especially well if your team cannot physically meet together. As with project management software, this software brings work style differences to the forefront. But the ability to see and record exactly how each member plans, drafts, and revises a collaborative text can be extremely powerful. The key is to structure the system so that you don't get lost in the many drafts and revisions. One way to do this is to develop a system for naming drafts that identifies the document, the collaborator, and the stage in the drafting process.

Screen-Sharing Software

Collaborators can "meet" across distances and also to share the same screen through software called **screen-sharing software.** For example, after one collaborator types, "I think we should change the beginning to start like this . . ." another responds, "Yes, let's do it," and both simultaneously see their conversation and resulting changes on screen.

Although such a system is a powerful tool for collaborators, currently only two computers can be linked in such a way. In addition, collaborators talk of the difficulty in sharing a screen with someone else: Who gets control of the screen first? What if you like a specific change and your collaborator immediately deletes it (when it hasn't yet been saved)? Such systems already exist; for example, Timbuktu™ and Aspects™ allow you to share screens during collaboration. For these systems to be used successfully, however, teams will need additional discussions at the relationship level: Who will be in control of these shared texts? How will members respond to changes other members make?

screen-sharing software
a computer technology that enables group members to link two computers and simultaneously see their conversation and resulting changes on screen

TECHNOLOGIES BEST SUITED FOR THE PUBLIC PRESENTATION STAGE

When you have finished your drafting tasks and are ready to present your project to your audience, computer technologies can also be helpful.

Presentation Support Software

The reality of the workplace is that you will make presentations, either to other members of your team or to outside clients. Numerous software programs can help you present your document, product, system, or solution publicly. **Presentation support software** often includes graphics programs and other desktop publishing software. In some cases, your team can print out its presentation and then make overheads for public display. If you have a computer projection system available for the presentation, the team can design a computerized presentation. The value of this computer technology

presentation support software
a computer technology that includes graphics programs and desktop publishing options that enhance collaborative group presentations

is that the electronic version of the project you have created can now be used directly in your public presentation.

During the initiation stage of your work, identify those members of the team skilled with programs such as these. At the public presentation stage, some roles may need to change as those members more skilled as graphic artists may take charge of formatting and arranging your public presentation.

Group Memory Management

group memory management
a computer technology that enables group members to store and retrieve documents by topic or date

All teams need some form of group memory of the many electronic notes, plans, drafts, and revisions generated. Nothing is worse than vaguely remembering the "perfect" idea, example, or statistic but not being able to recall it completely or remember who said it, when, and in what context. In addition, teams need systems that allow them to store and retrieve these documents by topic as well as by date. Hypercard™ is one widely available software product that can be used as a **group memory management.** As your team collaborates, you could develop a stack of hypercards that can be linked and cross-referenced very easily. You could, in effect, make each note, plan, draft, and revision a separate card.

At the public presentation stage, you could access these cards and project only those cards needed for your presentation. Or, if you remember a specific idea talked about at one meeting, and you recorded it on your meeting on a card, you could search all the cards and locate your idea, placing it in your final presentation.

Comprehensive Work Team Support

comprehensive work team support
a combination of computer technologies that enable group members to collaborate effectively through the initiation, execution, and presentation stages

After skimming through these collaborative technologies, you possibly have highlighted two or three that look attractive—perhaps your team has read these sections of this book and is now ready to request access to the most useful technology. Your combination of these systems would result in a form of **comprehensive work team support.** In the best of computer worlds, you would be able to link a system especially helpful at the initiation stage with one at the execution stage in turn with one at the public presentation stage. Unfortunately, we're not there yet. However, by talking about these technologies as part of your collaborative process, your group can structure its own comprehensive work team support system. By locating what's available on your campus to aid your collaboration, you can make informed choices about what computer tools to use at what stage.

Although the first vision of comprehensive team support was proposed by Douglas Engelbart in the 1960s, corporations are just now developing such systems that integrate several collaborative technologies into one package. One integrated system should be more powerful than several standalone systems.

The bottom line is that all of these collaborative technologies raise new sorts of questions and feelings in collaborators. The point is to pay attention

Initiation	Execution	Public Presentation
Face-to-face meeting facilitator	Project management software	Presentation support software
Computer conferencing software	Group writing software	Group memory management
Desktop videoconferencing	Screen-sharing software	Comprehensive work team support

FIGURE 6–4
User Approaches to Computer-Supported Teams

to these questions about styles of collaboration and how, when, and in what ways electronic support is appropriate. Always remember that when you link computers, when you share screens, and when you project your prose, the computer is no longer personal; it's interpersonal. It becomes a medium for sharing information at both the task level and the relationship level.

In Figure 6–4, we list again for you the computer technology approaches according to their potential use at the initiation, execution, and public presentation stages of your collaborative project.

COLLABORATIVE TECHNOLOGIES AND THE OPEN EXCHANGE OF IDEAS

As you decide which collaborative technologies you might employ, keep in mind that collaborating via interactive technologies involves changing the ways you have traditionally accessed and used information; it involves a change in the status of communication or of who owns information; and it can, of course, result in conflict. However, collaborative technologies, when designed and used properly, can offer each member of your team an equal voice. Collaborative technologies can also ease your working with others at sites outside your own country.

Part of structuring these technologies, indeed of structuring your entire collaborative experience, should be to encourage an open exchange of ideas. From diversity comes new ideas and solutions. When you collaborate, you are one among others. You no longer are the sole owner of an idea, question, process, or solution. You share your ideas and take on the ideas of others.

Earlier in this chapter we spoke of substantive or pseudo-conflict, and ways to defer consensus and explore additional options. But these options, these additional ideas, are never heard if one member stifles another. Open and lively debate over ideas enhances writing and collaboration. The key is to avoid interpersonal conflict. Collaborative technologies often allow team members to express their ideas without being silenced by traditional or cultural hierarchies. For example, research shows that in a face-to-face group setting women are more often interrupted than men (Edelsky). On the other hand, collaborative technologies potentially allow for all team members to contribute without the interruption that may come during face-to-face contact.

When you structure and use your technology, make sure that all members have access to it, that all members understand how to use it, that all members then actually use it, and that all members can call for changes in its use at any time. Never isolate members from each other via technology; use it to integrate and empower the entire team.

Therefore, when choosing and structuring a specific technology to facilitate your collaboration, remember to consider the following three issues:

1. **Open conversations:** How will the system promote open conversations; that is, how will the technology enhance collaboration and interaction among yourselves, other students, your instructor, your co-workers, and the larger community?

2. **Distribution of power:** How can you structure the technology so that each member has equal access to the system? Are there existing models of authority that will be affected by the technology? For example, if everyone has access to and uses the technologies, will a "boss" emerge within your group and your team members speak more freely? Or, will you need to appoint a leader to coordinate your efforts?

3. **Accessibility:** What system will all members be able to learn and use? How can you design a system that provides accessibility? Will this design allow all participants equal access to the technology?

Six Characteristics of Effective Teams

Throughout this chapter, you have read a great deal about how to approach collaboration at the relationship and task levels and the possible technologies to employ at the initiation, execution, and public presentation stages of your work. To reinforce these concepts, we leave you with the six characteristics that researchers use to explain how and why effective teams develop. Successful teams have the following characteristics:

1. A clear, elevating goal.
2. A results-driven structure.
3. Competent team members.
4. Unified commitment.
5. A collaborative climate.
6. Standards of excellence (Larson and LaFasto 8).

As you collaborate, use the suggestions in this chapter to get to know your team members, to clarify and visualize your goal, to identify the diversity and competencies represented by individuals on the team, to grow at the relationship level and establish a solid collaborative climate, and to establish high standards for your team's work. Through understanding how to work together, your team's interactions and products will be outstanding.

SUMMARY

This chapter provides you with a common language for initiating, executing, and presenting collaborative projects, and it presents technologies that can facilitate your collaboration. People collaborate to solve problems, create things, or discover new information. In order to collaborate effectively, people need to focus on both the relationship and task levels.

At the beginning of a collaborative project, you need to find team members, get to know each other, and share background assumptions about working with others. You also need to clarify your team's goal, choose an approach and schedule meetings, and generate a plan about how to proceed. Part of this plan should include accessing appropriate collaborative technologies to aid your work. You might pick from or be assigned one of several collaborative models, depending on your situation: the divide-and-conquer model; the specialization model; the dialogic model (marked by flexible roles); the sequential model (often characterized by a series of signatures of approval); and the synthesis model (formed with team members representing different views).

As you execute your project, you need to maintain trust by establishing ways to resolve pseudo-, simple, and ego forms of conflict. You can clarify your work styles by talking about control, credit, response, flexibility, and constraints on the project. And you can share information and assess your progress by using question prompts to help each other expand ideas, consider alternatives, and voice disagreements. Collaborative technologies that allow you to share texts and perhaps even share screens especially aid your work at this stage.

At the end of your project, you need to acknowledge and appreciate each member's work as well as package your work for presentation to your instructor or outside client. Any disputes during the process should now be resolved. Once again you should identify those members with special talents to format and present your work and use technologies to develop your presentation.

Collaborative work goes better if you participate in open discussions that include talk at both the relationship and task levels. Remember to listen carefully to other members and to change models and roles throughout the project when the need arises. Take time to identify your different learning and working styles, and capitalize on your diverse talents.

During your study of technical communication, you may frequently be asked to participate in collaborative writing activities. And, after you enter the workplace, you may be assigned to a collaborative writing team or in the future you may establish and supervise one. Keep in mind our suggestions for making this process an enjoyable and productive experience.

ACTIVITIES AND EXERCISES

1. Go to the library and read a chapter or two from one of the following sources in which popular writers, managers, theorists, and researchers talk about working and writing with others. Prepare a brief memo to your classmates in which you explain the important points in the reading.

Popular writers

Aburdene, Patricia, and John Naisbitt. "Collaborative Couples." *Megatrends for Women.* New York: Villard Books, 1992, pp. 165–87.

Schrage, Michael. *Shared Minds: The New Technologies of Collaboration.* New York: Random House, 1990.

Managers

Larson, Carl E., and Frank M. J. LaFasto. *TeamWork: What Must Go Right/ What Can Go Wrong.* Newbury Park, CA: Sage Publications, 1989.

Rosen, Ned. "How Groups Affect Their Members." *Teamwork and the Bottom Line: Groups Make a Difference.* Hillsdale, NJ: Lawrence Erlbaum Associates, Publishers, 1989, pp. 48–65.

Theorists and Researchers

Ede, Lisa, and Andrea Lunsford. *Singular Texts/Plural Authors: Perspectives on Collaborative Writing.* Carbondale, IL: Southern Illinois University Press, 1990.

Sproull, Lee, and Sara Kiesler. *Connections: New Ways of Working in the Networked Organization.* Cambridge, MA: The MIT Press, 1991.

2. Do the following individually, and then share this information with your team as part of building understanding and trust as you begin your project:

   a. Quickly write down the most recent writing project that you collaborated on either in college or at work (essay, speech, proposal, letter, instructions, memo, or computer documentation).

   b. When you collaborated, what aspects of the project (and the writing) were you most concerned with? List the first three things you think of.

   c. How much control did you have over the final collaborative product? How much control did others have?

   d. How were you given credit for the work that you did?

   e. What process did you follow when responding to the work of others?

   f. What procedures did you follow for resolving conflict?

   g. Did each member of your team understand how flexible you could be with pre-established formats/specifications/style guides?

   h. When did your team talk about constraints facing the project (deadlines, specific client or instructor needs)?

   i. What was the status of your project in the class? In the organization? Was the status different for different team members?

   j. What were your personal goals for being involved in the project? To simply get the job done? To get promoted? To get an *A*?

   k. How did you approach your part of the work?

   l. Could you see any personal characteristics (gender, age, personal background, cultural heritage, and so on) that affected the collaborative process for you and for your team members?

3. As a group, list all of the technologies available for your collaborative work (remember to include the phone, fax machines, copy machines, and tape recorders as well as computer technologies). Define each and state whether you think it would be best used at the initiation, execution, and/or public presentation

stages of your collaborative work. In a memo written to your classmates and instructor, discuss your findings.

4. Interview someone who is active in your future profession, asking that person to talk about how he or she:
    a. identifies possible collaborators,
    b. approaches the collaborative task,
    c. gets to know individuals' work styles,
    d. handles conflict, and
    e. reaches closure on a project.

    In a table or chart, display what you learned and how this matches your personal responses.

5. On August 2, 1992, the Coalition of Citizens Concerned with Animal Rights (CCCAR) sent a letter to Dr. Karen Wartala, Dean of the College of Agriculture, University of the Midwest, detailing their surveillance and investigation into the trapping and killing of birds in the experimental crop fields on its campus (Appendix A, Case Documents 3). This letter started a long process in which university personnel met to determine what they should do. In turn, they met with CCCAR representatives and worked to clarify the issues and their agreements.

    Collaborate with a partner to draft a document that includes the issues from both sides and the agreements made as stated in the two letters dated August 13 (from the CCCAR to Dr. Wartala) and August 22 (from Dr. Wartala to the CCCAR). Assume that one of you must summarize the position of the CCCAR and the other the University. The audience for this document will be people from both the CCCAR and the University.

    As part of your collaborative work, remember to begin at the relationship level. That is, what are your personal stands on the issue of trapping birds? What is the right thing to do: For the CCCAR? For the University? For both sides combined?

    Then, work at the task level and draft your document.

6. When three researchers studied groups of collaborators (staff, supervisors, and managers) at Exxon, they found that collaborators followed this process:

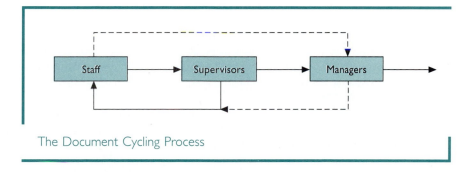

The Document Cycling Process

From this chart, list the process you think these collaborators followed:

Now read what the collaborators said about each other:

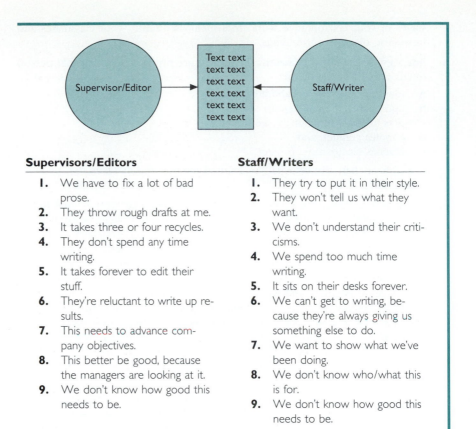

Supervisors/Editors	**Staff/Writers**
1. We have to fix a lot of bad prose.	1. They try to put it in their style.
2. They throw rough drafts at me.	2. They won't tell us what they want.
3. It takes three or four recycles.	3. We don't understand their criticisms.
4. They don't spend any time writing.	4. We spend too much time writing.
5. It takes forever to edit their stuff.	5. It sits on their desks forever.
6. They're reluctant to write up results.	6. We can't get to writing, because they're always giving us something else to do.
7. This needs to advance company objectives.	7. We want to show what we've been doing.
8. This better be good, because the managers are looking at it.	8. We don't know who/what this is for.
9. We don't know how good this needs to be.	9. We don't know how good this needs to be.

Source: James Paradis, David Dobrin, and Richard Miller. "Writing at Exxon, ITD: Notes on the Writing Environment of an R&D Organization." *Writing in Nonacademic Settings.* Ed. Lee Odell and Dixie Goswami (New York: Guilford Press, 1985). p. 301 (paraphrased to the plural). Adapted by permission of the Guilford Press.

Study these comments. What positive, future directions might you draw from them? Because you have participated in positive collaborative projects in the past, your supervisor asks you to develop a new process for collaborative writing projects at Exxon. Analyze the comments above and propose a new or revised collaborative process.

7. This case study focuses on the executive letter of the 1986 annual report of the Auldouest Insurance Corporation (pseudonym), described by one employee as "the most revised piece we do all year." This executive letter would go to over half a million policyholders.

The seven collaborative writers were the following:

■ Newly elected president (48, vocational-college graduate in computing, 26-year veteran).

Year	Message
1980	As a result of inflation and a series of summer storms, our experience did not measure up to our expectations and we suffered an operating loss for the year.
1981	The year's underwriting experience produced a loss. . . .
1982	These elements, coupled with increased loss frequency and severity, produced record underwriting losses for the industry and for [The Auldouest] Mutual Insurance Company.
1983	Even with improving economic conditions, [Auldouest] Mutual, as well as the casualty/property insurance industry, was once again hit with unfavorable underwriting results, a condition that has prevailed since the early 80's.
1984	Actually, the industry paid out in claims and operational costs 18 percent more than the amount of insurance premiums received. Our (the industry and [Auldouest]'s) task is to identify the reasons for such adversity and, where possible, to develop solutions.
1985	The underwriting loss was $18, 896, 203. . . .

Source: Geoffrey Cross, "A Bakhtinian Exploration of Factors Affecting the Collaborative Writing of an Executive Letter of an Annual Report," *Research in the Teaching of English* 24.2 (1990), pp. 173–203. Copyright 1990 by the National Council of Teachers of English. Reprinted with permission.

FIGURE 6–5
Disclosure of Underwriting Losses in Auldouest Executive Letters

- Senior vice president (49, liberal-arts-college graduate, 18-year veteran).
- Chief executive officer's secretary (high-school graduate, 34-year veteran).
- President's secretary (liberal-arts-college graduate, 19-year veteran).
- Vice president of corporate communications (46, high-school graduate with college coursework in public relations, 19-year veteran).
- Supervisor (37, university graduate in broadcast journalism, five-year veteran).
- Writer (23, summa cum laude liberal-arts-college graduate in English/communications, six months at Auldouest).

In terms of location in the 33-story Auldouest Building, the first 10 floors were for employees, and top-ranking officials were on the 33rd floor.

The collaborators began on October 13, 1986, and were expected to complete the letter in about five weeks, but the actual process required seven major drafts and the final letter was approved on December 29, 1986. This 77-day collaborative process exceeded its deadline by six weeks.

In a small group, read through the following collaborative process as well as data about the company (Figure 6-5), perceived audiences for the document (Figure 6-6), and the actual letter (Example 6-1). Then, develop a table or chart

Actual Audience	President	Senior Vice President	Chief Executive Officer's Secretary	President's Secretary	Vice President	Supervisor	Writer
Perceived Audiences for Each Writer and/or Editor							
Family/friends of the top executives	×						
Other chief executive officers	|ª	|	×		×	|ᵇ	|ᵇ
Other insurance staff	×	×	×		×	×	
Bankers					×		
Chamber of commerce	×						
Company representatives with clients		×					
Potential commercial policyholders	×	×			×		
Commercial policyholders	×				×	×	
Potential agents					×		
Agents	×	×	|		×	×	
Potential employees		×			×	×	
Potential domestic policyholders							
Domestic policyholders		×			|	×	
Media					×	×	
Public				|	×		

\|ª = most important; \|ᵇ = most important external audience. The Supervisor's most important audiences were the president, the CEO, and other CEOs. The writer's most important audiences were the President and the CEO.

Source: Geoffrey Cross, "A Bakhtinian Exploration of Factors Affecting the Collaborative Writing of an Executive Letter of an Annual Report," *Research in the Teaching of English* 24.2 (1990), pp. 173–203. Copyright 1990 by the National Council of Teachers of English. Reprinted with permission.

FIGURE 6–6
A Comparison of Actual and Perceived External Audiences of the Executive Letter

EXAMPLE 6–1 Executive Message

We are pleased to present this 1986 annual report. It reflects not only the year's achievements, but also the positive effect from prior years' actions. A result of these measures is that the combined assets of all Auldouest Companies now exceed $400 million.

The assets of Auldouest Mutual grew to $324 million in 1986, an increase of more than $34 million. The company's surplus (equity) account increased to $88 million, which represents a 23.7 percent growth from the preceding year. The combined loss and expense ratio decreased from 1985's 109.6 to 103.2.

Auldouest Life improved its presence in the marketplace by adding Universal Life to its portfolio. Assets increased to over $56 million, while the surplus account grew to more than $14 million, which represents a 20.5 percent improvement from the previous year. The amount of life insurance in force increased from 1985's $470 million to $601 million.

AULDCO, Auldouest's nonstandard company, continues the trend of positive results, with assets increasing 73.7 percent to $15 million. Its surplus account now exceeds $8 million, which represents a 41 percent increase. AULDCO currently conducts business in Michigan and Wisconsin.

AULDLEASE, a wholly owned subsidiary of Auldouest, has as its primary product the offering of private vehicle leases to individuals and small groups. At the end of 1986, AULDLEASE had assets of $5.5 million with a year-end net worth of $2.5 million.

During 1986, we initiated many changes in our operations—changes we believe will have both immediate and long-term positive effects. To name just a few: installation of a new IBM computer to enhance automated processing; construction of a new claims training facility; inauguration of additional training classes for employees and agents. Thus, we anticipate greater efficiencies and an even greater degree of professionalism in providing protection and service to our policyholders.

As we look to the future, the insurance industry faces numerous challenges, some old and some new. One issue, the need for tort reform, is an external factor which has been present for several years. Although 39 states enacted some type of tort reform in 1986, the issue is far from resolved. As insurers, we are concerned that unpredictable court verdicts and the awards associated with those vehicles will continue to negatively influence insurance availability and affordability.

The Tax Reform Act of 1986 is another external factor that will affect the affordability of insurance products. It is estimated that taxes for the property-casualty industry will be increased in excess of $7 billion over the next five years. The life insurance industry will also be adversely affected.

While the future may present some obstacles, the Auldouest Companies are prepared. We are confident that through cost containment programs, increased operating efficiency, and enhanced professionalism of employees and agents, all Auldouest Companies will continue to grow and prosper.

EXAMPLE 6–1   (*continued*)

Steven P. Norton	John H. L. Vogel
Chairman of the Board	President and Chief
and Chief Executive Officer	Operations Officer

Source: Exercise, including all figures and examples, comes from Geoffrey Cross, "A Bakhtinian Exploration of Factors Affecting the Collaborative Writing of an Executive Letter of an Annual Report." *Research in the Teaching of English* 24.2 (1990) pp. 173–203. Copyright 1990, the National Council of Teachers of English. Reprinted with permission. For a full description and analysis of this collaborative writing process and a summary of related research on group writing and conflict, see Geoffrey A. Cross, *Collaboration and Conflict: A Contextual Exploration of Group Writing and Positive Emphasis* (Cresskill, NJ: The Hampton Press, 1994).

of this case study in terms of what the collaborators did at the initiation, execution, and public presentation stages (and at relationship and task levels) of their work. Analyze and evaluate their process in terms of:

- The distribution of power.
- The audience(s) for the letter.
- The purpose(s) of the document.
- Expectations for the document.
- The collaborative model which this group followed.

Then, share your table or chart and your evaluation of this collaborative process with other groups.

The actual collaborative process went through three periods: a period of stability, a period of instability, and a resolution.

Period of Stability

- Corporate Communications held two departmental brainstorming meetings. The president had only been on the job six months. The vice president encouraged his subordinates, including the supervisor and the writer, to present fresh approaches and ideas.
- At their second meeting, collaborators learned that they were to ghostwrite the letter. They agreed that the letter should discuss a recovering Auldouest in the context of the troubled insurance industry.
- The supervisor wrote an annual report outline that presented this image of Auldouest. It included the statement, ''We are a forward-looking company dealing successfully with difficult problems.''
- The supervisor and vice president met with the president and CEO, who suggested only a few additions (namely, that statistics/numbers were to be placed at the beginning of the letter).
- The supervisor and vice president were elated that their first outline was approved. What had not surfaced was the new president's strong commitment to positive emphasis.

- The supervisor delegated the writing of the first draft to the writer, but the writer had not attended the letter-planning meeting with top management.
- After finishing her draft, the writer discussed it with the supervisor. The supervisor made suggestions, but didn't talk about her perceptions of the statement, "must understand that surplus is a sign of corporate health, not corporate greed."

Period of Instability

- The vice president reviewed the revised draft of the letter and rejected it, objecting to the phrase, "The public must understand." The supervisor challenged this decision. The vice president saw the public as the chief audience and felt the words were dictatorial (the vice president was not aware that the writer had written the letter and assumed that the supervisor had misunderstood his advice).
- Challenged by his subordinate, the vice president (along with the supervisor) consulted the senior vice president, who supported the vice president's judgment. In order to defend Auldouest's financial position, the senior vice president told the supervisor to include the number of recent bankruptcies of insurance companies and other information related to performance. The senior vice president also wanted the supervisor to discuss problems at the beginning of the letter.
- The supervisor communicated these needed changes to the writer, who was asked to revise.
- The supervisor and senior vice president reviewed the revised letter, and a clean draft was sent to the 33rd floor (executive offices).
- The executive secretaries incorporated several of their own editing changes, substituting words like "halt" for "stop."
- The CEO reviewed the draft and asked for more statistics.
- The letter went to the president who made suggestions that required a global change, emphasized certain audiences and purposes, ignored others, and altered the company's account of itself in 1986. The president communicated this new concept for the letter to the senior vice president. They decided that the senior vice president, the supervisor, and the writer would each write his or her own new version of the letter.
- The supervisor and the writer submitted new letters to the senior vice president, who chose not to use any parts of these letters in the final letter.

Resolution

- The senior vice president wrote a "success story" draft that ignored the interests of domestic policyholders but met the chief intentions of the president and CEO.

Burnett, Rebecca. "Substantive Conflict in a Cooperative Context: A Way to Improve the Collaborative Planning of Workplace Documents." *Journal of Technical Communication* 38.4 (1991) pp. 532–39.

WORKS CITED

Ede, Lisa, and Andrea Lunsford. *Singular Texts/Plural Authors: Perspectives on Collaborative Writing*. Carbondale, IL: Southern Illinois University Press, 1990.

Edelsky, Carole. "Who's Got the Floor?" *Language in Society* 10.3 (1981) pp. 383–421.

Engelbart, Douglas. "A Conceptual Framework for the Augmentation of Man's Intellect." In *Vistas in Information Handling,* ed. P. W. Howerton and D. C. Weeks. Washington, DC: Spartan Books, 1963, pp. 1–29.

Faigley, Leslie, and Miller, T. "What We Learn from Writing on the Job." *College English,* 44 (1982) pp. 557–69.

Johansen, Robert. "Use Approaches to Computer-Supported Teams." In *Technological Support for Work Group Collaboration,* ed. by Margrethe H. Olson. Hillsdale, NJ: Lawrence Erlbaum Associates, Inc., 1989, pp. 1–31.

Kiefer, C. F., and P. M. Senge. "Metanoic Organizations." In *Transforming Work,* ed. J. D. Adams. Alexandria, VA: Miles River Press, 1984, pp. 109–122.

Larson, Carl E., and Frank M. J. LaFasto. *TeamWork: What Must Go Right/What Can Go Wrong*. Newbury Park, CA: Sage Publications, 1989.

Parker, Glenn M. *Team Players and Teamwork: The New Competitive Business Strategy*. San Francisco: Jossey-Bass Publishers, 1990.

Wall, V. D., and L. L. Nolan. "Small Group Conflict: A Look at Equity, Satisfaction, and Styles of Management." *Small Group Behavior* 18 (1987) pp. 188–211.

Secondary Sources of Information: Libraries and Internet

A Continuing Case to Demonstrate Gathering Information ▪ Library Research ▪ Internet Research

Some of our students, perhaps more than we imagine, will actually contribute to the published, indexed literature. As authors of journal articles, technical reports, or conference papers, or as contributors to electronic databases of various sorts, they will need to make decisions about how to title, to abstract, and to list keywords for their work. Many more students will need to use indexes and computerized literature searches; they need to become efficient and intelligent users of the electronic literature.

Donnelyn Curtis and Stephen A. Bernhardt,
"Keywords, Titles, Abstracts, and Online Searches: Implications for Technical Writing,"
The Technical Writing Teacher 18.2 (Spring 1991), pp. 142–61.

Introduction

When you are charged with a writing or speaking task, you often have to research your subject thoroughly as part of your planning. This research enables you to discover the previous "conversations" about that subject—what solutions other researchers have been proposed, what tests they have run, what decisions they have made, what arguments they have posed, what "good reasons" they have used, and what problems they have encountered. While you may eventually "talk" to those researchers, in person, over the phone, or by regular or electronic mail, you'll usually start by reading information on your subject.

When you gather information this way, you are often not seeking specific data about your local situation. Instead, you are looking for already published information that can cast light on the questions you are asking. Such research, commonly referred to as **secondary research** because you are investigating the research that others have already done, can be undertaken through two different but interrelated means:

secondary research
using published research and scholarship to gather information; includes databases and computer networks such as the Internet

1. You may go to a university or company library and conduct the research, or
2. You may log on to a computer connected to a computer network—the largest of which is the Internet—and conduct **online research.**

online research
research conducted by using computer or electronic means, such as computer networks and databases

Both options offer massive amounts of information sources, but the trend is toward the latter. More and more secondary research is being conducted through electronic searches. This chapter provides an introduction to such investigations as well as to traditional library research.

A Continuing Case to Demonstrate Gathering Information

In this chapter and the one that follows, we refer to a continuing case to discuss gathering information. Let's assume that you work for Biomed Corporation, a company that produces computer software for the healthcare industry. You have been asked to do research on your customers' knowledge and attitudes about a new software product, Med-Ease Software. This product is designed to computerize the patient-management system of hospitals. With Med-Ease, hospitals can admit, track, and release patients, monitoring their treatments, their medications, their billing, and their insurance claims. This is a large product, and Biomed has installed it at City Hospital in a pilot testing program designed to verify the performance of the software. If the product works well for three months in City Hospital, it will be released to the market.

usability study
an investigation of how well a product meets its users' needs; conducted on-site or in the laboratory

You have been asked to begin a **usability study** of Med-Ease Software. You are told that the first step will be to develop information about the users' responses to the new product: How easy is it to use? How well do users like it? Are there differences in attitude among different types of users?

You know that before you contact Med-Ease users, you need to read the published studies of software use in other hospitals. Others may well have studied the issues that you are investigating. You predict that you can gain further insight into your own project by comparing what others have found to what you may find when you contact users. Careful secondary research is often done before you begin primary research as a way to avoid "reinventing the wheel." Other people's research can give you ideas and information you didn't have before, and it can often save you time by pointing out dead ends. In order to benefit by the work of other researchers, you will need to conduct both library and online searches for appropriate information.

Library Research

In this section, we cannot provide you with a detailed analysis of how to use the library. If you are unfamiliar with your library, ask your librarian or instructor to arrange a tour. Some colleges and universities provide courses for new researchers on how to use library resources. What we are providing you with here is an overview of how to conduct searches at most college and public libraries. Any library, and especially those at large research universities, can be intimidating the first couple of times you use it. As a consequence, you should read this chapter over before you go to the library to begin doing research. That way you will have a plan in mind and won't become overwhelmed. You should also realize that your interactions with the **reference librarians** will enable you to conduct research accurately and efficiently. Reference librarians help new researchers find their bearings and start working.

reference librarians
librarians trained to help researchers find needed information accurately and efficiently

The information you seek from libraries may be stored in a variety of media, including books, magazines, newspapers, printed journals, government documents, microfiche, audiotape, videotape, and compact disks. Most of the sources you seek can be found through one of two types of searches: a book search or a periodical search. The former represents the most traditional and commonly known library research. The latter encompasses a wonderful array of avenues and possibilities.

CLASSIFICATION SYSTEMS

Despite their size and the variety of their holdings, libraries are highly organized places. Librarians are experts in handling information storage and retrieval. One of the first keys to a successful search is understanding the library's method of organizing its holdings. A library may use one or both of the two standard systems for organizing its books and journals: the Dewey Decimal system or the Library of Congress system. Each system has a different way of identifying its holdings, and each classification system generates a **call number** to identify a particular holding.

call number
the coding device used to identify library holdings according to either the Dewey Decimal or Library of Congress classification system

Dewey Decimal Classification

At the heart of the **Dewey Decimal classification** system is a three-digit number. The first number identifies the 10 most general classes of holdings; the second has a narrower focus, and the third an even narrower focus. In Example 7–1, you can see the system's 10 basic categories, and these are the same in every library that uses the Dewey Decimal classification. The numbers identify the ranges of call numbers for books and journals of that topic. In this system, the 500s are reserved for natural science and mathematics, the 700s for the arts, and so on.

If we were going to work in a library organized by the Dewey Decimal system to begin our secondary research on Med-Ease, the computerized patient-management system used by City Hospital, we would first have to identify where other usability studies of patient-management software programs might be stored. When we look at the most general classification in Example 7–1, it seems likely that materials we might want would be found in the 600s—Technology (Applied Sciences). We would go to that part of the library and begin browsing just to get an idea of what might be available, but at this stage, we wouldn't have any concrete idea of which books would have what information or even where in the section the books and periodicals dealing specifically with software would be.

The Dewey Decimal system uses the second number in its classification pattern to narrow down further what is covered in each of its 000 to 900 general categories. Example 7–2 shows one of the general sections, 600 Technology (Applied Sciences), at the next level of detail. By looking at this level of specificity in the system, we see that the materials we are looking for might be in section 610 or 650. Section 610 is technology, medical sciences, and section 650 is technology, management.

Example 7–3 shows the third level of specificity in the Dewey Decimal system. At this level, we can see what materials are included in 610 (Technology, Medical Sciences) and 650 (Technology, Management). On the basis of

---

EXAMPLE 7–1   Dewey Decimal Classification System

000 Generalities
100 Philosophy & Psychology
200 Religion
300 Social Sciences
400 Language
500 Natural Science & Mathematics
600 Technology (Applied Sciences)
700 The Arts
800 Literature & Rhetoric
900 Geography & History

EXAMPLE 7–2   Dewey Decimal: 600 Technology

600–609—Technology (Applied Sciences)
610–619—Medical Sciences, Medicine
620–629—Engineering & Allied Operations
630–639—Agriculture
640–649—Home Economics & Family Living
650–659—Management & Auxiliary Services
660–669—Chemical Engineering
670–679—Manufacturing
680–689—Manufacture for Specific Purposes
690–699—Buildings

---

this information, we could narrow down the categories in which we are likely to find materials related to our usability study of the Med-Ease patient-management software. At this level of focus, it looks as if the 610s are not what we need. The 650s hold more promise. In fact, if we were to look more closely at section 658 (General management), we would discover that the next refinement of the system (not shown) gives us 658.9—Management of enterprises engaged in specific fields of activity. We could now look for books and serials beginning with this general number dealing with hospital management. Of course, at this point we do not have the authors and titles of specific works, and we don't know if any of these materials deal with usability studies of *hospital software.*

The library carries the Dewey Decimal system to many decimal points, narrowing down topics and identifying them with appropriate numbers.

EXAMPLE 7–3   Dewey Decimal Classification

610 Medical Sciences, Medicine

610—Medical sciences. Medicine
611—Human anatomy, cytology, tissue biology
612—Human physiology
613—General & personal hygiene
614—Public health & related topics
615—Pharmacology & therapies
616—Diseases
617—Surgery & related topics
618—Other branches of medicine. Gynecology and obstetrics
619—Experimental medicine

650 Management

650—Management & auxiliary services
651—Office services
652—Processes of written communication
653—Shorthand
654—Unassigned
655—Unassigned
656—Unassigned
657—Accounting
658—General management
659—Advertising & public relations

EXAMPLE 7–4 Dewey Decimal System

600	Applied Sciences
630	Agriculture and agricultural business
631	Farming
631.5	Crop production
631.58	Special cultivation methods
651.587	Irrigation farming
631.5872	By furrow system

Example 7–4 shows how the system is used to define holdings accurately and narrowly. Of course, it is not necessary for you to memorize this or any classification system, but it is important for you to have an idea of how libraries categorize their holdings. Such knowledge will enable you to create more effective searches.

Library of Congress System

Library of Congress classification
library system that uses a letter coding

Libraries may use the Dewey Decimal system, the **Library of Congress classification system,** and several others in combination or alone. Example 7–5 shows the most general classification level of the Library of Congress system, which is the same in every library that uses this system.

If we were looking for information for our Med-Ease study, using the Library of Congress system, we would probably consider holdings in **R** (Medicine) or **T** (Technology). At this level of specificity our search would be virtually useless, including as it would hundreds of thousands of materials. Because the Library of Congress system uses both letters and numbers to classify holdings, we would have to turn to the *Library of Congress Subject Headings,* a major reference guide found in most libraries, to see what is covered under both **R** and **T**. In this case, we would discover that R855–R855.5 designates Medical Technology. The *Library of Congress Subject Headings* would tell us that Medical Technology covers the following:

> Here are entered works on the techniques, drugs, and procedures used to deliver medical care and the systems within which such care is delivered (2993).

This is helpful information, but it would be impossible to look up every subject in the *Library of Congress Subject Headings* to get a call number. Without very powerful searching tools, you would have difficulty undertaking any secondary research if this general information was all you had to rely on. The rest of this chapter discusses a variety of such searching tools.

EXAMPLE 7–5   Library of Congress Classification System

A—General Works
B—Philosophy, Psychology, Religion
C—Auxiliary Sciences of History
D—History: General and Old World
E–F—History: America
G—Geography, Anthropology, Recreation
H—Political Sciences
J—Political Science
K—Law
L—Education

M—Music
N—Fine Arts
P—Language & Literature
Q—Science
R—Medicine
S—Agriculture
T—Technology
U—Military Science
V—Naval Science
Z—Library Science

BOOK SEARCH

Most books can be found by using either the traditional or the electronic card catalogue. The **traditional card catalogue** consists of hundreds of drawers of cards arranged alphabetically and separated into three major sections: *author, title,* and *subject.* If you know the author or the title of the book you are seeking, you can find it listed in the appropriate section of the card catalogue. If you have a topic in mind, such as "computer software," but don't know of any particular title or author, you can begin your search in the *subject* section. Finding useful information through a subject search is the most common strategy but also the most challenging of all library research strategies. A broad topic such as "computers" will turn up hundreds of cards to sort through. Usually the first card under such a broad category tries to direct you to a narrower topic (e.g., "computer software," "computer hardware," or "impact of computers"). In the early stages of a search, you may not know which of these subcategories to examine, and you can spend a lot of time in unrewarding searches. To do an effective subject search, you must decide what words or phrases best represent the breadth and depth of the information that you desire on your subject. We'll explain how to choose these *keywords* shortly.

traditional card catalogue
drawers of cards arranged alphabetically and separated into three major sections: author, title, and subject

   Example 7-6 shows typical card catalog entries for author, title, and subject. These entries provide you with considerable information. Often you will be able to tell from looking at an entry whether or not the book is useful to you. By using the cards thoroughly, you can save considerable time when you finally go to the shelves. For example, if you only want current materials, you can eliminate all books written before a particular date from your consideration. Likewise, you can determine what language the book is written in, whether or not it has a bibliography, and if it has illustrations—all useful information as you consider which materials you want to gather.

EXAMPLE 7–6   Author, Title, and Subject Cards from a Card Catalogue

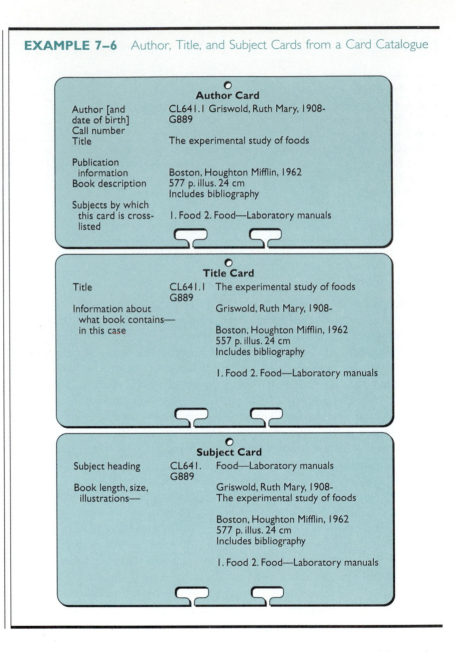

Author Card

Author [and
  date of birth]   CL641.1 Griswold, Ruth Mary, 1908-
Call number   G889
Title   The experimental study of foods

Publication
  information   Boston, Houghton Mifflin, 1962
Book description   577 p. illus. 24 cm
                          Includes bibliography

Subjects by which
  this card is cross- 1. Food 2. Food—Laboratory manuals
  listed

Title Card

Title   CL641.1 The experimental study of foods
                          G889
Information about   Griswold, Ruth Mary, 1908-
  what book contains—
  in this case   Boston, Houghton Mifflin, 1962
                          557 p. illus. 24 cm
                          Includes bibliography

                          1. Food 2. Food—Laboratory manuals

Subject Card

Subject heading   CL641.   Food—Laboratory manuals
                          G889
Book length, size,   Griswold, Ruth Mary, 1908-
  illustrations—   The experimental study of foods

                          Boston, Houghton Mifflin, 1962
                          577 p. illus. 24 cm
                          Includes bibliography

                          1. Food 2. Food—Laboratory manuals

**electronic card catalogue
system**

organized along the same
principle (author, title, and
subject) as the traditional
catalogue system but con-
tained on a library's computer system

Subject searches can be more successfully completed at a library that
offers an **electronic card catalogue system.** Although systems may differ
in key ways, they will be organized on the same principle as the traditional
catalogue: author, title, or subject. Let's say that because you want to use
secondary materials in your usability study of Med-Ease patient-management

---

EXAMPLE 7–7   Sample Electronic Catalogue Entry

Title: Software for health sciences education ⟨computer file⟩:
                        an interactive resource/designed and developed by the
                        Learning Resource Center, The University of Michigan
                        Medical Center
Author/Editor:
Edition: 2nd ed.
Published: Ann Arbor, Michigan: The Center, c1989
Location: Bio-med Learning Resource Center Library
Call No.:   W18 CMM1989
Status: Not checked out

---

software, you decide to search first for books on the topic. Using the library's computer terminal, you type in the instructions to search for each of the words you think might yield information. In this case, you've given your problem some thought and you decide to look for materials on the following subjects: computer, software, usability. Most electronic library search systems ask that you type something that looks like this: **s=computer** to indicate that your subject is computers.

In this case, your subject search is both successful and not. You have turned up 5000 entries on the subject of computers, 153 entries on software, and 0 on usability. You decide to try to search for the subject management, but it turns up 3,617 entries. Fortunately, you don't have to stop here. On most electronic card catalogues, subject searches are greatly enhanced because you can use a more thorough method of searching. **Keyword searches,** as they are called, take advantage of the library's finely differentiated classification systems. Keyword searches allow you to combine subject names and narrow your search accordingly.

In our example, let's assume that because you have access to an electronic card catalogue system, you try several combinations of keywords to see what you get. The first time you try *management software.* After you browse through the screens listing their titles, you decide that none are what you had in mind. Next you try *hospital software* but that fails to turn up any entries. Finally, you try *medical software* and find three entries:

1989   Software for health sciences ⟨data file⟩

1988   Laboratory diagnosis of acid-base disorders

1986   The Code Blue Professor

The first title looks like a possibility to you, so you choose it on the screen, and the computer gives you more information. Example 7–7 indicates what kind of information comes up on the screen. It looks a lot like the

keyword searches
used in subject searches; allows users to combine subject names and narrow search effectively

actual cards in the old catalogue, and so you are prepared for the information on it.

As we noted above, keyword searches in a traditional card catalogue will be a bit more cumbersome, but they can still be rewarding. Here you have to try each of your keywords separately. That is, you would look up *computer*. Under the entry for *computer* you look to see if there are subcategories for *software* and/or *usability*. The difficulty of this method, as compared to the electronic search, comes when you need to flip through the dozens—maybe hundreds—of cards listing *computer* as a subject title. As you search the card catalogue, take careful note of the headings for cross-listings at the bottom of the cards. These cross-listings can give you ideas about other subjects you might wish to search.

Serendipity

Once you find some call numbers identifying books that may be useful to you, you will head for the stacks, or shelves, to gather these materials. Knowing as you do something about classification systems, you realize that related materials will be classified similarly to the specific materials you have identified. Therefore, when you are in the stacks and you have located your book, look around on the same shelf and above and below your book. Also, looking through the bibliography that might appear in one book you have found will lead you to other sources. Often **serendipity**—being in the right place at the right time—can help you discover additional related materials not turned up in your search. The library is systematically organized, and the librarians may have identified related materials that escaped your search. Serendipity does not substitute for systematic examination of the card catalog, but it can play a helpful role in your research as you browse through related materials and discover something particularly useful.

serendipity
being in the right place at the right time

PERIODICAL SEARCH

Although library book holdings can be vast, the amount of information available through periodicals is almost too large to imagine. Through electronic advances in information storage and retrieval, the modern library can provide literally millions of sources of information through periodical searches. Once you begin to learn the ins and outs of periodical searching, you will rarely face the problem of too few sources. Instead, you will more likely seek ways to eliminate sources before you are overwhelmed in information, analyses, opinions, and data.

As you search for information in periodicals, you will find a variety of useful publications. Generally, research journals, particularly those that report scientific and technical studies, will offer you the latest data and interpretations of that data available in your field. Researchers at university or industrial sites report on their studies—including the research question asked, the method used to collect data, and the findings—in such journals. For example, in your search for material on user attitudes about healthcare computer software products, you might find results of similar user testing in computer science research journals, or you may find recommended methods of conducting user testing in psychology research journals.

Trade journals, periodicals that address practitioners in a field, usually suggest how to improve workplace practices. Although not all libraries carry trade publications, these journals are designed to help practitioners interpret and apply the recommendations and results that have been reported in research journals. For example, in your project for Biomed Corporation and City Hospital, you might find that hospital association trade journals offer suggestions on how to use new software products, and computer science trade journals offer recommendations on how to test the impact of such products.

Popular magazines both inform and entertain their readers and can give you a glimpse at how society in general might view your subject. For example, in your search for information for healthcare computer software programs, articles in popular magazines might indicate general social attitudes toward hospital care. Finally, newsletters produced by associations and corporations offer a wide variety of information. Usually newsletters update readers on events and people within a specific corporation or professional group. Perhaps you'll need to read back issues of the City Hospital newsletter to understand the climate of the hospital before you question Med-Ease users.

As with the book search, you can follow both traditional and electronic channels to search for periodical information. Either way you will follow one of two avenues: first, if you know the particular article you wish to find, you read the printed listing of your library's periodical holdings, find the call number, and proceed with your search. If you don't already have a title and journal in mind, then you must begin subject searching.

Indexes and Abstracts

If you do not have electronic means at your disposal, you will search the traditional way by manually paging through reference books to find the sources you need. The two most commonly used types of reference books for this type of search are **indexes,** which contain listings of citations organized by title, subject, and author, and **abstracts,** which include, in addition to the bibliographic information provided by indexes, brief summaries of the periodical articles.

For example, if you are searching for articles about stock market trends, you may use the *Business Periodicals Index* or *The Wall Street Journal Index.* If you are searching for articles about alternative energy sources, you may find them in *Energy Research Abstracts.* Two of the most commonly used indexes are ones that cover the popular press: *Reader's Guide,* which

indexes
reference collections that contain lists of citations organized by title, subject, and author

abstracts
reference collections that contain bibliographic information and brief summaries of sources

---

EXAMPLE 7–8   Selected Abstracts, Indexes, and Bibliographies of Indexes

Abstracts and Indexes

Traditional Periodical Indexes and Abstracts

Aerospace Database	Energy Research Abstracts
Animal Behavior Abstracts	Engineering Index
Applied Mechanics Review	Environment Abstracts
Applied Science and Technology Index	Index Medicus
Astronomy and Astrophysics Abstracts	Metals Abstracts
Biological Abstracts	Microbiology Abstracts
Ceramic Abstracts	Nuclear Science Abstracts
Chemical Abstracts	Nursing Studies Abstract
Chemical Engineering Abstracts	Oceanic Abstracts
Computer and Control Abstracts	Plant Breeding Abstract
Computer Literature Index	Pollution Abstracts
Ecology Abstracts	Science Citation Abstract
Education Abstracts	Social Science Citation Index
Electrical and Electronics Abstracts	Wildlife Review

Computerized Periodical Indexes and Abstracts
Arts and Humanities Citation Index
Biological and Agricultural Index
Computer Data Base
Food Service and Technology Index
Microcomputer Index
National Newspaper Index
Sociological Abstracts

Bibliographies of Indexes

Abstracts and Indexes in Science and Technology: A Descriptive Guide
Information Sources in Engineering
Information Sources in Science and Technology
Science and Engineering Literature
Scientific and Technical Information Sources

---

indexes hundreds of magazines and journals, and the *New York Times Index,* which indexes the *New York Times* newspaper. Many specialized indexes exist as well. What you choose will depend on the subject you are researching and how technical you need to be.

Example 7–8 lists some print and online specialized abstracts and indexes. It also gives some bibliographies of indexes and abstracts relating to science and technology. These reference books provide lists of available sources of information. In that sense, they are indexes of indexes and abstracts of abstracts.

EXAMPLE 7–9 Sample of *Psychological Abstracts*

Heading	Psychopharmacology
Authors	38374. Abel, E. L.; Bilitzke, P. J. & Cotton, D. B. (C.S. Mott
Title	Ctr for Human Growth & Development, Detroit, MI)
Site of research	**Alarm substance induces convulsions in impra-**
Journal	**mine-treated rats.** *Pharmacology, Biochemistry & Behav-*
Date of publication	*ior,* 1992 (Mar), Vol 41(3), 599–601.—Male rats were
Abstract	injected with imipramine (0–30 mg/kg) and tested in the
	forced-swim test in either fresh water or water soiled by
	other rats, which presumably contained an alarm sub-
	stance. Imipramine did not affect the behavior of rats in
	fresh water. More than half of the Ss given the combination
	of imipramine (30 mg/kg) and stress from alarm substance
	had clonic convulsions. Adrenalectomy did not affect this
	relationship, indicating that corticosterone is not involved,
	although imipramine and stress cause increases in plasma
	corticosterone levels. The convulsive effect of combined
	stress and imipramine may be due to their common ac-
	tions at central noradrenergic and serotonergic receptors.

Example 7-9 shows the kind of information you will find in an index. Just as it is essential for you to know how to read cards in the catalog, it is also important for you to know what information is available in an index. As you see from Example 7-9, you apply your subject search techniques to find material of interest to you. The index entries provide authors, titles, abstracts, and they direct you to the correct journal. Indexes frequently abbreviate the titles of journals, as Example 7-9 indicates. A key to these abbreviations is given at the beginning of each index.

OTHER REFERENCE MATERIALS

Although much of your research will grow from books and periodicals, there are a number of other potentially fruitful avenues to explore. Reference librarians will be able to tell you what sources are available at your location. Even if you think that you will only be using books or periodicals, it often is a great time saver to check in with the reference desk and ask for the librarian's advice as you begin to search. Often, he or she will direct you to additional sources of the type we describe below.

EXAMPLE 7–10   Selected Government Documents

Monthly Catalog of US Government Publications
GPO Publications Reference File
Government Reports Announcements and Index
Census of Population and Housing
County and City Data Book
National Retail Trade
Census of Agriculture
Dept. of Defense Hazardous Materials Information
National Economic, Social, and Environmental Data Bank

Government Documents

US government documents
published by state and national government agencies on a variety of topics

Many libraries are registered repositories for **US government documents** that cover topics ranging from economic forecasts, to environmental studies, to declassified intelligence reports, to census data, to a wide variety of other topics. The federal government publishes more than 25 million different documents yearly aimed at both specialized and general audiences. You can find reports from departments and agencies including the Departments of Defense, Interior, and Labor, the US Information Agency, Environmental Protection Agency, NASA, the Copyright Office, and the Central Intelligence Agency.

Libraries that serve as repositories of federal documents often employ a special reference librarian. If your library is such a repository, ask at the reference desk where to find the librarian in charge of the collection. Example 7–10 lists a few additional sources of government documents information.

Many state government agencies also provide publications. If your state has a land-grant university, it also has an Extension Service, which publishes a wide variety of information on agriculture, rural sociology, food science and nutrition, and gardening. These and other materials, often housed separately in the library, are designed for both lay and specialized audiences. Check with your librarian to see how these are indexed.

dictionaries
reference sources that define words

encyclopedias
reference sources that contain short, descriptive entries on topics

handbooks
reference sources that contain guidelines, standards, and procedures in particular fields

Reference Works

Other sources of information include specific disciplinary **dictionaries, encyclopedias,** and **handbooks.** There are both general and specialized versions of these reference materials (see Example 7–11). When you know very little about a topic, general reference works will help you get started. Later, when you have a clearer idea of your subject and its specialized vocabulary, you may wish to look at more specialized reference materials. Encyclopedias

EXAMPLE 7–11   Selected Reference Works

CRC Handbook of Chemistry & Physics
CRC Standard Math Tables
Dictionary of Advanced Manufacturing Technology
Dictionary of Ceramic Science & Engineering
Engineering Acronyms & Abbreviations
Electronic Engineers Handbook
Encyclopedia of Chemical Technology
Encyclopedia of Fluid Mechanics
Encyclopedia of Materials Science & Engineering
Encyclopedia of Physical Science & Technology
Encyclopedia of Polymer Science & Engineering
Engineering Mathematics Handbook: Definitions, Theorems, Formulas, Tables
Handbook of Industrial Toxicology

Handbook of Mechanics, Materials, & Structures
IEEE Standard Dictionary of Electrical & Electronic Terms
McGraw-Hill Dictionary of Scientific & Technical Terms
McGraw-Hill Encyclopedia of Science & Technology
Machinery's Handbook
Metal's Handbook
Perry's Chemical Engineers' Handbook
Standard Handbook for Electrical Engineers
Standard Handbook of Engineering Calculations
Standard Handbook of Industrial Automation
Standard Handbook for Mechanical Engineers

and handbooks are particularly useful in this way because they frequently list a bibliography of other sources on the topic at the end of each entry.

The abundance of reference guides should make it clear that there are many sources of information within the physical walls of a single library building. But those walls of wood, concrete, and steel need not serve as the barrier for your research now that you can also explore the electronic library. The rest of this chapter deals with electronic sources available in libraries and through computer networks.

The Electronic Library

Most libraries will provide you with access to search for your source either online or through a **CD-ROM** system. Online searching connects you to **databases** containing millions of sources of information. Most online databases, such as the *Knowledge Index, Chemical Abstracts Online,* or *BRS/ After Dark,* can be searched through a keyword system similar to that used on the electronic card catalogue discussed above. Each system, however, has its own **syntax** or way in which words are put together to form phrases or sentences; therefore, you will need assistance when trying to use these systems for the first time. The major advantage of searching online is that the databases are continually updated so that you will have access to the

CD-ROM
online research sources that display databases on various topics; CD-ROM searching enables the researcher to scan holdings on a laser disk

database
research term referring to lists and collections of information on various topics; usually refers to online research via the computer

syntax
in secondary research, the way words are put together to form phrases or sentences in a particular database

most current information available through library research. A disadvantage can be the cost and availability of the online system. You may be charged to use certain online services. These systems are not just searching for what is available in your local library. Many perform worldwide searches of databases held in various sites.

CD-ROM searching enables you to scan the holdings of an index that is stored on a read-only laser disk capable of holding thousands of citations. Although the searching strategies may be similar to those in electronic searches, CD-ROM systems are relatively inexpensive given the amount of

---

EXAMPLE 7–12   One Library's Online Research Service—Newsletter Excerpt

Library Offers New Search Service

In addition to the evening search services such as the Knowledge Index and the Chemical Abstracts Service (CAS), the Andrew S. Schuler Educational Resources Center also offers FirstSearch to the Clarkson community.

FirstSearch is an online search service which has databases that are used to find books and journal articles on a wide variety of subjects. . . .

To search for journal articles in a particular field, users would be able to access a host of other indexes involving the subject they are interested in. Some of the databases most relevant to Clarkson include:

- Article First—indexes 11,000 scholarly journals in academic disciplines from 1990 on.
- Contents First—search by journal title, then display the contents pages for recent issues of the same 11,000 journals.
- Periodicals Abstracts—indexes articles from 900 top general and academic journals and also transcripts from 30 television news programs. This database includes abstracts, with coverage from 1989 on.
- Newspaper Abstracts—indexes 25 national and regional newspapers. . . . Includes abstracts, coverage from 1989 on.
- PAIS Decade—indexes latest 10 years of material on public affairs, including international relations, environmental topics, health care financing and others. Includes abstracts.
- PsychFirst—indexes articles published in 1,200 psychology journals in the last 3 years. Includes abstract. . . .

Unlike the evening search services, FirstSearch is available during the daytime hours. . . . ERC patrons who wish to use this service should ask the reference librarian on duty. . . . The reference librarian can also walk new users through their first FirstSearch.

Source: *ERC News* 9.1 (1993). Andrew S. Schuler Educational Resources Center, Clarkson University, Potsdam, NY. Reprinted with permission.

data they can provide. And although they are not updated as easily and quickly as online databases, many CD-ROM systems are updated on a monthly or quarterly basis. Also, because of the tremendous capacity of compact disks (CDs) to store a variety of formats, you may be able to find and read the full text of the source you are seeking. For example, your library may subscribe to a CD-ROM service that supplies disks with the full text of several major daily newspapers, such as the *New York Times* or *The Wall Street Journal.* In other cases, the CD might hold an actual video of a speech or event.

To illustrate some very useful electronic search services offered by many university libraries, Example 7–12 provides an excerpt from one library's newsletter describing an extensive online database. In addition to the descriptions of all of the services provided, note the final paragraph. As this passage implies, online searching usually must be scheduled and assisted by reference librarians.

In general, periodical searching provides the greatest potential for secondary research. And over time, the means of doing such searches will continue to evolve. Already, it is possible to search library holdings using your own computer connected to a computer network, such as the Internet. The following section provides an overview of this network.

Internet Research

The potential for individuals to have vast databases available at their homes and offices is being realized through a large web of computer networks known as the **Internet.** This system isn't merely a network of computers; it is a network of networks, each of which is composed of a group of linked computer systems in organizations such as universities, government agencies, or research labs. No one knows how big the Internet community really is, but in 1993, the best guess is that it included about 1.3 million computers, 8,000 computer networks, and linked about 10,000 users! Hundreds of university and government library card catalogues are available through Internet, and most federal agencies are on Internet.

Internet
international electronic computer information system; consists of a large web of computer networks

For example, let's assume that University Z has a computer network that allows all computer users on campus to access the links of the library's database of its holdings. An *electronic database* is a collection of information stored on computer that can be organized and sorted in a variety of ways. Let's also assume that University Z has several other research centers on campus, each of which contains an electronic database on a particular topic, for example a database for a University Z's Cancer Research Center. Anyone on campus with a networked computer and a user ID and account can access these databases. That means you could conduct a general library search from a computer in a dorm room or residence hall.

Now, imagine that University Z's network, with all its vast holdings, is connected to hundreds and hundreds of other similar networks worldwide

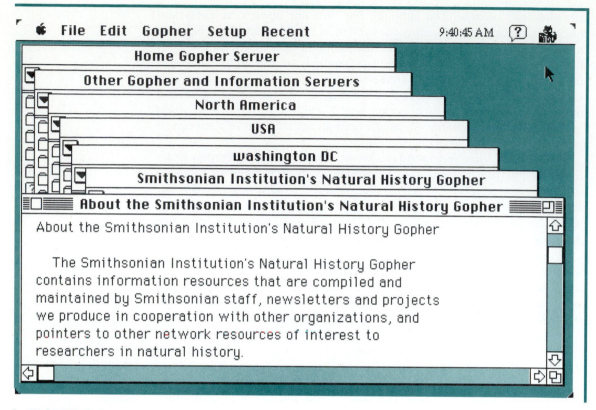

FIGURE 7–1
Overview of an Internet Search

with all of their vast holdings. Add to universities the growing numbers of corporations, nonprofit organizations, and individuals from all over the world, and you have the Internet.

Using Gopher or other search tools you will find out about shortly, you can find much relevant material on the Internet. For example, if you were looking for information about computer applications that might provide information about the Med-Ease usability study, you could find several promising leads. These leads include ongoing discussion groups, called **Listserv lists,** which are conducted on the Internet, as well as details of research done at hospitals listed on the National Science Foundation public Gopher site. Searches on the Internet are potentially confusing, because once you enter that electronic domain, it is sometimes hard to know where in the

Listserv lists
ongoing discussion groups conducted on the Internet

EXAMPLE 7–13   Surfing the Internet

Here is a sampling of what you can find on the Internet. For more information consult Krol or Laquey and Ryer.

- Historical documents: Bill of Rights through to Martin Luther King's "I Have a Dream" speech
- Contemporary speeches: Copies of President Clinton's speeches are available daily
- Books: Many books are available in their entirety
- Electronic addresses: Of almost all scientists in the world as well as faculty, staff, and students whose institutions are on Internet
- Research results from the Department of Agriculture
- Updates on the Hubbell Space Telescope
- Free software
- Weather conditions, as well as windsurfing and skiing conditions around the world
- Recipes, jokes, and reviews
- UPI news stories
- Conference and job announcements
- Reports on current National Science Foundation grants

Source: Ed Krol, *The Whole Internet User's Guide and Catalog* (Sebastopol, CA: O'Reilly & Associates, 1992).

---

world (literally) you have transported yourself electronically. Fortunately, many interfaces on the Internet allow you to have a visual map of where you have traveled.

Figure 7-1, for example, gives you an overview of how a researcher at the University of Minnesota traveled along the Internet in search of specialized information—in this case, information on natural history. By selecting from among available information servers throughout the world, this researcher was able to step through **cyberspace,** the electronic environment in which all this communication takes place, to find the Gopher server for the Smithsonian Institution.

cyberspace
the electronic or computer environment

Because of the utter vastness of the Internet, you need to use a set of tools that will help you find your way through the web with some order and efficiency. The rest of this chapter explains some of these tools. If you would like to see what Internet's resources include and how to navigate it, you might want to consider looking at Laquey and Ryer's, *The Internet Companion: A Beginner's Guide to Global Networking,* and Krol's, *The Whole Internet User's Guide and Catalog* (see Example 7-13).

INTERNET RESEARCH TOOLS

Conducting research via the tools described below represents only one use of the Internet. This chapter does not explore other very powerful uses, such as electronic mail and bulletin board services. In order to be successful in Internet searches, you must know how to access and use the Internet, which is beyond the scope of this book. Just as we suggest you consult with the reference librarian before beginning your search at the library, we suggest you ask your teacher to put you in touch with your campus' academic computing personnel. They can tell you what is available at your school, and they can assist you in the technical details of getting on the Internet. However, once you learn how to use the Internet, the following explanations can serve as an introduction to the network's research capabilities.

WorldWide Web and Mosaic

If you and your computer suddenly had the ability to connect to thousands of databases all over the world, how would you ever begin to find what you want to find, when you don't have any idea what is in any of the databases? If you were looking for information about studies of the usability of computer software, where would you begin to search? Maybe some of those thousands of databases have information about the topic, and maybe most don't. If you had to access and search each database separately, it might take you weeks and months to find even a single useful entry. This is a particularly critical issue in dealing with the Internet, which is growing rapidly. Also, although there is a group called the Internet Society that oversees the network, there is no one "in charge" of it, per se.

Fortunately, network software designers at several locations have developed systems that make Internet searches much more efficient. Two of these systems are known as the **WorldWide Web** (WWW), developed at the European Laboratory for Particle Physics, and **Mosaic,** developed at the National Center for Supercomputing Applications at the University of Illinois. These two systems enable you to use **hypertext** (that is, nonlinear) links to search through databases throughout the network. WWW is a hypertext system that can enable a user to search multiple databases via keywords. Mosaic sets up connections between certain bodies of information that have been entered into it. Over time more and more bodies of information will be integrated into Mosaic, as expressed in the following quotation:

WorldWide Web and Mosaic
systems that make Internet searches more efficient

hypertext
nonlinear links used to search through databases

> A user of *Mosaic* who is searching for information on a particular topic will be presented with a document on the topic that may include several words or phrases printed in blue. Clicking on the phrases with a mouse will present the user with other materials on the Internet related to the phrase.
>
> The blue phrases, which become red once they have been clicked on, are known as "hot links" and may connect to other documents, photo-

graphs, video sequences, or satellite images stored on computers anywhere in the world. There is little delay in calling up the related materials because all the links are pre-established so that no searching is needed. (DeLoughry A23)

Gopher

Developed at the University of Minnesota, **Gopher** is a tool designed to make electronic communications on the Internet easier. Gopher provides a series of easy-to-use menus that enable you to browse through whatever is stored in a Gopher site anywhere in the world. Because Gopher was provided free on the Internet, it is widely distributed over the world. As a result, Gopher is an effective tool for accessing information from a staggering number of databases (hence the play on words in its name—"go fer"). Gopher sites make online databases at their location available to other Internet users. As a result, when you ask Gopher to search all Gopher sites for a piece of information, your search encompasses all Gopher sites around the world and all their databases.

The useful thing about Gopher is that it enables you to search different databases at different locations without having to contact each location and connect with it. With Gopher you are exploring information that may be housed at different locations, but location is irrelevant to you once you are in "Gopher Space." Gopher has organized the information at numerous sites and placed it all under topic headings. For example, if you wanted to see if anyone else had examined patient-management software at a hospital, you could use Gopher to search all its databases for this information. Such a search might take you to a variety of libraries without your having to contact any one of them. Before there was such a service, you had to connect electronically with each library and explore it separately, or, as Krol describes the changes:

> Think of the pre-Gopher Internet as a set of public libraries without card catalogs and librarians. To find something, you have to wander aimlessly until you stumble on something interesting. This kind of library isn't very useful, unless you already know in great detail what you want to find, and where you're likely to find it. A Gopher server is like hiring a librarian, who creates a card catalog subject index. You can find something by thumbing through the subject list, then showing the card to the librarian and asking "Could you help me get this please?" If you don't find it in one library, you can electronically walk to the next and check there. (Krol 191–92).

Archie, Veronica, and Other Tools

One of the great things about the Internet is that you can use it to send and receive files to anyone else on the Internet. Although individual users may be able to send files (say, a file containing your resume to a prospective employer who needs it immediately) through electronic mail, those same users may request and receive files from a number of **file archives** through-

Gopher
a computer tool that provides easy-to-use menus to browse through the holdings and databases at any Gopher location or site

file archives
collections of computer files

EXAMPLE 7–14   Selected Internet Library Sources

Resources listed in **The Whole Internet User's Guide and Catalog**

■ CARL (Colorado Association of Research Libraries)—Provides library catalog and "a wide assortment of online text, indices of model school programs, online book reviews, facts about the metropolitan Denver area, and a database on environmental education."

■ Internet Accessible Library Catalogs and Databases—Contains several hundred online library catalogs.

■ Library of Congress Records—Contains the Library of Congress catalog and records.

■ Reader's Guide to Periodical Literature—Contains citations to articles printed in most general interest magazines printed in the US.

■ RLIN (Research Libraries Information Network)—One of the largest online "catalogs of catalogs," encompassing most major research libraries in the United States.

Source: Ed Krol, *The Whole Internet User's Guide and Catalog* (Sebastopol, CA: O'Reilly & Associates, 1992).

Archie and Veronica
computer tools that allow Internet researchers to search master lists of file archives that reside at any Gopher site

out the world. **Archie,** a system designed at McGill University in Montreal, Canada, and named for the comic book character, enables you to search through a master list of files that may reside at any Gopher site connected to the Internet that has files (containing papers, software, graphics—anything that could be stored digitally) that are free to be copied and transferred to individuals on the Internet.

If you need to find articles that describe, say, research into computer software usability, and you want to see if there are any repositories of papers that are available to the public on the topic, you can use Archie to do it. Archie is a database of all such archives on the Internet.

If Archie fails you, then you can use **Veronica** (which stands for Very Easy Rodent-Oriented Net-Wide Index to Computerized Archives). Veronica works easily with Gopher (thus the term, Rodent-Oriented), and like Archie allows you to search all Gopher sites. Developed at the University of Nevada at Reno, Veronica lists every menu item available from every Gopher site. So, instead of having to search one Gopher site at a time for information, with Veronica you have a centralized listing of all information available from all Gopher sites.

Other tools exist (e.g., WAIS, the wide area information service, and Jughead), and others are being developed and refined. The only way they will become useful is for you to experiment with them. Internet is available to corporations and individuals outside of universities for a fee, and more and more corporations are taking advantage of this data highway. As a consequence, if you have the opportunity to use Internet while in college, you will be gaining a valuable skill for the workplace. You'll need help getting set up to do Internet research, and the very nature of the system almost requires you to practice searching for information through some informed trial-and-error sessions. Only through this type of practice can you begin to grasp the scope of potential sources available through the network. Example 7–14 gives you only a few of the sources listed in Krol's *The Whole Internet User's Guide and Catalog.*

SUMMARY

The suggestions and guidelines presented in this chapter are only introductions to a vast world of secondary research. Whenever you are beginning to learn new research techniques, whether primary or secondary, you should work with experienced researchers who can help you both learn and refine your skills and guide you through the pitfalls and challenges of research. First-time library and electronic network research can leave you overwhelmed and confused. Learning how to become proficient at both can take some time, but the payoff is potentially enormous.

As a technical communicator, with secondary research, the type of information gathering you are most likely to do, you are not generating new data about your specific situation as you are when you do primary research. Instead, you are seeking research already done by others and published in any of a variety of documents. Secondary research can be undertaken by going to a library and conducting the research, or by logging on to a computer connected to a computer network and conducting online research.

Although libraries are highly organized places, a staggering amount of information is available in a variety of formats, including books, magazines, newspapers, printed journals, government documents, microfiche, audiotape, videotape, and compact disks. As a consequence, you need good search techniques to find what you want in this mountain of information. You can find most of the sources that you seek through one of two types of searches: a book search or a periodical search. Books are organized through one of two classification systems, Dewey Decimal or Library of Congress systems, and can be searched through electronic or traditional card catalogs. Periodicals are also organized through either of the two classification systems and can be searched through a range of indexes and abstracts.

Although much library research covers books and periodicals, other types of publications can also be useful: government documents as well as a variety of types of reference books, such as dictionaries, encyclopedias, and handbooks.

In addition to searching through printed materials, most libraries provide electronic access to search for your source either online or through a CD-ROM system. Online and CD-ROM searching connects you to databases containing millions of sources of information. The other major way to research information electronically is through the Internet. Several additional tools help you access information from electronic libraries all over the world, including WorldWide Web, Mosaic, Gopher, Archie, and Veronica.

ACTIVITIES AND EXERCISES

1. The Du Pont documents (Appendix A, Case Documents 4) describe a product liability case related to the issue known as "failure to warn." Many product liability cases are the result of industrial and commercial accidents. The following lists areas in which such accidents have occurred.

   Industrial accidents

   manufacturing: steel, aluminum, auto, petroleum, aerospace, chemical

   construction: commercial, residential, industrial

   food service

   agriculture

   forest products

   power utilities

   Consumer product accidents

   automobiles

   motorcycles

   boats

   household appliances

   food

   medical care

   computers

   lawn, garden, and farm equipment

   pesticides/insecticides

   airlines

   Choose one of these areas (or find another one not listed) and conduct secondary research on accidents in that area:

   a. Find the citations for five books that include information about accidents in the area you choose. List the bibliographic information for each entry.
   b. Search the popular press indexes and list the references to five articles about two or more accidents within the area you've chosen. In doing so, identify not only the entry but also in what index or abstract you found the entry and whether or not the sources that contain the entries are available in your library.
   c. Search technical, scientific, business, and/or legal indexes/abstracts about one of the accidents in the area you've chosen. List five citations, the indexes or abstracts that you use, and the location and availability of the entries.
   d. If you have undertaken parts B or C and *have not used an electronic means of searching,* find different citations for five books or articles relating to accidents in the area you've chosen by using an electronic searching technique. List the five citations and write a brief step-by-step description of the electronic search that you (and the librarian?) used to find them.

2. Read over the Fact Sheet on AZT in Appendix A, Case Documents 2. Since the publication of that fact sheet, other treatments for people living with HIV/AIDS

have been discussed, tested, developed, and/or rejected. Conduct either a traditional periodical search or an electronic search to find the latest information on medical treatment for HIV/AIDS. Find at least two very recent articles on the disease and its treatment. Write a brief description of their content and how you found them for your instructor and classmates.

3. Bioremediation, as explained in Appendix A, Case Documents 3, was the main treatment used by the Exxon Corporation to clean up the oil spill in Prince William Sound, Alaska. Assume that you are an engineer working for a competing oil company and have been asked by your supervisor to gather information in three areas:

- the ultimate success of bioremediation in the *Exxon Valdez* spill area;
- other procedures and measures that Exxon used to supplement bioremediation; and
- procedures and measures used in other oil spills since the *Exxon Valdez* spill.

Using whatever means you and your librarian decide, gather enough information to write a one- to two-page memo to your supervisor. You will need at least three sources (make sure that they are the very latest) for each of the three areas assigned. Be sure to refer to your sources in your memo and list them at the end of the memo.

4. Choose a topic from your major that interests you, and complete the following:

   a. A library search for the latest information on that topic. Ultimately you want to find two or three relevant sources on the topic. Keep notes on your search procedures: where you started, where you went next, what indexes you used, how you chose keywords, and such. Then, in an oral presentation to your classmates, describe your library search from beginning to end. You might also include advice, warnings, and helpful hints about library searches in general based on your experience.

   b. An electronic search for the latest information on that topic. Follow the instructions and assignment as described in **a**.

5. With a partner from your class, search out information on the Internet on your campus. You might need to interview your computing services director. Find out if your campus has Internet (and if not, when they plan to acquire it), how to access it, to whom it is available, and such. Talk to one of the reference librarians on your campus as well and ask them how they are planning to deal (or already dealing) with electronic libraries. Prepare your findings in a brief memo to your instructor.

WORKS CITED

DeLoughry, T. J. "Software Designed to Offer Internet Users Easy Access to Documents and Graphics." *The Chronicle of Higher Education* 39, Issue 44, 1993, p. A23.

Krol, Ed. *The Whole Internet User's Guide and Catalog.* Sebastopol, CA: O'Reilly & Associates, 1992.

Laquey, T., and J. C. Ryer. *The Internet Companion: A Beginner's Guide to Global Networking.* Reading, MA: Addison-Wesley Publishing, 1993.

Library of Congress Classification Outline. 6th ed. Washington, DC: Library of Congress Office for Subject Cataloging Policy, 1990.

Library of Congress Subject Headings. 16th ed. Washington, DC: Library of Congress Office for Subject Cataloging Policy, 1993.

Primary Sources of Information: Surveys, Interviews, and Observations

The Interpretation of Information ∎ Primary Research Tools

What is research? For most, it is a search for knowledge, for understanding, for meaning—and, in some cases, it is a search for the ability to predict, and thus, control.

> Mary Beth Debs, "Reflexive and Reflective Tensions:
> Considering Research Methods from Writing-Related Fields,"
> *Writing in the Workplace: New Research Perspectives,* ed.
> Rachel Spilka (Carbondale: Southern Illinois University Press, 1993) p. 239.

Introduction

primary research
research methods used to create or discover new knowledge, such as interviews, questionnaires, observations, and experiments

In Chapter 7, we discussed secondary research—how to obtain information from existing sources. When you conduct secondary research, you uncover and interpret the results of others. This chapter covers **primary research,** the research methods designed to help you create or discover new knowledge. Before discussing primary research methods, let's review by way of examples the purposes of conducting research.

Your purpose may be as simple and direct as a need to learn about a new process or product, which would require you to find published reports on the topic and to interview experts. If your company wants to develop a multimedia training and documentation system, you might be charged with researching the products and recommending the best system for purchase. If your boss wants you to write a report for upper management explaining your company's research on artificial intelligence, you might have to find, organize, and interpret expert knowledge for a less knowledgeable audience. To do this, you would have to conduct interviews to learn about the status of artificial intelligence in your company. Or, if your company has developed a complex software system, you might be part of a team given the task of studying how easy or hard the system is to use. To determine the system's effectiveness, your research team might want to question users through survey questionnaires and interviews, simulate actual use and observe the product or process in action, or watch real customers using the system in their day-to-day work.

All of these activities are different types of information development, which generally combine secondary and primary research. The information you find through secondary research or create through primary research often provides you with the "good reasons" that you need to persuade your audience, as we discussed in Chapter 5. The research will help you describe the actual situations and propose the ideal situations that help you solve problems for your audience, as we discussed in Chapter 2. Although the methods of developing information are many, this chapter helps you become familiar with three primary research methods: surveys, interviews, and observations.

The Interpretation of Information

Regardless of your purpose or your method for developing information, you must realize that such activities are not neutral investigations of fact, particularly when you generate new knowledge through primary research. Your activities are much more likely to be interpretive actions based on your perceptions of fact. That is, the research you do is likely to lead to conclusions that are not absolutely verifiable facts, but instead are educated inferences based upon careful and disciplined interpretations of things you treat as facts. (We will deal with interpretation of information, discovery of "facts," and understanding of inferences in Chapter 9). Whenever you conduct an experiment, write a survey question, or interview someone, you are communicating from a particular cultural, political, or organizational point of view. Likewise, the responses you get from others through surveys, interviews, or observations are imbued with particular points of view. Therefore, you should always try to recognize your inherent biases and try to uncover the biases of others with whom you develop information.

You can see this challenge if you go back to the examples at the beginning of this chapter. If, in your research on the products and manufacturers of multimedia training and documentation systems, you interview experts, you will find that these experts have values and standards developed over time and through experience. Two experts may differ greatly on their professional opinion of which multimedia system is best—one expert may value the longevity of the system and one may value low cost. Or, when doing interviews for your report on the status of artificial intelligence in your company, you may be reminded that your division depends on funding from upper level management. You will need advice as to how to inform this audience about the basic research done within your company's research and development labs if some of that research has failed or gone over estimated cost or deadlines. Finally, in surveying users of your new software system, you might find that the responses of real customers may vary greatly depending on their cultures. For example, if your product is used in international sites by users for whom English is a second language, you may find that while the software is flawless, the accompanying documentation is perplexing to these users. Therefore, the "facts" you gather must be interpreted and understood within the social context of the workplace.

Primary Research Tools

The rest of this chapter presents several different types of primary research tools. Each section introduces specific research strategies that will help you develop information. Remember that most of the research tasks you are assigned will require a combination of primary and secondary research.

Throughout this section we refer to the continuing case we used in Chapter 7 to illustrate secondary research techniques. To refresh your memory, let's assume that you work for Biomed Corporation, a company that produces computer software for the healthcare industry. You have been asked to do research on your customers' knowledge and attitudes about a new software product, Med-Ease Software. This product is designed to computerize the patient-management system of hospitals. With Med-Ease, hospitals can admit, track, and release patients, monitoring their treatments, their medications, their billing, and their insurance claims. This is a large product, and Biomed has installed it at City Hospital in a pilot testing program

usability study
an investigation of how
well a product meets its
users' needs; conducted
on-site or in the laboratory

designed to verify the performance of the software. If the product works well for three months in City Hospital, it will be released to the market.

You have been asked to begin a **usability study** of Med-Ease Software. You are told that the first step will be to develop information about the users' responses to the new product: How easy is it to use? How well do they like it? Are there differences in attitude among different types of users?

SURVEYS

survey questionnaires
a primary research tool, often called a survey, that
asks sources to complete a
question-answer form; of-
ten conducted through the
mail or over the phone

If you need to find out what certain people know about a process, product, or concept, or how those people think or feel about those things, you may be able to develop this type of information through **survey questionnaires.**

To discover the answers to the questions you have asked in your Biomed-City Hospital task—How easy is Med-Ease Software to use? How well do users like it? Are there differences in attitude among different types of users?—you could employ several different research methods. For example, you could interview selected users personally, or you could observe users at work on the system. Both of these methods could yield interesting information that would help you and others improve your company's products—and later sections of this chapter explain both methods. But here we are going to assume that you want to get data from a population too large to interview. Although you may have time to interview or observe a handful of users, you need to find a way to question a large number of users—enough users so that their responses will be representative of all users at City Hospital. The best way to do this is to develop and administer a printed questionnaire to survey the users. Example 8–1 provides an overview of this task.

EXAMPLE 8–1   Survey Research for Biomed Corporation:
Med-Ease Software Use at City Hospital

Research Questions
How easy is Med-Ease to use?
How well do users like it?
Are there differences in attitude among different types of users?

Research Goal
To ask these questions of a representative sample of the entire population of users at City Hospital

Research Method
Survey questionnaire

Choosing a Design

The first step in developing your survey is to decide its general design: Will it be a **longitudinal survey** or a **cross-sectional survey?** A longitudinal survey allows you to ask questions of the same or similar group over a period of time. For example, college administrators who want to survey student attitudes may decide to give questionnaires to the class of 1999 in each of its four years at the college. If the administrators ask the same questions each year, their surveys will allow them to see trends in the attitudes of the students. The administrators will gain insight into the ways the student attitudes change over time.

When you use a cross-sectional survey, you ask a set of questions only once because you want to get a current picture of the phenomena under study. Imagine a cross-sectional survey to be a snapshot of the target group you are studying. For the City Hospital survey, you decide that you need to conduct a cross-sectional survey because your company needs to learn about the current state of Med-Ease and its use (Example 8–2).

longitudinal survey
a primary research tool that allows the researcher to ask questions of the same or similar group over a period of time

cross-sectional survey
a primary research tool that allows the researcher to ask questions of members of a heterogeneous group about a current phenomena

---

EXAMPLE 8–2   Survey Design for Study of Med-Ease Software Use at City Hospital

Design Options

   1. Longitudinal study of users from beginning to the end of their training
   2. Cross-sectional study of users near the end of their training

Choice

Cross-sectional study

Rationale

Because management wants to evaluate user attitudes and knowledge of the product and not the progress of their training, the one-time cross-sectional survey near the end of their experience was chosen

---

Choosing a Sample

Once you decide on the general design, you need to focus on the people you will survey. In some cases, you may be able to survey every single person that comprises the group you want to study. You may, for example, be able to survey everyone in the sophomore class of a small private college. But it would be impractical to try to survey all sophomores at a large state university. The former may include 250 students; the latter, 10,000! Clearly, at the state university, the best you could hope to do would be to survey a **representative sample** of the sophomore class (the **general population**). You might identify that representative sample so that it contains the same proportions

representative sample
a selection of some of the people involved in the subject or topic being researched

general population
all of the people involved in the subject or topic being researched

of female and male students, students age 18 to 25 and students over age 25, students living on campus and those living off campus, or any other identifying characteristics you feel essential to your task. Collecting statistics about a representative sample in the ways we describe in this chapter often allows you to generalize your findings to the entire population, for example, all the members of the sophomore class.

Furthermore, there may be times when you want to survey only certain subgroups. In one study, for example, you may wish to question all female employees over the age of 40 who have incomes over $40,000, 10 years or more of service with the company, and no more than a high school education. Just by limiting the type of person you wish to question, you limit the possible number of respondents. Whether or not you can survey every single person in the subgroup depends on the time you have, the amount of money you can spend producing and distributing the survey, and the availability of everyone in the subgroup.

Although it is not the purpose of this book to explain the subtleties of probability sampling, it is useful to note that you can achieve a representative sample by randomly choosing a relatively small group of respondents. For example, public opinion polls of presidential races are usually based upon a sample of 500 to 1000 respondents, chosen randomly from a variety of regions across the country. In other words, the opinions of only 1000 randomly chosen citizens can represent the opinions of all potential voters across the nation!

You can choose respondents randomly several ways. If you are conducting a survey of a relatively small population, the simplest way to get a random sample is to assign each potential respondent a number and then consult a random number table (which can be found in most statistics textbooks) or run a random number generation program on a computer. If, for example, the first random number is 57, then you choose person number 57 to be included in your sample. You continue this process until you have chosen the number of potential respondents you wish to have.

At City Hospital, you discover that although there are 1,500 employees, only 400 use Med-Ease in their work. Of those 400, 50 are managers who regularly receive reports generated by the system, but who do not actually use the system to do their jobs. The remaining 350 employees are spread evenly across five different departments at the hospital. You decide to choose 100 employees randomly from those five departments (see Example 8–3).

This does not mean, however, that you can necessarily count on 100 users responding to your survey. In large surveys, often only 10 percent of those who receive a questionnaire actually fill it out and return it. It is difficult to generalize with confidence when you get a small return rate, and you might need to encourage the people surveyed to respond by offering to send them the results of your survey or by stressing the importance of their feedback. In a closed population like City Hospital, you may achieve up to

---

EXAMPLE 8–3   Survey Sample for Study of Med-Ease Software Use at City Hospital

Total Population

350 nonmanagerial users evenly spread across five different departments

Sample

20 users from each department for a total of 100 potential respondents

Method Chosen

Random number generator to pick respondents from numbered list of users

---

75 percent response rate, however. Either way, you must choose your sample knowing that only a percentage will respond.

Finally, you must consider how you will distribute your survey to your sample. If you send questionnaires through the mail, you must anticipate the costs involved in sending out a large number of surveys. Even if there are no mailing costs, you must consider the time and effort involved in copying and preparing a large number of surveys.

Determining the Form of Questions and Answers

The questions you ask take a variety of forms. You can allow your respondents to write anything they wish to write (this is called an **open question**) or you can write questions that force respondents to answer in specific ways (these are called **closed questions**).

open questions
in a survey questionnaire, the questions that allow respondents to answer anything they wish

closed questions
in a survey questionnaire, the questions that force respondents to answer in specific ways

Open Questions   The *advantages* of using open questions are that they are easier to write, and, because you are not determining the answer, you may be able to get responses that accurately reflect the thinking of the respondents. Ask open-ended questions when you want to learn about things that you cannot yet imagine. For example, if you do not feel you can stipulate all of the problems that a respondent may perceive in the workplace, you may ask the following at the end of a closed-question survey:

> *Are there any other work-related problems that you would like to identify? Please explain in the space below.*

The answers you get to such a question may both surprise you and provide you insight that your closed questions would not have allowed.

The *disadvantages* of working with open questions are that:

1. It is time consuming to have to read the answers, and
2. The validity of those answers is dependent on how articulate the respondents can be.

The latter point can be crucial. If your respondents use vague, unclear language, you may not be able to verify their meanings. Accordingly, if a small minority of the respondents are quite articulate, then you run the risk of weighing more heavily the opinions of that minority without a clear indication of how the majority feels about the topic. However, you can do a *content analysis* of the responses to all open-ended questions and then *code* these responses to find out how many people responded in certain ways.

A content analysis is a particular research approach that allows you to determine the relative frequency of themes or actions in a collection of texts by counting the number of times they occur. Content analysis is often done to determine what's going on in the mass media, for instance, the amount of violence in children's cartoons. To do this kind of analysis, you must develop a coding system. Assuming you had already determined your sample—to look at cartoon violence—you would have to determine next what constituted violence (Were threats included? Did it only count when someone was hurt?) and how you were going to measure instances (Did a fight count as only one example, or did each time someone hit another count as an act of violence?). Content analysis works only when the code is clearly spelled out and two different people looking at the same example can get the same results.

In terms of the Med-Ease project, a content analysis of the open-ended questions might help develop a list of what people considered work-related problems with the product and how frequently these problems were identified. You might also be able to determine people's attitude toward the problems and product. Remember, that you will need to develop a code for determining how to classify responses so that your count is accurate.

Closed Questions   Closed questions can take many forms. Some of the most commonly used are *yes/no questions, check-offs,* and *scaled questions.* **Yes/no questions** allow the respondent to answer only yes or no to your question:

> Do you use the multitasking function when word processing?
>
> Yes   No

Check-off questions present the respondent with a list of options to choose from. You may ask a respondent to choose one or more items from a list by placing a check next to each choice:

> *Which of the following features have you used to do your work?*
>
> _____*Spreadsheet.*
> _____*Word processor.*
> _____*Database manager.*
> _____*Multitasking functions.*
> _____*CD ROM.*
> _____*Online help screens.*

yes/no questions
in a survey questionnaire, the closed questions that force respondents to give a yes or no answer

check-off questions
in a survey questionnaire, the closed questions that force respondents to choose from a list of options

Scaled questions ask respondents either to place their answers in a list of categories or to rate their response on a continuum of possibilities. There are three commonly used types of scaled questions:

scaled questions
in a survey questionnaire, the closed questions that force respondents to place their answers in a list of categories or rate their responses on a continuum of possibilities

1. *Nominal*

    When you ask respondents to identify a category, you are using a **nominal scale.** Nominal scales are best used when you want to identify your respondents in very specific ways. *Demographic information,* the vital statistics of general populations, can be easily collected with closed questions and nominal scales. Common nominal scale questions ask respondents to identify their sex, religious affiliation, level of education, and so forth.

    nominal scale
    scaled questions that ask respondents to identify a category

    What is your religious affiliation?

    > _____*Buddhist.*
    > _____*Catholic.*
    > _____*Islamic.*
    > _____*Jewish.*
    > _____*Protestant.*
    > _____*Other:* _____
    > _____*No religious affiliation.*

2. *Ordinal*

    When you ask respondents to place their answer in a rank order, you are using an **ordinal scale.** Ordinal scales are useful in identifying when respondents feel strongly one way or another about a response. In the example that follows you would probably find it useful to determine how many respondents are either highly motivated or poorly motivated. If a majority of your respondents were moderately motivated, you might consider motivation not an important characteristic among this population:

    ordinal scale
    scaled questions that ask respondents to place their answers in rank order

    When you consider your drive to succeed as a student, would you consider yourself:
    highly motivated   moderately motivated   poorly motivated

    or

    Identify the extent to which you agree or disagree with the following statements:
    I am motivated to do well in school.
    strongly agree   agree   no opinion   disagree   strongly disagree

3. *Interval*

    When you ask respondents to provide answers that are intrinsically numeric, such as age or income, you are using an **interval scale.** Using an interval scale allows you to place responses in categories when specific

    interval scale
    scaled questions that ask respondents to provide answers that are intrinsically numeric, such as age or income

statistics are not necessary or when respondents are reluctant to offer very specific information, as in the example below:

What is your income per year?

　　　_____*$10,000.*
　　　_____*$10,000 to $19,999.*
　　　_____*$20,000 to $29,999.*
　　　_____*$30,000 to $39,999.*
　　　_____*$40,000 to $49,999.*
　　　_____*Above $50,000.*

Another type of interval question asks respondents to count things that they do. Here respondents, who do not remember whether they save files three, four, or five times, can still give useful information:

How many times a day do you save the files that you are working on?

　　　_____*Zero.*
　　　_____*1 to 5.*
　　　_____*6 to 10.*
　　　_____*11 to 15.*
　　　_____*15 to 20.*
　　　_____*More than 20.*

What types of questions should you use for the City Hospital survey? To answer this, go back to the questions that are driving this research (see Example 8–1). You want to gauge the attitudes of the users toward the new system, and you want to identify different types of users; therefore, you decide to use a combination of nominal and ordinal questions. The nominal questions will enable you to establish a **profile of users** by asking them to identify factors including their age, years with the company, sex, education, and job function. The ordinal questions will give you insight into their attitudes. Finally, you decide to add an open-ended question at the end of the survey to enable respondents to bring up issues not addressed by the closed questions. Example 8–4 provides an overview of the question types you chose.

profile of users
the identity factors gained from nominal questions

Constructing Questions

Writing effective questions for surveys is the most important and often the most difficult part of the entire enterprise. Your goal is to write *valid* and *reliable* questions.

valid questions
survey questions that accurately ask what the researcher intends, without any misunderstanding between the researcher and the respondent

Validity and Reliability   **Valid questions** accurately ask what you intend them to ask. If you write a question that you believe asks one thing while the respondents believe the question asks something else, then you have not written a valid question. For example, if you asked respondents if the Med-Ease software made their job "easier," you imagine that the software

Nominal questions for developing user profile, for example:
Please circle the correct answer.

1. You are:

    male   female
    18–25 years old
    26–45 years old
    46 or older

2. Your highest level of education?

    high school graduate
    associate's degree in college
    bachelor's degree
    master's degree
    doctorate

Ordinal questions on attitudes, for example:

5. The menu system enables me to move to the functions that I need to do my job.

    strongly agree
    agree
    no opinion
    disagree
    strongly disagree

6. The help screens can be found quickly.

    strongly agree
    agree
    no opinion
    disagree
    strongly disagree

7. The help screens do not provide enough information to solve my problems.

    strongly agree
    agree
    no opinion
    disagree
    strongly disagree

Open-ended question:
Is there anything else we should know about the new Med-Ease system? Please write your response in the space below:

could enable your users to complete tasks quicker, but your respondents might interpret "easier" to mean completing the tasks with fewer mistakes. **Reliable questions** ask the same question every time. If you write a question that appears to change its meaning over time, then it is not reliable. For example, you would have asked unreliable questions if at the beginning of the survey, you asked how the Med-Ease software enabled users to complete "tasks" and meant data entry, but later in the survey used "task" to mean billing patients.

Although it is difficult to write perfectly valid, reliable questions, you can approach this goal by doing the following:

1. Use clear, concrete language and define terms. The more specific your language, the greater the chances that respondents will interpret a question as you hope they will. Note the vague language in the following example:

   Is the multitasking feature good?
   Always   Sometimes   Never

   Now look at a more specific version of the question. Note how the second question defines the vague term, *good.*

   Does the multitasking feature enable you to complete more tasks in a workday than you could without multitasking?
   Always   Sometimes   Never

   It is possible to write a question that appears to be completely clear to you but will not be easily understood by your respondents. For example, if you are working for a computer company, you may use computer terminology every day. However, when you attempt to survey customers, you may find that many of them are not familiar with those terms. In the examples above you are asking about *multitasking.* You know what that means and some of your potential respondents will know, but will *all* of them know? Maybe not. Instead of using the jargon, define the feature briefly:

   Would you prefer multitasking, a system that allows you to run more than one program at a time (e.g., makes printouts while you continue to use the spreadsheet program)?
   Always   Sometimes   Never

   As you can see, multitasking is not hard to define and by defining it you increase the validity of your question.

2. Recognize your biases and try to limit them. It does you no good to write a question that pushes respondents toward an answer that you may like them to give. If you want a valid response, you need to attempt to curb your biases. You would be pushing a respondent by asking "Wouldn't you agree that our system, with its multitasking capabilities,

is a step beyond the competitors?'' You will get a more valid response by asking

Does the multitasking feature allow you to get more work done in less time?
Always   Sometimes   Never

Yet, even this version casts your product in a positive light. Therefore, you may increase the validity of your survey by asking a similar question in the negative:

Does multitasking impede your work?
Always   Sometimes   Never

By including a balance of questions like these, you can check the validity of your responses. If respondents are answering consistently, that is, positively to one question *and* negatively to the other, then you can feel more confident in the validity of your questions. If, however, you are finding that your respondents are saying that multitasking allows them to get more work done in less time *and* that multitasking impedes their work, then you've got a validity problem. For example, your respondents may not know what ''impede'' means. You may have to revise your entire approach to the issue you are investigating.

3. Limit each question to one idea. Be sure to only ask one question at a time. For example, the following question is actually two questions:

Do the multitasking and multimedia features impede your work?
Always   Sometimes   Never

Respondents may not be able to answer this question if they believe that one feature impedes their work while the other doesn't.

4. Avoid untrue absolutes in questions and answers. Be sure that if you ask your respondents to choose an absolute (yes/no, always/never, and so forth) these absolutes are ''true'' to the given situation. For example, look again at one of the sample question/answer sets we analyzed above:

Does the multitasking feature enable you to complete more tasks in a workday than you could without multitasking?
Always   Sometimes   Never

Could your multitasking feature ever really enable users to complete more tasks in a workday? This question would be effective if rephrased to ask respondents how much they agree with statements.

The multitasking feature allows me to complete more tasks in a workday than I could before multitasking.
Strongly Agree   Agree   Neutral   Disagree   Strongly Disagree

Remembering these four guidelines will help you design reliable and valid questions.

self-administered questionnaire
printed survey form sent through the mail

telephone survey
survey method in which a caller asks respondents questions over the phone and records answers by marking an answer sheet

Presenting Questions

Once you begin to write questions, you also need to decide what form the questionnaire will take. The most common form is the printed, **self-administered questionnaire** that is sent to respondents through the mail. Another common form is the **telephone survey.** Using this technique, a caller asks the questions of the respondent and records the answers by marking an answer sheet. When you are developing a printed, self-administered questionnaire, you must pay attention to the format of the document. When administering the questionnaire by telephone, you must consider your verbal presentation as interviewer.

Self-administered Questionnaire Format   Because you want respondents to be motivated enough to fill out the questionnaire, you should attempt to make the document as brief as possible. Yet, at the same time you want it to be easy to read. Remember, format can affect the respondent's willingness to answer your questions. Will respondents be less likely to fill out the survey if they see that it extends for three, four, five, or more pages? Maybe. Will they be less likely to fill it out if it appears to be one page of small type and dense copy? Perhaps. Try to make the form both brief and readable.

Usually, you will need to provide brief but clear *instructions* at the top of the page. Sometimes you may also want to provide a brief *rationale* — an explanation of why the respondent is being asked to fill out the survey.

Finally, you must consider the order in which you present the questions on the survey sheet. Although there are no hard and fast rules on this, you may wish to present simpler questions first, followed by questions that may take more thought and reflection. The reasoning behind this pattern is simple: if respondents are hit immediately with challenging or difficult questions, they may be less inclined to complete the survey. But if you write a survey that is easy to start, then the respondents build a kind of momentum that can carry them through to the end of the questionnaire.

For the City Hospital survey, you decide to begin with a brief rationale, followed by simple instructions. The questions will begin with the profile questions, followed by the attitude questions, and end with the open-ended question (see Example 8–5).

Telephone Survey   An alternative to sending printed forms to respondents and hoping they send them back is for you to call potential respondents on the phone to conduct a survey interview. The advantage of this method is that you may achieve a higher response rate. It's much harder to hang up on a person conducting a telephone survey than it is to throw away a questionnaire. However, the disadvantage is that you may not be able to reach as many potential respondents as you could through the mail. Although it is feasible to send 250 survey forms through the mail, it may be far too

EXAMPLE 8–5   Self-Administered Survey Format Survey Format for Study of Med-Ease Software Use at City Hospital

Rationale

To enable Biomed Corporation to learn more about your attitudes toward the new Med-Ease system, please take a few moments to fill out this questionnaire.

Instructions

For each question, please circle the best answer.

Order of Questions

1.   Profile questions.
2.   Attitude questions.
3.   Open-ended question.

time-consuming to attempt to call 250 respondents on the phone. Of course, if you can develop a team of interviewers, a phone survey may be the way to go—if you can train all of the interviewers to conduct the survey in a similar fashion. Training all interviewers to ask the same questions, in the same way, and to record responses in the same way adds to the reliability and validity of your survey. Whether you do them all by yourself or you create a team of callers, you should abide by a few key guidelines:

1.   Make the calls at the most opportune time. If you are calling respondents at home, you don't want to catch them when they are busy. For example, if you are expecting that your respondents are at work during the day, then call in the evening—but you do not want to interrupt their evening meals nor do you want to disturb respondents with late-night phone calls. Or, if you are calling respondents at work, try to call early in the day when they are not in the middle of lengthy tasks or rushing to lunch or home.

2.   Identify yourself and your purpose immediately. Many potential respondents will be initially suspicious of your call. Many will wonder if you are selling something and be tempted to hang up quickly. However, given the opportunity, many people are quite willing to express their opinions publicly. Therefore, you need to have a crisp, brief opening that makes it clear that you are not selling but researching. If you are calling respondents at work, be sure to give them an accurate estimate of how long the survey will take and give them the opportunity to reschedule the survey for a more convenient time.

3.   Use a polite, business-like tone. A professional, conversational tone is the least threatening. When introducing yourself, your purpose, and the survey, do not read from a prewritten text. If your pitch sounds canned,

it comes off as less personal, and some people will be less likely to respond. At the same time you want to be sure to say the same thing to every respondent; therefore, you should have a prearranged introduction that you know well enough not to read. When you get to the questions, however, you will most likely need to read them. This is expected and acceptable to most people.

4. Make the questions as brief as possible and the answers easy to choose. Because the phone survey respondent cannot reread a question and the range of possible answers, you need to write questions and answers a listener can easily grasp. For example, because it may be too cumbersome to expect respondents to choose from five possible answers, such as strongly agree, agree, no opinion, disagree, strongly disagree, you may want to limit the possible answers to agree, disagree, and no opinion.

5. Decide how to handle alternate answers. Many respondents may not want to be forced into any limited answer scheme. They may want to explain their opinions in some complexity. You must decide beforehand how you plan to handle that type of response. Do you throw out such responses, not counting them as valid? Do you listen for something that fits your answer scheme and then say, for example, "So you are saying that you disagree?" hoping that the respondent will verify your interpretation? You can decide how you will handle this issue, but the key is to be consistent across every interview.

Determining the Results

pilot survey
a preliminary version of a survey sent to a small, representative sample (about three percent) to test the survey's usefulness and validity

In order to increase your chances of getting useful and valid results from your survey, you should first run a **pilot survey.** That is, once you have produced the questionnaire, you should try it out on a small, but representative, sample (perhaps about three percent of those you intend to survey) to test its performance before reproducing it and administering it on a large scale. A pilot study can help you refine your questions, whether they are in written or interview form.

Once you have piloted your survey, refined it, and administered it to the full sample, you will need to organize your results. With most closed-question surveys, the primary job is tabulating the quantifiable results. Simply put, you must count up the answers: how many, for example, answered "strongly agree" for a particular question, how many answered "agree," and so forth. Once you have counted all of the answers, you have tabulated the **raw scores** of the survey.

raw scores
the final count of all answers gathered during a survey

statistical analysis
a mathematical interpretation of the raw scores of a survey

Although the raw scores may tell you something useful or insightful, usually your goal is to develop some sort of **statistical analysis** of those raw scores. Such analyses can provide even more insight into the issues you are investigating. There are two types of statistical analyses that can be undertaken in survey research: *descriptive statistics* and *inferential statistics.*

Descriptive statistics provide manageable ways of presenting quantitative descriptions. There are a variety of ways to gather descriptive statistics depending on whether you want to describe single variables or the ways variables are connected. One of the most familiar kinds of descriptive statistics involves data reduction. If you sent out a survey with 100 questions to 200 people, you would have 20,000 answers with which to deal. **Data reduction** allows you to reduce mountains of data to manageable summaries.

Inferential statistics allow you to handle data in a different way. Often the research you do will involve a sample drawn from a larger population. You aren't particularly interested in the sample, but you do want to know what the sample tells you about that larger population. A number of statistical techniques allow you to make such inferences. These techniques allow you to determine the **statistical significance** of your findings. Any type of statistical analysis can become quite complicated, and inferential statistics are particularly complex. In this chapter, we discuss only the simplest descriptive statistic: **frequency.**

Frequency is merely the percentage of responses to a given question. If, for example, 20 out of 50 respondents answer "strongly agree" to a particular statement and another 20 answer "agree," that means that 80 percent of the respondents agree or strongly agree with the statement.

See Example 8–6 for some of the raw scores from the survey at City Hospital. To practice tabulating other descriptive statistics, try Exercise 1 at the end of this chapter.

INTERVIEWS

Interviews are the most commonly used research method in the workplace. From informal conversations around the coffee maker to formal depositions in the presence of lawyers, more people gain information through interviews than through any other research activity. Oftentimes in the workplace, you will gather information from other people quickly and informally, through a short phone call or over electronic mail. However, in this section of the chapter, we introduce you to the most challenging situation—formal interview strategies. Some of these strategies will prove useful in more informal situations as well. Interviews differ from surveys in that the primary research instrument is not a questionnaire but the interviewer. Therefore, interviews are primarily a function of interpersonal communication.

Although both interviews and surveys are based upon question and answer strategies, there are a couple of key differences between the methods. First, surveys are designed to be delivered to a large number of potential respondents, whereas the time demands of interviews usually allow for far fewer respondents. Second, most surveys employ closed questions, thus limiting the respondents' possible answers. On the other hand, interviews are ideal for probing follow-up questions, allowing for lengthy, possibly

---

descriptive statistics
a type of statistical analysis used to describe single variables or connections between variables

data reduction
a type of descriptive statistical analysis used to reduce raw scores to manageable summaries

inferential statistics
a type of statistical analysis used to determine what a sample indicates about a larger population

statistical significance
the result of inferential or descriptive statistical analysis

frequency
the percentage of responses to a given question; the simplest descriptive statistic

interviews
a primary research tool, the most commonly used in the workplace, where the research instrument is the interviewer

EXAMPLE 8–6   Raw Scores: Selected Survey Results of Study of Med-Ease Software Use at City Hospital

Total number of respondents: 72 (out of 100 surveyed)
Please circle the correct answer.

1. You are:

    male = 24   female = 48
    18–25 years old = 15
    26–45 years old = 20
    46 or older = 37

2. Your highest level of education?

    high school graduate = 56
    associate's degree in college = 10
    bachelor's degree = 5
    master's degree = 0
    doctorate = 1

Ordinal questions on attitudes

5. The menu system enables me to move to the functions that I need to do my job.

    strongly agree = 12
    agree = 10
    no opinion = 5
    disagree = 40
    strongly disagree = 5

6. The help screens can be found quickly.

    strongly agree = 2
    agree = 8
    no opinion = 5
    disagree = 50
    strongly disagree = 7

7. The help screens do not provide enough information to solve my problems.

    strongly agree = 2
    agree = 10
    no opinion = 3
    disagree = 25
    strongly disagree = 32

unanticipated answers. Interviewees, in other words, are often allowed far more freedom in constructing their answers and therefore may provide unexpected insights. And because interviews enable you to begin investigations before you know the parameters of the issue, they are often employed before constructing survey questionnaires in which the issues and the answers are limited by the researcher.

This is not to say that all interviews are completely unstructured; in fact, few are. Generally, there are two types of interviews: *open* and *directed.* **Open interviews,** in theory, are completely unstructured, meaning that the interviewer imposes no limits on an interview. In reality, an open interview is based upon one or more starter questions that give the interview a general focus. Then the interviewer allows the interview to flow according to the responses of the interviewee.

Directed interviews are the counterpart to self-administered survey questionnaires. The interviewer brings a set of specific questions and attempts to stick to those questions regardless of the responses of the interviewees. More often than not, the best form of interview *combines* features of the open and directed interviews. These interviews impose a structure on the interviewee but allow for the interview to move into unanticipated but relevant areas. This process attempts to use time efficiently while allowing for unexpected discoveries.

For example, in the City Hospital investigation, you may decide that you need to supplement your survey with several in-depth, directed interviews with users in each department. These interviews would focus on the users' attitudes about the new system but still allow them to raise issues through the one open-ended question at the end of the survey. Because an interview allows for follow-up questions that can probe a particular issue, you may be able to investigate key points in great depth. When you combine this depth with the breadth of issues you may cover with the survey, you will gain a complex view of the attitudes you are investigating. Example 8–7 provides an overview of your interview goals at City Hospital.

Before actually sitting down with your interviewees, you need to prepare by scheduling the interview, determining the site, choosing a recording method, and writing your questions.

Scheduling

If you are conducting interviews in a workplace setting, usually the interview will interfere with the normal work patterns of the interviewees. Generally, you should make it clear to the interviewee that you would like to schedule the interview at the most convenient time for him or her. Because this request may be your first contact with the potential interviewee, you want to set a tone that encourages cooperation.

At the same time, you want to schedule a time when the interviewee is reasonably rested and alert. Interviews right before the workday ends may

open interviews
unstructured interviews in which the interviewer imposes no limits

directed interviews
structured interviews in which the interviewer asks a set of specific questions; counterpart to the self-administered survey questionnaire

---

EXAMPLE 8–7   Interview Research for Biomed Corporation:
Med-Ease Software Use at City Hospital

Research Questions

How easy is Med-Ease to use?
How well do users like it?

Research Goal

To explore in-depth the attitudes of two users in each of the five departments

Research Method

Interviews

---

not allow you to get the best from the interviewee, who may be watching the clock and may be reticent to provide in-depth information because of the time constraint. Typically, a time earlier in a person's workday is better than later.

Be sure to inform the interviewee of the expected length of the interview. Interviews that last between 15 and 45 minutes are typical. Any interview over an hour is pushing both your endurance and that of the interviewee. You both need to remain alert and focused throughout the interview.

Determining the Site

The ideal site for an interview is both convenient for the interviewee and reasonably free from distractions or interruptions. If possible, do some leg-work beforehand to inspect the best interview site. The worst thing that could happen is that you agree to a site—for example, the interviewee's workplace—where the interviewee is continually interrupted with telephone calls and visits by co-workers. You could conduct the interview in a conference room or even in a quiet nearby coffee shop.

Choosing a Recording Method

Always ask the interviewee beforehand whether or not you can use an electronic recording device. Typically, interviews are tape recorded. Occasionally, they may be videotaped, although it is the rare interview that demands a visual record. Furthermore, video cameras can be very intimidating to an interviewee, so you should generally avoid this technology unless there is a compelling reason to use it.

If tape recording is allowed, be sure that you have a reliable power source, extension cords if necessary, and more than enough tapes. Arrange the microphone and tape recorder so that it can clearly record your voices, but do not place it so prominently that it may intimidate the interviewee. Few people like to have a microphone stuck in their faces.

It's always a good idea to take notes even though you are recording an interview; your recorder could malfunction, or you may not have the time and resources to transcribe your tapes completely. If your interviewee does not want to be tape recorded, then you must rely on your notes. In such a case, knowing shorthand is helpful. It is also a good idea to have your questions written on a sheet or notepad that is separate from the pad you are writing on. The less flipping of pages you have to do the better.

Writing Questions

Most interview questions will be designed to allow the interviewee to speak at length. (If you only need one word, yes/no type answers for the entire interview, then you may be better off sending a questionnaire to the interviewee instead.) Therefore, most interview questions will be open-ended to some extent. For example, if at City Hospital you wanted to ask employee Jane Smith about her general attitude toward the new Med-Ease software, you want to avoid a question like, "Do you like the new software?" This merely asks someone to answer yes, no, or maybe. You open up a potentially more fruitful channel by asking, "What do you like about the new software? What do you dislike about the new software?" The interviewee's response may give you several issues to probe in more depth.

Often, you will go into an interview not with specifically written out questions, but instead with a list of topics that you want to explore. Your goal is to ask questions about those topics in such a way that encourages the interviewee to speak in some detail. Then you can move the direction of the interview by asking specific follow-up questions.

For example, at City Hospital you decide to segment your interview into sections that correspond to the major components of the Med-Ease software system. For each of those segments, you may begin with a question such as, "What do you like about the insurance claims manager system? What do you dislike about the insurance claims manager system?" and follow-up with more specific questions based on the interviewee's initial response (see Example 8–8).

Finally, you may want to pre-plan a sequence of questions or topics so that the most difficult or most controversial comes either during the middle of the interview or near the end. Often, you will want to build up to the difficult part of the interview. Once you get an interviewee in a reflective pattern of thinking, you may be able to probe the tough issues better than if you hit the interviewees with the tough questions right at the start.

At City Hospital you anticipate such a situation. You know that one component of the software system has been plagued by bugs. You suspect that has frustrated many of the workers; therefore, you decide to ask about

---

EXAMPLE 8–8   Plans for Interviews on Med-Ease Software Use at City
                  Hospital

Schedule

Most interviews will be held from 10:00 to 10:45 A.M.

Site

The interviews will be held in private conference rooms within each department's
offices.

Recording Methods

All interviews will be tape recorded.

Questions

The questions will focus on five topics, each of which asks users to discuss their likes
and dislikes regarding a different part of the Med-Ease software system.

---

that section last. By that time, you will have explored the other components
that you assume are working without errors, and you may have enabled the
interviewee to express opinions about the other four components within a
less emotional context. Of course, such a plan could backfire. It may be that
you will find so much hostility that the interviewee will want to begin
complaining right from the start. There is little you can do in such a situation
but try to direct the interviewee's anger into answers that are as informative
and constructive for you as possible.

Conducting the Interview

Regardless of how open or directed you make them, all interviews should
have a simple structure: a beginning, middle, and end. As the interview
begins, provide a brief overview of your purpose and your plan so that the
interviewee knows what to expect. The middle of the interview comprises
the sequence of questions and topics that you wish to cover. At the end of
the interview, ask the interviewee if there is anything he or she wishes to
add before finishing. After that, you might restate your rationale, indicating
what will be done with the information gathered in the interview. Finally,
you may wish to set up a follow-up interview, if necessary.

   You must recognize that throughout the interview you are engaged in
a **dyadic** (two-person) exchange that involves power and status to some
extent. As the interviewer, you have one form of power: you ask questions.
If you also have hierarchical status—that is, if you outrank the interviewee
in the organizational hierarchy—then you possess even more of the power
in the dyad. Sometimes, of course, you may find yourself in the exact opposite
situation. Either way, a dyad that is unbalanced can negatively influence the

dyadic
a two-person exchange of
information, as in an in-
terview

outcome of the interview. The key is the degree of respect that you show the other. If you act in a professional and respectful manner—never trying to intimidate, always displaying professional respect for the other—then you increase your chances of acquiring valid information during the interview.

Clearly, one of the ways you give and gain respect is through your verbal communication—the way you describe your purpose, present your questions, and delve into issues in follow-up questions. But another important consideration is your **nonverbal communication:** Have you dressed appropriately for the interview? Do you present an alert but relaxed posture (not so alert as to make the other nervous but not so relaxed as to make the other wonder whether you are paying attention)? Do you show genuine interest in the words of the interviewees by looking at them when they speak, by listening to them carefully, by providing appropriate nonverbal gestures (nodding, smiling, etc.)? Never forget that in interviews you embody the research instrument. Your entire performance as interviewer, not just the questions you ask, is the means by which you gather information.

One of the key elements during your performance will be the degree of control you exert during the interview. At one end of the control spectrum there is the unstructured, open, informal interview in which you exert almost no control. At the opposite end is the self-administered questionnaire containing nothing but closed questions. Usually, interviews fall somewhere in the middle of this spectrum: you want to cover certain topics, but you want to explore aspects of those topics that you cannot predict. The key here is the balance between keeping the interviewee on track but still allowing for the unexpected.

One way that you can maintain this balance is to allow interviewees to wander a bit, then pull them back toward the topic at hand. Although some of those wanderings may prove fruitful, some may not. Therefore, you need always to be ready to refocus the interviewee on the topic at hand when you sense the interview meandering into irrelevancy. Of course, this type of problem may only occur when you are interviewing a talkative person. Sometimes you will encounter someone who is reticent, a person who, for whatever reason, does not say much of anything. This type of person usually poses the toughest situation for the interviewer. How do you prompt reticent interviewees to speak more without determining what they will say (that is, without putting words in their mouths)?

Two simple techniques may help you out in this situation. First, don't be afraid to use silence. If a person provides too brief an answer, don't quickly jump in with another question. Wait a moment. Keep looking at the person as if you are still expecting him or her to say more. Silence in a dyad usually sends the message that the silent one expects the other to speak. However, don't let silence lapse into a standoff; you may need to go on to the next question and then circle back to the question that provoked the silence.

nonverbal communication

the communication signals and cues conveyed by features other than words themselves, such as dress, facial and hand gestures, posture, and volume and tone of voice

A second technique is for you to restate a couple of key words that the interviewee just used in such a way as to indicate that you expect some further elaboration on those words. For example, read the interview excerpt presented in Example 8–9. Notice that in this situation, the interviewer has not pushed the user to say anything even though the first response indicated no problems and the last response began to pinpoint some problems. Don't overuse this technique, as it can become tedious to the interviewee, but restating key words can be quite useful when you know the interviewee has some valuable information to offer.

Ultimately, this is the balance you seek in any interview: you want to prompt people to tell you what they think without determining what they think. Exercise 2 at the end of this chapter will allow you to test your ability to respond to a variety of interviewees, from the reticent to the hostile.

OBSERVATIONS

observations
a primary research tool that gathers information about people and their activities as they go about their jobs or about products, devices, and environments

When you interview people or ask them to fill out a questionnaire, you are asking them to step out of their natural environment for a few moments and reflect on a given topic. Although these methods are often fruitful, there are times when you may want a different kind of information. Instead of asking people to tell you about their thoughts and actions, sometimes you may want to watch them act and listen to them talk as they go about their jobs. You can do this in one of two ways: you can do observations of *natural work* or you can make use of *simulated work* and observe those simulations.

Moreover, at times you need to observe *things:* bridges, roads, buildings, natural events, equipment, lab experiments, and so on. Often these things are found in a site or location that you must visit. Of course, people often construct, operate, or control these things. Most of the strategies we discuss next can be used to observe people and things.

Natural Work

natural work
what happens naturally in a group, at a site, or on the job

One way to observe is for you actually to enter a group of people and try to watch what happens naturally in that group, or to visit a site and observe that environment and facilities. For example, we can learn about a foreign society by asking the members of that society to tell us about it; however, we can also learn about that society by visiting it, living in it, and recording what we see and hear. Anthropologists often do the latter. They study different cultures by observing those cultures first hand. Margaret Mead lived in Samoa to learn about Samoan culture. Other anthropologists have entered and observed a variety of cultures from remote aboriginal societies to suburban and inner-city communities.

Just as an anthropologist studies a community, so too may you need to study the actions of others in a working situation to understand better how

---

EXAMPLE 8–9   Prompting Responses in an Interview

Excerpt: Interview at City Hospital

Interviewer: Can you tell what it is that you like about the spreadsheet component of Med-Ease?

User: Hmmm. Well . . . Not sure, really. Guess I like it all right. (Silence.) I can use it all right once I get into it.

Interviewer: Get into it?

User: Yeah, you know, once I figure out which commands get me into what I want to do.

Interviewer: Once you figure out what . . . (voice trails off)

User: Yeah, figure out. Actually, I always forget which is the first key to hit to get the menu to appear at the top of the screen. Then I can never quite remember if I want to first go into the Workbook or into File or one of the other sections. But once I figure that out, I'm OK.

---

they communicate, why they communicate, what the barriers to communication in that community are, and what communication strategies succeed with members of that community. You may want to observe experts performing a task so that you can learn the task and, in turn, communicate it to others who wish to learn the task. You may want to observe how people use something you have designed so that you can determine the worth of that design. Or, you may travel to a site to observe, measure, and access the facilities in which products are made.

In your research at City Hospital, you want to understand better how the first users of the new software system feel about the new system. So far, you have surveyed and interviewed users of Med-Ease software, but you haven't actually watched users in action. If you add that kind of information to the results of your surveys and interviews, you may develop a well-rounded perception of the attitudes and abilities of the users and the strengths and weaknesses of your company's product.

Of course, such research has its disadvantages. First, you must be careful not to assume that the few people, the small amount of work, or the specific site you are observing represents all of the people, all of the work, or all the sites in a given population. Because you are entering someone else's workplace and watching him or her work, you cannot set the agenda. For example, if you are able to observe several City Hospital employees using the software as they do their jobs, you may see only a small facet of the kind of work that is done every day. Or, if you observe specific facilities or locations, you may not be able to generalize to all such facilities or locations. For example, if you observe the transportation flow at one intersection during rush hour, you may not be able to assume that the same number of cars will pass through the next intersection at the same rate. Or, even though one bridge is unsafe because of eroding soil and frequent flooding, other bridges may not be affected in the same way.

Second, as an outsider, your presence may disrupt the normal activities that you want to observe. You may stifle or distort the words and actions of those under observation.

Third, you may misinterpret what you see. For example, a lab experiment might seem to go quite well during your observation, but the final product may deteriorate after you leave.

simulated work
work that is recreated within a laboratory setting

Simulated Work

What can you do when you know that you need to observe people at work but it is impossible to gain entrance into their workplace? What can you do when you need to predict the production rate of a future product or the durability of a certain material or device? You may be able to simulate the work that people do, the steps in production, or the material or device in use. For example, in the early 1980s a group of researchers from IBM Corporation, headed by John Carroll, wanted to study how people in offices learned how to use word processing programs. At the time they could not actually go and visit real offices where employees were learning how to use computers, so they constructed a simulated office, complete with all of the noises and interruptions that are typical in most offices. In this situation, Carroll and his team were able to observe and study people working.

You may have already studied or even created computer simulations to test the feasibility of creating and using potential products. Manufacturers, consumer interest groups, and even magazines such as *Consumer Reports* test the durability and safety of everything from cars to electric can openers, often through simulation.

The disadvantage of this approach is that the work the participants are doing isn't their real work; instead, it is work that the researchers *envision* real people doing. Simulating the manufacturer of a new product may provide you with an idea about the materials needed and the timing of each stage, but the simulation may not indicate the time needed to train employees. There may be significant differences between the way people work in a simulation and the way they work in their natural environments. Furthermore, setting up realistic simulations can be expensive and time-consuming. In the following section we suggest ways to conduct effective observations as a means of developing information.

Preparing to Observe

Once you have decided that you'd like to observe some phenomenon, you need to undertake four general steps:

1. Decide specifically what activities or facilities you want to observe. For example, at City Hospital, you have determined that you need to watch

several individuals in each department using the Med-Ease system. You would like to observe some or all of the participants using every component of the system, but you cannot stipulate that beforehand. What if no one ever uses, say, the spreadsheet component? To force someone to use it would distort the data that you seek to collect. After all, you want to study actual use and actual attitudes.

2. Determining if those activities or facilities are observable in a natural setting. Sometimes you know what you'd like to observe, but, for whatever reasons, it is not feasible actually to do the observations. At City Hospital, you realize that all of the work you want to observe is conducted openly and regularly and, therefore, it is quite feasible to observe work as it is being done.

3. Request entry into the site where you want to observe. Never assume that others will automatically understand your need to enter their worlds and observe them or their employees doing their jobs. On the contrary, most people see observers as intruders and disrupters, or worse, spies! When you request entry, you want to be prepared to explain in detail your purpose and your methods. Make it clear that you will try to be as unobtrusive as possible. On the other hand, don't over-promise. The participants should realize that you will be somewhat intrusive by your very presence. Also, remember that most people will associate observation with evaluation. If you are not there to evaluate someone's performance, then by all means make that clear to those you observe. Your presence in that case may seem less intimidating and less of an imposition.

4. Choose one or more data recording methods. When you negotiate your entry, you should discuss how you will record what you see and hear. In some cases, you will not be allowed to bring any external recording device to the site, other than your memory. In such cases, be prepared to record, either in writing or on tape, what you saw and heard immediately after leaving the site.

In most situations you will be allowed, at least, to write observations, known as **field notes.** One commonly used style of field notes stipulates three different types of notes:

1. **Observational notes** are hastily scribbled descriptions of what you see going on as you watch.
2. **Theoretical notes** are statements that express your opinion about or interpretation of what you see.
3. **Methodological notes** are reminders to yourself to do some further research work, to ask a certain question tomorrow, or to remember to talk again to a certain person (Shatzman and Strauss).

field notes
written record of observations

observational notes
written record of what an observer sees

theoretical notes
written record of an observer's opinions or interpretations

methodological notes
written record of an observer's reminders about additional research needed

You need to distinguish among these three types of notes because you want to try to separate a description of what you see (an observational note) from an analysis of what you think it means (a theoretical note) from a statement about your future actions in the setting (a methodological note).

In many situations, you may be allowed to tape record conversations just as you would tape record an interview. Or, you may find a situation that will allow for videotaping the activity under study. In the latter case, you must realize, of course, that this technology may be the most intimidating recording technique of all. In any case, you want to try to ensure that the presence of any means of recording distorts the study only minimally.

At City Hospital, you are granted entry to selected offices for a limited time on several different days. You arrange to be allowed to write field notes and tape record any discussions that go on while you are observing. You obtain a battery-powered, voice-activated, mini-tape recorder so that you can easily record whatever is said in your presence, with the permission of the City Hospital and its employees, of course (see Example 8–10).

Observing

When you are in the field or on site, be sure to do the following:

1. Make your agenda known. As we noted above, you can alleviate tension and suspicion about your presence by making your purpose clear. Generally, you observe work because you don't have a clear understanding of how it is done. When you tell the participants that you are there to learn from them, often they will accept your presence and cooperate freely. If you suddenly appear unannounced and unexplained and begin furtively taking notes as you watch them, you will obviously arouse anxiety and hostility.

2. Choose effective vantage points. Try to position yourself in places where you can observe freely without getting in the way of the normal activities. In offices this is often easy to do. In an industrial site, when you are observing to learn a new process, you must be careful not to obstruct activity because you want to get close and watch a particular action.

3. Capture as much as you can. First-time observers often worry that they cannot capture everything that is happening in a setting. Don't worry. You can never capture the full complexity of any activity, even if you are videotaping it. All you can do is try to record as much as you can and assume that if you miss something important, it will happen again in some form. If it is important, you are usually afforded more than one opportunity to record it.

Sometimes, just watching does not enable you to learn what you need to learn. You need more information: often you need the participants to explain what it is they are doing and thinking; however, in the course of

---

EXAMPLE 8–10   Observation Plan: Observing Med-Ease Software Use at
City Hospital

Activity to Be Observed

Use of the Med-Ease system

Participants

Two employees from each of the five departments

Length of Observations

One hour in the morning and one in the afternoon for each employee

Recording Methods

Field notes and tape recordings

---

their normal work, they do not typically articulate what they are thinking. Therefore, if you only watch them work, you will never capture their thoughts and explanations.

How can you break into the normally silent thoughts of those whom you observe? You do this when you ask someone to fill out a questionnaire or when you ask someone a question in a formal interview. But how can you do this while observing people working in their natural settings? One method, the **think-aloud protocol,** enables you to achieve this goal but requires that you intrude quite dramatically in the normal activities of the participants. It enables you to get a glimpse of individuals' thought processes as they work.

> **think-aloud protocol**
> an observation tool in which the person observed states continual thoughts while performing a task

In order to explore a person's thought process, you can ask the person to state continually what is on his or her mind while reading and/or performing a task. But because speaking aloud while working is quite difficult, you should choose a limited task, something that a person could complete within a half hour, and tape record the person's running commentary. When the participant falls silent, a few prompts, such as, "What are you doing now?" should get the person talking again. But be careful that you don't interfere when the person is really concentrating (for example, to avoid a dangerous situation, such as electrocution). At City Hospital you may want to have an employee think aloud while sitting at a computer terminal and completing several tasks. Your goal will be to gain some insight into how the user perceives the task and the usefulness of the software.

Be Careful Out There!

The suggestions in this chapter serve as an introduction to primary research. Only after much practice can you or anyone become a skilled researcher. So, therefore, be very careful when you undertake any primary investigations.

Remember, primary investigations of people, processes, and products are fraught with an array of difficulties. In particular, if you are involved in research into the lives and habits of others, never forget that you must always treat each person you are studying with care and respect. If your interviewee or participant requests anonymity or confidentiality, you must respect that request. If you want people to help you, you must value them.

SUMMARY

Many technical professionals are likely to conduct research in order to develop the information needed to carry out their jobs. As we saw in Chapter 7, often this research starts with or is supplemented by secondary research, which uncovers and interprets the research of others. Primary research helps you create or discover new knowledge and includes the techniques discussed in this chapter: surveys, interviews, and observations.

Surveys are questionnaires designed to ask the same questions to a large number of potential respondents. Longitudinal surveys allow you to ask questions of the same or similar groups over a period of time, while cross-sectional surveys allow you to get a current picture of the phenomena under study. If you cannot question everyone involved, you should choose a representative sample. Often a balance of open and closed questions will provide you with excellent information, if your questions are valid and reliable. Surveys can be presented to the respondents through self-administered forms or over the phone.

More people gain information through the face-to-face question and answer sessions, or interviews, than through any other research activity. Interviews differ from surveys in that the primary research instrument is not a questionnaire but the interviewer. Therefore, interviews are primarily a function of interpersonal communication. Interviews can be open, directed, or a combination of both structures. Planning a successful interview demands careful scheduling, determining the site, choosing a recording method, writing questions, and, finally, conducting the interview, with careful attention to verbal and nonverbal communication and respect for the interviewee.

Instead of surveying or interviewing, sometimes you must find your information by observing people engaged in a particular task or facilities used at particular sites. Such observations can be natural or simulated. Your field notes should consist of observational notes, theoretical notes, and methodological notes.

Again, whenever you conduct any type of research, you are communicating from a particular cultural, political, or organizational point of view. Likewise, the responses that you get from others are also imbued with particular points of view. Your goal in doing primary research is not only to gather information, but also, as we will discuss in more detail in the next chapter, to try to recognize your inherent biases and try to uncover the biases of others with whom you develop information.

ACTIVITIES AND EXERCISES

1. Consult a statistical textbook or one of the books on survey design listed at the end of this chapter and calculate the following types of descriptive statistics: mean, median, and mode. Using the raw data in Example 8–6, compute the

frequency, mean, median, and mode of each item. In a memo to your classmates and instructor, discuss when it is most appropriate to use each.

2. Write follow-up questions to provoke more information or to help the interviewee focus for each of the following interview situations. (Note the interviewer is *Q*, the interviewee is *A*.) Your instructor may ask you to role-play this activity.

    ### Situation 1
    Q: Have you ever used the spreadsheet function of the software?
    A: You know, when I use the word processor, I never know how to print when I'm done. I mean, I have looked it up in the manual but I always have trouble finding it. That's another thing. Those manuals! Who wrote them? I can never find anything in them and then when I do I can't understand it! Anyway, I don't print out documents all that often so when I do I have forgotten how to do it and even if I remember the first print command, then I get completely confused if it shows me the screen with all of the printing choices. You know, which printer am I using, and all that.

    ### Situation 2
    Q: Do you find anything difficult about the word processing program?
    A: Well, no, not really.
    Q: No?
    A: No.

    ### Situation 3
    Q: Would you say that you are satisfied or unsatisfied with the spreadsheet program?
    A: Oh, well, I'd say that I am very satisfied with it. I have used it extensively and I would say that there is very little bad about it, other than the fact that I have trouble finding the correct embedded commands in the menus. Several times I have erased data that I had entered because I thought that I was moving data from one column to the next when actually I must have been deleting data. Stuff like that has happened to me several times. Lost a bunch of work, too. But, no. . . . You know, everyone makes mistakes in this business. So, I'm pretty satisfied with the spreadsheet.

    ### Situation 4
    Q: I want to thank you for taking this time to meet with me. I don't expect this interview to take long, so if we could just. . .
    A: You know, I am sick and tired of management sending their lackeys down here to snoop on us and then giving us grief about not getting our work done! You want to come in here in the middle of the day to ask me questions and then just walk out of here while I get stuck working over my lunch break because my boss wants to know why I haven't gotten my morning's work done. And because your boss told my boss that I had to talk with you. Let me tell you something—with all due respect, I don't feel like giving up my lunch break for you!

3. Discover a problem that involves technology on your campus. For example, first-year students might have difficulty learning a new software package in a computer lab, the telephone system in the dormitories might be expensive or inadequate, the air conditioning or heating system in the library might fail frequently, or the

computer program used to schedule rooms and classes might not accommodate both student and instructor preferences. Use your own experience or that of your classmates to discover such problems. An open discussion for a few moments in your class will probably generate a number of ideas.

Then, form a team of three or four students from your class and prepare to investigate one such technical problem on your campus. Choose one primary research method discussed in this chapter that will best help your team understand the nature of the problem (at this point, don't worry about the solution). Design your research method using the advice in this chapter, and describe in a brief memo how you would administer, conduct, or distribute that method or tool. For example, if you decide to conduct an interview, prepare not only your questions but also a description of the place, the recording method, and the people whom you would interview. If you design a questionnaire, don't just write your questions, but also determine your sample. If you wish to observe, decide not only what you are looking for but also whether that situation should be natural or simulated and what types of notes you will need to take.

4. Look at the Aetna Casualty and Surety Company v. Jeppesen & Company case in the appendix (Appendix A, Case Documents 4). This case involved the death of an airplane crew and all passengers as the result of errors in the graphic depiction of the airport instrument approach chart. Assume that you work for Jeppesen Company and you have been given the task of interviewing the people in the communication department to find out what went wrong in the development and editing of the charts. Your goal is not to find a guilty party who can be blamed but to determine the editorial processes that allowed this error to occur. The company wishes to make sure this problem never occurs again.

You realize that the people you will be interviewing will be defensive and nervous. Develop and write out what you will say in your initial communication with the department members in which you prepare them for your interview. Also, write out what questions you will begin your interview with and which questions you will use to try to pinpoint the place where the editorial process failed.

WORKS CITED

Carroll, John M.; Robert L. Mack; Clayton H. Lewis; Nancy L. Grischkowsky; and Scott R. Robertson. "Exploring a Wordprocessor." in *Effective Documentation: What We Have Learned from Research,* ed. Stephen Doheny-Farina. Cambridge: MIT Press, 1988, pp. 103–26.

Carroll, John M., Penny L. Smith-Kerker; Jim R. Ford; and Sandra A. Mazur-Rimetz. "The Minimal Manual." *Effective Documentation: What We Have Learned from Research,* ed. Stephen Doheny-Farina. Cambridge: MIT Press, 1988, pp. 73–102.

Debs, Mary Beth. "Reflexive and Reflective Tensions: Considering Research Methods from Writing-Related Fields." In *Writing in the Workplace: New Research Perspectives,* ed. Rachel Spilka. Carbondale IL: Southern Illinois University Press, 1993, pp. 238–52.

Shatzman, L., and Anselm Strauss. *Field Research.* Englewood Cliffs, NJ: Prentice Hall, 1973.

Information Management: Evaluating, Organizing, Summarizing, and Documenting

Evaluating Information ▌ Organizing Information ▌ Documenting
Information: Crediting Sources

Reading the world always precedes reading the word, and reading the word implies
continually reading the world. . . . In a way, however, we can go further and say
that reading the word is not preceded merely by reading the world, but by a certain
form of *writing* it or *rewriting* it, that is, of transforming it by means of conscious,
practical work.

> Paulo Freire and Donald Macedo, *Literacy:*
> *Reading the Word and the World* (South
> Hadley, MA: Bergin & Garvey, 1987), p. 35.

Introduction

Information is around us in abundance. The real challenge in dealing with information is finding ways to make it useful. In Chapters 7 and 8 we looked at how to conduct primary and secondary research. In this chapter we look at ways to manage the information generated by that research. When you gather information or create new knowledge, you need to evaluate, organize, summarize, and document it. Information management isn't an easy task, but it is a crucial one. Your documents will be more useful and your arguments more effective if you organize and keep track of the information you convey to your audiences.

Evaluating Information

As you read or listen to the information you gather from primary or secondary sources, you will need to assess your sources' credibility, or *ethos;* you need to determine whether your sources use *pathos* to appeal to your values and emotions; and you need to decide what "good reasons," or *logos,* your sources offer in their own arguments. The lessons you learned in Chapter 5 should help you analyze the persuasive strategies your sources use to convince you to listen, read, and accept their arguments.

In this section, we offer you additional help in evaluating the *logos* of your sources. Recall for a moment the *because-clauses* that we discussed in Chapter 5. A new highway should be put through downtown *because* of improved safety, better traffic flow, and better aesthetics. A certain paint should be used on automobiles for good reasons *because* it resists chipping, for instance, and *because* it is inexpensive and easy to use. And, recall that these because-clauses support the ideal situations that you saw in Chapter 2 when we discussed communicating to solve problems. The *ideal situation* is to have foreign language study within the engineering curriculum, but at the moment it is not required or encouraged *(actual situation),* so you are proposing how foreign language might be included in the curriculum *(purpose) because* other schools are doing it, *because* that might make students into more flexible, informed, and culturally aware engineers, and *because* it might give students an edge on the job market. Looking at ideal and actual situations enables you to solve problems for your own audience; solving these problems through persuasive because-clauses ensures that your audience will respond to you. You can use these same tools to evaluate the information you gather from your sources.

In particular, you need to determine whether your sources back up their good reasons, their because-clauses, with facts, inferences, or opinions. Of course, what appears to the source as fact might seem opinion to you, and so on. Your evaluation will affect how you present the material to your own audience and how you interpret the importance, relevance, and value of the material. Your decisions about what information to include will be based on your determination of whether the information seems verifiable, logically inferred from sufficient evidence, or based on expert opinion. No information comes without context, interpretation, and bias—both your sources' and your own.

FACTS

If you gather and present information that can be verified or proven by personal observation or research in qualified sources, that information comes as close to **fact** as we can get. We add that caution because the context in which the fact is presented and even which fact is selected adds subjectivity or potential bias to any objectivity your sources or you might attempt to achieve. For example, by using a standard unit of measurement, a source might say "the computer screen is 4 inches from top to bottom," but the choice to use inches, rather than millimeters, and the decision to convey that dimension rather than the height of the terminal itself reflect the source's personal choices, goals, perspectives, or values, and affect you and your own audience. Given this qualification then, keep in mind that a fact is universally agreed upon or can be verified by anyone repeating the observation, experiment, or calculation. For those reasons, you may more readily accept as "fact" a source's description of an actual situation than of an ideal situation. And, you need to look at your sources' because-clauses to see how many of them could be described as factual.

fact
a piece of information that can be verified by observation, experiment, or calculation

When reading or hearing what appears to be a fact, look carefully at every word. For example, if your source refers to an actual situation such as, "The University of Minnesota trapped and killed 10,000 birds in a three-month period," you will need to ask whether this number reflects those birds trapped in experimental agricultural fields only; whether the number reflects sparrows, blackbirds, grackles, starlings, cowbirds, or other types; whether the source refers to only the Twin Cities branch of the University or includes coordinate campuses; and what three-month period was involved. Generally, qualifying words, particularly of time and place, add to the credibility of any statement (e.g., "Ten thousand blackbirds and starlings were trapped and killed in the seven experimental fields, numbered 1001–1007, at the Twin Cities branch of the University of Minnesota between March 1, 1993 and June 2, 1993.").

Finally, when you believe you have encountered a fact, look carefully at your source. The statement about the number of birds trapped in University fields might be more credible if offered by the person in charge of counting the number of birds than by an unqualified observer. It would also be more credible if confirmed by more than one person and if published in a research report rather than stated in a phone call.

INFERENCES

inference
an assumption or theory that can be supported by observation or experiment but that cannot be fully verified

Inferences are usually theories and hypotheses, statements backed up by some observations or experiments but not completely verifiable. Much of the information you gather from your primary and secondary sources, many of your sources' good reasons and because-clauses, will probably be inferences, or those logical conclusions and statements of probability, supported by as many facts as the observer or experimenter can provide. For example, the statement, "Nevertheless, extensive laboratory studies suggested that enhanced bioremediation might be applicable to stranded oil on the beaches of PWS [Prince William Sound] and the GOA [Gulf of Alaska]," reflects the source's inference that bioremediation would be effective. The laboratory studies might indeed predict that the conditions in Prince William Sound would be conducive to bioremediation. However, look carefully at each word within the statement: What does the source mean by "extensive"? By "enhanced"? Who did the laboratory studies? How many studies were done? In what environment? Even though the inference may seem logical or reasonable, you will need to evaluate your source's background, experience, values, organizational affiliation, and so forth before presenting the inference to your own audience.

OPINIONS

opinion
a personal interpretation of information

Finally, at times you might encounter what seems an **opinion,** a personal attitude, in the information you gather and must interpret. That opinion might appear in your source's description of an ideal situation, or one of your source's because-clauses might seem an opinion to you. To say that trapping and killing birds in University of Minnesota experimental fields is "a waste of animal life" is an opinion. What is a waste to some might be a necessity to others. If the birds that are killed are destroying valuable research seeds and if they are used to feed the injured raptors being cared for by the veterinary school, another person might have a different opinion about the University's actions. However, do not dismiss opinions entirely from the information you gather but instead evaluate the *ethos* of your source and the information upon which the source has based the opinion. A strongly stated opinion from an expert on a topic might carry more weight than an inference offered by a less qualified person.

Organizing Information

Some of the tools provided in Chapter 4 were designed to help you organize information. Now that you've learned how to gather and evaluate information, we offer you some additional help in organizing that information. As you gather data for a report, memo, talk, letter, or any other situation, you are likely to become overwhelmed with information. You may be faced with a variety of sources of data, all of which offer far more than you actually need to construct your message. One of the most difficult tasks you face is prioritizing this information. But because you have not yet written your document, you might be unsure of what you actually need and what you can discard. At this point, you need to select and organize your information — to make some sense of it all, to decide what is important and useful and what is not. One of the best ways to do this is to begin to outline what you have, what you want to save, and what you still need to find. In this chapter, we explore outlining again, this time to show how outlining can help you select and organize the information you have gathered from primary and secondary sources.

We all organize the information we collect in a variety of ways. Sometimes we do things as simple as stack books and papers in groups. If, for example, you were writing about engineering disasters, one pile might be for the information about the collapse of the Hartford, Connecticut, Civic Center, and another pile might cover the explosion of the space shuttle *Challenger*. These piles physically organize information. But this stacking method doesn't allow for very specific organizing. What if one book or one set of notes contains information about five different topics that you may want to use in five different places? Usually we handle this problem by describing our organization. For example, when some sources overlap, we might find ourselves affixing notes to the covers of the books and papers indicating where the information contained in them goes. This need to keep track of data and cross reference it is one reason Post-it™ notes are so popular.

OUTLINE DEVELOPMENT

A far more sophisticated way to organize information is to develop outlines. Using outlines as you plan your documents can help you manage information.

EXAMPLE 9–1   Informal Outline for a Report on Portable Computers

Intro: the benefits of portable computers and the increasing miniaturization of computing hardware.
Notebook, subnotebook, palm-top, wallet-sized computers.
Key developments in the miniaturizing components:

CPU/chip.

Hard drives.

Power packs.

Floppy drives.

Screens: b/w, color (active, passive).

Keyboards and pointing devices.

Recommendations.

outline
any of a variety of frameworks or skeletons for organizing data

An **outline,** or skeleton, of your document represents the basic structure upon which you will elaborate in creating the full document. Outlines can be as simple as an informal list of ideas and information or as complex as a formal, fully developed, enumerated, formatted list that shows every topic as well as all of its subtopics. Example 9–1 shows a simple, informal outline for a report on portable computers. This informal outline uses one formal element: indentations to indicate a list of subtopics.

Example 9–2 shows a more formal version of the same information found in Example 9–1. This outline makes use of several formatting devices: listing, indenting, and decimal numbering. This outline is a bit more detailed than the informal one. Sometimes when you follow the convention of numbering sections and subsections, you discover that some topics require multiple levels. That is, the decimal numbering convention reminds you to consider the levels of detail needed to cover a topic sufficiently. The nature of the information itself often suggests how it should be structured.

EXAMPLE 9–2   Excerpt of a Formal Outline for a Report on Portable Computers

1. Introduction
  1.1 The benefits of portable computers
  1.2 The increasing miniaturization of computing hardware
2. Types of portable computers
  2.1 Notebook
  2.2 Subnotebook
  2.3 Palm-top
  2.4 Wallet-sized
3. Key developments in the miniaturizing components
  3.1 CPU/chip
    3.1.1 486SX, 486DX
    3.1.2 486SL
  3.2 Power packs
    3.2.1 Ni-cad batteries
    3.2.2 Ni-mh batteries
    3.2.3 AC power adapters
  3.3 Screens
    3.3.1 B/W
    3.3.2 Color
      3.3.2.1 Active matrix
      3.3.2.2 Passive matrix . . .

Figuring out the best time to develop an outline will have something to do with the nature of the task you face, your timeline, and your (or your team's) writing process. Should you begin your research with an outline? Should you wait until you are ready to write your paper to develop an outline? Should you jot down provisional outlines throughout the entire process of research and writing? The simple answer is that usually you will do all of the above. From your initial thoughts on a topic to the finished document some days, weeks, or months later, you will go through a variety of topics and sources, each of which is an attempt to organize your work at that stage.

In this chapter you will see brief outlines showing various organizational patterns. Do not feel as if you have to develop such complete outlines in the early stages of research and writing. Let your organization develop as you go, as your information, audience, and purpose work together to shape your document. The most important aspect of your writing is the finished document, and outlines are merely tools to reach that product. Therefore, outlines should be useful. To impose an outline structure on ideas and information that are only in the early stages of development would be a waste of time. Complicated and perfectly constructed outlines can be a thing of beauty, but of little use.

ORGANIC OUTLINES: DEVELOPING INFORMAL WORKING LISTS

As you collect information about a topic, you will begin to see similarities among certain detailed bits of information, and you may want to create an **organic outline** to organize those similar details into groups. Likewise, as you develop provisional groups of details, you may see similarities among separate groups and then begin to group the groups. This process really isn't as confusing as it sounds. These initial attempts to organize the information are a time to experiment, to try categorizing pieces of information and seeing patterns within the mass of details you collect through your research.

organic outline
an informal collection and grouping of pieces of information

For example, let's assume that you work as a biomedical engineer in the cardio-thoracic surgery department of a research hospital. This department performs and studies differing types of heart surgery. You are asked to write a report on the current status of research and development of artificial hearts—devices designed to replace human hearts either temporarily or permanently. The report will provide your colleagues with an overview of the field so that your supervisors can begin to choose which types of devices should be used in your department. You begin your research by assuming that you will investigate different devices, and you sketch out the following simple outline:

1. Intro: overview of the field—different types of pumps.
2. Pump 1: describe.

3. Pump 2: describe.
4. Etc.
5. Conclusion: summary of the field.

After several days of work in medical libraries, you soon discover that the field is divided among several radically different concepts of heart pumps. You discover that different biomedical experts interpret certain human heart processes and heart diseases in different ways and that these interpretations have led them to develop very different pumps. As you learn more, your plans for the report evolve. Originally, you planned a simple analysis of several different devices. Now you begin to develop a detailed comparison/contrast outline that includes analyses of human physiology, heart disease, and the differing artificial hearts. Eventually, you decide that your conclusion will be evaluative, giving your opinions on the most effective type of device for your department. At some point in the process, you would have developed a fairly detailed outline that might continue to change but that would be more formal than your initial one. Your outline may never look perfect, with all of the decimals lined up perfectly, but it will be most useful to you if it contains sufficient detail.

FORMAL OUTLINES: SETTING EXPECTATIONS AND FULFILLING THEM

formal outline
an orderly pattern for organizing information

When you create a **formal outline,** you have to decide what comes first and what comes next. You can't just put down information haphazardly in your document, and you should use your outline to establish the structure you will use to present and organize your information. As you create a formal outline, you can organize your information using a variety of conventional patterns recognized in technical and business cultures.

When you use these patterns, you are communicating some very useful information to your readers. You are telling them you know the rules for the orderly presentation of information, and you are allowing them to trust their expectations. When they use your document, they will recognize its format, either consciously or unconsciously. As a result, they will have confidence in your document and will find it easier to read because of its familiar structures and patterns.

For example, let's assume you are writing a memo reporting an accident at an aluminum plant. Example 9–3 shows how you begin a section of the memo. This beginning sets up certain expectations for your readers. They will continue to expect your account to follow in a chronological order until signals are given that the chronology is ended and another organizing principle will be used in its place.

If another section of the same memo contains the language displayed in Example 9–4, a reader will have a different set of expectations. This passage sets up a comparison/contrast pattern that readers will implicitly

EXAMPLE 9–3 Excerpt from Memo on an Aluminum Plant Accident:
Chronological Ordering

At the start of the 4 P.M. shift, the furnace electrician put in a request to cut power
to the crane. At approximately 4:20 P.M., the maintenance supervisor called up to the
furnace room and stated that power would be cut at precisely 4:30 P.M. At 4:35 the
electrician. . .

recognize; readers will expect this pattern to continue until you signal a
change of pattern is necessary.

Knowing some standard patterns of organization is important because
they are widely recognized in the workplace. And because your readers
already understand how these patterns organize information, your writing
will gain credibility as a result of meeting their expectations. Examples 9–
5 through 9–9 illustrate some common ways of organizing information as
you create your outlines.

Classification/Partition

By placing items into different groups that share characteristics, you are
classifying information. If you are asked to investigate the range of available
personal computers for the employees of your management consulting firm,
for example, you could organize the information by classifying the types of
computers available and the features of each type. Example 9–5 shows one
way this classification could be done.

If you wished, you could extend the classification pattern further. Under
each subcategory (e.g., 2.7 input devices), you could easily list additional
sub-subcategories (2.7.1 keyboard, 2.7.2 mouse, 2.7.3 touch screen), and, if
you wished, you could continue to partition each of those sub-subtopics
further.

classification
a method of organizing information by grouping it according to shared characteristics

Comparison/Contrast

Now let's assume that you don't merely want to identify the types and features
of personal computers. You would like to evaluate them and recommend one

comparison/contrast
a method of organizing information by grouping it to juxtapose shared and contrasting characteristics

EXAMPLE 9–4 Excerpt from Memo on an Aluminum Plant Accident:
Comparison/Contrast

My investigation has revealed two differing accounts of the incident. On the one hand,
the electrician says. . .

On the other hand, the maintenance supervisor says. . .

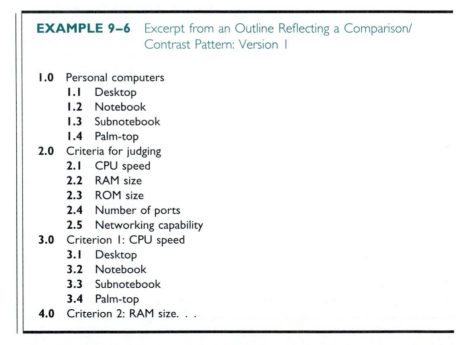

EXAMPLE 9–5   Excerpt from an Outline Reflecting a Classification/
Partition Pattern

1.0   Personal computers
   1.1   Desktop
   1.2   Notebook
   1.3   Subnotebook
   1.4   Palm-top
2.0   Desktop features
   2.1   CPU
   2.2   RAM
   2.3   ROM
   2.4   Ports
   2.5   Networking
   2.6   Monitor
   2.7   Input devices. . .

type over another for your employees in the firm. To do this task you will compare and contrast the different types using a set of criteria that you consider to be most important. Example 9–6 compares each type of computer under each criterion. An alternate but common way to organize this comparison/contrast information would be to repeat the list of criteria under each type of computer.

EXAMPLE 9–6   Excerpt from an Outline Reflecting a Comparison/
Contrast Pattern: Version 1

1.0   Personal computers
   1.1   Desktop
   1.2   Notebook
   1.3   Subnotebook
   1.4   Palm-top
2.0   Criteria for judging
   2.1   CPU speed
   2.2   RAM size
   2.3   ROM size
   2.4   Number of ports
   2.5   Networking capability
3.0   Criterion 1: CPU speed
   3.1   Desktop
   3.2   Notebook
   3.3   Subnotebook
   3.4   Palm-top
4.0   Criterion 2: RAM size. . .

EXAMPLE 9–7 Excerpt from an Outline Reflecting a Comparison/
Contrast Pattern: Version 2

1.0 Personal computers
   1.1 Desktop
   1.2 Notebook
   1.3 Subnotebook
   1.4 Palm-top
2.0 Criteria for judging
   2.1 CPU speed
   2.2 RAM size
   2.3 ROM size
   2.4 Number of ports
   2.5 Networking capability. . .
3.0 Desktop
   3.1 CPU speed
   3.2 RAM size
   3.3 ROM size
   3.4 Number of ports
   3.5 Networking capability. . .
4.0 Notebook
   4.1 CPU speed
   4.2 RAM size. . .

Example 9-7 illustrates what an outline based on that organizational pattern would look like. The major advantage of version 1 (Example 9-6) is that it places the features to be compared near each other. The major advantage of version 2 (Example 9-7) is that it allows you to present each type of computer in its entirety before going on to the next. You will have to decide which pattern works better for the kinds of information you need to present.

Spatial Patterns

Let's assume that you recommended the purchase of notebook computers for your firm, but you realize that none of the employees has ever used these devices. With the delivery of the computers, you decide to provide information about the key physical characteristics of each unit; that is, you organize the information about the computers spatially. You'll need to decide where to start and end your communication using this type of pattern: top to bottom, right to left, outside to inside, front to back, and so forth. Example 9-8 illustrates what an outline following this pattern of organization might look like. Other spatial patterns are equally possible and effective.

spatial pattern
a method of organizing information by arranging it spatially—describing something from top to bottom, left to right, front to back, etc.

EXAMPLE 9–8   Excerpt from an Outline Reflecting a Spatial Pattern

1.0   Physical layout of your new notebook computer
2.0   Screen
    2.1   Open/closed positions
    2.2   Adjustment positions
3.0   Keyboard
    3.1   Multi-purpose keys
    3.2   Function keys
4.0   Ports
    4.1   Mouse
    4.2   External monitor
    4.3   External keypad
    4.4   Internal modem
    4.5   Serial
    4.6   Parallel. . .

---

chronological
a method of organizing information by arranging it in the order it occurs in time

Chronological/Step-by-Step

Look back to Example 9–3 and reread the example of the aluminum plant accident report. That passage was organized chronologically—a pattern that proceeded by presenting what happened first, second, third, and so on. A similar type of organizational pattern common to the world of technical communication is the step-by-step pattern illustrated in Example 9–9. This pattern is commonly used when the document you create needs to be organized around the actions needed to complete a process or produce some product.

The patterns for organizing information contained in this chapter represent only a few common ways that you can shape the information you collect through research. There are others that you will come across when writing in technical fields. For example, you may often need to develop problem/solution organizational patterns with the data you collect. Such patterns first explain the problem and then describe a solution or alternate solutions that must be evaluated. We will say more about this pattern when we look at how to create reports.

COMPLEX INFORMATION: COMBINING OUTLINE PATTERNS

Many times you bring various types of information to bear on a report or argument, thus requiring you to use more than one organizational strategy. You may need to organize the information in one part of a report in one pattern, and another part may need to be organized differently. Although

---

EXAMPLE 9–9   Excerpt from an Outline Reflecting a
Step-by-Step Pattern

1.0   Starting to use your new notebook computer
2.0   Opening the screen
3.0   Using the keyboard
    3.1   Using the multi-purpose keys
        3.1.1   (step 1)
        3.1.2   (step 2). . .
4.0   Installing software
    4.1   Word processor
        4.1.1   (step 1)
        4.1.2   (step 2). . .

---

such a practice is common and often necessary, you must guide your readers through each pattern and provide a transition to the next.

For example, some of the organizational patterns shown in the outlines dealing with different types of computers (Examples 9–5 to 9–9) could be combined. Certainly the classification/partition analysis and the comparison/contrast analysis could be rolled into one report. Likewise the spatial and step-by-step analyses could also be combined.

Although these combinations seem to make sense, don't forget that when readers see the beginning of a pattern, they will expect to see that pattern played out to its conclusion. When you combine patterns, you run the risk of confusing readers. Readers can handle these changes, but you must craft the transitions carefully. The purpose of learning these patterns of information organization is not so you can construct perfect outlines. Rather, it is to provide you with ways of organizing information that are efficient, effective, and easily recognized by audiences in the workplace.

Summarizing Information: Developing Summaries and Abstracts

Another useful tool to help you manage information is the **summary** or **abstract.** When you collect information from secondary or primary sources, you will often summarize or reduce the essence of that information to a paragraph or two of notes. When you decide what secondary research will be valuable for your project, you may read the abstracts published at the beginning of research articles or within collections (look back at Example 7–8 in Chapter 7 for a list of abstracts and indexes).

More often than not, when you do any kind of detailed research, you will present the results of that research in a shortened form. After all, if your research generates pages and pages of notes and data from dozens of books

summary
a way to manage information by reducing it to its main points

abstract
a shortened form of a complete article that stands alone but clearly describes the article's content and structure

and articles, your written and oral reports serve, in part, as extended summaries of all of that information. And, within those reports, you will need to provide shorter versions of the information you are presenting in carefully constructed *abstracts* or *summaries*. We cover summaries next, as a general technique for reducing your information to a manageable form; we then cover reading and writing abstracts.

SUMMARIES

Summaries serve several functions in your research and communication. In order to manage the information you gather through primary and secondary

---

EXAMPLE 9–10   An Article on Botany and Ecology

Many **plants** that are **pollinated by animals**—examples would be anything pollinated by ants or the proverbial birds and bees—**develop their flowers in three fairly discrete phases.** First, there's an **initial phase;** during this period the plant produces a **small but increasing number of flowers each day.** Next comes a **short peak phase** in which most of the **total number of flowers are produced.** And finally comes a **final phase** when the plant produces a **small and decreasing number of flowers** each day. Recently many evolutionary ecologists, such as Dan Janzen, Al Gentry, Gary Stiles, and Carol Augsperger (right down the hall), have been concentrating their efforts on studying the effects of this **mass-flowering pattern**—on how pollinators like bees and ants are attracted to plants and how and why they move from plant to plant. These people are showing that many kinds of floral visitors (for example, the ants) **prefer to eat individual plants that produce really large numbers of flowers per day,** presumably because these individual plants are **easy to spot** by the pollinators (and are very tasty to them) and because it takes **less effort** for the pollinators to do all their pollinating when flowers are so close together. In contrast, when daily flower production is low, during the first and last phases of the flowering pattern, the plant through its flowers might not be able to attract enough pollinators from other flowers. Consequently, this mass-flowering pattern poses a dilemma: When flowers are few and far between, the bees or ants fly or walk around and visit many different trees, since flowers are scarce; that's good in that cross-pollination occurs (and cross-pollination makes for a more vigorous next generation), but it's bad in that the birds and ants spend a lot of their time and energy visiting various trees and therefore don't make as many stops as usual. On the other hand, when flowers are many, the pollinators can save their energy and work more efficiently on one plant; that's certainly more efficient, but **it cuts down on the cross-pollination because the pollinators are sticking to one location.** So what's a plant to do?

In our study, Terry Haas and I tried to answer two questions: During what phase of the flowering pattern does outcrossing occur? And what factors promote outcrossing? Specifically, we've examined the effect of different floral abundances on the proportion of flowers that are self- and cross-pollinated on an individual plant. What's more, we've documented the flowering pattern of the catalpa tree (*Catalpa speciosa*, Bignoniaceae).

research, there are times when you will record that information word for word. At other times you will record only the essence of that information in a summary. Also, when you write or speak, you often offer your audience short summarizing statements *throughout* a communication to help that audience remember and appreciate your arguments. Also, you might conclude your argument with a summary of all your good reasons or main points. Moreover, your introduction to a document or speech often previews the structure of your communication by providing a kind of summary of your main points.

We return to summaries later in this chapter when we distinguish between quotation, paraphrase, and summary style when referring to and citing specific sources. Here we demonstrate some techniques for summarizing the information you encounter in your research or for summarizing the main points in your own documents. Example 9–10 repeats the passage that you studied in Chapter 5 (Example 5–4) to demonstrate summarizing techniques.

When you write a summary of information you gather, first identify the main points and the important supporting evidence. You can either underline this information or record it on a note card or in a computer database that you create for your project. You cannot include all of the source's examples, but you might record the most revealing or universal example in your summary. You will encounter some words and phrases in the original that convey the meaning to such a great extent that you cannot change them without losing that meaning. Record those words and phrases as direct quotations and note the page number on which they occurred in the original source.

Your use of the information will also determine what you capture in your summary. For example, let's say that you are gathering information on cross-pollination of flowering plants by animals. You are particularly interested in what happens to cross-pollination when animals encounter plants in the peak or second flowering stage. The passage in Example 9–10 appears in one of the articles you read. We have placed in boldface the main points you would probably record for your summary.

You'll notice that the summary in Example 9–11 begins with a short explanation of the three phases of flowering and contains the specific name

---

EXAMPLE 9–11   Summary of Example 9–10

Plants pollinated by animals develop their flowers in three phases:

Phase one: a few flowers are produced.

Phase two: almost all the total number of flowers are produced.

Phase three: only a few flowers are produced.

This phenomena is called a "mass-flowering pattern" (p. 3). Pollinators prefer to eat the plants in the second phase because they can find them easily and stay in one spot. However, less cross-pollination occurs because the pollinators are not moving around.

EXAMPLE 9–12   An Article on Botany and Ecology

Many plants that are pollinated by animals—examples would be anything pollinated by ants or the proverbial birds and bees—develop their flowers in three fairly discrete phases. First, there's an initial phase; during this period the plant produces a small but increasing number of flowers each day. Next comes a short peak phase in which most of the total number of flowers are produced. And finally comes a final phase when the plant produces a small and decreasing number of flowers each day. Recently many evolutionary ecologists, such as **Dan Janzen, Al Gentry, Gary Stiles, and Carol Augsperger** (right down the hall), have been concentrating their efforts on studying the effects of this mass-flowering pattern—on **how pollinators like bees and ants are attracted to plants and how and why they move from plant to plant.** These people are showing that many kinds of floral visitors (for example, the ants) prefer to eat individual plants that produce really large numbers of flowers per day, presumably because these individual plants are easy to spot by the pollinators (and are very tasty to them) and because it takes less effort for the pollinators to do all their pollinating when flowers are so close together. In contrast, when daily flower production is low, during the first and last phases of the flowering pattern, the plant through its flowers might not be able to attract enough pollinators from other flowers. Consequently, this mass-flowering pattern poses a dilemma: **When flowers are few and far between, the bees or ants fly or walk around and visit many different trees, since flowers are scarce; that's good in that cross-pollination occurs (and cross-pollination makes for a more vigorous next generation), but it's bad in that the birds and ants spend a lot of their time and energy visiting various trees and therefore don't make as many stops as usual.** On the other hand, **when flowers are many, the pollinators can save their energy and work more efficiently on one plant; thats certainly more efficient, but it cuts down on the cross-pollination because the pollinators are sticking to one location.** So what's a plant to do?

In our study, Terry Haas and I tried to answer two questions: **During what phase of the flowering pattern does outcrossing occur? And what factors promote outcrossing?** Specifically, we've examined the **effect of different floral abundances on the proportion of flowers that are self- and cross-pollinated on an individual plant. What's more, we've documented the flowering pattern of the catalpa tree** (*Catalpa speciosa,* Bignoniaceae).

for the phenomena and some essential information about the effects of this phenomena on cross-pollination. Because you are concentrating on the second phase of the mass-flowering pattern, you record just one aspect of cross-pollination.

On the other hand, if you need to identify the research questions asked about this "mass-flowering pattern," the researchers asking the questions, and the dilemma that has caused some of these questions, you would have marked different parts of the same passage (boldfaced parts in Example 9–

EXAMPLE 9–13   *Summary of Example 9–12*

Researchers Dan Janzen, Al Gentry, Gary Stiles, and Carol Augsperger have asked: "How pollinators like bees and ants are attracted to plants and how and why they move from plant to plant?" (p. 3). In asking these questions, a dilemma surfaced: if pollinators find lots of flowers clustered, as during the peak flowering phases of many plants, the pollinators stay in one spot, and less cross-pollination occurs. But if pollinators find few flowers clustered, they move around a lot but they don't make as many stops, and less cross-pollination occurs.

In this article, Terry Haas and A. G. Stephenson have asked two questions: "During what phase of the flowering pattern does outcrossing occur? And what factors promote outcrossing?" (p. 4). They have researched: "the effect of different floral abundances on the proportion of flowers that are self- and cross-pollinated on an individual plant" on the catalpa tree (p. 4).

12). The summary of these researchers' questions and the dilemma that raised these questions appears in Example 9–13.

When writing a summary of your own words and ideas, for an introduction or conclusion to a long report, for example, you can use the same techniques.

ABSTRACTS

An abstract often stands alone or represents a document in a list or index and helps a reader decide whether to read, skim, or skip the document or to pass it on to others. Therefore, an abstract must be understood by audiences who do not have the document at hand. Imagine yourself in a library searching for articles on a research topic and finding an abstract in an index or electronic database. As a reader and researcher, you will want that abstract to describe the full article clearly enough so that you can decide whether it will be useful to find and read it. If the abstract is effective, its writer has achieved that stand alone quality. Moreover, you will be required to write abstracts for the formal reports you produce on the job. Your readers will use your abstract to make the same kinds of decisions you would in your research: to read, skim, skip, or pass it on.

Abstracts provide readers with a glimpse of an entire communication or document, such as a report, in shortened form. In general, they follow the style and the order of the document. You will encounter several types of abstracts in your reading and will be asked to write a variety of abstracts on the job. In this section, we review the three most common types of abstracts.

EXAMPLE 9–14   A Limited, Descriptive Abstract

This report summarizes the recent changes in the waste management system that services the city of Roseville. The report first reviews the problems the city experienced with the old system, then narrates the installation process of the new system, and finally proposes the expected advantages of the new system.

Limited, Descriptive Abstracts

limited, descriptive abstracts
an abstract of 50 or fewer words that describes the purpose, scope and organization of a document

Limited, descriptive abstracts typically do not exceed 50 words and describe the purpose, scope, and organization of a document in a few sentences, giving readers an idea of what the document is about. Descriptive abstracts do not offer conclusions and recommendations contained in the document or offer any final evaluations about the topic covered in the document. By describing the topic and organization of the document, descriptive abstracts help the audience decide whether to read that document or not. Example 9–14 shows an example of a limited, descriptive abstract.

Notice that the abstract in Example 9–14 uses transitions ("first," "then," and "finally") to indicate the organization of the document described. The verbs within the abstract relate to the document itself ("summarizes," "reviews," "narrates," and "proposes") rather than to any conclusions about the new waste management system. Thus, the words *describe* the document.

Complete, Descriptive Abstracts

complete, descriptive abstracts
an abstract of more than 50 words that describes the entire contents, including each part and facet, of a document

Complete, descriptive abstracts are much longer than limited descriptive abstracts (about twice as long) and describe the entire contents, including each major part or facet, of the document. However, as with limited, descriptive abstracts, these abstracts also describe the purpose, scope, and organiza-

---

EXAMPLE 9–15   A Complete, Descriptive Abstract

This 50-page analytical report summarizes the recent changes in the waste management system that services the city of Roseville, the closest northern suburb to St. Paul, Minnesota. The changes covered include the new sewer system installed in the 20-block area bounded by Larpenteur Avenue, Lexington Parkway, Snelling Avenue, and County Road B2. The report includes a complete listing of independent contractors licensed for home trash collection in the suburb and the recycling contractor approved by the Roseville council. The report first reviews the problems the city experienced with the old system, including clogged or limited drainage and overloaded sewer lines. The report then narrates the installation process of the new system between April 1993 and September 1994. Finally, the report describes the expected financial and safety advantages of the new system.

EXAMPLE 9–16   An Informative Abstract

The city of Roseville, the closest northern suburb to St. Paul, Minnesota, had experienced problems with its waste management system after a period of growth and new housing development between 1960 and 1980. The city council reviewed the possibility of installing a new waste management system in November and December of 1992, which was approved by the voters in January 1993, and construction was completed between April 1993 and September 1994. The new waste management system includes a new sewer system, type 678B7, installed in the 20-block area bounded by Larpenteur Avenue on the south, Lexington Parkway on the east, Snelling Avenue on the west, and County Road B2 on the north. The effectiveness of this sewer system was tested by the engineering firm of Walzer and Gross in late September 1994 and found to be 95 percent effective in transporting household and business waste. The firm concludes that the previous problems with clogged or limited drainage and overloaded sewer lines could be further relieved by installing new sewer lines north of County Road B2 in the summer of 1995. In August 1994 the Roseville council contracted with five independent contractors to remove trash from households and businesses and published this list, with price options, for all Roseville residents in local newspapers. Recyclable waste is picked up curbside in containers provided by residents twice a month and includes paper, glass, aluminum, and plastic. The city has raised property taxes by 2 percent for 1994 and will raise them an additional 2 percent for 1995 to cover the cost of the system, as approved by voters in January 1993, and has remained within budget. As confirmed by Walzer and Gross, water quality in Roseville has proved to be 10 percent higher than other nearby suburbs since the installation of the system.

tion of the document without offering the findings or conclusions about the topic the document discusses. A complete, descriptive abstract of the document on the Roseville waste management system is shown in Example 9–15. Notice again the use of transitions and verbs to describe the document itself.

Informative Abstracts

Informative abstracts provide the audience with an idea of the purpose and organization of a document, but they also give specific findings, evaluations, and conclusions about the topic covered in the document. They describe the topic, the methods used to research that topic, the experiments set up to test some aspect of the topic, the findings of that research or those experiments, and the conclusions based on research or experiments. Although the audience may elect to go on and read the entire document based on the information within this kind of abstract, informative abstracts can also serve as substitutes for the document. The audience knows enough about the specifics in the document and the significance of the findings to learn something new about the topic. An informative abstract of the document on the Roseville waste management system is shown in Example 9–

informative abstract
an abstract that describes the purpose and organization of a document as well as evaluations, findings, and conclusions about the topic covered

ment on the Roseville waste management system is shown in Example 9–16. Notice that the verbs within the abstract ("proved to be," "found to be," "tested," "confirmed," and so on) reflect action and results pertaining to the waste management system.

Again, you will encounter all three kinds of abstracts as you do secondary research in the library, and, because you will have to write abstracts for your own documents, you will likely need to turn back to this section several times as you work your way through this text.

SUGGESTIONS FOR WRITING ABSTRACTS

The following guidelines should help you write more effective abstracts.

1. If possible, write your abstract *after* you have written the full document. This practice provides two benefits. First, it helps you avoid discussing a topic in the abstract that has not actually been included in the document. Second, it enables you to construct an abstract that follows the structure of the document.
2. For a first draft of your abstract, scan the document and pull out passages that best characterize the overall document, the major conclusion of the document, and each section of the document. If you have written your document on a computer, this can be done quite easily.
3. If you are writing a long abstract, choose the most significant details and find a way to include them in the abstract in a way that will make them understandable. Although it is easy to pull out the key details of a document, it is far more difficult to make those details meaningful outside of the context of the full document. Remember, in general, abstracts should be able to stand alone.
4. Once you have pulled out key passages from the full document, put them together and read them through from beginning to end, be sure each passage moves smoothly to the next. Add transitions. Edit the abstract so that it reads as a cohesive statement, not just a series of disconnected sentences. Make sure you have chosen the most appropriate verbs for the type of abstract you are writing. Also, be very sure that you delete any words or statements referring to points you covered in the document but don't need to mention in the abstract.
5. The best test to see if you've written a good abstract is to have someone read it who has not read your full document. If the abstract reads as a cohesive whole to that person, then you are on the right track.

Another aspect of managing the information you collect through research is identifying the source of the information clearly and accurately. The whole point of doing research is to obtain and analyze the knowledge of others to help you inform your readers and solve a problem. You must always remember, however, that the knowledge you gain from your sources is not coming from you but from them. That is, you must **cite** your sources. Failure to do so is **plagiarism**. Plagiarism in school results in failure in a course and in disciplinary action; plagiarism on the job leads to a severe reprimand or dismissal. You must always be accurate and honest in giving other people credit for their ideas and work.

However, on the job you are often asked to collaborate with others to produce a document or you might use existing internal company reports and manuals (called boilerplates) to produce a new company document. It may be difficult to assign credit to each of your collaborators or to these internal company sources. Be conscientious, and, when in doubt, ask your supervisor. The guidelines we offer in this next section will help you properly credit your primary sources and your secondary library sources.

QUOTATION, PARAPHRASE, AND SUMMARY STYLES

As you gather information from both primary and secondary sources, you will need to decide when to quote from your source, when to paraphrase, and when to summarize what they have said. The following guidelines will help in your decision and ensure that you give proper credit to your source.

Quotation Style

When you quote from a source, you are conveying to your audience what that source said or wrote *word for word*. Generally, you should use a **quotation** if the words are particularly interesting, startling, or important to your topic; will catch the audience's attention; or contain essential details or statistics. For example, let's say that you encountered the following source and passage:

> In the summer of 1991, there were about 55,000 microcomputers and terminals in use in US elementary and secondary schools for instructional purposes. By the end of the year, there were more than 100,000. We think that there will be more than 200,000 microcomputers in the schools by the end of 1992, and even our conservative estimate would be that the number of microcomputers will grow to 500,000 by 1994.
>
> Thomas Q. Jansen and Neil L. Smith. "Computers in the Schools." *Journal of Applied Communication Research* 21.1 (1991): 123.

If you find the passage essential to your document, you need to quote directly much of the information, and might do so as follows and still use your own transitions to introduce the quotes:

Documenting Information: Crediting Sources

cite
the process of acknowledging the sources of a quotation or information

plagiarism
the failure to acknowledge the sources of information

quotation
a passage taken word for word from another's document

According to 1991 statistics, "there were about 55,000 microcomputers and terminals in use in the US elementary and secondary schools for instructional purposes" (Jansen and Smith 123).

The numbers grew to 100,000 by the end of 1991. Moreover, researchers Jansen and Smith predicted that "there will be more than 200,000 by the end of 1992," and "that the number of microcomputers will grow to 500,000 by 1994" (123).

Of course, whenever you use others' exact words, you must give them credit in one of two ways within your text: by mentioning the source's name or some other identifying label in your own sentence or by including the source's name or label parenthetically in your sentence. The exact page number upon which the quotation is found in the original must also be included. The passage above uses MLA style (discussed in detail below) and therefore relies on the name of the source, not the name and publication date as called for in other styles. Because it is clear that all quotations come from the same source, only the page number needs to be included parenthetically after the first time the source name is mentioned.

Whenever possible, work the quotation in smoothly into your own words and introduce the quotation in some way ("Researchers Jansen and Smith"; According to 1991 statistics"). Use ellipses if you leave out any words ("In . . . 1991, there were about 55,000 microcomputers and terminals in use in U.S. elementary and secondary schools . . ."), and use square brackets if you insert your own words into the quotation ("By the end of the year [1991], there were more than 100,000 [microcomputers].").

Paraphrase

paraphrase
to restate in your own words ideas and information taken from another's document

If your source's words themselves are not as important as the ideas, or if the words from the original source would be difficult or unfamiliar to the reader, you might elect to **paraphrase** those words. When you paraphrase, you need to maintain the essence or main ideas of your source. A paraphrase is faithful to the structure of a source, listing most major and relevant minor points, in their original order and with approximately the same emphasis. The following would be an acceptable paraphrase of the Jansen and Smith passage:

Approximately 100,000 small personal computers were used to teach grade school and junior high school students in this country by the end of 1991. By 1992, researchers predict there presumably will be more than 200,000 and about 500,000 by 1994 (Jansen and Smith 123).

Notice that when you paraphrase you still must indicate the source and page number of the passage you are paraphrasing.

Summary

If you just need to convey the information, facts, or ideas, but you don't need the words or phrases themselves from your source, you can summarize them. A summary, often much shorter than a quotation or paraphrase, captures the gist of a source or some portion of it, boiling it down to a few words or sentences. The following passage summarizes the Jansen and Smith material:

> The number of microcomputers and terminals in US schools will more than double by the end of 1992 and be 500,000 by 1994 (Jansen and Smith 123).

In general, your decision to quote, paraphrase, and summarize depends on your audience's interests and needs and the importance of the passage to your document.

The key to crediting your sources accurately is to identify which words and ideas are derived from your sources. You must cite your sources not only when you quote them directly but also when you paraphrase them or restate their information in your own words. There are a number of ways for you to give credit to your sources. Your company may supply you with a complete style guide, including preferred citation style. Whichever one you choose or are required to use in your class or in the workplace, you must remember to be thorough and consistent.

CITATIONS AND WORKS CITED SECTIONS

You are probably familiar with the practice of footnoting, in which references mentioned in the text are documented at the bottom of a page or at the end of a chapter. Generally speaking, most of the documents you create will not have footnotes but will rely on brief in-text **citations** to identify references. At the end of a document, or in the case of a long document at the end of a chapter or part of it, complete references are given in a **Works Cited** section or a bibliography. Probably the best way for you to grasp this convention is to look at samples of the practice in Examples 9–17 and 9–18. These examples follow the American Psychological Association or APA style guide (see the APA style section that follows).

Notice that Example 9–17 has citations for both a direct quotation (Roth, 1991, p. 10) and for a paraphrase (Grant, 1992; Person, 1993). Notice as

citation
a formal reference to another's document in your own text

works cited
a compilation of the complete references to all citations in a document

---

EXAMPLE 9–17 In-text Citation

The key to creating a usable multimedia development lab is to gain the ability to digitize every audio, video, and mixed media input available. This can only be done with "a device that accepts a full range of data standards and that is not yet feasible" (Roth, 1991, p. 10). Some experts believe, however, that a universal standard—a common way to create and digitize all information—is close at hand (Grant, 1992; Person, 1993).

EXAMPLE 9–18　Excerpt from a Works Cited Page (APA Style)

Works Cited

　Grant, J. (1992). The coming revolution in digitization. *CD Journal, 10,* 157–159.

　Person, B. (1993). Even the dog's bark can be digitized. *New Digital World, 2,* 120–121.

　Roth, A. (1991). Digitization: New impact on the world. *New Digital World, 1,* 101–109.

---

well the format for each reference is the same: the author's name is followed by the date of publication and this information is put in parentheses. This convention tells the reader that he or she can find complete information about the sources in the Works Cited section of the document, a portion of which is shown in Example 9-18.

When you list multiple references in Works Cited, list them alphabetically by last name of the first word in the citation. That first word is usually an author's or editor's last name, but it may be the first word in the title of an article for which no author is identified. Also, when you list two or more citations by the same author, place them in order of the publication date with the most recent listed first followed by the next most recent, and so on.

ACADEMIC DOCUMENTATION STYLES

Whenever you are documenting your sources in academic setting, you should always verify the proper format for that setting. Different disciplines use different formats, and the key to being correct and accurate is to follow the format that is most appropriate. Below are listed a few examples from three of the most widely used formats, Chicago B style favored by the natural and social sciences, the APA (American Psychological Association) style used in the social sciences, and the MLA (Modern Language Association) style used in the humanities. These styles are academic documentation styles; your company probably will provide you with its preferred documentation style in a style guide. This list of examples is extremely limited, and it won't answer all the questions you might have as you create your Works Cited section. It will provide you with a few of the most common formats, however, to get started.

Chicago Style B

Book

Yates, J. 1989. *Control through communication: The rise of system in American management.* Baltimore, MD: Johns Hopkins Univ. Press.

Edited Book
Spilka, R., ed. 1993. *Writing in the workplace: New research perspectives.* Carbondale, IL: Southern Illinois Univ. Press.

Essay in an Anthology
Selzer, J. 1993. "Intertextuality and the writing process: An overview." In *Writing in the workplace: New research perspectives.* Ed. by R. Spilka, 171–180. Carbondale, IL: Southern Illinois Univ. Press.

Journal Article (paginated by volume)
Winsor, D. A. 1990. Engineering writing/writing engineering. *College Composition and Communication* 41, 58–70.

Ervin, E., and D. L. Fox. 1994. Collaboration as political action. *Journal of Advanced Composition* 14:53–71.

Article (author unidentified)
A business that defies recession. 1982. *Business Week,* Oct. 25, 30–31.

Newspaper Article
Whalen, T. 1948. Is digital the wave of the future? *Boston Beacon,* Oct. 10, A2.

Government Publication
US Department of Interior. 1987. *Underwater archeological sites.* Washington, DC: Government Printing Office.

APA Style

Book
Yates, J. (1989). *Control through communication: The rise of system in American management.* Baltimore, MD: Johns Hopkins University Press.

Edited Book
Spilka, R. (Ed.) (1993). *Writing in the workplace: New research perspectives.* Carbondale, IL: Southern Illinois University Press.

Essay in an Anthology
Selzer, J. (1993). Intertextuality and the writing process: An overview. In R. Spilka (Ed.), *Writing in the workplace: New research perspectives* (pp. 171–180). Carbondale, IL: Southern Illinois University Press.

Journal Article (paginated by volume)
Winsor, D. A. (1990). Engineering writing/writing engineering. *College Composition and Communication, 41,* 58–70.

Ervin, E., & Fox, D. L. (1994). Collaboration as political action. *Journal of Advanced Composition, 14,* 53–71.

Article (author unidentified)

A business that defies recession. (1982, October 25). *Business Week,* pp. 30–31.

Newspaper Article

Whalen, T. (1948, October 10). Is digital the wave of the future? *Boston Beacon,* p. A2.

Government Publication

US Department of Interior. (1987). *Underwater archeological sites.* Washington, DC: Government Printing Office.

MLA Style

Book

Yates, JoAnne. *Control Through Communication: The Rise of System in American Management.* Baltimore, MD: Johns Hopkins UP, 1989.

Edited Book

Spilka, Rachel, ed. *Writing in Workplace: New Research Perspectives.* Carbondale, IL: Southern Illinois UP, 1993.

Journal Article (paginated by volume)

Winsor, Dorothy A. "Engineering Writing/Writing Engineering." *College Composition and Communication* 41 (1990): 58–70.

Ervin, Elizabeth, and Dana L. Fox. "Collaboration as Political Action." *Journal of Advanced Composition* 14 (1994): 53–71.

Essay in an Anthology

Selzer, Jack. "Intertextuality and the Writing Process: An Overview." *Writing in the Workplace: New Research Perspectives.* Ed. Rachel Spilka. Carbondale, IL: Southern Illinois UP, 1993. 171–80.

Article (author unidentified)

"A Business That Defies Recession." *Business Week* 25 Oct. 1982: 30–31.

Newspaper Article

Whalen, Thomas. "Is Digital the Wave of the Future?" *Boston Beacon* 10 Oct. 1948: A2.

Government Publication

United States Department of Interior. *Underwater Archeological Sites.* Washington: GPO, 1987.

For any citations that do not match these, you should consult the appropriate style guide. You will need to get used to noticing what is underlined (or italicized), what appears in capital letters, what is abbreviated, what is followed by a comma or period, and so on. Example 9–19 lists several different style guides.

EXAMPLE 9–19 A Selection of Style Guides

American Chemical Society. *American Chemical Society Style Guide: A Manual for Authors and Editors.* 2nd ed. Washington, DC: American Chemical Society Publishing, 1986.

American Institute of Physics. *Style Manual for Guidance in the Preparation of Papers.* 4th ed. New York: American Institute of Physics, 1990.

American Psychological Association. *Publication Manual of the American Psychological Association.* 3rd ed. Washington, DC: American Psychological Association, 1983.

CBE Style Manual. 5th ed. Bethesda, MD: Council of Biology Editors, 1983.

The Chicago Manual of Style. 13th ed. Chicago: University of Chicago Press, 1982.

Gibaldi, Joseph, and Walter S. Achert. *MLA Handbook for Writers of Research Papers.* 3rd ed. New York: Modern Language Association of America, 1988.

The GPO Style Manual. 28th ed. Washington, DC: US Government Printing Office, 1984.

LIMITS ON CITATIONS: PROTECTING CONFIDENTIALITY WHEN APPROPRIATE

Sometimes you will be reporting information that does not come from published sources but from the primary research that you gather in interviews, surveys, and observations. Upon occasion, there may be very good reasons for you to not identify the source of this information. You may have been granted an interview with someone only under the condition that the interviewee's identity would remain confidential, that is, known only to you as the interviewer. For example, if you were to investigate employee complaints about a new policy, you might decide that none of the employees interviewed would be identified. That way you could hear their complaints, and you could gain information that might be for the overall good of the company. If you did not grant such confidentiality, the key issues might never be revealed.

Although under such circumstances you would not reveal your sources, you *must* make it absolutely clear that the information you are reporting does not come from you but from unnamed sources. You must not present such information as if it is made up of your ideas when it is not.

Overall, collecting information under a promise of confidentiality is tricky business, and you must take extreme caution in undertaking such an enterprise. Sometimes the smallest detail can identify a source to a knowledgeable reader. If that happens, you have put your sources in jeopardy. Other times it may be inappropriate to grant confidentiality. You could put yourself in an untenable position if you were to grant confidentiality only to discover information about the danger inherent in some product or design or some failure to take necessary precautions. If you don't plan to guarantee confidentiality, don't promise it. Above all, if you are prepared to offer confidentiality, be sure you can deliver it.

SUMMARY

This chapter offers you some methods of managing information you collect through primary and secondary research. One of your first considerations is evaluating the material you encounter. Is the source presenting what appear to be facts, inferences, or opinions? What qualifying words add credibility to the information?

Your second consideration is to organize the information you find. As you gather information for a document you have to choose an appropriate strategy for prioritizing and arranging it. One of the best ways to choose that strategy is to use an outline to lay out an effective organizational pattern to your work. Some common organizational patterns are comparison/contrast, classification/partition, chronological/step-by-step, and spatial. These patterns are useful in that they are easily recognized by most readers of professional and technical writing. In addition, these patterns are often combined in complex presentations.

A third task you face as you deal with information is summarizing it. Within many documents you will need to provide shorter versions of the information you are presenting. You provide those versions by carefully constructing summaries or abstracts. Typically, most abstracts appear in one of three forms: (1) the limited, descriptive abstract that describes the purpose and significance of a report in a few sentences; (2) the complete, descriptive abstract that describes the entire contents of a report in as few words as possible; or (3) the informative abstract that provides an overview of the purpose, significance, scope, and important conclusions of a report.

Finally, you need to decide whether quoting, paraphrasing, or summarizing the information will be most useful to you and your audience. And, of course, you need to document your sources. Three common ways to document information in an academic setting are to follow the MLA, Chicago B, and APA documentation style guides. Your company will probably provide you with its preferred style. Finally, sometimes a researcher doing primary research must provide confidentiality to the human sources of information. Such practices must be exercised judiciously and carefully.

ACTIVITIES AND EXERCISES

1. Write a paragraph on the preservation of coastal habitats using information from the sources below. Choose parts of these sources to quote, paraphrase, or summarize as you see fit. Keep in mind whether the source is stating what appears to be a fact, an inference, or an opinion. Use your own thoughts to make connections between your sources and draw your own conclusions as needed.

Source 1

Burger, Joanna. "Factors Affecting Distribution of Gulls (Larus spp.) on Two New Jersey Coastal Bays." *Environmental Conservation* 14 (1987): 59–72. Taken from p. 59:

Gulls of the genus *Larus* are important components of coastal and marine ecosystems, and hence can constitute a factor of substantial environmental significance. Yet their relative contribution to coastal avifaunas, and the basic factors affecting their distribution along coasts, are rarely studied. Because gulls are so common along coasts, and tend to frequent areas that are used extensively by people, they are usually considered to be undisturbed by the presence of human beings . . . .

Marine biologists who are interested in birds usually census marine birds over the oceans, or shorebirds along the coasts. Most censuses do not report on the numbers of all species found along coasts. Shorebirds and gulls are often omitted from these seabird censuses (Burger). However, data are available from the shelf waters off the US (Powers), Gulf of San José in Argentina (Jehl *et al.*), several bays in England (Prater), Mollendo in Peru (Hughes), and Cape Town in South Africa (McLachlan *et al.*). In those studies gulls and terns together accounted for between 12 percent (Weston, England) and 94 percent (Wash, England) of the birds. In general, gulls and terns comprised 30–60 percent of the inshore avifauna (Burger).

Source 2

Thorhaug, Anitra, and Beverly Miller. "Stemming the Loss of Coastal Wetland Habitats: Jamaica as a Model for Tropical Developing Countries?" *Environmental Conservation* 13 (1986): 72–85. Taken from p. 75:

Rapid and accelerating development of coastal-based industry, infrastructure, and urban expansion is occurring throughout the developing world, of which many nations are tropical and have extensive coastlines. Owing *inter alia* to the World Conservation Strategy (ICUN-UNEP-WWF, 1980), increased attention is being paid by developing nations to coastal-zone management for long-term conservation of critical coastal habitats (in the tropics especially mangroves, marshes, seagrasses, and coral reefs). Tropical nations are implementing new policies and regulations to protect coastal habitats and their associated fisheries from industrial and other "development," which erodes away the natural resource-bases if left unmanaged (Fig. 1).

To overcome the ill effects of rapid industrial, urban, and tourist expansion along the 420 miles (672 km) coastline, with its accompanying transmigration of inland population towards coastal cities, increasing demophoric pressures, and over a million tourists per year, the island nation of Jamaica has provided a model for the concept of zero-loss policy of their coastal wetlands.

Source 3

Sumeriz, Philip. "Environmental Reconciliation of Opposing Views: The New Course." *Journal of Environmental Thought* 27 (1991): 47–90. Taken from p. 47:

Surely, those with any sense will see the importance of preserving our coastal habitats. Each species contributes to the overall ecological balance of a habitat in numerous ways that we can only begin to fathom. Directly or indirectly, each species is delicately connected to all of the other species in a habitat. If we had known what we know now twenty years ago, maybe our cause would be less critical. But we must act with what we know now.

What is needed is a consolidation of the many sides of the ecological debate. On one side are the industrialists, who are concerned more with the preservation of their products than with the niches of the seagulls and seagrasses. On the other side are the ecologists, who are concerned with allowing each species to play its role in nature undisturbed. On some of the other sides of the issue are those who would seek ways to reconcile the views of the industrialists and ecologists. This should be our course.

(**Source:** Exercise and third passage created by Ronald L. Stone and used with permission of the author. First and third passages reprinted with the kind permission of Elsevier Sequoia, Lausanne, Switzerland, publishers of *Environmental Conservation*.)

2. Let's assume that you are working for a health agency in Minneapolis, Minnesota, and you are asked by your supervisor to write a report on the status of AIDS in the state. In researching this topic you have gathered the documents concerning AIDS in Minnesota contained in Appendix A, Case Documents 1. Develop an informal outline that classifies all of this information.

3. Read "AZT Step Fact Sheet" (Appendix A, Case Documents 1). Assume that you are going to draw information from this document for a report on AIDS treatments. Are there sections of the "Fact Sheet" that you could be presented in any of the following organizational patterns: chronological step-by-step, spatial, classification/partition, comparison/contrast? Develop outlines for each pattern that you can see in that information.

4. Write a limited, descriptive abstract and an informative abstract of the "AZT Step Fact Sheet" (Appendix A, Case Documents 1). In addition to your abstracts, explain in a brief memo the relative importance or unimportance of what you stated/left out.

5. Your research on writing processes in the workplace has led you to include information from the sources listed below. Revise this list so that it meets either APA style, Chicago style B, or MLA style standards. If necessary, consult an APA, Chicago, or MLA handbook for guidance.

Carl G. Herndl, Barbara A. Fennel, Carolyn R. Miller: "Understanding failures in organizational discourse: The accident at Three Mile Island and the shuttle *Challenger* disaster," Textual Dynamics of the Professions: Historical and Contemporary Studies of Writing in Professional Communities. Charles Bazerman and James Paradis, editors, University of Wisconsin Press, Madison, WI, pages 279–305, 1991.

Lee Odell, Dixie Goswami, Anne J. Herrington, Doris Quick (1983): "Studying Writing in Non-Academic Settings" found in New essays in technical and scientific communication: Research, theory, practice. Paul Anderson, R.

John Brockmann, and Carolyn R. Miller, editors, Baywood Publishers, Far-mingdale, NY, pages 17–40, 1983.

Lee Odell: "Beyond the text: Relations between writing and social con-text." Writing in nonacademic settings. L. Odell and D. Goswami, editors, Guilford Press, New York, NY, pages 249–280, 1985.

Christine Barabas, Technical writing in a corporate culture. Ablex Publish-ing, Norwood, New Jersey 1990.

Lisa Ede, Andrea Lunsford: (1990) Singular Texts/Plural Authors: Perspec-tives on Collaborative Writing. Southern Illinois University Press, Carbon-dale, Illinois, 1990.

6. In Appendix A, Case Documents 2, you will find an "Executive Summary" to "A Report Detailing the Effectiveness and Safety of Bioremediation in Prince William Sound." This "Executive Summary" is really an informative abstract. What features does it contain that are typical of an informative abstract? Does it differ in any way? Now rewrite the abstract so that it becomes a complete, descriptive abstract.

7. Go to your library and find current statistics on the reported HIV positive and AIDS cases in your state. Then, find a recent article on the treatment of AIDS using AZT. Finally, see if you can find some specific information on the costs of AZT versus the benefits of AZT. Write a one- or two-page summary of your findings.

8. Go to your library and find information on the response of environmental groups to the *Exxon Valdez* oil spill. Review the case documents in Appendix A (Case Documents 2) on the spill and write a comparison/contrast piece on the different points of view you discover in your reading and research.

9. Select one of the liability cases described in Appendix A, Case Documents 4. Assume that you are working for a corporation similar to the one most heavily involved in the liability case.

    Your supervisor would like a brief (about one page) summary of the case, its consequences, and its implications for your company. You will need to decide what appears to be factual in the case, what is inference, and what is expert opinion. You will also need to decide when to quote from the document, when to paraphrase, and when to summarize.

    After you have completed the summary of one case, write for your supervi-sor a brief (one- to two-page) summary of all the liabilities cases in Appen-dix A, Case Documents 4. You will have to make the same evaluation and judgments called for in your first summary.

Technical Editing and Style

Editing ▮ Substantive Editing ▮ Copyediting ▮ Proofreading ▮ Style

I am in love with language. . . . It has been a curious discovery for me that working . . . with the many ways of writing English, almost all prose, has given me a strong sense of the unity behind the infinite variety of English. It is a unity you can only find through a conscious exploration of the language as it serves many different purposes. If precision and exactness are central to the writing of a poem, so are they to the framing of a business memo, the working out of an application for a job, or . . . the writing of a speech.

Lauris Edmond, "Imagining Ourselves,"

Thinking about editing and style is not likely to be something you look forward to when you work on a document. You may think about these elements in terms of the red marks you may have received on a paper turned in for a college writing class. However, there is much more to editing and style than visions of nitpicking detail; they are integral components of successful writing projects and an essential part of revising. Failure to pay attention to these elements of effective writing can make your project more difficult to read at best and at worst can result in injury to that person relying on your words. Moreover, as one who communicates about technology, you have an obligation to create accurate and easily understood documents.

One way to begin thinking about editing and style is to consider the elements that comprise them. **Editing** includes attention to style, grammar, punctuation, and usage. You edit to ensure that your document has the level of correctness and accuracy necessary for all technical documents. Throughout this editing process, you address issues of style. **Style** is the way in which you express yourself, as opposed to the content or substance of your words. When you think about style, you must consider the overall effect of the words and sentence structures you choose. Because you may have had limited experience dealing with these issues, we will look at style separately.

Grammar, punctuation, and usage should be familiar concepts to you. Once you are certain that you are on the right track with a paper in terms of its content, organization, and style, then you will have to turn your attention more fully to grammar, punctuation, and usage. We use the word **grammar** to refer to the rules that govern how words fit together in sentences. **Punctuation** refers to the rules governing the correct use of punctuation marks; **usage** refers to following accepted practice in the choice of words.

To determine to what extent you might need to brush up in any of these areas, you might wish to take the diagnostic test in Appendix B. If you score less than 80 percent on it, you should review basic grammar, punctuation, and usage by trying the exercises provided to you by your instructor or the ones that follow the diagnostic test, reading some of the additional sources we have listed in Appendix B, or taking an appropriate class. If English is not your first language, you may wish to consult with the English as a Second Language (ESL) faculty at your college or university for additional advice. In this chapter, we discuss those aspects of editing and style that are most essential to technical communication.

Introduction

editing
the process of making a document more accurate, effective, and readable by ensuring that the grammar, usage, style, and punctuation as well as the organization of a document are accurate and appropriate

style
the overall effect of word choice, sentence and paragraph structures, and tone in a document

grammar
the rules that govern the way words are placed and used in sentences

punctuation
the rules governing the correct use of punctuation marks

usage
the accepted practice of word selection

Editing

The goal of all editing is to create a more accurate and readable document. **Readability** refers to the ease with which a reader can make use of a document. By readability we are not referring to the rather artificial analyses provided by some word processing and writing programs that count the number of words per sentence or the average number of syllables in a

readability
the ease with which a reader can make use of a document; includes accuracy, appropriateness, and completeness of a document

substantive editing
an evaluative task that occurs in the editing process in which the writer examines the document as a whole as well as its form, content, organization, and style

copyediting
an editorial task that occurs after a document has been substantively edited in preparation for typesetting; focuses on completeness, accuracy, and correctness

proofreading
an editorial task that occurs after the last draft has been prepared for typesetting but before it has been printed; the last time to catch all errors

document and then derive the reading level at which the document is aimed. Neither do we mean that a readable document is the same as a document whose content is easy or low level. Rather readability refers to a level of complexity, design, and style appropriate for the document and the readership. Editing then is a process designed to take a document from its first completed draft through to its final accurate and readable form. Depending on where you are in the writing process, your editing task will differ, and depending on where you work, you will find yourself responsible for different tasks. In some companies, much of the editing is covered by special departments that do nothing but proofreading and copyediting. Most companies, however, require that writers shepherd a document through several distinct edits that may involve a range of readers and checks on a number of different document elements.

In your preparation of technical and scientific documents, you will generally perform at least substantive editing and copyediting. During these processes, for example, you will make sure that references in the text (such as, "See Figure 1") match up with the things to which they refer. You will eliminate unacceptable errors: sentence fragments, misspelled words, subject-verb disagreement, or incomprehensible sentences. You will also perform proofreading throughout the writing process, but particularly in the final stages. We'll look at substantive editing, copyediting, and proofreading separately in the sections of this chapter, but in actual practice you will often engage in all three as you work through the writing process.

Substantive editing is an evaluative task that occurs early in the editing process. Substantive editing requires that you look at a document as a whole and make necessary changes in form, content, organization, and style. When you do substantive editing, you often delete or add material to your document, check the strength of your argument or evidence, make sure you have conveyed your message accurately, or change your document based on how well it "tested" with your user.

Copyediting and **proofreading** are related tasks and are often confused with one other. Copyediting occurs after a document has been substantively edited, after you've made certain that the document addresses its audience and has the correct form, content, and organization. Part of your task at this stage is to make sure that the document is correct in terms of grammar, usage, and punctuation, and that the format and style are accurate and consistent. Another goal of copyediting is preparing a document for typesetting or desktop publishing. Copyeditors make sure that all the parts are there, that everything matches, and that the document is accurate and correct.

Proofreading occurs after the final draft has been prepared but before copies of the document have been printed. It's the last chance for you to make sure that the copy is error free, that the page breaks occur at all the right places, and that the document will look as you intended. Your credibility and often your readers' safety will depend on your thorough and careful editing and proofreading.

Substantive Editing

When you edit substantively, you are looking at the big picture, at how content and meaning are conveyed in the text. Substantive editing is the process you use when you are working with an early draft and making sure that it has all the information you need and that it is aimed at the correct audience. Early in this book, we noted that a technical document could have a variety of purposes and audiences. These ranged from documents that informed experts within an organization and documents that described products to clients, to documents that communicated with regulatory groups. When you write a document, you perform a careful analysis of the audience and purpose of your work. When you edit a document, you will need to return to that earlier analysis.

Once the first draft is complete, you can begin giving the document a substantive edit. Example 10–1 illustrates how substantive editing can make a document clearer, more appropriate for its audience, more accurate, and more easily understood, often in fewer words. When you edit substantively, you consider how well the document meets the audience and purpose for which it was designed. You look at the document from a number of perspectives: organization, style, correctness, appeal, clarity, and so forth.

ANALYZING YOUR AUDIENCE, PURPOSE, AND USE

Although you have already determined the audience, purpose, and use of your document before you or your group began writing it, you need to look at your draft during the editing process with those same points in mind. Writing Strategies 10–1 can remind you about what you want your document to accomplish and who will be reading it. Once you are satisfied that you have kept in mind throughout the document who will be reading your document and for what purpose, you will need to evaluate it in terms of its content, organization, format, and style.

EVALUATING YOUR DOCUMENT

Evaluating how successfully your own document addresses its audience, how well it serves its purpose, and how well it is designed for the uses to which it will be put is difficult to do. We are all attached to our own words. That's why in collaborative writing environments, it is often helpful to ask one of the people in the group who is not tied to the form and content of the current draft to do the first editing tasks. If you are going to edit your

---

EXAMPLE 10–1   Substantive Editing of Documents

Look at the differences between the two versions of the same manuscript below. Although the unedited manuscript contains spelling, usage, grammar, and punctuation errors, it also leaves out essential information for its audience of medical care providers. The edited manuscript has been edited and is easier to read and understand. Identify the problems with the unedited manuscript. Indicate what changes have been made and why you think they were made.

Unedited Manuscript

The Minnesota Obstetric Management Inititave is a systems analyses, organized by time, consisting of a questionnaire. It is a research tool; not a quality assurance tool yet. This study wanted to know if the cases were managed appropriately or not. MOMI is the study of hospital births and medical record analysis. The study is a sampling of the 37,921 births in Minnesota between 1988 and May 1989. It is a random sampling of 5000 births which hopefully serves as a real reflection of this community. The statistics presented here are not final or publishable, but you can see trends in terms of the study and eventually the data will be published in JAMA. (114 words)

Edited Manuscript

The Minnesota Obstetric Management Initiative (MOMI) is a systems analysis, organized by time. MOMI examines hospital births and medical records, using a 32-page questionnaire to gather data. Designed to determine whether pregnancy cases were managed appropriately or not, the study covers a random sample of 5,000 births selected from the 37,921 births in Minnesota between May 1988 and May 1989. As such, it provides an accurate reflection of pregnancy in the state. MOMI is a research tool, not a quality assurance tool. The statistics presented here are not final but identify trends and eventually will be published in the *Journal of the American Medication Association* (108 words)

---

own work, try to let it sit a day or two before beginning your evaluation process so you can gain some perspective on what you've done.

Because it's easier to edit someone else's work, we have given you a sample text to evaluate in Example 10-2. Look at Appendix A, Case Documents 4, *Aetna Casualty and Surety Co. v. Jeppesen and Co.,* to refresh your memory about the context of the example. Let's assume you have written a memorandum to the communications department informing them of the outcome of the lawsuit against Jeppesen and Co. and asking them to make changes in their editing procedures to prevent future litigation.

Before you begin editing, you perform an analysis of the existing draft. You know that your audience consists of experts in document design, that they will read the entire memo, and that they may be hostile or defensive about the matter because their work has resulted in both a fatal accident and a lawsuit. You remind yourself that the goal in writing the memorandum

---

WRITING STRATEGIES 10–1   Analysis Checklist

Before you begin editing your draft, answer again the following questions about its audience, purpose, and use:

- What is the purpose of this document? What problem are you helping your audience solve?
- Who will use this document? Do you have a single audience with multiple needs? A simple, homogeneous audience? A multiple audience?
- Does your audience consist of transmitters? Decision-makers? Advisors? Implementers?
- How knowledgeable is that audience about the content of the document? What is the audience's organizational role?
- How familiar is the audience with this format?
- What is that audience's attitude toward you? Your message? What is the audience's attitude in the act of reading?
- How will the document be read? For reference? Straight through? As a quick reminder on the shop floor?
- What experience has the audience had with the subject? In school? On the job?
- Are you creating an audience or reader within your document?
- How can you establish your *ethos* and that of your organization or company? How can you create a reliable, trustworthy persona? What is an appropriate voice for that persona?
- What aspects of *pathos* should you employ? How can you appeal to your audience's most basic, heartfelt attitudes and values?
- What "good reasons" should you provide for your arguments in the documents? How should you express these reasons in *because-clauses*? What is the *logos* or logic of what you must communicate? Should you argue by definition, from consequence, from value, by comparison and contrast, by citing authority, and/or by countering objections?
- What kind of document is it? A memo? A manual? A feasibility report?

---

is not to have them all quit but to tell them the outcome of the lawsuit and to ask them to institute a new editorial process to make sure oversights such as the one that caused the accident don't happen again.

You will need to decide if the memo has all of the information necessary for your audience to understand the issue facing them. You also want to organize the information effectively because your memo will be read by people who work in communication, and it's important that you retain your credibility as their manager. Also, because you are sure that your audience will be defensive, you want the style to be formal but friendly enough to convey support for the staff along with a clear indication that the staff will have to make the changes you indicate. With these considerations in mind, substantively edit Example 10-2. After you have finished your edit, notice

EXAMPLE 10–2 First Draft of Memo

30 May 1993
      TO: Communications Department Staff
   FROM: Jane Lilly, Manager
  SUBJECT: Lawsuit Involving Communications Department

By now you know that Jeppesen lost a major law suit because of the charts you produced of the airport instrument approach to Las Vegas. On November 15, 1964, a Bonanza Airlines plane crashed in its approach to Las Vegas, and all the people died in the crash caused by your faulty charts. Aetna Casualty and Surety Co. was required to pay a large wrongful death settlement. Aetna sued our company to recover the damages. The court found that the crew members were not negligent in relying on the defective chart because this chart was so much unlike other charts we've created.

You know that our products involve the setting forth of FAA specifications in tabular form accompanied by graphic approach charts. Usually, we have two views of the proper approach. One is a plan, depicted as if one were looking down on the approach segment of the flight from directly above. The second chart depicts a profile, presenting a side view of the approach with a descending line depicting the minimum allowable altitudes as the approach progresses. Well, you got all the data right, but you really messed up on the charts. The plan chart showed minimum altitude at the distance of 15 miles from the Las Vegas Airport as 6,000 feet. The profile chart does not extend beyond three miles and shows a minimum altitude as 3,100 feet. You created two views that were the same size but the scale of the plan is five times that of the profile. This isn't like other charts you have created for other airports in which the profile and plan charts had the same scale.

The courts found that the information you conveyed was completely correct, but that your Las Vegas charts "radically departed" from the usual presentation of graphics on our other charts and as a consequence the graphics rendered the chart unreasonably dangerous and a defective product.

From now on we can't have any more mistakes like this. I want you all to double check all charts making sure that not only are the data correct but that the same scale applies to all charts. I then want you to go through all existing charts and develop a revision schedule to make sure that any problems we might have with existing products are cleared up. I want these changes to be effective immediately.

Let me know if there's a problem.

the changes we made in Example 10-3. The use of headings and white space help to make the document friendlier to the eye. Because the author anticipated a hostile audience, she revised the text to eliminate the blaming tone and to replace the word *you* with the word *we*. Look over this revision carefully and see what other changes you can identify. How does it resemble and differ from your own revision?

MAKING YOUR SENTENCES WORK

Substantive editing requires that you pay special attention to the building blocks of your document: the sentences that convey your meaning. Each of your sentences should be structured to reinforce meaning. When we look at style shortly, we will look at some other ways of improving your sentences.

EXAMPLE 10–3   Draft of Memo After First Editing

30 May 1993
          TO: Communications Department Staff, Jeppesen and Co.
      FROM: Jane Lilly, Manager
   SUBJECT: Resolution of the Lawsuit and Development of New Editing Procedures

Resolution of the Lawsuit

As many of you know, Jeppesen recently lost a major law suit stemming from the charts we produced of the airport instrument approach to Las Vegas. On November 15, 1964, a Bonanza Airlines plane crashed in its approach to Las Vegas, with 100 percent fatalities. Aetna Casualty and Surety Co. was required to pay a large wrongful death settlement. Aetna sued us to recover the damages. The court decided that the crew members were not negligent as they had relied on a chart much unlike other charts we've created. That means that we have to bear the ultimate legal responsibility for what happened, despite the fact that our error was unintentional.

Specific Nature of the Defect in the Charts

The materials we create set forth FAA specifications in tabular form accompanied by graphic approach charts. Usually, we have two views of the proper approach. One is a plan that depicts the approach segment of the flight as if one were looking down from directly above. The second chart shows a side view of the approach with a descending line depicting the minimum allowable altitudes as the approach progresses. In this case, the data in the tabular form were accurate, but there was a problem of scale between the two approach charts. The plan chart showed minimum altitude at the distance of *15 miles* from the Las Vegas Airport as *6,000 feet*. The profile chart did not extend beyond *three miles* and showed a minimum altitude as *3,100 feet*. The two views were the same size but the scale of the plan was five times that of the profile. The pilots couldn't anticipate this difference because our charts created for other airports have profile and plan charts at the same scale.

   This was the basis of the lawsuit against us. The courts found that the information conveyed in the charts was completely correct, but that the Las Vegas charts "radically departed" from the usual presentation of graphics on our other charts. As a consequence, the graphics rendered the chart unreasonably dangerous and a defective product.

New Procedures

I know that the defects in our charts were accidental and that we all wish we had discovered the problem under other circumstances. However, in order to minimize the chance of this kind of problem ever happening again, we're going to need to develop some new editing procedures. I have appointed an editing committee whose responsibility will be to look at our work both in terms of accuracy and consistency. I've asked them to begin by double checking all charts to make sure that the data are correct and that the same scale is used on all charts. I would like all of you to schedule a time when you can meet with this new group and help it develop a series of procedures to guard against any other serious defects that we might inadvertently introduce into our work.

   Let me know if you can think of any other procedures and safeguards we might want to introduce at this same time.

At this point, however, we'd like to direct you to a number of techniques you can use to polish your sentences.

Emphasizing the Main Point

There are a number of ways to emphasize your main point, and when you evaluate your sentences you will want to consider where you place your main idea, how you handle coordination and subordination, how you get your sentences off to a fast start, and whether your use of modifiers supports your main ideas.

Placement of Core Ideas   The main point you hope to make should be in the subject and predicate of the independent clause making up the core of each sentence. Less important parts of the sentence—subordinate or dependent clauses and phrases—should be used for additional and less central information.

Look at the following sentence:

> A *program has been initiated* to encourage the nesting of sparrow hawks in the plot areas as a further means of biological control of birds.

In this sentence, the subject and predicate, which represent the core of the sentence, are italicized. Unfortunately, in this case the writer has used the core of the sentence for less important information and placed the most important idea in subordinate phrases. What is actually most important is that sparrow hawks are being encouraged to nest in the plot as a means of biological control of birds.

The following sentences show two ways the sentence might be rewritten, placing the most significant elements in the core (again, subject and predicate have been italicized):

> The *nesting of sparrow hawks* in the plot areas *has been encouraged* as a further biological control of birds.

> As a further means of biologically controlling birds, *the nesting of sparrow hawks* in the plot areas *has been encouraged.*

If the writer wants to put even more emphasis on biological control, the sentence could be rewritten as follows:

> *Birds can be biologically controlled* if we encourage sparrow hawks to nest in plot areas.

The sentences above have the important idea in the sentence located centrally. Restructuring sentences in this way helps readers determine what the important ideas of a document are.

Coordination and Subordination   It's also important to make sure ideas in the sentence are **coordinated** or **subordinated** to enhance clarity and proper emphasis. If you put your main ideas in a subordinate position, your

coordination
refers to the linking of two equal ideas, usually with a coordinating conjunction

subordination
refers to the linking of two unequal ideas so that the relationship between them is clear

readers may be confused by your sentences. What should be subordinated and what should be central will depend on what you are trying to convey. Consider the following sentence:

> Al Abramson, who is a former employee of the company, owned the patent for the first model of this machine.

This sentence is structured to emphasize the fact that Al Abramson owned the patent—that he is a former employee is subordinated to that main idea. The sentence could be written differently, however:

> A former employee of the company, Al Abramson, owned the patent for the first model of this machine.

In this sentence, the reader is directed to a different main idea: that a former employee of the company owned the patent. That it was Al Abramson is less important and subordinated.

Sometimes your sentence includes a number of points all of equal significance. In this case your task is not to subordinate one idea to another but to coordinate them properly. The cluster of sentences below need to be revised so that the ideas contained in them can be coordinated. As they are, the sentences are repetitious, choppy, and ineffective:

> Do not install the appliance near water. Do not place it near a washbowl, bathtub, kitchen sink, or laundry tub. Do not consider putting it in a wet basement or near a swimming pool.

Below the sentences have been combined and their ideas coordinated. The sentence, with its central idea about where to avoid placing the appliance, is more effective:

> Do not install the appliance near water such as a washbowl, bathtub, kitchen sink, or laundry tub, and do not place it in a wet basement or near a swimming pool.

Often coordinating the ideas of several short, related sentences can cut down on the wordiness of your document. It can also create more effective and more convincing sentences.

Vague Pronouns and Slow Starts

Writers often give away the power in their sentences by placing a **vague pronoun** in the place of the main subject or getting their sentences off to a slow start. When you edit, look for sentences that begin with *It, There,* and other pronouns to see if you could begin the sentences more directly and quickly. Consider the sentences below:

> It is important to wear a respirator when cleaning with solvents.

> It is clear that tuberculosis has been brought under control in the United States but has not been eliminated.

> There is an important reason for getting a tetanus shot: fatal lockjaw.

vague pronouns
refers to the use of *it*, *there*, and other pronouns to begin a sentence in place of the main idea

These sentences are not as effective as they might be because they squander the power of the core sentence by using vague pronouns as the main subject or delay the reader getting to the important part of the sentence. Below, we've rewritten these sentences, eliminating the vague pronouns and moving the content into the sentence core to make them more effective:

Wear a respirator when cleaning with solvents.

or

You should wear a respirator when cleaning with solvents.

Tuberculosis has been brought under control in the United States but has not been eliminated.

or

Although under control, tuberculosis has not been eliminated in the United States.

Averting fatal lockjaw is an important reason for getting a tetanus shot.

or

Getting a tetanus shot prevents fatal lockjaw.

You do not have to remove every *it is* or *there are* you come across in your writing. On some occasions, you would change the meaning of a sentence if you eliminated the pronoun reference, as in the case below:

It's an imperfect system, but it's better than none.

An imperfect system is better than none.

In other cases removing the vague pronoun reference doesn't do much to make the sentence more effective, as in the following examples:

There are several reasons why you should set the timer on your VCR.

Several reasons exist why you should set the timer on your VCR.

However, your document is generally stronger when you put the main idea of a sentence in its core, and one way to do that is to eliminate as many vague pronoun references and slow starts as possible. If you are working with a computer, you can use some of its features to help you identify these problems in your document, and you can edit those that are ineffective. Use the *find* or *replace* command in your word processing software to locate each ''it is'' and ''there is'' in your text and examine each one as it is identified.

modifiers
words, phrases, and clauses that describe other elements in sentences

Modifiers are words, phrases, and clauses that describe other elements in your sentences. Because technical and scientific documents make heavy use of modifiers, you will need to make sure they are used accurately in your writing. You do not want modifiers to confuse a reader looking for your main idea. One way to check accuracy in your use of modifiers is to look at their placement. In the sentences below, notice how the meaning changes when the modifier is moved:

> *Even* the adapter may become warm when it is being used.
>
> The adapter *even* may become warm when it is being used.
>
> The adapted may become warm *even* when it is being used.

Sometimes modifiers are used in sentences but are not clearly linked to a referent. Such modifiers are called dangling or misplaced modifiers, and they don't say what you think they do. In fact, the result of dangling modifiers is often unintentionally humorous at your expense. Examine the pairs of sentences below. The first sentence in each pair has a dangling modifier underlined; in the second sentence, the error is corrected.

> *Running down the street,* the trash can tripped the police officer.
>
> Running down the street, the police officer tripped over the trash can.
>
> *Looking northward,* the Huron Islands can be seen.
>
> Looking northward, you can see the Huron Islands.
>
> *Although ten years old at the time,* my father decided that I was still too young to accompany him on the Canadian fishing trip.
>
> Although I was ten years old at the time, my father decided that I was still too young to accompany him on the Canadian fishing trip.

Notice that we have fixed these sentences by moving the modifier next to the subject to which it refers or moving the subject into the modifying phrase. During substantive editing, a close look at your modifiers will help ensure clarity and conciseness.

LOOKING FOR THE BEST WORDS

Not all of your editing will be done at the sentence level; some will be done at the word level. As with editing sentences, your goal is to create the most effective and precise communication. Often your success in creating useful documents will depend on how well you select and arrange specific words.

Action Verbs

One way to write effective documents is to choose verbs that convey action precisely. When you edit, look at each sentence and ask yourself if the verb conveys the precise meaning you intend. The following sentences illustrate how imprecise verbs can obscure the action of your sentence. In the rewritten forms, the less precise verbs are replaced by strong, precise **action verbs** (in italics):

action verbs
verbs that convey precise, rather than general, meaning

An electrical surge caused the destruction of the hard drive.

An electrical surge *destroyed* the hard drive.

Gears can injure fingers.

Gears can *crush* fingers.

Whenever possible avoid the use of the verb *to be (is, am, be, was, were, being, been)*, using in its place more precise action verbs, as in the following sentences and their revisions:

Failure to respond to the complaint would be in violation of university policy.

Failure to respond to the complaint *violates* university policy.

The facts are in support of the judge's recommendation.

The facts *support* the judge's recommendation.

Replacing weak verbs with strong ones will make your writing both more precise and more interesting.

Concrete Nouns

In addition to selecting the right verbs, you will need to pay attention to the nouns in your sentences. In some cases, problems with the nouns create problems with the verbs as well. Generally speaking, try to avoid the use of **nominalizations** in your writing (Figure 10–1 gives you some examples of how nominalizations are formed). Nominalizations are verbs that have been made into nouns and can often be identified by their endings: *tion, ment, ance, ing,* and *ence.* Not all words that end with these suffixes are nominalizations; for example, the word *nation* ends in *tion* but isn't a nominalization. When you find nominalizations in your sentences, try to turn the noun back into a verb and move the main idea of the sentence into the subject position. Examine the pairs of sentences below to see how this might be done. The nominalizations have been italicized.

nominalizations
verbs that have been made into nouns

Incorporation of the latest electronics technology makes this keyboard versatile.

This versatile keyboard incorporates the latest electronics technology.

Performance of the tape player is better with fresh batteries.

The tape player performs better with fresh batteries.

Confinement of criminals seldom results in *rehabilitation.*

Criminals who are confined are seldom rehabilitated.

Notice that in the second sentence of each pair, the nominalization has been replaced as the subject and has been recast as a verb. Word processing programs can help with revision on the word level by identifying nominaliz-

Verb	Nominalization
condemn	condemnation
abandon	abandonment
house	housing
perform	performance
emerge	emergence

FIGURE 10–1
Common Nominalizations

ations for you. As you revise, have the computer search for the word endings that mark nominalizations.

You can also make your words concrete by lowering the level of abstraction. For example, the following two lists begin with the most abstract term and end with the most concrete:

List 1	*List 2*
computer program	destructive birds
word-processing programs	sparrows
Microsoft Word program	2,500 sparrows per year
Microsoft Word 5.1 for the Macintosh	2,500 sparrows between March 1993 and March 1994

Ask yourself precisely who, what, where, when, how much, what amount, what degree, what size, what percent, and what type, and reflect your answers in the most concrete way.

Parallelism

Sometimes as we revise and combine sentences we lose track of the relationship among the elements in a sentence, and we can create structures that aren't parallel. **Parallelism** means that similar items are expressed in a similar manner; lack of parallelism makes a sentence much harder to read. Consider the two sentences below: the first lacks parallelism in its structure, but the revision expresses similar actions—retains/relies—in a similar manner:

parallelism
expressing terms in a similar manner with respect to such elements as tense, verb form, and mood

> The memory back-up power means the electronic memory *retains* its contents and *is relying* on either the AC adapter or the batteries.

> The memory back-up power means the electronic memory *retains* its contents and *relies* on either the AC adapter or the batteries.

Making sure that sentences have parallel construction helps the reader interpret the information correctly.

Positive Phrasing

Readers will become more confident about your choices and suggestions if your words carry a positive tone. Also, people misunderstand and make mistakes interpreting and acting upon negatively worded messages. The first choice in the following list emphasizes the negative and the second choice the positive:

First Choice	*Second Choice*
do not accept	reject
not many	few
not reliable	unreliable
not efficient	inefficient

Changing your whole sentence is sometimes more effective in emphasizing the positive. Look at the sentences below and the revisions that follow each:

The project will not be complete on time, and so we will not be able to stay within the budget.

The project has been delayed, and the budget must be extended.

Because safety tests are not complete, we cannot recommend adopting the new system.

Until safety tests are completed, we must postpone adopting the system.

Of course, when warning your readers about potential hazards, the negative carries the strongest possible message:

Do not operate this equipment without a surge protector.

As you will see in later discussions about warnings, other design features such as color, boldface, and layout can help readers comprehend the messages.

EDITING FOR ORGANIZATION

Part of the job you face when you edit substantively is making sure the organization of your material is clear and fits audience expectations. Although you can organize material many ways in a document, sometimes the conventions of the genre in which you are writing limits your choices. At other times, you will have to apply what you have learned about how people read and use documents to organize your work most effectively.

Following the Conventions of the Genre

In many cases, you will be asked to create a specific kind of document such as a request for proposal (RFP), a memorandum, or a feasibility report. As you work with these specific documents, you will need to keep in mind that each of these genres has specific components, a recognized order, and an acceptable length. When you are asked to write a particular kind of document, you are often provided with a guide for the appearance of the document, its audience, its length, and its due date. You should look at this guide carefully and follow it exactly. When you write in an established format, you can obtain successful examples of other people's work to serve as a template for organizing your own. Always follow the order specified in the genre because that is the order your audience expects and is most familiar with. If you violate that order, your work will run in opposition to your audience's expectations and is more likely to fail.

Using Recognized Patterns for Organizing Materials

Research by cognitive psychologists has shown us that readers bring to a document a set of organizing principles or expectations. By having your document fit into that set of principles, you increase the likelihood that your document will be read accurately and easily. In all cases you must edit your document with this advice in mind: the structure of the document must support its meaning and your argument.

As we noted in Chapter 9, a variety of organizing principles apply to documents: comparison/contrast, spatial patterns, chronological/step-by-step, and combinations of these patterns. When framing an argument, as we discussed in Chapter 5, you might organize according to your *logos* or the logic of the argument: defining, describing consequence, determining value, comparing/contrasting, citing authority, and countering objections. When you edit for organization, your task is to make sure that information is organized in the pattern most familiar to readers of your document and most effective to your argument. You also need to make certain that there are no missing parts in the document; for example, you must have a beginning, middle, and end.

As you edit, verify that each section of your document has the same components. For example, this textbook is composed of chapters. Each chapter is organized on a similar template. You might want to look at the table of contents to see if you can determine what the component parts of each chapter is. In a similar way, when you edit your document, you will want to determine what the component parts of each unit are and make sure all are present.

transitions

words that emphasize links between thoughts within or between sentences by indicating relationships in time, place, cause and effect, contrast, or similarity

Providing Transitions Between Thoughts

Regardless of the organization you choose, guide your reader through your document with **transitions**—or words that link thoughts. The following passage illustrates how transitional words help a reader see the relationship among sentences or paragraphs. In some cases, a word or phrase is repeated to help the reader keep track of the main topic:

> The new computer system replaced System 7.0 only after all virus protection had been updated. *When* [transition] the *system* [repetition] was in place, personnel had to be trained on the new codes. These *codes* [repetition] challenged even the experienced users, a *challenge* [repetition] that management failed to understand in planning the amount of time needed for the complete changeover. *As a result* [transition], the training period was extended for three months.

Although repeated words remind the reader of the topic discussed, transitions indicate relationships in time *(after, before, until, while, during),* place *(above, below, outside, to the right),* cause and effect *(because, as a result, due to),* contrast *(however, on the other hand, instead, although),* and similarity *(moreover, furthermore, likewise, as).*

EDITING FOR FORMAT

Editing for format means checking the appearance of individual pages and sections as well as the appearance of the entire document. When you edit for organization, you make sure the content is complete and arranged in the most familiar and accessible way. When you edit for format, you make the organization of content visible and make sure that textual devices are consistent.

Genres generally have some format characteristics associated with them, as you will learn in the last section of this book. When you edit for format, you will have to make sure that your document follows the conventions of the genre. Writing Strategies 10–2 provides a checklist of elements to consider when you look over the format of your document.

When you analyze a document before editing its format, you are looking for ways to make it consistent and more readable throughout. The format of a successful document doesn't call attention to itself. Rather, it unobtrusively guides the reader through the document. As you edit, try to match format with meaning. For example, typeface should not call attention to itself at the cost of a document's readability. Warnings, cautions, and statements of danger may well be put in bold or even require the use of color in a document, but the use of color, bold or italic letters, or unusual typefaces should never be used just for themselves. Headings should be used only when they are necessary to guide readers through a document and make specific information accessible.

WRITING STRATEGIES 10–2   Editing for Format

Check Your Document for Effective, Appropriate, and Consistent Use of:

- Typeface and type size
- Paragraph divisions
- Margin widths
- Columns
- Bold, italic, underlining
- Headers and footers
- Headings and subheadings
- Space between lines, with headings, with figures
- Genre conventions for layout
- Style manual guidelines
- Indentation and white space
- Numbers and bullets
- Labeling of figures and boxes
- Right, left, and center-justified text
- Measurements and mathematical formulas

Copyediting

Often, after substantively editing, you turn to copyediting, although these two kinds of editing can occur at any time and are not necessarily discrete activities. Copyediting is a more limited kind of editing work, more narrowly focused to look at grammar, punctuation, spelling, tables, and graphs to make sure your document is correct, consistent, and complete.

CORRECTNESS

When you examine a document for **correctness,** you verify the spelling, grammar, and punctuation of each word and sentence. You also make certain that all quotations are accurate, that bibliographic references are exact, and that dates and locations cited are correct. The content of technical and scientific documents must be absolutely accurate. A company's legal liability or a person's life may depend on correct data or information. Therefore, you should recheck all data for accuracy.

correctness
accuracy of citations, bibliographic references, punctuation, spelling, and grammar in a document

CONSISTENCY

When copyediting for **consistency,** you make sure a variety of elements appear appropriately throughout the entire document. For example, look to see that numbers either are consistently written out or presented in numeric form. In scientific and technical documents, rules exist for when to write out a number and when to display it as a figure. For example, some styles

consistency
similarity in display or description; includes consistent use of numeric and written out numbers, units of measurement, hyphens, and stylistic considerations

dictate that all numbers that are 10 or above appear as figures, although one to nine are written out. Numbers preceding units of measurement, such as 2 inches, 10 centimeters, 120 miles, 20 Hz, usually appear as figures regardless of their quantity. Be sure to follow the standards in the discipline in which you are writing.

If you do not have written guidelines, achieve consistency within sentences by using figures for all numbers in a sentence if one of those numbers is 10 or above, as in the following example:

> In our personnel office, 3 managers and 17 support staff are housed.

If any numbers in the sentence are followed by a unit of measurement and therefore are expressed as figures, other numbers can remain written out if they are less than 10, as in the following example:

> The nine tables we ordered each measured 5 feet by 3 feet.

When numbers and words are joined to modify a noun, called a unit modifier, hyphens help the reader follow your sentence, as in the following examples:

> Seven 5-inch screws were missing from the assembly kit.
> Two-car garages have become popular in the suburbs.

Notice that *two* in the last sentence is written out, not only because it begins a sentence but also because *car* is not a unit of measurement. No matter what rules you are following, your job as a copyeditor is to make sure that numbering is consistent.

You will also have to attend to stylistic decisions. For example, many writers of documents used by an international audience no longer feel comfortable using the word *American* to refer to people from the United States. People from Canada and Mexico are also Americans and find the use of the term to refer only to people from the United States exclusionary. Style is a rather complex consideration, and we explore it more fully below.

While copyediting, you will need to decide to upon one consistent set of terms to be used throughout your document. For example, you will need to decide on such issues as hyphenation. In your document, is *online* to be one word or to be hyphenated as *on-line?* Copyediting also requires that you determine that labels, titles, and references to figures appear in the same form.

completeness
the totality of a manuscript, making sure that all parts are present, that all figures referred to in the text actually exist, and that cross references and sources are complete

Completeness

When copyediting for **completeness,** make sure that all the parts of the manuscript are present and that all figures and illustrations referred to in the text actually exist and are labeled with the same numbers as their refer-

---

WRITING STRATEGIES 10–3   Proofreading Checklist

In addition to any errors you can find in spelling and punctuation, proofread for the following:

- Hyphenation, especially at the ends of lines
- Extra spaces between words, between lines
- Words that run together
- Failure to italicize or boldface where called for
- Transpositions of letters and words
- Broken fonts
- Inappropriate page breaks, especially headings that appear as the last line on a page, the last word of a paragraph on a separate page, etc.
- Unaligned type
- Crooked margins
- Inconsistent indentations

---

ences. All cross references and all bibliographic references should be checked.

Proofreading

Proofreading is often the final check of your document and occurs before you print your last draft or before you have a printer print your document. Your goal is to produce a clean copy ready for printing. Proofreading is your last chance to make sure that the document will look and read as you hope it will. Writing Strategies 10-3 will give you an idea of what to look for as you proofread.

You can use several techniques while proofreading. For example, you can work with a dead copy, that is, an earlier, copyedited version of the document. Using a dead copy will allow you to make sure that all the errors caught in earlier editing have been corrected as specified. You can also proofread with a colleague. One person can read the text aloud, including all punctuation and spacing, while the other follows along on a second copy. This process guarantees that two sets of eyes and ears are attending to the text. If you do proofread with a partner, change off the role of reader and the role of listener frequently so that each person stays sharp.

Style

Style may be a feature of technical communication to which you may have given little thought, but it is a critical component in the creation of effective documents. When you work on style, you adjust your document to fit its audience, purpose, and use. If you fail to identify and make use of the correct style, your readers may ignore your document or discount its content.

Unlike most grammar questions for which there is a generally accepted, correct answer, style requires judgment calls on the part of the writer. Many styles exist, each of which can be correct in a given context. The challenge for the technical writer—or for any writer, for that matter—is to select a style that is appropriate for the context in which it will be used. This section offers some guidelines to help you make correct style choices.

CHOOSING THE RIGHT STYLE

There are a wide variety of styles to choose from when you create a document. Style deals with *how* you transfer your ideas to speech or paper, rather than to the technical content of the material. Generally, we think of style as a continuum from formal to informal, as shown in Figure 10-2. Informal style, the kind you might use in a letter to a friend, is marked by a relaxed tone that may include such elements as contractions and slang.

Formal writing and speaking are what you do when you create scholarly or professional communications. Formal styles are used by specialists or experts in a field who convey information to peers or to an educated but nonspecialist audience. Vocabulary in formal style is specialized and has precise meanings. Writers and speakers attempt to create an objective or impersonal tone, in direct contrast to the personal and subjective tone of informal styles. Contractions or slang are avoided, and the sentences are usually longer and more complex. However, a formal style does not have to be stuffy or rely heavily on passive voice. In fact, documents in formal style are often interesting to read because their creators put a special premium on clarity, sentence variety, and active voice.

As you will remember from Chapter 3, when you compose technical documents you create a persona, an image of yourself as author, to whom the reader relates in your documents. To a large extent, it is the style you select that creates this image. Ideally, you want to create the impression that you are competent, trustworthy, accurate, and that you have the reader's needs in mind. Once you choose a style, all the elements of your document, from your salutation to your closing, must reflect it so that your work will be consistent.

ADJUSTING SENTENCE STRUCTURE AND VOCABULARY

When writing for highly trained and/or highly educated audiences, you use a different style than you would select in writing for less well-trained or less well-educated audiences. Of course, you would not write in a condescending manner to less educated audiences, nor would you use stuffy, boring prose for the more educated groups. Instead, your choices about vocabulary, sentence

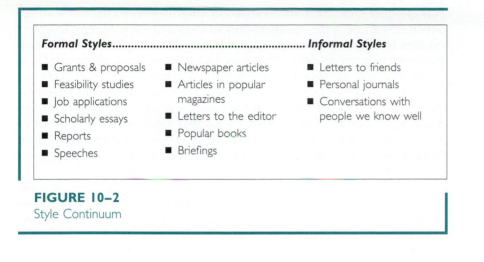

FIGURE 10–2
Style Continuum

complexity, and overall document design must fit the nature of the audience.

Generally speaking, you can increase the complexity and technical nature of your writing as the expertise and education levels of your group increase. This is not a hard and fast rule, however. One important part of your style evaluation should be to examine your use of sentence structure. Your document should show an appropriate variety of sentences for the topic and audience, and each sentence should be carefully crafted to emphasize its meaning.

Varying Sentence Structure

As you probably remember, sentences are either **simple, compound, complex,** or **compound-complex** in structure. Each of these kinds of sentence is effective in certain situations, and you will need to use a variety of sentence structures in your documents to convey your ideas effectively. You can often use more complex and compound-complex sentences and a greater variety of sentence structures for audiences who are highly educated. Stick with simple and compound sentences when your audience is less able to navigate the subtleties of language. If you have trouble recalling what distinguishes these sentence structures, you may want to look over one of the grammar and editing books we list in Appendix B.

Choosing Appropriate Vocabulary

When considering style, you need to determine what vocabulary—which words and terms—you need to address your audience. For example, you would not use discipline-specific terminology when writing to a highly educated audience who lacked expertise in your topic. Example 10–4 contains

simple sentence contains a single independent clause and no dependent clauses.

compound sentence contains two or more independent clauses joined by a coordinating conjunction, a correlative conjunction, or a conjunctive adverb

complex sentence contains both an independent and a dependent clause

compound-complex sentence contains two independent clauses and one or more dependent clauses

passages that each use a formal style but differ in vocabulary and sentence structure because of the audiences they address. Note that the legal judgment consists of one long sentence. Formal legal writing would not be the most effective choice of style for even a very educated audience unless they were attorneys or law students. This style is marked by many prepositional phrases *(in view; in using; in clear, plastic containers)* and expressions familiar to those who work with the law *(could not be held liable; failing to give warning; failure to show).* A formal legal style like this one is also marked by terseness. There is no embellishment, no use of extraneous adjectives. Examine the other passages in Example 10-4 to see how vocabulary and sentence structure varies.

Often you need to "move between" formal styles; that is, you may use a formal style to address both a well-informed and a novice audience on the same subject. Again, your vocabulary (as well as your sentence structure) will vary, as demonstrated in Example 10-5.

DECIDING ABOUT ACTIVE AND PASSIVE VOICE

active and passive voice
active voice clearly identifies the agent of the action, while passive voice does not

As you write, you will have to choose between **active and passive voice.** In active voice, the subject performs the action expressed by the verb, while in passive voice, the subject receives the action (Example 10-6). Be careful not to confuse passive voice with past tense. Generally speaking, you will want to avoid use of the passive voice and concentrate on active voice in your documents. Active voice is recommended for most technical writing because it tends to make your style much more direct and explicit.

As you can see from Example 10-6, active voice sentences generally have strong verbs and more clearly indicate who or what is the agent of action in the sentence. Passive voice sentences are usually longer and have a form of the weak verb *to be* in them. If you don't know the agent of action, or if you want the reader's primary focus to be on the action itself or the object of the action, passive voice is certainly appropriate. However, because our goal is precision in language, consider using the active voice whenever possible. Ethical and practical reasons for preferring active voice also exist, because active voice clearly indicates *who* is responsible for the action.

In the first sentence below, the passive voice obscures the person responsible for maintaining a safe working environment. In the second sentence, responsibility is clearly identified by the active voice:

> The steam pressure control must be turned off by the boiler room engineer before any maintenance of the turbine can be undertaken.

> The boiler room engineer must turn off the steam pressure control before the engineer can undertake any maintenance of the turbine.

The clear assignment of responsibility in technical documents can often be the difference between safety and injury or costly litigation. As you edit,

EXAMPLE 10–4   Variations in Vocabulary and Sentence Structure

Look at the following examples taken from the case documents in the Appendix A. Each is an example of formal style, but not all are alike because the documents in which they appear have different audiences and purposes.

After you look at each example carefully, jot down what assumptions you believe the authors had about their readers and how those assumptions are manifested in examples of vocabulary, sentence complexity, and organization. Determine which example uses the most specialized vocabulary and explain why that is appropriate. Which example assumes the most highly trained or educated audience? Which has the longest, most complex sentences? Which seems to be written for lay audiences and which for content specialists? Make a list of other elements that you believe characterize these as examples of formal style.

Description

AIDS stands for Acquired Immune Deficiency Syndrome, a viral infection that causes a breakdown of the body's immune system. A person infected with Human Immuno-deficiency Virus (HIV) is vulnerable to various diseases and infections. Some of these diseases can be treated, but there is no cure for HIV infection.

Informational Release

The purpose of our bird control program on the campus plots is to protect the experimental plantings that have been initiated for development of new crop varieties. The seed produced on these plots is extremely valuable because it represents the product of genetic crosses that have been carried out over a period of years. With each plant selection there may be only a few seeds produced so it is essential that they be protected.

Legal Judgment

Retail seller of insecticide/nematicide could not be held liable for failing to give warning to buyer about dangers inherent in using poison which appeared like water and was packaged in a translucent one-gallon container similar to plastic milk jug, in view of plaintiff's failure to show that retailer had reason to know or duty to discover by exercise of reasonable care product-connected danger complained of or that retailer should have known that buyer would not appreciate possible harm involved in using toxic pesticide which was packaged in clear, plastic container and looked like water.

be alert for passive constructions and replace them with active constructions whenever possible and appropriate.

USING INCLUSIVE LANGUAGE

Language is a powerful tool that can make your audience feel included or excluded. You do not want to alienate any members of your audience inadvertently because of poor language choices, by displaying sexism toward

---

EXAMPLE 10–5 *Moving Between Formal Styles*

The legal judgment in Example 10–4 is written for a specialized audience. Rewrite this passage so that it remains in a formal style but so that it could be easily understood by an educated lay audience. You should consult the complete case in Appendix A before you begin your revision.

Original Passage

Retail seller of insecticide/nematicide could not be held liable for failing to give warning to buyer about dangers inherent in using poison which appeared like water and was packaged in a translucent one-gallon container similar to plastic milk jug, in view of plaintiff's failure to show that retailer had reason to know or duty to discover by exercise of reasonable care product-connected danger complained of or that retailer should have known that buyer would not appreciate possible harm involved in using toxic pesticide which was packaged in clear, plastic container and looked like water.

Possible Revision

Here's what a revised version might look like:

The hardware store that sold a farmer insecticide, which was inadvertently consumed and poisoned a worker on his farm, cannot be held liable for failing to warn the buyer of the dangers inherent in using a colorless poison packaged in what resembled a plastic milk carton. There was no reason for the seller to think that the buyer would fail to understand the danger implied by using a toxic pesticide that looked like water and that was packaged in a clear, plastic container.

---

them, or by using discriminatory language. Such language is not only unacceptable; it is also insensitive and can constitute a form of harassment or discrimination.

inclusive language
the conscious choice of language that recognizes and includes diversity

In this section of the chapter, we alert you to some ways to recognize exclusionary and discriminatory language while developing a clear idea of how to use **inclusive language**.

EXAMPLE 10–6 *Active and Passive Voice*

Passive Voice	**Active Voice**
The woman was attacked by an assailant.	An assailant attacked the woman.
A decision was reached by the committee.	The committee reached a decision.
Surgical equipment must be checked by the support staff assigned to the operating theater at the beginning of their shifts.	Support staff assigned to the operating theater must check surgical equipment at the beginning of their shifts.

Sexist Language

Sexist language excludes individuals or treats them unequally on the basis of their gender. You should avoid sexist language in your communications. Sometimes that's difficult because you may not always be conscious of how others might read your words. What we offer here are some ways to recognize sexist language and make your writing more inclusive.

While you write and edit professional documents, be sure to avoid the generic pronoun *he*. Once accepted as meaning all people, male and female, research now shows that audiences have difficulty envisioning female subjects when writers and speakers use *he*.

To avoid using *he*, you can recast sentences in the plural:

Original: An engineer must understand not only theory but also application when he works with his clients.

Revision: Engineers must understand not only theory but also application when they work with their clients.

You might also avoid the pronoun altogether:

Revision: Engineers must understand not only theory but also application when working with clients.

You could also shift the person of the pronoun to the first person (*I or we*) or the second person (*you*), or substitute *one*:

Revision: As an engineer you must understand not only theory but also application when you work with your clients.

Revision: As an engineer one must understand not only theory but also application when one works with clients.

You could also use *he or she*, or *she or he* to avoid exclusive constructions but avoid repeating these phrases too often as they become cumbersome.

Revision: An engineer must understand not only theory but also application when he or she works with clients.

In addition to examining pronouns, you must also check your nouns to make sure they are inclusive. Gender-marked terms, such as *manmade* or *chairman*, should be changed to terms that include both genders (*artificial* and *chair* or *chairperson* are acceptable substitutes, for example).

Also, you need to edit out any suffixes that mark a noun as exclusively female or male. Be aware feminine suffixes often imply lesser, special, or diminutive, or simply discriminate between genders when not necessary. For example, we now use *flight attendant* rather than *stewardess* and *waiter* or *server* for both genders rather than *waitress* for females. Generally, suffixes such as *-ess* and *-ette* should alert you that the word might be marked as feminine and should be changed.

You should also avoid identifying jobs or activities as belonging to one gender or the other. If one gender has traditionally filled a role, avoid noting the other gender's entry into that role as an exception. Also, do not distinguish people by their gender (no more than you would do so by their race or religion). To do either makes it difficult for your audience to imagine the job or activity as being open to all. The following examples illustrate how exclusive language can discriminate by gender:

Original: The *lady* manager introduced us to her staff.

Problem: Implies that most managers are men, and calls attention to her gender as her distinguishing characteristic.

Revision: The manager introduced us to her staff.

Original: The *doctor* must use *his* talents to diagnose while the *nurse* must use *hers* to nurture and heal.

Problem: Implies a clear division of job and talent by gender; that is, only men can be doctors, and only women nurses.

Revision: Doctors must use their talents to diagnose while nurses use theirs to nurture and heal.

Original: I would like to introduce you to our two new employees. *Dr. Collins* comes to us from the *physics graduate program* at MIT, a *man* of much distinction already, and *Nancy* is *mother* to three and obviously a very busy *gal.*

Problem: Identifies the male by his formal or professional title and the female by her first name, showing less respect for her. Also notes professional accomplishments for the male but places the female in her traditional private role. Finally, considers the male an adult and the female a youth.

Revision: I would like to introduce to you our two new employees. Professor Collins comes to us from the physics graduate program at MIT, and Professor Dailey from a nine-year position as an aerospace engineer at NASA.

In Figure 10-3, we list some words and phrases that are considered exclusive on the basis of gender and their acceptable alternatives. Although our list is not inclusive, it should begin to give you a good idea of what to look for as you do your final editing of your document.

Other Exclusive or Discriminatory Language

You must also check your language to make sure that you have not excluded or discriminated against other groups—perhaps on the basis of race, religion, age, or special appearance. In some cases, it's a matter of choosing the term that the group prefers. For example, the terms *African American* and *Asian American* are now the terms acceptable to those whose ancestry is so noted. Those people who are living with AIDS prefer to be called just that—or *people with AIDS*—rather than *AIDS victims* or *people with the "disease"*

Exclusionary Terms	Alternatives
fellow man	other people, humans
forefathers	ancestors
kinsman	relative
layman	ordinary person, nonspecialist
man	human, person, individual
to man (verb)	to work, to operate, to staff
manmade	artificial
manpower	personnel
spokesman	representative, speaker, spokesperson
workmanship	skilled work
workman	worker
statesman	leader
chairman	chairperson, chair, head, moderator
clergyman	member of the clergy
congressman	member of Congress, Congressional representative
fireman	firefighter
foreman	supervisor
mailman	letter carrier, postal worker
salesman	salesperson, salesclerk
coed	student
poetess	poet
actress	actor, performer

FIGURE 10–3
Exclusionary Terms and Alternatives

of AIDS. Older people prefer to be called *senior citizens* rather than *the elderly.*

People with disabilities prefer not to be called *the handicapped* ("The handicapped need ramps.") but rather *people with handicaps or disabilities* ("People with disabilities often require ramps."). Those with poor or no eyesight, those who have lost their hearing, those who get around in wheelchairs, each have definite preferences. They generally prefer to be called physically challenged, hearing impaired, or visually impaired, rather than handicapped, deaf, or blind. If you have a question about what to call someone, be sure to ask what he or she prefers.

In general, remember to use your language choices to be inclusive. Make jobs and activities open to all, and to invite your audience to see themselves within your words.

SUMMARY

This chapter has covered a great many aspects of editing and style. You have learned that editing greatly increases the readability of your documents, and that substantive editing includes a careful look at your document as a whole—its form, content, organization, and style—to make sure that it meets the needs of your audience.

To conduct a substantive edit of your document effectively, you need to revisit your analysis of audience, purpose, and style. Substantive editing includes evaluating your document to see how successfully it addresses your audience's needs. Substantive editing also means checking to see if your sentences work, if they emphasize the main point, if they include proper coordination and subordination, if vague pronouns and slow starts have been eliminated, and if modifiers are used accurately.

When doing a substantive edit, you also need to make sure you have used the best words in your document. Editing for action verbs, concrete nouns, parallelism, and positive phrasing strengthens your document. Substantive editing also includes editing for organization, whether it be following the conventions of the genre, using recognized patterns for organizing materials, or providing transitions between thoughts. Editing for format is the final goal of substantive editing.

Copyediting includes a narrowly focused look at grammar, punctuation, spelling, and visual display of data to make sure that your document is correct, consistent, and complete. Proofreading is the final check of your document for any errors or inconsistencies you may have overlooked in previous edits.

Checking a document for consistency in style is also a critical part of editing. Choosing the right style means gauging how formal or informal you should be and adjusting sentence structure and vocabulary to your audience. You should also check your sentences for use of active and passive voice. Finally, editing enables you to ensure that you have used inclusive language so that you do not exclude members of your audience because of their sex, age, race, physical challenge, and so forth.

Editing and checking for style are not simple tasks, but they are ones that you cannot afford to ignore or to slight as you allocate time to the tasks involved in creating a polished document. Too many things, including your reputation, the safety of the document's user, and your company's legal liability, depend on your careful attention. Like other aspects of the writing process, editing has ethical, political, legal, and moral considerations involved with it. Good editing also can mean the difference between whether or not a text will be suitable for a global audience.

ACTIVITIES AND EXERCISES

1. Performance appraisals are common in business, government, and academic settings. They become part of an individual's personnel file and serve as a basis for decisions about promotion and salary increases. These documents also serve as a road map for a person's future work and are used to guide the career of

the person evaluated. Therefore, performance appraisals must be clearly written. The performance appraisal below is not as effective as it should be. It is long, repetitious, and inadequately organized. Revise this appraisal, reducing it significantly and organizing it logically.

Professor Alexandra is an extremely impressive and outstanding young scholar whose progress toward tenure has been and continues to be consistently excellent. She is good natured, energetic, enthusiastic. The people around her all respond well to her positive attitude. Professor Alexandra's abilities and skills as an administrator and leader in the department are excellent. Her desire to identify and seek out an increase in her administrative responsibilities as well as her other qualities are demonstrated in her outstanding performance as chair of the graduate committee for the past twelve months. During this time she served as next-in-charge for the department head for two weeks as well, performing her work excellently.

Professor Alexandra continues to sharpen her scholarly and research skills by attending conferences and scholarly meetings on a regular basis several times a year. She has attended 3 conferences and 2 workshops in this evaluation period and has given a total of 5 scholarly presentations. She is extremely well read and is up to date in her field as well as up to date in a number of collateral areas. She also knows many of the more established scholars in her field. She does thorough research and is well prepared when it comes to advising students which makes her a much sought after advisor among graduate and undergraduate students. She has a keen and perceptive intellect and is especially quick at identifying and solving both research and administrative problems. She is well prepared to handle the complex tasks of administration which includes making personnel decisions easily and without letting them confuse or upset her.

Professor Alexandra is very articulate in both her written and oral presentations, and she expresses herself clearly and forcefully. She is well liked and has good rapport with her colleagues, students, and with the civil service staff in the department, and she is well respected for her good judgment, easy-going nature, and fairness. Professor Alexandra continues to make significant contributions to new knowledge in her field, publishing more articles in the past year than her colleagues in similar stages in their careers and being nominated for several scholarly awards. She was also nominated for the distinguished teaching award in both the department and in the college on the basis of her quality advising and her teaching evaluation. Professor Alexandra is a young professor of the quality that we rarely find. Her scholarship, her teaching, her visibility in her field of research as well as her administrative and leadership abilities suggest that she will be ready for early promotion before others who came into the department the same year that she did. She is extremely well qualified for administrative duties, and I will recommend her with great enthusiasm for selection as one of the college's nominees as outstanding young professor. Her scholarly and teaching career as well as her administrative future is extremely promising. She has the potential for a highly successful and positively regarded career in the academic world.

2. Choose two journals in your field. In the front or back of each journal should be listed information about manuscript submissions. From this information, create a short style guide or checklist that could be used to help a person copyediting a manuscript for submission to these journals. For example, the *Technical Communication Quarterly* requires manuscripts to have an informative abstract of 50–70 words, to have titles centered on top of first page of text, and to indicate first-level headings by boldface or all capital letters.

3. Identify the sexist, discriminatory, or exclusive language in the following and edit to make them inclusive:

The girls in the main office always provide support for the salesmen in the field.

The groups represented in the survey were policemen, firemen, housewives, and waitresses.

Each man has to weigh the consequences of his own actions.

Engineering is a challenging occupation for any boy, or for any girl, for that matter. The math involved demands a youthful mind, the spirit of the hunter or sportsman.

We must take special care to help our handicapped clients realize the alternatives they might miss in our building designs. Those confined to wheelchair, those who cannot see, the mentally retarded, those suffering from any abnormal condition need these design features pointed out to them.

For a woman, she possesses a competitive spirit and exceptional endurance.

Every user needs to learn our graphics and word-processing software through examples and exercises. Whether it's the housewife recording her favorite recipes, the young boy designing computer games, the young girl describing her dreams of romance in a new ''electronic'' diary, the older person writing a letter to his grandson, or the woman listing emergency babysitters to step in when she's at work, each person needs to learn the value of the software in his own particular way.

Document Design and Packaging

Document Design and Readers' Goals ▌ Design Techniques to Engage Readers ▌ Design Techniques to Help Readers during Reading ▌ Techniques to Help Readers after Reading ▌ Developing a Style Sheet ▌ Evaluating Document Design Decisions ▌ Considering Differences When Designing a Document

By revising its forms, Citibank reduced the time spent training staff by 50 percent and improved the accuracy of the information that staff gave to customers.
"Plain English Pays," *Simply Stated*
(Washington, DC: American Institutes for Research, February 1986.)

Since the British Government began its review of forms in 1982, it has scrapped 27,000 forms, redesigned 41,000 forms, and saved over $28,000,000.
Robert D. Eagleson, *Writing in Plain English*
(Canberra, Australia: Australian Government Publishing Service, 1990).

Southern California Gas Company simplified its billing statement and is saving an estimated $252,000 a year from reduced customer inquiries.
"Gas Utilities Switch to Plain Language," *Simply Stated*
(Washington, DC: American Institutes for Research, March 1986).

Introduction

I n this chapter we will teach you how to make wise choices regarding a document's design and final packaging. **Design features** include everything from white space—that empty space in the margins and around headings and figures—to the use of color and the choice of binding. In going through this textbook, you might notice that we've used figures and examples throughout and that there is room in the margin for our definitions and your own notes. Other than that, you might not notice much. That's because design features are supposed to be invisible, or at least unobtrusive. Their purpose is to make reading easier, not to call attention to themselves. For that reason, you may not know some of the most common aspects of design at this point and probably little of the terminology; this chapter begins to remedy that by making you more aware of design features in general, especially those you might use to enhance the readability of your technical documents.

But before we delve into specific techniques, you need to know why these techniques are vital to a document's strength. Document design techniques are not just decorations; good document design helps busy readers, readers who basically do not want to read.

design features
text features such as white space, color, and binding that enhance the readability of a document

Document Design and Readers' Goals

Why are you reading this text? What are you using it for? What are your goals as a reader or user of this text?

To envision your own readers' goals in the technical documents you'll write and design, think of your readers as users of texts. That is, readers use texts in order to:

- Learn something;
- Do something; or
- Learn how to do something

Reading to learn something includes tasks where readers are using a document to extract and retain information that they will "call up" later and use again. In this type of reading, readers use study strategies such as previewing and reviewing in order to get the information to stay in memory. Most college reading tasks are reading-to-learn tasks (e.g., studying a textbook to prepare for a quiz).

Reading to do something includes tasks where readers are using documents to get information to help them complete a needed task. This is when readers look up information, use the information to complete a task, and then perhaps forget the information. Most workplace reading tasks are reading-to-do tasks because most documents in the workplace are manuals, flyers, labels, and forms (e.g., filling out a reimbursement form or ordering parts for a machine).

Reading to learn how to do something includes tasks where readers are using documents to learn how to complete a task that they are likely do again or need to know how to do later. Reading to learn means that readers locate information, use it, and then record how they used it so that they will not have to spend time relearning how to do something. Many college and workplace reading tasks include reading to learn how to do something (e.g., completing laboratory procedures or assembling a spreadsheet or databases). Essentially, this is the primary purpose for your reading this textbook; you should be learning how to do something. In the case of this chapter, you should learn how to use document design techniques to help readers.

Of course, whether your reader uses your document to learn, to do something, or to learn how to do something, you are helping your reader solve a problem—to order new parts, to get reimbursed for expenses, and so forth. To illustrate these three major goals for reading, you can complete the following exercise in Figure 11-1. The first column in this grid presents 10 contexts or scenarios for reading. In the second column, decide whether the reader's main goal is to learn something, to do something, or to learn how to do something. Then, working with a partner, compare your responses and list a document design technique that you have seen or would use to help a reader better understand and use a text in this scenario.

As shown in the short exercise in Figure 11-1, in every reading context, a reader uses a document actively to do or learn to do something: to solve a problem or to complete a task. In the workplace, we rarely have the pleasure of using technical documents only to learn something. Cognitive psychologists call this active use of documents **situated learning;** that is, readers participate actively in the process of understanding a document. Patricia Wright, an applied cognitive psychologist specializing in document design, has created a six-step sketch to show how readers (users) interact with technical documents. Readers interact with technical documents before reading, during reading, and after reading.

situated learning
active use of documents to do or learn to do something

Before Reading

1. Formulate question, not necessarily a precise question.
2. Find location of potential answer.

During Reading

3. Comprehend text, but reading may be highly selective.
4. Construct action plans or use information to make a decision.

After Reading

5. Execute action plans or implement decisions.
6. Evaluate outcome of actions or decisions (Wright, "Writing Technical Information," p. 340).

Context for Reading	Reader's Goal: (1) To learn (2) To do (3) To learn to do	Document Design Technique to Help the Reader/User
Filling out a tax form		
Learning the best ways to avoid exposure to HIV		
Comprehending one's rights in a legal contract with a landlord		
Reading an advertisement about exercising equipment		
Locating an emergency phone number		
Synthesizing texts in order to construct an argument about product liability laws		
Making a decision to vote for or against the building of a nuclear power plant in your county		
Assessing the current status of tuberculosis control efforts in your state		
Using a manual to solve a telecommunications problem		
Following directions to assemble a gas grill		

FIGURE 11–1
Goals for Reading

   We can choose any of the reading contexts listed in Figure 11–1 and follow the scenario through Wright's six steps. For example, if you need to assess the current status of tuberculosis control efforts in your state, you rarely jump into reading; rather, you formulate a question (though not necessarily a precise one) and search for the location of a potential answer.

   In the Appendix A, Case Documents 1 in this textbook, we have included a document entitled, "AIDS Public Policy in Minnesota." Imagine that you are a reader asking, "Given that people living with AIDS frequently suffer from tuberculosis, what efforts is Minnesota taking to control the spread of tuberculosis in AIDS patients and in the non-HIV-infected population?" Find and study this document now and implement these six steps as you search for an answer, comprehend the text, and execute any action plans or decisions.

As you went about the preceding task, you probably noticed the title of the document immediately, "Tuberculosis and AIDS." This title gave you a good idea that this document would help you answer your question. As you continued to skim through the document, you probably located quickly the second boxed heading, "Current Status of Tuberculosis Control Efforts in Minnesota." In this section, you no doubt focused on the table reporting the number of new tuberculosis cases, but reading below the table, you saw that control efforts seem to consist primarily of seeking grants to support treatment services. You want information on more concrete efforts to control the spread of tuberculosis. Did you stop reading there and decide you needed to turn to a different document? Or, did you persist and discover the "Recommendations" section that includes a list of concrete measures?

Now, study a different document from the Appendix A, Case Documents 2 or 3—the documents concerning the trapping and disposal of birds in University agricultural areas, or documents from the *Exxon Valdez* oil spill— and envision a complete reading scenario, going through all six of Wright's steps. What questions might readers come up with? How will readers find the location of a potential answer? And so on. Basically, what will readers do before, during, and after reading? What document design techniques in the document help or hinder readers as they search for answers, read the texts, and execute actions?

As a developer of texts, you need to focus on document design techniques that help readers use texts before, during, and after reading. You can use the design techniques in the following sections to help readers access and use your information at all three stages.

Design Techniques to Engage Readers

When all else is considered equal in printed materials, we know that physical attractiveness can tip the balance for readers. As Martha Andrews Nord and Beth Tanner state, "Attractive documents look as if the producers value the information enough to care about its appearance and suggest that the same care has gone into writing the text as was invested in its design" (p. 224). Look across the texts on any bookshelf or the many handout packets that you have purchased during your college career. What packaging features motivate you to use a textbook or other document?

WRITING STRATEGIES 11–1   Reader Goals and Design Techniques I

Reader's Goals before Reading	Writer's Document Design Techniques
To formulate a question and become motivated to open a document and locate an answer	Focus on packaging:
	Attractive binding
	High-quality paper
	Balanced and consistent layout
	Interesting graphics
	Legible cover and print
	Appropriate use of color
To find the location of a potential answer	Focus on the larger organization:
	Table of contents
	Indexes
	Structural headings
	Tab markers and dividers

---

navigational tools
devices for guiding a reader through a document, including tables of contents, indexes, tabs, and headings

We also know that readers need **navigational tools** or guideposts to help them find the location of a potential answer. Numerous studies have found that the most effective readers are those who use navigational tools such as tables of contents, indexes, and headings to aid their search for information. Thus, even in the shortest document (e.g., a one-page memo or a short form), you should provide document design techniques that allow readers to locate information quickly and use that information effectively.

You can motivate your readers to open a document and locate a potential answer by focusing on packaging as well as the document's larger organization. Writing Strategies 11–1 presents a reader's main goals before reading as well as the document design techniques you can use to engage readers.

ENGAGING READERS

You can engage readers with a document by considering your readers' needs and focusing on attractive packaging and the structural layout of the document. Although there are numerous document design techniques for this stage of reading, here are some starting techniques and examples in each case.

Attractive Binding

If readers need to keep a document open to a specific page, provide a spiral binding. To enable readers to locate a document quickly on a bookshelf, if possible, provide identifying information on the document's spine or binding.

High-Quality Paper

If you want your resume to be noticed and not tossed, use bond or other high-quality paper. If you are printing a document on both sides of the paper, choose high-quality paper so that the text doesn't bleed through to the other side.

Structural Layout

Begin major sections of a document with a consistent layout (note the first pages of every chapter in this text). When using figures and tables, place these in appropriate positions on the page and label them consistently.

Graphics and Visuals

Use the suggestions that follow in Chapter 12 to choose the right visual display of data. Remember that readers will be drawn to visuals; use them to engage readers. Use a common visual to unify a group of documents such as tutorial and reference texts or field and laboratory reports.

Cover

Although there are numerous materials available for creating covers for documents, choose one that showcases your document without detracting from it. Make the title on the document large, legible, and clear.

Color Scheme

Choose a color scheme that works for your readers. A bright color scheme for a train schedule may not work for forms or labels. Unify a set of documents with a similar color. Do not use multiple colors to be flashy; this detracts rather than engages readers. Consider these guidelines:

- Use only four to six color codes.
- Use color conventions from a reader's work world (e.g., red for danger, blue for water), but be aware of cultural differences when you choose a color.
- Pick harmonious colors and avoid conflicting colors.
- Never use colors that compromise legibility.
- Never depend on color alone for any critical distinction (Horton, "Visual Literacy"; "Pictures Please").

Use caution when you use color and avoid it if color-blind readers might be confused or endangered. To test your understanding of color blindness, answer the questions in Figure 11–2.

Example 11–1 shows the cover of a document that describes how to conduct emergency medical treatment for children. Notice the specific document design techniques used by the writers to engage readers:

Color blindness can limit people's ability to perceive information displayed in certain colors. To test your understanding of color blindness, answer the following questions:

1. What fraction of men have trouble distinguishing red or green?
   100%   80%   8%   0.8%   0.4%   0%
2. What fraction of women have trouble distinguishing red or green?
   100%   80%   8%   0.8%   0.4%   0%
3. As we age our ability to perceive one color diminishes. What is that color?
4. The center of our vision, where our vision is most acute and where we read text, is less sensitive to one color than others. Which color is that?

Answers are at the end of this chapter.

Source: Adapted from William Horton, "Understanding Color Blindness," *Technical Communication* 39.3 (1992), pp. 447–51. Reprinted with permission from *Technical Communication,* published by the Society for Technical Communication, Arlington, VA.

FIGURE 11–2
Understanding Color Blindness

EXAMPLE 11-1   Cover for an Emergency Medical Treatment Document

916363-00 ∗ $7.95 ∗ EMT, Inc.

EMERGENCY MEDICAL TREATMENT:

CHILDREN

TECHNICALLY
REVIEWED BY THE
**NATIONAL
SAFETY
COUNCIL**

A HANDBOOK OF WHAT TO DO IN AN EMERGENCY TO KEEP A CHILD ALIVE UNTIL HELP ARRIVES.

NOT BREATHING • NO PULSE • CHOKING • BURNS • BLEEDING
• SEIZURES • POISONING • DROWNING • HEAD, NECK, BACK INJURY
• ACCIDENT PREVENTION • BROKEN BONES

By Stephen Vogel, M.D. & David Manhoff

- The spiral binding enables the document to stay open once the reader locates the appropriate section.
- Although you cannot "feel" this document, it is made of card stock (heavy paper) with a waxed coating. Thus, it's durable, and if it gets dirty, the reader can wipe it off easily.
- The tab markers on the right side makes the structural layout of the document automatically visible from the front cover of the document.
- The visual on the front cover automatically cues readers as to the purpose of the document.
- The color scheme signals the subject (green); technical accuracy of the information (green); and the seriousness of the subject (bold and black).

Now, before reading further, evaluate the effect of *this* textbook's binding, the quality of the paper, the overall layout, and so on, on you as a reader. Does it motivate you to open the text and use it to learn to communicate effectively?

HELPING READERS LOCATE POSSIBLE ANSWERS

Once readers become motivated to open and use a document, they need to locate possible answers quickly and efficiently. To aid readers, you need to focus on the document's larger organization. Essentially, you need to divide the text into manageable units and signal relationships between these units. We suggest the following techniques.

Table of Contents or a List of Headings

For larger documents, include a table of contents. Many word processing programs include a table of contents utility. Even for short documents (one or two pages), providing a list of the headings at the top helps readers to determine whether they have located the appropriate section. This list also functions as a preview of the document and guides readers' understanding of the information. When developing a table of contents or a list of headings, follow suggestions for developing lists given later in this chapter.

As an example, note the contents for the Centers for Disease Control's *Fact Book FY 1991* in Example 11–2. This example also shows this document's two-page structural organization.

Indexes

For longer documents, you need to include an index that allows readers to find specific information quickly. Readers locate information via a good index far more often than from a table of contents. Many word processing programs include an indexing utility.

CDC *Fact Book FY 1991*

CENTERS FOR DISEASE CONTROL

U.S. DEPARTMENT OF HEALTH & HUMAN SERVICES • Public Health Service • Centers for Disease Control

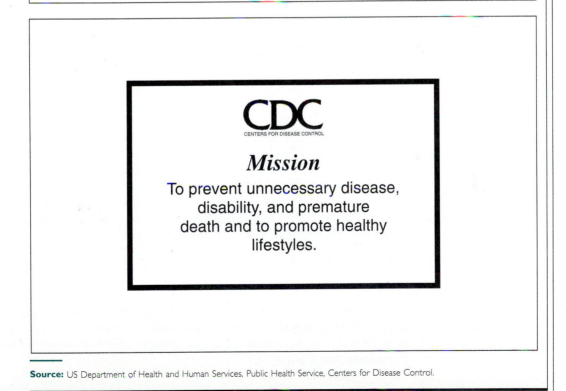

CDC
CENTERS FOR DISEASE CONTROL

Mission

To prevent unnecessary disease, disability, and premature death and to promote healthy lifestyles.

Source: US Department of Health and Human Services, Public Health Service, Centers for Disease Control.

Structural Headings

Structural headings identify the largest chunks of information in a document. Running **headers** and **footers** appear in small type at the top or bottom pages and repeat information such as topic, page number, and date. Magazines and newspapers generally use **standing heads**—a separate line above the text and set off by a different typeface or graphic feature—to identify regular departments such as editorials. You can also use icons as **structural headings;** for example, the icon below might be used to indicate a section about time limits.

You can also guide readers by using **hanging heads** (or headings placed in margins) so that readers can skim and skip text.

Tab Markers or Dividers

Separate sections of a larger document by using **dividers or tab markers.** These function as structural headings and quickly guide readers to appropriate sections. (Note the dividers in Example 11–1.)

Again, before reading further, apply each of these suggestions to this textbook. That is, evaluate this textbook's table of contents, index, and use of structural headings. To test these elements, write down the time that it takes you to locate information about collaboration, ISO–9000 standards, and questionnaires. Then, using Figure 11–3, go back and list the document design techniques that aided your search in each case.

As you completed this short exercise, what document design techniques helped you the most? The least? What techniques listed above are not used in this text, perhaps due to cost constraints or traditional textbook specifications?

headers
navigational aid in small type running at the top of a page and repeating information such as topic, title, page number, or date

footers
navigational aid in small type running at the bottom of a page that repeats information such as topic, title, page number, or date

standing heads
separate line of text above the main text and set off by a different typeface or graphic feature to identify different components of a document

structural headings
headings indicating the major structures or components of a document; can be marked by headings or icons

hanging heads
headings placed in the margins to facilitate a reader's skimming through a text to find what he or she is seeking

dividers or **tab markers**
dividers often are sheets of paper inserted between sections; tabs are heavy markers attached to the right edge of a page; both guide readers to appropriate sections of a document

Design Techniques to Help Readers during Reading

Once readers have found the location of a possible answer, you need to help them read selectively and easily in order to make a decision. Writing Strategies 11–2 lists a reader's goals at this stage and the document design techniques that you can use.

Locate the Following:	Time to Locate the Information	Document Design Techniques That Aided Your Search
Information on how to initiate a collaborative writing project		
Information on writing with standards such as ISO–9000		
Information on designing a questionnaire as a secondary research tool		

FIGURE 11–3
Design and Location Exercise

WRITING STRATEGIES 11–2   Reader Goals and Design Techniques II

Reader's Goals during Reading	Writer's Document Design Techniques
To read selectively	Focus on guiding readers by using:
	grid layout
	white space
	headings and subheadings
	margins
	color
	visual elements (boxes, rules)
To read with ease and use information to make a decision	Focus on typography:
	case
	style (typeface)
	size (point)
	spacing (leading and kerning)
	length
	highlighting

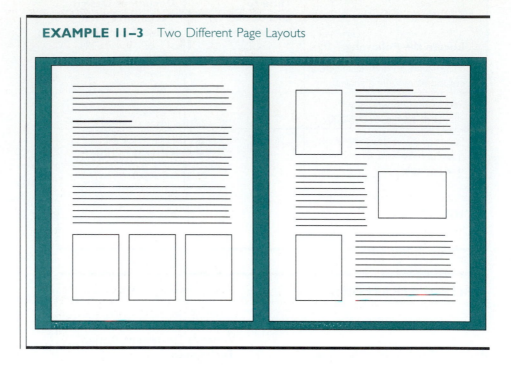

EXAMPLE 11–3   Two Different Page Layouts

HELPING READERS READ SELECTIVELY

To help readers read selectively, you need to visualize the layout of a page of your document. For this stage, you have a variety of techniques to influence how readers interact with the text.

Grid Layout

Imagine your page as a grid, a blueprint that you can repeat on each page for consistency. Does the purpose for your document warrant a traditional one-column format, or should you think about using a two- or three-column format? Where should you place visual display of data in this **grid layout**?

grid layout
use of the page as a grid or blueprint upon which to make design decisions such as number of columns and placement of graphics

folio
two-page spread used in page design

Example 11–3 shows two layouts. Working with a partner or in small groups, discuss the elements of each layout that you prefer and why. In this case, both layouts might be preferred in different cases. For what purpose and audience might the first layout be used? The second? You are not limited to the standard 8½" by 11" page for your layout; you can use a two-page spread called a **folio.** Example 11–4 presents a two-page spread or a folio from the emergency treatment document. Continue to refer to this two-page spread as we discuss white space, margins, color, and other visual elements.

EXAMPLE 11-4   Folio from the Emergency Medical Treatment Document

Not Breathing/No Pulse

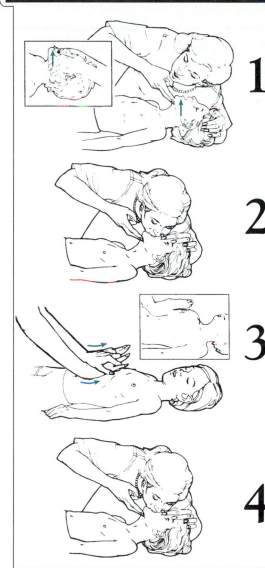

Gently tap or shake child. Ask: "Are you OK?" If child does not respond, shout for help.

1 Open airway, check breathing.

Lay child on back. If you must roll child on back, keep (support) head and neck in straight line. Tilt head back gently, lift chin slightly. If you suspect neck or back injury, pull open jaw without moving head (see inset). Look, listen, feel for breath (3-4 sec.). If not breathing, or you are in doubt, start rescue breathing (#2). If breathing, place child on side (unless head, neck or back injury).

2 Give two slow breaths.

Pinch nose. Cover child's mouth with yours. Give two slow, *gentle* breaths (1-1¹/₂ seconds each) into child's mouth. Allow chest to rise and fall between breaths. NOTE: *Watch chest.* If chest does not rise and fall after 2 breaths, retilt head, lift chin up and try again. If airway is blocked, go to picture #3. If chest does rise and fall, *check pulse* as in picture #5.

3 Something in windpipe.

Straddle child's thighs. Place the heel of one hand just above the navel, but well below lower tip of breastbone. Place your other hand directly on top of your first hand with fingers pointing to head. Press upward into stomach with up to 5 quick thrusts. Open mouth by grasping tongue and lower jaw between thumb and fingers, and lifting. Only if you see object, gently sweep index finger (hooking motion) deeply into mouth at base of tongue to remove from throat.

4 Repeat two slow breaths.

Tilt head back. Lift chin. Pinch nose. Cover child's mouth with yours and give two slow, *gentle* breaths. Watch chest rise and fall. Repeat #3 and #4 if necessary.

EXAMPLE 11–4 *(continued)*

NOTE: The use of latex gloves and mouth barrier with one-way valve (for rescue breathing) is recommended.

5 Check for pulse on side of neck.

Press 2-3 fingers into neck just to side of Adam's apple. Feel for pulse 3-4 seconds. If no pulse, start chest compressions immediately (#6) along with rescue breathing. If child has pulse, but is not breathing, continue one breath every 3 seconds for one minute (20 breaths). **Call 911/ambulance.** Return to child, recheck breathing/pulse (3-4 sec.). Continue one breath every 3 seconds until child breathes on own or ambulance arrives. Roll child onto side if breathing resumes (unless head/neck/back injury suspected).

6 Place heel of hand below mid-breastbone.

Follow rib cage to where it meets in the center of the lower part of chest. Place entire heel of hand 1-2 finger widths above lower tip of breastbone, just below mid-breastbone.

7 Push down on chest 1"-1$^1/_2$" (2.5 - 3.8 cm) - 5 times.

Straighten you arm, lock elbow and push straight down on chest 5 times (a rate of 100 per minute). Let chest relax completely between downstrokes, without removing hand from chest.

8 Give 1 breath.

Tilt head back. Lift chin. Pinch nose. Give 1 slow, gentle breath.

9 Continue with 5 chest compressions then 1 breath for 20 cycles (one minute). Call 911/ambulance.

Alternate 5 compressions and 1 breath for 20 cycles (1 minute). Call for ambulance and quickly return to child. Recheck pulse/breathing (3-4 sec.). If there is a pulse but no breathing, give 1 breath every 3 seconds (20 per minute). If no pulse, continue 5 chest compressions then 1 breath until child breathes on own or ambulance arrives. Recheck breathing/pulse every few minutes. NOTE: Allow chest to rise and fall between breaths, let chest relax completely between downstrokes. Roll child onto side if breathing resumes (unless head/neck/back injury suspected).

White Space

Imagine this life-saving strategy explained in a series of paragraphs rather than in this layout. In Example 11–4, note the ample use of **white space.** White space guides readers through structure, prevents fatigue, and creates aesthetic appeal. You should devote at least 50 percent of a page to white space, never less.

  You can also add white space to your text by considering your spacing and creating additional space between paragraphs and between lines. For documents such as letters, you can center the text on the page vertically; that is, don't let the letter fill only the top half of the page. Adjust and add white space to enhance the letter's organization and make the content easier to read. Look at Example 11–5, which shows one letter lacking white space and one with ample white space. Each letter contains the same amount of text.

white space
blank or negative space on the page

Headings and Subheadings

Use of **headings and subheadings** also creates white space. More importantly, efficient readers skim headings and subheadings to locate and use information. To signal relationships between major sections and subsections, use two or three levels of headings. For example, look at the three levels of headings used below:

headings and **subheadings**
navigational devices indicating major and subsections of a document and their relationship to each other

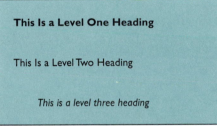

This Is a Level One Heading

This Is a Level Two Heading

This is a level three heading

As you can see, a level one heading is boldface and starts at the left margin. A level two heading also starts at the left margin but is not boldfaced. A level three heading is indented and set in italics.

  Generally, a level one heading should be in boldface print. It can be either left-justified or centered on the page.

margins
white space at the top, bottom, and sides of a page

left-justified margin
all lines of text line up along the left margin

right-justified margin
all lines of text line up along the right margin

Margins

Ample **margins** also help create needed white space on a page. A **left-justified** margin means that all lines begin in a straight line at the left. Readers are used to reading left-justified text. A **right-justified** margin means that all lines line up along the right margin. Right-justified margins help readers identify discrete chunks of text. A **ragged right margin** (that is,

ragged right margin
lines of text do not line up along the right margin, although they line up with the left margin

---

EXAMPLE 11–5 Two Letters Positioned on a Page

---

not justified) guides readers to continue reading and avoids unnatural spacing between words in a line of text that often appears with right-justified margins. What type of justification is used in Example 11–5? In this short paragraph on margins? Why should this help readers?

Use of Color

Use of color can be an effective document design technique throughout the reading process. You should use color to:

- Link objects and different colors to distinguish objects.
- Focus attention on important information or action readers should take.
- Speed search by color-coding symbols in a complex diagram.
- Express a range of values such as temperatures on a map.

Again, note the use of color in Example 11–4. Why is each step presented in color?

Visual Elements for Highlighting

Lines, borders, circles, and other shapes emphasize, isolate, and separate information to help readers locate and understand text. You can emphasize crucial information with a line or **rule** to attract attention. Use **boxes** to isolate and separate information. Borders and circles also help to isolate text for emphasis. Note the rules used at the top of Example 11-4. For what purpose have boxes been used in this document?

Information such as warnings, caution statements, and legal contracts should be highlighted so that readers can find it easily and consistently. Shading this information with an appropriate color (as determined by ANSI standards for danger, warning, and caution), placing a box around the information, and/or identifying it with an appropriate icon will help readers to locate important information.

HELPING READERS READ WITH EASE

To take the burden out of reading, you need to make conscious decisions regarding the **typography** of your document. Typography refers to the look and legibility of letters; this includes their case, style, size, spacing, and length on the page. Look at Example 11-6 as we discuss typography.

Case

Perhaps you're most familiar with **case** (lower and upper case letters). Lower case letters have **ascenders,** or parts of the letter that rise above the line (e.g., *h* and *f*), and **descenders,** or parts of the letter that drop below the line (e.g., *y* and *p*). Ascenders and descenders create a distinctive shape for a word. Draw a close line around any word in this paragraph, and you'll see that your line goes up and down around these ascenders and descenders. These irregular outlines help readers recognize words and read with ease.

Thus, contrary to what you might think, ALL UPPER CASE WORDS ARE EXTREMELY DIFFICULT TO READ; THEY PROVIDE NO IRREGULAR OUTLINE TO SPEED UP READING. Moreover, users of online documents report that reading documents with all upper case words makes them feel as if they are being shouted at.

Therefore, use all upper case letters for WARNINGS or other information that must be intimidating. Otherwise, use upper case letters only as capital letters to begin sentences or for titles and headings.

Typeface

Typeface or **font** refers to a collection of type that has a unique name and set of distinctive characteristics. A typeface or font comes in two styles, **serif** and **sans serif** (French for without serifs). Serifs are the little "feet" attached to the ends of letters. These extensions pull a reader's eye across the line of text. A typeface such as Palatino is a serif typeface; that is, it has these little extensions:

This is Palatino font.

---

rule
line used in the text to emphasize information or to draw the reader's attention

boxes
borders that separate or isolate information in a document

typography
the look and legibility of letters, including their case, style, size, spacing, and length on the page

case
the upper or lowercase letters in a document

ascenders
parts of the letter that rise above the line (e.g., h and f)

descenders
parts of the letter that drop below the line (e.g., y and p)

typeface or **font**
refers to a collection of type that has a unique name and set of distinctive characteristics

serif and **sans serif**
two styles of type; serifs are the little "feet" attached to the ends of letters; sans serif means without serifs

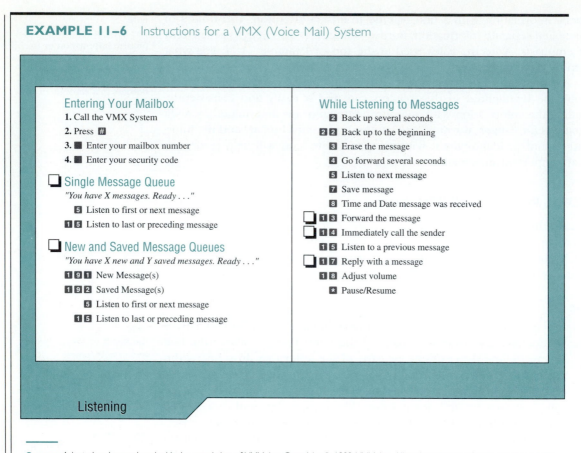

EXAMPLE 11–6   Instructions for a VMX (Voice Mail) System

Entering Your Mailbox
1. Call the VMX System
2. Press **#**
3. ■ Enter your mailbox number
4. ■ Enter your security code

Single Message Queue
"You have X messages. Ready . . ."
  5 Listen to first or next message
  1 5 Listen to last or preceding message

New and Saved Message Queues
"You have X new and Y saved messages. Ready . . ."
  1 9 1 New Message(s)
  1 9 2 Saved Message(s)
  5 Listen to first or next message
  1 5 Listen to last or preceding message

While Listening to Messages
  2 Back up several seconds
  2 2 Back up to the beginning
  3 Erase the message
  4 Go forward several seconds
  5 Listen to next message
  7 Save message
  8 Time and Date message was received
  1 3 Forward the message
  1 4 Immediately call the sender
  1 5 Listen to a previous message
  1 7 Reply with a message
  1 8 Adjust volume
  * Pause/Resume

Listening

A typeface such as Helvetica is a sans serif typeface; sans serif typefaces do not have extensions:

This is Helvetica font.

Base your decision on what is familiar to your readers. People in the United States most often use serif typefaces, and those in Europe use sans serif typefaces (White).

Size or Point

You'll hear people refer to **type size, point,** and **font size** somewhat interchangeably. For your purposes, they all basically refer to the size of the print. Technically, a **point** is the unit for measuring font size: one point is

type size, point, font size
often used interchangeably, these terms refer to the size of the print in a text

point
unit for measuring font size; one point equals 0.014 inch

equivalent to 0.014 inch, and 72 points equal one inch. For titles and headings, use a larger point size such as 12, 14, or 18 point. For text, standard point size ranges from 9 to 12 point. If you're using columns, the narrower the column, the smaller point size you should use, so that your reader encounters a sufficient number of words before jumping to the next line of text.

Spacing

Spacing is the distance between letters and between lines of type on a page. People use two terms to describe spacing: leading and kerning. **Leading** refers to the space between lines of text; generally, shorter lines need less space between them. Type fonts with short ascenders and descenders need more space between lines so the text doesn't look too heavy and dense. **Kerning** refers to the space between letters in a word. Some typefaces create extra space between letters and may be harder to read.

spacing
distance between letters in a line of text and between lines of type on a page

leading
the space between lines of text

kerning
the space between letters in a word

Line Length

Line length is the width of a column of text. There are at least two ways to judge proper line length. You can count off and set about 50 characters per line. Or, you can set an alphabet and a half of your chosen type style, that is, the length of the letters *a* through *z* if set in a row. Longer lines make it harder for readers to retain information and return to the beginning of the next line of text.

line length
the width of a column of text

Highlighting

Highlighting is a way you can emphasize text by using **boldface** or *italics* or by underlining. Keep in mind that excessive use of any of these techniques wears on readers. The more you boldface, the less readers will notice and retain. Therefore, use these techniques wisely to highlight what really needs emphasis.

highlighting
emphasizing text by using **boldface** or *italics* or by underlining

Now, looking at Example 11–6, work with a partner or in small groups and discuss the following questions:

- What case did the writer use in the document?
- Has the writer used a sans serif or a serif typeface?
- What point size would you guess the headings and text are in the document? How effective is the point size?
- Are line lengths appropriate in the document? Leading or space between lines?
- How is text highlighted?
- What other aspects of typography do you notice in the example? How effective are they for the readers' use?

Continuing to work with your partner or group, locate a paper that you're each working on for this or another course. Exchange papers and

---

WRITING STRATEGIES 11–3   Reader Goals and Design Techniques III

Reader's Goals after Reading	Writer's Document Design Techniques
To execute action plans or implement decisions	Focus on providing:
	lists
	examples
	glossaries
	appendices
To evaluate the outcome of actions or decisions	Focus on providing:
	checklists
	boxes for recording information
	quick reference cards

---

evaluate the document's typography. Remember to evaluate the typography in terms of the purpose of the text and the audience that will be reading it. Typographical techniques that may have been fun for you to create may not be best for the purpose and audience of the text. Or, if no points are distinguished visually, offer suggestions for visually clarifying the organization and substance of the text.

Techniques To Help Readers after Reading

After readers have read your document, they need to execute action plans or implement decisions based on what they have read. Writing Strategies 11–3 lists goals and techniques for this stage of reading.

HELPING READERS EXECUTE ACTION PLANS

You can help readers implement their decisions by providing lists, examples, glossaries, and appendices.

Lists

lists
chunks of information clearly separated from the rest of the text

By their format alone, **lists** are chunks of information clearly separated from the rest of the text. Because of the additional white space around them and their structural order, they automatically increase comprehension. They are especially helpful for executing action plans. Document design specialists suggest that you:

- Indent the list so that numbers or bullets stand out in the left margin.
- Avoid setting lists in type smaller than the body unless you want the reader to consider the information less important than the rest of the text.
- Punctuate consistently (Nord and Tanner).

You can also place cues before list items:

- ☐ A checkoff box if each item must be performed.
- A bullet if list items are equally important.
1. A number if list items should be considered in a certain order (Carliner).

Examples

Throughout this textbook, we have worked to provide examples of what we describe. Examples are valuable for helping readers to understand what you mean and to execute action plans. Provide examples whenever possible, and separate these with visual elements to highlight them for readers.

Glossaries

As you know, **glossaries** define information presented in the main text. In the workplace you will often understand more technical details than will your readers. Use glossaries to elaborate on difficult terms used in your document. Readers can refer to your glossary during or after reading as a means to implement decisions.

glossaries
lists that define terms presented in the main text; usually appear at the end of a document or section

Appendices

Appendices provide supplemental information without cluttering your document. Note the appendices used in this textbook. What chapters would this information interrupt if placed in the body of the text? You can place examples or other information in appendices so readers can access the information to execute action plans.

appendices
section at the end of a document in which additional material not directly needed in the text can be placed

HELPING READERS EVALUATE ACTIONS

Rarely do writers seem to provide areas where readers can take a moment to evaluate the outcome of their actions or decisions. Three techniques that you can use are checklists, boxes, and quick reference cards.

Checklists

Checklists enable readers to identify necessary items or actions. For example, if you need to assemble a gas grill, you'll need to refer to a checklist to see if all of the parts are available or to ensure all steps are taken in order.

Boxes

Boxes separate or emphasize a portion of text. You can include key points that summarize the text and place them in boxes alongside the text. After reading, users can refer mainly to the boxes to get a gist of the text. You can also provide boxes as specific places for readers to record information after they have read the text. In this way readers have a space for evaluating the outcome of their actions or decisions.

Quick Reference Cards

You can place the most important steps, hints, or points on a **quick reference card** for readers to refer to after they have used a document. Although quick reference cards are often included with computer manuals (see Example 11-7), readers appreciate shortcuts for reading, and quick reference cards can function as a summary or reference to the most important points in a document.

Developing a Style Sheet

One of the best ways to plan for consistency throughout your document is to create a style sheet or style guide. Although a **style sheet** can contain the word and sentence level choices we discussed in Chapter 10, a style sheet can also be your plan for the visual look of your text. Style sheets are especially helpful whenever you contribute to a group writing project. What document design techniques will you use to help readers before reading, during reading, and after reading? What spacing, margins, justification, typeface, and point size will you use? How will you build consistency into your headings, subheadings, and lists?

Whether you're writing individually or collaboratively, if your writing task requires a lot of formatting or has several sections, it helps to design a style sheet. To begin, keep your style sheet simple. Writing in a consistent typeface with consistent margins and spacing will help.

To create a style sheet, make planning decisions about which techniques you will use and why, as shown in Writing Strategies 11-4. By creating a style sheet, you will spend less time in the future making decisions about style. You can develop your style sheet on paper, but we suggest using the style feature in a word processing program to develop a style sheet; that way you can access it each time you begin a new document.

EXAMPLE 11–7 Microsoft Quick Reference Card

Microsoft Word Quick Reference Guide

If you want to	Do this
Add a footnote	Click insertion point where you want footnote to go. Press **E**. Click **OK**. Type footnote text in footnote window.
Add a header or footer	Under **Document**, choose **Open Header** … OR **Open Footer** … Type header/footer text in header/footer window. Close window.
Boldface text	Select text. Press **-[Shift]-B**.
Center text	Press **R**. Select Centered paragraph alignment icon on ruler.
Change line spacing	Press **R**. Select desired line spacing icon on ruler.
Change text	Press **H**. Type text you want to change in **Find What:** box. Type what you want to change it to in **Change To:** box. Click **Start Search**.
Change to plain text	Select text. Press **-[Shift]-Z**.
Check spelling	Press **L**. If necessary, click **Start Check**.
Close a file	Press **W**. If necessary, save changes to open file.
Copy text	Select text. Press **C**. Selected text remains in document and is copied to clipboard.
Cut text	Select text. Press **X**. Selected text is cut from document and is copied to clipboard.
Define styles	Press **T**. Make font and format selections from pull-down menus while in dialog box. Click **OK**. (Requires Full Menus.)
Find text	Press **F**. Type text you want to locate. Click **Start Search**.
Format a paragraph	Press **M**. Make desired selections in dialog box. Click **OK**. (Requires Full Menus.)
Go to page number	Press **G**. Enter page number in dialog box. Click **OK**.
Italicize text	Select text. Press **-[Shift]-I**.
Left-justify text	Press **R**. Select Flush Left paragraph alignment icon on ruler.
Number pages	Under **Document,** choose **Open Header** … OR **Open Footer** … Click insertion point in header/footer window where you want page number to go. Click page number icon. Close window.
Open a new file	Press **N**.
Open an existing file	Press **O**. Select file to open. Click **Open**.
Outline a document	Press **U**. (Requires Full Menus.)
Paste text	Click insertion point in document where you want text to go. Press **V**. Text from clipboard is copied to insertion point.
Print a file	Press **P**. Make desired selections in dialog box. Click **OK**. Follow directions on screen.
Quit Microsoft Word	Press **Q**. If necessary, save changes to any open files.
Save as new file	Under **File**, choose **Save As...** Enter document name in dialog box. Click **Save**.
Save an existing file	Press **S**.
Set margins	Press **R**. Move indent markers on ruler to desired settings.
Show/Hide the ruler	Press **R**.
Underline text	Select text. Press **-[Shift]-U**.
View the page layout	Press **I**.

Source: Ann Hill Duin and Kathleen S. Gorak, *Writing with the Macintosh ® Using Microsoft ® Word.* Cambridge, MA: Course Technology, Inc. Reprinted with permission.

WRITING STRATEGIES 11–4   Style Sheet Decisions

	Document Design Technique	**Style Sheet Decisions**

Focus on these techniques first	grid layout
	white space
	margins
	headings and subheadings
	color
	visual elements
	
	typography:
	
	     case
	
	     style (typeface)
	
	     size (point)
	
	     spacing (leading, kerning)
	
	     length
	
	     highlighting	

Focus on these techniques second or third	binding quality of paper layout graphics cover color table of contents indexes headers and footers tab markers and dividers lists examples glossaries appendices checklists boxes for recording info. quick reference cards	

When you design a document, you have some idea of its intended use; you have some idea about your target for readers' subjective feelings and attitudes about the document. But how do you measure reader satisfaction with your document? How do you know if your document design decisions are correct? How do you know if the techniques you've chosen have resulted in a higher quality document?

The drawback of evaluating documents is that they take time; the benefit is that the information obtained is extremely valuable for guiding revision and future document design decisions. For example:

> When a technical publications group at AT&T began evaluating documents and focused on streamlining the process of developing technical documentation, they reduced the cost of documentation by 53 percent, reduced documentation production time by 59 percent, and increased the number of projects completed by 45 percent. (Edwards, MG45–MG46)

> In England, when the Department of Customs and Excise revised its lost-baggage forms used by airline passengers, they cut a 55 percent error rate down to a mere 3 percent. ("Plain English Pays," 1, 4)

Research aimed at measuring quality in document design can be classified by two dominant approaches: direct, or "criterion-reference," measures and indirect, or "prediction," measures (Schriver). These two approaches serve different purposes when evaluating document design.

DIRECT MEASURES (CRITERION MEASURES)

Direct measures are those that study how readers use, read, or rate a document and then judge that activity against a criterion. Direct measures include: (1) **concurrent methods,** designed to collect data about how people think, feel, or respond as they use a document, and (2) **retrospective methods,** designed to collect data as readers reflect on their experience after reading or using a document (Schriver).

direct measures
ways to study how readers use, read, or rate a document and to rate those activities against a criterion; includes concurrent methods and retrospective methods

For example, if you designed instructions for assembling a gas grill and then wanted to evaluate your document, you might give your instructions to readers/users and record how they used the instructions. You would then judge that activity against a criterion such as the maximum time that it should take users to assemble the grill. Although this concurrent method would provide information about how users respond as they use your document, you might also ask users to reflect later on their experience, a retrospective method.

concurrent methods
methods designed to collect data about how people think, feel, or respond as they use a document

As another example, if you designed a formal report with the bulk of your findings in appendices, you might give the report to readers and record how often they flipped to the appendices to study specific results. A day or two later, you might also ask readers to record what they remembered most from your report.

retrospective methods
methods designed to collect data as readers reflect on their experience after reading a document

INDIRECT MEASURES (PREDICTION MEASURES)

indirect measures
measures designed to predict a reader's ability to comprehend or use the document, such as quality metrics

quality metrics
a four-phase indirect measure of a reader's ability to use a document; ends with a regression analysis

Indirect measures are those designed to predict a reader's ability to comprehend or use the document (Schriver). We will cover **quality metrics** in this chapter.

Quality Metrics

To develop a quality metric, you need to carry out a four-phase plan:

- In Phase 1 you gather evidence from existing research about how readers use a specific type of document (e.g., a tax form, a proposal, or a brochure), especially noting the document design features readers appreciate most.
- In Phase 2 you modify a document according to the document design features that readers appreciate most. You then test these documents with readers/users to determine whether the features improve understanding, attitudes, and ease of use.
- In Phase 3 you use the results from this testing to identify a set of features (i.e., a style sheet) that will be used for the prediction measure.
- In Phase 4 you conduct a **regression analysis** to determine what weight each feature should have when predicting the comprehensibility or usability of the document (Schriver).

regression analysis
a method of data analysis in which equations are used to show the relationships among variables

Although this four-phase plan may appear too complicated to use as you develop technical documents, it may be extremely worthwhile to follow when designing documents for the corporation or other business in which you may be working.

EVALUATING A PHONE BILL

Let's apply the evaluative measures we've discussed to the redesigning of a telephone bill. Deborah Keller-Cohen and her colleagues studied the collaborative effort as a linguist, a graphic designer, and an executive manager developed and tested new telephone bill solutions to improve customer comprehension and satisfaction with a very familiar, yet complex document. Example 11–8 shows the original telephone bill.

Based on what you have now learned about document design techniques, how might you redesign this telephone bill? This group's main concerns were to make relationships among the charges clearer, to increase legibility, and to simplify the language. The collaborative team followed this process when redesigning it:

1. They made initial changes that they felt customers desired and that the company could implement. They then tested their changes through face-to-face interviews with customers.

EXAMPLE 11–8 Original Telephone Bill

```
              ⊙ Midwest Bell    P.O. Box 1234   Anywhere, Midwest 01234


                                               123 456-7890 000
                                               DEC 10, 1984
                                                  PAGE    2

MIDWEST BELL - CURRENT CHARGES

FLAT RATE SERVICE
MONTHLY SERVICE     DEC 10-JAN 09                      19.11
DIRECTORY ADVERTISING                                  2.75
LOCAL AND ZONE USAGE-SEE DETAIL PAGE                   6.85
OTHER CHARGES AND CREDITS-SEE DETAIL PAGE              1.92
ITEMIZED CALLS-SEE DETAIL PAGE                         2.12
TAXES:  (FED      .87)   (STATE      1.16)  (LOCAL    1.16)   3.19


          MIDWEST BELL TOTAL CURRENT CHARGES           35.94


BILLING INQUIRIES - CALL TOLL FREE 1-123-4567
IF MOVING OR PLACING AN ORDER FOR SERVICE - CALL TOLL FREE 1-123-0000



                                                     CONTINUED

              KEY TO CALLS ON OTHER SIDE
```

Source: Deborah Keller-Cohen, Bruce Ian Meader, and David W. Mann, "Redesigning a Telephone Bill," *Information Design Journal* 6.1 (1990), pp. 45–66. Used with permission of the authors.

2. Based on the interviews in stage 1, they developed three possible designs for the bill. They tested these designs against the original bill by using three methods with customers:

- A comprehension task to assess which version was easier to understand (e.g., one multiple-choice question asked how many zone calls were made).
- An attitude survey to see which version customers preferred (e.g., customers were asked to indicate how easy each bill was to understand).
- **Focus groups** in which customers talked about their preferences.

3. They developed a composite telephone bill based on their findings in stage 2 (see Example 11-9). They tested this bill against

focus groups
a method of gathering data by bringing together groups of people around a specific topic or issue

EXAMPLE 11-9 Composite Telephone Bill

◉ **Midwest Bell**

P. O. Box 1234
Anywhere, Midwest 01234

| **MIDWEST BELL** **CURRENT CHARGES** | 123 456-7890 0000 Dec 10, 1984 | Page 2 |

Monthly Charges	Local Services.....................................Dec 10 thru Jan 9			19.11
	Directory Advertising			2.75
Local Calls	Calls over 50 call allowance - 23 at 8.2¢ each			1.89
		Calls	Minutes	
Zone Calls	Day Rate	9	38	3.62
	Evening Rate	3	16	1.04
	Night/Weekend Rate.......................................	2	6	.30
Itemized Calls	See page 3			2.12
Other Charges and Credits	We increased our monthly service rate by 64¢ per day beginning Dec. 7. The new rate applied to Dec 7 thru Dec 9 and was not billed at that time.			1.92
Taxes	Federal: .87 State: 1.16 Local: 1.16			3.19

| **MIDWEST BELL CURRENT CHARGES** | **35.94** |

This amount is included in the total amount due on page 1

For questions about your bill call 555-5555
If moving or placing an order for service call 555-4444

KEY TO CALLS ON OTHER SIDE

Source: Deborah Keller-Cohen, Bruce Ian Meader, and David W. Mann. "Redesigning a Telephone Bill." *Information Design Journal* 6.1 (1990), pp. 45–66. Used with permission of the authors.

the original bill, again using a comprehension task, an attitude survey, and focus groups.

This collaborative group essentially followed the first three phases of quality metrics.

Alone or with a partner, evaluate the document design techniques used to design this final composite telephone bill. In light of the evidence that readers do not read a good bit of the material with which they are regularly confronted, would this redesigned phone bill attract your attention? What are *your* preferences as a reader?

Whatever method you choose—concurrent methods, retrospective methods, or quality metrics—be sure to evaluate your document's design. We cannot stress this enough. Once you do, you will be shocked at how much you learn, and you will be able to apply what you learn to your future document design efforts.

EXAMPLE 11–10 Three Visual Symbols

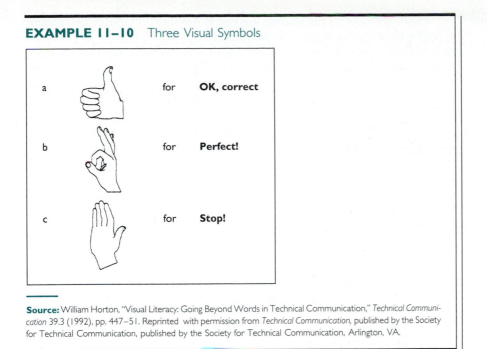

a   for   **OK, correct**

b   for   **Perfect!**

c   for   **Stop!**

Source: William Horton, "Visual Literacy: Going Beyond Words in Technical Communication," *Technical Communication* 39.3 (1992), pp. 447–51. Reprinted with permission from *Technical Communication*, published by the Society for Technical Communication, published by the Society for Technical Communication, Arlington, VA.

Keep in mind that every suggestion in this chapter may not hold true for someone from another culture. Additionally, in some cases, a document design feature that one gender prefers may not be preferred by the other gender. Visual symbols or gestures that are common and well understood in one culture may miscommunicate or even offend when used for another culture.

Example 11–10 depicts three frequently used gestures and the meanings they convey in the United States. However, these gestures may be understood very differently if used in another culture:

1. Although widely used in the United States and northern Europe to signal approval or to hitchhike, this gesture using the thumb has vulgar meanings in Australia, some Mediterranean countries, and much of Africa.
2. The circle gesture has a variety of meanings other than the one common in the United States. In Japan, for instance, it means money—the circular opening representing a coin. In France it means zero or worthless. In South America, it is an obscene gesture.
3. The palms-out gesture, used by traffic cops in the United States and Great Britain to tell motorists to come to a halt, has a scatological meaning in Greece. (Horton, "Visual Literacy," p. 450).

Considering Differences When Designing a Document

Although you may run slightly less risk of misinterpretation when making typographical decisions, in the information age in which we live, assume that at some point in time your document will likely be read by someone from another culture. In short, don't put US blinders on. Consider your larger audience.

SUMMARY

This chapter describes how you might design documents so that busy readers can quickly locate what they need, comprehend the information, and refer to it later if they need to. By envisioning readers' goals and by thinking of readers as users of texts, you can predict whether a reader mainly wants to learn something, do something, or learn to do something when using your document.

You now have a list of document design techniques to use to motivate readers to pick up your text: packaging features such as an attractive binding, the quality of paper, a balanced and consistent layout, interesting visual display of data, a legible cover and print, and the appropriate use of color. Once readers become motivated to open and use a document, they need to locate answers to their questions quickly and efficiently. You can help readers locate information by providing a table of contents and/or index and using structural headings (headers, footers) as well as tab markers or dividers.

To help readers read selectively, you now know how to visualize the grid layout or page design of your document. Remember to create ample white space by using headings and subheadings and margins. Focus readers' attention through the appropriate use of color and visual elements such as lines, borders, circles, and other shapes. To help readers comprehend the information, you need to make conscious decisions regarding the typography of your document; that is, the case, style, size, and spacing of letters as well as the length of a line on the page. To highlight specific text, use boldface, underlining, and italics; however, remember not to overuse highlighting features.

After users have read your document, they need to execute action plans or implement decisions based on what they have read. You can help readers implement their decisions by providing lists, examples, glossaries, and appendices as well as checklists, boxes for recording information, and quick reference cards.

One of the best ways to plan for consistency throughout your document is to create a style sheet in which you make planning decisions about which document design techniques you will use in a document and why. Focus on page design and typography first, and then determine techniques for helping readers before and after reading. By creating a style sheet, you will spend less time in the future on document design decisions.

To evaluate your document design decisions, you can use direct or indirect measures. It's most important to watch real people using your document and to collect information (data) about how they use it. You then use this information to redesign your document, and to determine which document design techniques you should be using as you develop a style sheet.

Last, keep in mind that document design techniques appropriate for one culture may not be as appropriate for another culture. Although some techniques may work across cultures (e.g., the use of headings and lists), some cultures may prefer one technique to another.

---

1. Take another look at the design features of this textbook or any other textbook you have available. Flip through the text and generate a list of the design features, such as the type style used, the distribution of white space, the display of lists, the nature of headings and margins, the binding method, the use of color, and the placement of figures. That is, first, focus on the *larger message:* How is the text organized visually so that you get a clear sense of where information is located? Make a list of features that help you get a sense of the overall organization of the text. Second, focus on the *smaller message:* What smaller features, such as headings, bullets, and figures or examples help you to find specific information quickly and use that information? Again, make a list of these specific features. Now, share your list with another person in your class. What did someone else notice that you didn't even see? What document design features do you most appreciate? What document design features are not present in this text that you wish were there to guide you?

2. Go to the library and read a chapter or two from one of the following sources in which technical communicators talk about document design:

   Barnum, Carol M., and Saul Carliner, eds. *Techniques for Technical Communicators.* New York: Macmillan Publishing Company, 1993.

   Horton, William K. *Designing and Writing Online Documentation.* New York: John Wiley & Sons, 1990.

   Swann, Cal. *Language and Typography.* London: Lund Humphries, 1991.
   Any of the articles in the *Information Design Journal.*
   Any of the *Proceedings* from the yearly conference of the Society for Technical Communication, Arlington, VA.

   Prepare a brief oral presentation on your findings for your class.

3. Explore your current word processing program to identify the document design features available to you. What document design techniques are available for you to use? Bring a list of these and compare word processing programs with other students. Chapter 12 includes a section on computer graphics. In this section, we note that because computer tools for communicators are being developed and improved so rapidly, any list of such tools is quickly outdated. At your campus or workplace, identify the computer graphics tools available for your use.

4. Add headings and subheadings to a document that you wrote for another course. Note how this process helps you to identify areas where your content and organization were weak. After you have created headings and subheadings, choose other document design techniques that would improve this document.

5. Locate a text that you or someone else have written for work. Evaluate the text's document design using one of the evaluation methods discussed in this chapter.

6. As you work near others perhaps in a computer lab, ask them what they consider "visual repellents," that is, what document design decisions annoy them as readers. For example, do certain typefaces annoy them? Do they think that college writers should avoid left- and right-justifying texts? Bring your findings to class for discussion.

7. Locate your phone bill or any other day-to-day functional document (forms, labels, whatever). Redesign and test the revised version(s) according to the steps followed by Keller-Cohen and her colleagues.

8. Choose any of the case study documents included in this textbook and record exactly how you approach and use the document:
   - Before reading;
   - During reading; and/or
   - After reading.

   Do you follow Wright's six-step sketch as discussed in this chapter? Why or why not? What document design changes would you recommend be made to the document?

9. In 1992 the University of the Midwest was informed by the Coalition of Citizens Concerned with Animal Rights (CCCAR) that the University was trapping and killing large numbers of birds in the experimental crop fields on its campus. After much discussion, the University developed the position statement entitled "Bird Control on the Experimental Plots," which is included in Appendix A, Case Document 3.

   Working with a partner, redesign this entire document using Figure 11-4. As you redesign the text, record your document design process. After you have redesigned the text, get opinions from outside readers regarding your changes and record those opinions.

10. This exercise focuses on the Exxon Company's report entitled "How Tiny Organisms Helped Clean Prince William Sound" in Appendix A, Case Documents 2. Make suggestions for redesigning this document for engaging readers:
    - Before reading;
    - During reading; and/or
    - After reading.

   Record your suggestions in the grids in Figures 11-5, 11-6, and 11-7. Then, bring your completed grids to the next class for discussion.

Document Design Decision	Specific Changes in the Document	Opinions from Outside Readers

FIGURE 11–4
Plans for Redesigning a Text

Reader's Goals before Reading	Writer's Document Design Techniques
To formulate a question and become motivated to open a document and locate an answer	Focus on packaging: attractive binding quality of paper balanced and consistent layout interesting graphics legible cover and print appropriate use of color
To find the location of a potential answer	Focus on the larger organization: table of contents indexes structural headings tab markers and dividers
Changes Suggested for the Exxon Document:	

FIGURE 11–5
Grid for Redesigning a Report I

Reader's Goals during Reading	Writer's Document Design Techniques
To read selectively	Focus on guiding readers through:
	grid layout
	white space
	headings and subheadings
	margins
	color
	visual elements (boxes, rules)
To read with ease and use information to make a decision	Focus on typography:
	case
	style (typeface)
	size (point)
	spacing (leading and kerning)
	length
	highlighting

Changes Suggested for the Exxon Document:

FIGURE 11–6
Grid for Redesigning a Report II

Reader's Goals after Reading	Writer's Document Design Techniques
To execute action plans or implement decisions	Focus on providing: lists examples glossaries appendices
To evaluate the outcome of actions or decisions	Focus on providing: checklists boxes for recording information quick reference cards

Changes Suggested for the Exxon Document:

FIGURE 11–7
Grid for Redesigning a Report III

WORKS CITED

Carliner, Saul. "Lists: The Ultimate Tool for Engineering Writers." In *Writing and Speaking in the Technology Professions: A Practical Guide,* ed. D. Beer. New York: IEEE Press, 1991, pp. 53–56.

Edwards, Arthur W. "A Quality System for Technical Documentation." *Proceedings of the 36th International Technical Communication Conference.* Washington, DC: Society for Technical Communication, 1989. pp. MG45–MG46.

"Gas Utilities Switch to Plain Language." *Simply Stated.* Washington, DC: American Institutes for Research, March 1986.

Horton, William. "Pictures Please—Presenting Information Visually." In *Techniques for Technical Communicators,* ed. Carol M. Barnum and Saul Carliner. New York: Macmillan Publishing Company, 1993, pp. 187–218.

Horton, William. "Visual Literacy: Going Beyond Words in Technical Communications." *Technical Communication* 39.3 (1992), pp. 447–51.

Keller-Cohen, Deborah; Bruce Ian Meader; and David W. Mann. "Redesigning a Telephone Bill." *Information Design Journal* 6.1 (1990), pp. 45–66.

Nord, Martha Andrews, and Beth Tanner. "Design That Delivers—Formatting Information for Print and Online Documents." In *Techniques for Technical Communicators,* ed. Carol M. Barnum and Saul Carliner. New York: Macmillan Publishing, 1993. pp. 219–72.

"Plain English Pays." *Simply Stated.* Washington, DC: American Institutes for Research, February 1986.

Schriver, Karen A. "Quality in Document Design: Issues and Controversies." *Technical Communication* 40.2 (1993), pp. 239–57.

White, Jan. *Graphic Design for the Electronic Age.* New York: Watson-Guptill, 1988.

Wright, Patricia. "The Instructions Clear State . . . Can't People Read?" *Applied Ergonomics* 12 (1981), pp. 131–41.

Wright, Patricia. "Writing Technical Information." In *Review of Research in Education,* ed. E. Z. Rothkopf. Washington, DC: American Educational Research Association, 1987. pp. 327–85.

Answers to Questions about Color Blindness

1. About 8 percent of men have trouble distinguishing red and green.
2. Only 0.4 percent of women suffer from the same malady.
3. Blue, primarily because the lens of the eye yellows with age.
4. Blue. The center of the retina is covered with a yellow patch that filters out blue light.

Source: Adapted from William Horton, "Understanding Color Blindness," *Technical Communication* 39.3 (1992), pp. 447–51. Reprinted with permission from *Technical Communication,* published by the Society for Technical Communication, Arlington, VA.

CHAPTER

12

Visual Display and Presentation

Selecting a Visual Display ▮ Preparing Data for Visual Presentation ▮
Creating Tables ▮ Creating Graphs ▮ Accurate Representation of
Data ▮ Creating Photographs ▮ Creating Drawings and Diagrams ▮
Creating Maps ▮ Creating Charts ▮ Using Computer Software and
Hardware to Create Visual Displays

Graphical excellence is that which gives to the viewer the greatest number of ideas
in the shortest time with the least ink in the smallest space. . . And graphical
excellence requires telling the truth about the data.
Edward R. Tufte, *The Visual Display of Quantitative Information*
(Cheshire, CT: Graphics Press, 1983), p. 51.

I n Chapter 11, you learned how to design and package your documents effectively. Part of that design involves the **visual display of data,** which we will look at here. When your written and oral presentations contain data, you need to think about how to interpret that data so that they are meaningful and accessible to your audience. Quite often what you are displaying visually is the evidence that supports an argument. Because visual displays of data either support or take the place of text, you should plan these visual displays *as* you plan your text, rather than leaving them for last.

Generally, you use a visual display of data to accomplish one or two of the following tasks:

- Set off or emphasize important information or evidence.
- Make complex or detailed information accessible.
- Make the abstract more concrete.
- Make the concrete universal.
- Symbolize a structure or organization.
- Condense large amounts of data.
- Show relationships or trends among data.
- Compare and contrast the size or magnitude of data.
- Show what something looks like.
- Show what percentages or proportions are assigned to the parts of a whole.
- Demonstrate how to do something.
- Illustrate how something is organized or assembled.

To understand the **rhetorical function of visual display,** you need to understand first the difference between a prose statement that contains data and a visual display of those data. In the following prose statement, the readers would be so busy trying to keep the numbers in mind that they would have a difficult time seeing relationships.

> The University of the Midwest Agricultural Experiment Station traps sparrows, blackbirds, starlings, and grackles to protect valuable seed in its fields. In 1960, 6305 sparrows, 581 blackbirds, 1086 starlings, and 163 grackles were trapped. In 1970, 2317 sparrows, 1468 blackbirds, 400 starlings, and 131 grackles were trapped. In 1980, 3843 sparrows, 502 blackbirds, 1104 starlings, and 325 grackles were trapped. Finally, in 1990, 3396 sparrows, 395 blackbirds, 2160 starlings, and 412 grackles were trapped.

In Figure 12–1, the audience can see quickly that although more sparrows are trapped than any other kind of bird, the number of sparrows trapped dropped sharply in 1970 from its high in 1960.

In the text of a document containing this information, an audience would need an explanation of that contrast and other changes, such as the increase in blackbirds trapped in the same year that fewer sparrows were trapped and the sharp increase in the number of starlings trapped since 1970. Dis-

---

Introduction

visual display of data
placing data in a graph, chart, drawing, and so forth to support or take the place of text

rhetorical function of visual display
the specific purpose of visual display, including the clarification of relationships among pieces of data and the trends they suggest

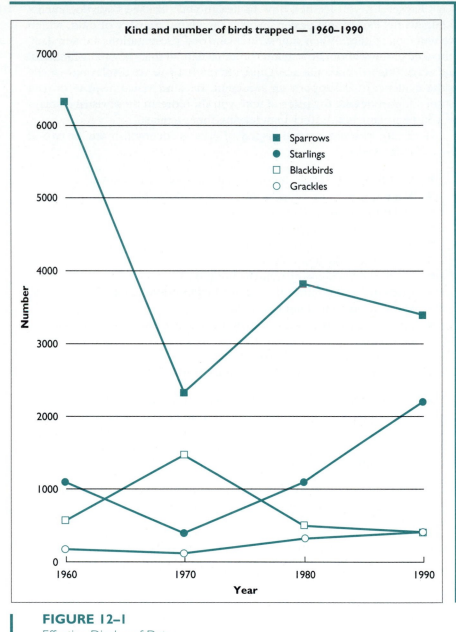

FIGURE 12–1
Effective Display of Data

playing data visually helps the audience see quickly the trends the text must identify, interpret, and explain.

Although visual displays are very effective, they cannot be dropped into your document without carefully preparing your audience to read and understand them. More concrete visual displays of data, such as tables and photographs, require surrounding text that helps an audience decide what is essential information and what is extraneous. More abstract visual displays of data, such as line graphs and drawings, often require less explanatory text but restrict the audience to those experienced in decoding these symbols. You need to introduce all visual displays of data, relate them to your subject, and help your audience interpret them. You need to mention the visual display by its title and number in the text the first time that your audience will need the information referred to, and you need to place the display as close to where it is mentioned as possible.

To be effective, text and visual displays of data must be fully integrated. For example, you might refer to specific data and then refer the audience to the visual display of those data:

> The total number of birds trapped in university fields in 1990 was 1591 compared to 2034 in 1960 (see Table IV).

Or, you might call your audience's attention to the function of the display the first time they would find it useful.

> Table IV displays the number of birds by type trapped in university fields in 1960, 1970, 1980, and 1990.

You may need to explain how or when to read your visual display. For example, you might say:

> To compare the number of sparrows, blackbirds, starlings, and grackles trapped in university fields from 1960 to 1990, see Table IV. Data are given for the first year only in each decade; see the bottom line of the table for the total number of birds trapped.

Finally, because you may be using your display of data to persuade your audience—to give evidence or good reasons—you may need to draw some conclusions about your data. For example, you might say:

> The data in Table IV demonstrate that since 1960 the university has success-fully decreased the number of birds trapped in its experimental fields.

Remember our advice about introducing, discussing, and interpreting your visual displays as we discuss each type of display in this chapter.

In order to decide which visual display to use, you need to understand generally which visual display works best in what circumstances. Use Writing

Selecting a Visual Display

Strategies 12–1 to help you select what form of visual display will best serve your audience's needs.

You will also need to select your type of visual display of data according to your purpose as well as according to the experience, technical understanding, language, and cultural background of your audience.

SELECTING A VISUAL DISPLAY ACCORDING TO PURPOSE

Even though you can use Writing Strategies 12–1 to get started on selecting and creating a visual display of data, the purpose of your display will help you determine the type of visual display you will need. For example, if you have to explain the mechanical and chemical processes used in office copying machines, Writing Strategies 12–1 tells you that drawings "show the physical components of mechanisms, objects, or organisms." But you will still have to decide what kind of drawing you should use. Writing Strategies 12–2 gives you some additional criteria to use once you are clear about the specific purpose of your document. To support your office copying machine description, you might select an action drawing if you want to display the interaction of the machine parts, an exploded drawing if you want to display the parts as assembled or attached to each other, or a symbolic drawing if you want to display the chemical interactions.

SELECTING A VISUAL DISPLAY ACCORDING TO AUDIENCE

Of course, your audience will also determine the type of visual display you select to present your evidence or support your argument. When considering a visual display, you can often focus on your audience's experience with the topic, their technical understanding, and their language and culture.

Experience

An audience that has little experience with your subject will need a simple and obvious visual display. A table with just two or three column and row headings and related data is easily understood by most audiences. A line graph with two lines plotted against familiar units of measurement such as time and dollars is a familiar display you can use to introduce an unfamiliar subject.

More experienced audiences can handle detailed or more abstract visual displays, such as exploded drawings that show how the parts of a mechanism fit together. An audience that has worked with a similar mechanism can handle the detail of a more complex or symbolic display such as a schematic.

As you think about each display discussed in this chapter, you should consider its appropriateness for your purposes in terms of how much experience your audience has had with your subject.

WRITING STRATEGIES 12–1   Visual Display and General Use

Tables

- Give a large amount of data or detailed information in a small space
- Show the characteristics of objects, ideas, or processes
- Aid item-by-item comparison
- Display exact numeric values
- Give exact values for comparison

Graphs

- Show the relationships, trends, and patterns in two or more sets of data
- Support forecasts and predictions
- Help the audience interpolate or extrapolate
- Present complex information symbolically
- Interest the audience in the data or add credibility to words

Charts

- Show the components, chronology, or steps of a whole organization, process, mechanism, or organism
- Display the interrelationships between the components, stages, or steps

Photographs

- Show the actual appearance of an object or organism
- Add realism of detail, tone, texture, and color to a display or discussion
- Display features of objects or organisms difficult to draw, such as emotions
- Prove something is real or exists

Drawings and Diagrams

- Show the actual appearance of an object or organism excluding unnecessary detail
- Show the physical components of mechanisms, objects, or organisms
- Show generic or nonspecific objects or organisms
- Display objects that might have existed in the past or might in the future
- Show views that would be impossible to see otherwise

Maps

- Offer geographical information on data and objects
- Show locations in relationship to each other
- Show distribution of objects over a particular location
- Prepare the audience for travel

WRITING STRATEGIES 12–2   Linking Purpose to Specific Visual Display

Visual Display	Purpose
Graphs	
Simple bar	Display approximate values
Multiple bar	Compare approximate values
Stacked or area	Show how values contribute to total
Pie or 100 percent bar	Represent percentages or portions of a whole unit
Deviation	Display values above and below zero (negative/positive, loss/gain, hot/cold)
Line	Demonstrate or predict trends
Combination bar/line	Relate approximate values to trends
100 percent area	Show trends in components of a whole
Pictorial	Communicate abstract or unfamiliar values through symbols
Drawing	
Line	Show the generic or selective features of an unfamiliar or complex object
Action view	Explain an action or process of an object or organism
Phantom or translucent	Offer an inside view of an object or organism
Exploded view	Show how to assemble an object
Schematic or symbolic	Symbolize a system to a specialized audience
Charts	
Block	Reveal relationships between parts or categories of a system
Organizational	Show relationships between parts of an organization
Flow	Show interrelated decisions or steps in a process

Technical Understanding

The degree of **technical understanding** your audience has should also affect your selection of a visual display. The more technical understanding the audience has, the more abstract and symbolic your display can be. General or inexperienced audiences will recognize common or familiar symbols and units of measurement only. A general audience would understand a pictorial graph of the number of birds trapped only if the birds were distinguished by color or tone and not by particular features such as wing span or markings.

Audiences with some degree of technical understanding, such as technicians or mechanics with a high school or two-year college education, can handle more detail and symbols or icons common to their trade or job. These audiences would also be familiar with certain types of visual displays; for example, the electrician would be educated to read schematics of circuits.

Professional audiences—engineers, scientists, systems analysts, computer programmers, and others—can understand visual displays using symbols and abstractions alone. These audiences can read everything from combination graphs to flow charts easily when the information and types of display are common in their profession.

technical understanding
specialized knowledge about a product or process that an audience brings to a document

professional audiences
audiences composed of highly trained professionals, such as scientists, computer programmers, engineers, physicians, and systems analysts

Language and Culture

The audience's ability to read English and their cultural background also should influence your choice of visual displays. Generally the less comfortable the audience is with English, the more you need to limit your words and rely on widely accepted symbols, pictures, or icons. When possible, determine your audience's ability to read English. The majority of your audience should fit into one or two of the following categories:

- Reads English well.
- Has difficulty reading English.
- Reads English as a second language.
- Must rely on a translation of English.

If your words are going to be translated, use the most common, least ambiguous words and back up these words with widely understood symbols. On the other hand, when your audience has difficulty reading or does not read English at all, you will have to rely on the symbols alone to carry the message. The audience that reads well can handle most combinations of numbers and words.

The **cultural background** of the audience also determines their ability and willingness to read your visual display. For example, colors in various cultures carry different meanings. In Western cultures yellow can carry a negative connotation, while in some Eastern cultures it symbolizes courage. Images of parts of the body and gestures as symbols should be selected with your audience's culture in mind. A waving hand might represent the end or

cultural background
the shared understanding of symbols, shapes, gestures, and colors within a given culture

"good-bye" in Western culture, but in other cultures can signal the beginning or "hello." Finally, a culture that reads right to left would have difficulty reading a table designed for a Western audience that reads left to right.

Preparing Data for Visual Presentation

Before you create your visual display, you need to make sure your data are complete and accurate, and you will want to organize them usefully and meaningfully. In order to do that, keep the following guidelines in mind:

- Include all data available that pertain to your topic, even the data that might contradict final recommendations.
- Collect your data from a large enough **random sample** to support your recommendations and interpretation.
- Indicate all the possible interpretations of your data.
- Make sure that the data really support what you are saying.
- Clearly indicate the **units of measurement** of your data.
- Condense the data to a manageable form (e.g., $11,000 can be expressed as 11 with a column head or axis label of "dollars in thousands").
- Indicate, if needed, the following:

random sample
a representative sample selected from a table of random numbers or computer program for random selection

units of measurement
universal standards by which data are measured, such as inches, Hertz, meters, pints, or miles

Mean: the arithmetic average obtained by dividing a sum by the number of items added.

Median: the middle number in a series containing an odd number of items, or the number midway between the two middle numbers in a series containing an even number of items.

Mode: the number or value that appears most frequently in a series.

In this chapter, examples of visual displays come from two sources: technical and scientific publications and the documents on bird trapping that appear in Appendix A, Case Documents 3, of this text. The bird trapping displays were created with Microsoft Excel™, a spreadsheet computer program.

Creating Tables

Tables enable you to offer large amounts of information in a small space for comparison and interpretation, and, with their rows and columns of numbers or characteristics, are particular and concrete. An audience will not have to read symbols and deal with abstract lines and curves. Complex tables (more than two or three column and row headings) are usually reserved for audiences with experience in the topic or with a high degree of technical understanding.

As with all visual displays of data, you should integrate your table into the text by introducing, discussing, and interpreting it. When creating a table, organize data into groups and write short descriptive labels for each group. Order the table in the most appropriate way for your audience and purpose: alphabetically, geographically, chronologically, highest to lowest, smallest to largest, and so forth. Include all your data and do not ask your audience to interpolate or extrapolate. Compare your data vertically, rather than horizontally, and place columns that must be compared next to each other. **Columns** in a table have vertical references, while **rows** have horizontal references. **Stub heads** classify row heads and therefore have vertical reference (see Figure 12-2 for the typical arrangement of a table).

Finally, if you remember the following guidelines, your audience can read your table more easily:

columns
vertical structure of information in a table

rows
horizontal structure of information in a table

stub heads
categories or classification of row headings in a table

- Select standard units of measurement; use common abbreviations for them, and indicate them either in column and row or line headings or beside each number.
- Align the numbers along the decimal point or along the units column.
- Align prose along the left-hand margin, or center prose within the column.
- Round off numbers if possible.
- Convert fractions to decimal points and limit numbers to two decimal points.
- Use parallel grammatical forms for words.
- Box the table, but avoid excessive rules (seldom are vertical rules necessary). Use spaces to help cluster and separate groups of information.
- Use footnotes, indicated by lowercase letters and numbers as **superscripts,** for clarification.
- Place the source of the table below the table.
- Turn large tables 90 degrees and place the top at the binding of the document.
- Continue long tables by writing "continued" at the bottom of the first page and repeating the full title, "continued," and full column headings on each page that follows.

superscripts
numbers or letters raised above the line used to identify footnotes

Finally, remember that displaying data is a rhetorical act. When you create a visual display, you are persuading your audience that you have gathered your data thoroughly, that you have interpreted the data correctly, and that you can use those data logically to support your argument. Figures 12-3 and 12-4 show two displays of the same data. The figures tell different stories, however, and helping an audience read them provides different challenges for the writer. The first table (Figure 12-3) might persuade an audience protesting the trapping of birds that numbers have declined since

Stub Head	Single Column Head[a]	Multiple Column Head[b]		Single Column Head
		Subhead	Subhead	
Line head	111.1	222.2	333.3	444.4
Line head	11111.1	22.2	33.3	44.4
Subhead	111.1	2.2	3333.3	44.4
Subhead	111.1	2222.2	333.3	4.4
Line head	11.1	2.2	33.3	444.4
Column average or total	111.1	222.2	33.3	44.4

Source: XXXXX

[a] Footnote

[b] Footnote

FIGURE 12–2
Parts and Labels for a Typical Table

Kind of Bird	Year			
	1960	1970	1980	1990
Sparrow	6305	2317	3843	3396
Blackbird	581	1468	502	395
Starling	1086	400	1104	2160
Grackle	163	131	325	412
Average per year[a]	2034	1079	1444	1591

[a] Rounded off to the next highest number.

Source: University of the Midwest, Agricultural Experiment Station.

FIGURE 12–3
Kind and Number of Birds Trapped—1960–1990: Version One

the 1960s, or it might help experimental station researchers analyze how effective the traps have been in capturing each kind of bird over four decades. The second table (Figure 12-4), on the other hand, could be used by those protesting the trapping to emphasize the totals per year and per kind. Certainly a total of 8,135 birds trapped in one year or 15,861 sparrows trapped

| | Kind of Bird | | | | Total per |
Year	Sparrow	Blackbird	Starling	Grackle	Year
1960	6305	581	1086	163	8135
1970	2317	1468	400	131	4316
1980	3843	502	1104	325	5774
1990	3396	395	2160	412	6363
Total per kind	15,861	2946	4750	1031	

Source: University of the Midwest, Agricultural Experiment Station.

FIGURE 12–4
Kind and Number of Birds Trapped—1960–1990: Version Two

in four years seems more significant than 2,034 birds per kind trapped in one year. However, notice that because Figure 12-4 asks the audience to read horizontally, as well as to follow the natural vertical pattern, the audience might miss the ''Total per year'' column.

Creating Graphs

All graphs translate numbers into pictures—or bars, lines, areas, or pictorials. The numbers are plotted as a series of points on a coordinate system. The **vertical axis,** called the **ordinate** or **Y axis,** begins with zero and often indicates quantity; the **horizontal axis,** called the **abscissa** or **X axis,** often indicates time, item, or category. Therefore, graphs usually illustrate changes in one amount (such as quantity) in relation to changes in another measure (such as time). In the following sections, we discuss common types of graphs. Because graphs have many features in common, here are some general guidelines for making them:

- Always title the graph with the largest font or type used in the graph and center the title at the top.
- Make the axis lines heavier than the grid lines but not as heavy as curves. You can omit the Y axis line in graphs with horizontal grid lines.
- Start at zero, unless a large amount of the grid is unnecessary; then clearly indicate a break in the grid but retain the zero base line.
- Avoid using two amount scales (such as temperature on a left vertical axis and humidity on a right vertical axis).

vertical, ordinate, or **Y axis**
on a graph, the vertical axis that begins at zero and usually indicates quantity

horizontal, abscissa, or **X axis**
on a graph, the horizontal axis that usually indicates time, item, or category

- Label each axis (horizontally, if possible) and indicate the unit of measurement used if the unit of measurement does not follow each number.
- Place a figure every second or fourth grid ruling. Use standard intervals of 10, 20, 30, and so forth.
- Use small tick marks inside the scales if grid rulings are not used.
- Distinguish between curves or lines, bars, pieces of the pie, and so forth by tone, color, or symbol listed in a key or legend.

In general, graphs, particularly line graphs, are more abstract than tables; that is, graphs are more generic, more symbolic, and more likely to show the structure or organization of the data than a table would. Therefore, if a graphic is suited to your purpose—for example, to display trends or to forecast—then be aware that your audience must have had some experience with your topic and with this type of visual display to understand the display and be persuaded by it. Graphs are not always interchangeable. Each type of graph lends itself to specific purposes.

BAR GRAPHS

Bar graphs demonstrate magnitude or size of several items or emphasize difference in one item at equal time intervals. Each bar represents a separate quantity, and multiple bars may be grouped and displayed horizontally or vertically. Vertical bars best indicate height and depth, and horizontal bars best indicate distance, length, and time.

Although bar graphs are not the best choice when showing trends, they work well when your data consist of distinct values over a specified period of time and when the values are noticeably different. Bar graphs can also show how data change from place to place, and 100 percent bar graphs can show composition as a whole. Bars can also be stacked to show totals or combined with line graphs to help an audience understand difference along with trend. Be aware, though, that the darkest bars in your display may seem larger to your audience and attract their attention; this phenomenon might help emphasize a point in your argument, but you need to play fair with your audience and make sure that your data really support your point.

When planning any bar graph, keep in mind the following guidelines:

- Make the bars the same width and label each bar.
- Make the space between bars one-half the bar width.
- Make the vertical or Y scale at least 75 percent as long as your horizontal scale.
- Make the longest bar extend nearly to the end of its parallel axis.
- Consider adding exact numbers or extending the gridlines to help relate the bars to the values.

- Consider using a **legend** or **key,** and distinguish among your bars by varying pattern or color in multiple bar graphs.

legend or **key**
information included to explain devices such as color or patterns on a graph

Simple Bar Graphs

Again, although graphs are more abstract than tables, almost any audience can grasp the point of a simple bar graph quickly. For example, the bar graph in Figure 12-5 shows how many sparrows were trapped in university experimental fields over a certain period of time. The audience can quickly and easily contrast the years in which a great many sparrows were trapped (e.g., 1955 and 1960) and the years in which few were trapped (e.g., 1970 and 1985).

If your data are too close to compare by the size of the bar alone, you can place exact numbers over each bar (for example, 1965 and 1980 in Figure 12-5). You also need to be aware that the farther a bar is from the vertical axis, the more difficult it is for the audience to see differences accurately. However, Figure 12-5 would work well if you wanted to emphasize for either a general or a technical audience that the number of sparrows trapped in 1955 was far greater than those trapped in more recent years.

Multiple Bar Graphs

You can group several bars in a multiple bar graph to show the magnitude of different variables or items in relation to each other over distinct periods of time. However, you need to limit the groups to three or four bars. Your audience generally needs a legend that lists the variables and shows the pattern or color that distinguishes them.

In Figure 12-6 the kinds of birds trapped are displayed using bars of various patterns. Because few blackbirds and grackles are trapped over this time period, these bars are quite small. To help the audience see them, they are darker. Grid lines allow the audience to gauge the number of each kind of bird trapped more easily. The great numbers of sparrows trapped becomes obvious immediately because of the contrast in bar height. You can make a graph three-dimensional to help your audience better see the smaller bars, but you need to remember that using this technique increases the visual size of bars and may misrepresent your data.

A horizontal bar graph is made by turning a vertical bar graph and scales on its side or 90 degrees to the right. This placement usually eliminates the need for a key. If possible, the longest bar should be placed at the bottom of the graph for balance. Some of the data from Figure 12-6 are displayed horizontally in Figure 12-7.

In a horizontal bar graph, you have the room to label the bars themselves. However, labeling at the end of the bar lengthens the look of the bar and therefore changes the audience's impression of its magnitude. The horizontal

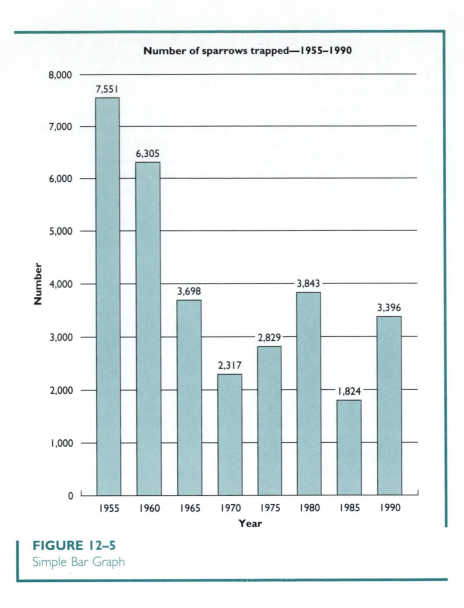

FIGURE 12–5
Simple Bar Graph

bars in Figure 12-7 could also have been labeled at the left of the Y axis, between the date and the axis line. A horizontal bar graph helps an audience see increasing or decreasing values more easily. However, the audience would have more difficulty seeing the differences between items—in this case, the kinds of birds.

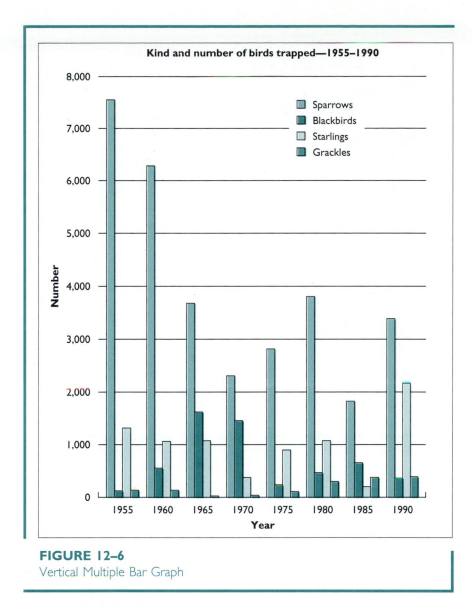

FIGURE 12–6
Vertical Multiple Bar Graph

Stacked Bar Graphs

Stacking the bars, either vertically or horizontally, in a graph can inform your audience of the size or magnitude of items and also let them see how items contribute to a whole quantity and how that quantity changes over time. The magnitudes of the different items or components in a bar are differentiated by patterns or tones. For example, in Figure 12-8, an audience

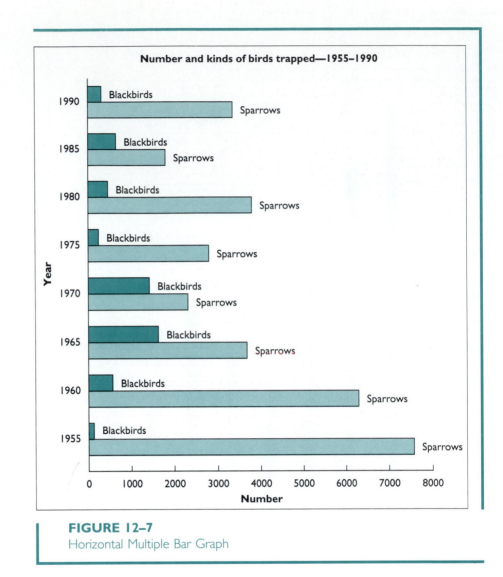

FIGURE 12–7
Horizontal Multiple Bar Graph

can see clearly how many grackles, starlings, blackbirds, and sparrows were trapped in a given year and also how the total number of birds trapped changed from year to year in this stacked bar graph. Because it would be difficult to read those totals, we have placed the numbers above the stacked bars. This aid is also important because in this graph the numbers on the Y axis are compressed to save space. The Y axis now carries a label that reads "Number (in thousands)" so that the audience can translate 1 into 1,000, and so on.

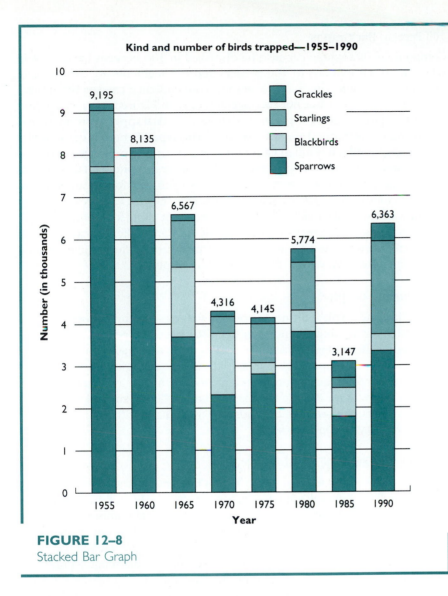

Kind and number of birds trapped—1955–1990

Legend:
- Grackles
- Starlings
- Blackbirds
- Sparrows

Number (in thousands)

Year

FIGURE 12–8
Stacked Bar Graph

Finally, notice in stacked bar graphs how difficult it is to compare the upper segments (the grackles, starlings, and blackbirds), as these bars have no **common base line.** It's impossible to tell how many blackbirds were trapped in 1955, for example. Stacked bar graphs are intended to serve as a means for comparing *totals* rather than individual elements. Although lines could be added to link each segment, it's wise to limit the number of items to two or three.

common base line
in vertical and horizontal bar graphs, the shared starting point of each bar; this feature is missing from stacked bar graphs

100 Percent Bar Graphs

Components of a whole can also be displayed in 100 percent bar graphs in which each bar extends to 100 percent and the segments are measured in percentages. When you use these graphs, your audience can see the relative size of each segment but not the magnitude, as in Figure 12–9, in which it is easy to figure out that the greatest portions of birds trapped were sparrows, but the numbers trapped are not displayed. This type of graph is quite similar to a pie graph, which is much easier for a general or inexperienced audience to interpret. However, as with the stacked bar graph, individual segments do not line up along a common scale, so it's difficult for an audience to compare, for example, the number of blackbirds trapped each year with any accuracy. Reading data according to their position along a common scale is often the most accurate **perceptual task** that an audience can perform (Cochran, Albrecht, and Green, p. 27), so be aware of the difficulties your audience might have with a 100 percent bar graph.

Deviation Bar Graphs

A deviation bar graph indicates **opposite values**—negative/positive, loss/gain, and so forth—that extend on either side of a central point. For example, if the night shift, midnight to 9:00 AM, takes three readings of the river level on the hour and computes an average, the readings might be expressed in Figure 12–10. Zero equals normal height, and the hourly averages are either above or below zero. Placing the exact numbers above and below the bars helps overcome the problem the audience might have with reading data not positioned along a common scale.

Combination Bar/Line Graphs

You can easily combine a bar graph with a line graph. Notice that in Figure 12–11, the bars show the annual purchase of alternative-fuel vehicles (AFVs) by the federal government, and the line shows the total number of AFVs the government projects it will have purchased by 1998. The line, rather than stacked bars, gives greater emphasis to this number than a stacked bar graph would. However, line graphs, as we will see in the following section, are more abstract than bar graphs; they demand that the audience interpret symbols and understand the structure and organization, rather than the particulars, of the data.

LINE GRAPHS

Line graphs are most useful in showing trends or movement in a long series of data or several series of data, particularly if your audience must **interpolate** (fill in) or **extrapolate** (extend) that series and if your audience does not need to know the exact values. Line graphs show cause and effect—what

perceptual task
intellectual effort performed by an audience to retrieve and interpret data from a graph or other visual display

opposite values
values such as positive/negative, gain/loss, increase/decrease that can be displayed on a deviation bar graph

interpolate and **extrapolate**
means of understanding trends in data by filling in gaps (interpolating) or projecting trends (extrapolating)

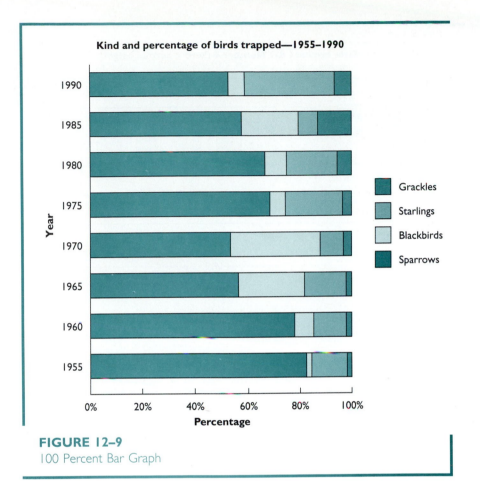

FIGURE 12–9
100 Percent Bar Graph

happens to one value (dependent value) as another value (independent value) changes.

Because line graphs are more abstract than bar graphs and can accommodate and symbolize complex data, you should be sure that your audience has enough experience and technical knowledge to understand your display. Keep in mind the following guidelines:

- Place the dependent variable on the Y (vertical) axis and the independent variable on the X (horizontal) axis.
- Keep vertical and horizontal axes proportionate.
- Make sure the slope or steepness of the line accurately indicates the trend indicated by the data.

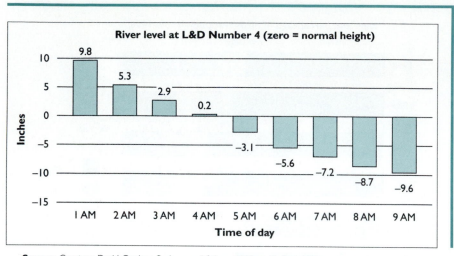

Source: Courtesy, David Gardner, Business and Science Writers, St. Paul, MN.

FIGURE 12–10
Deviation Bar Graph

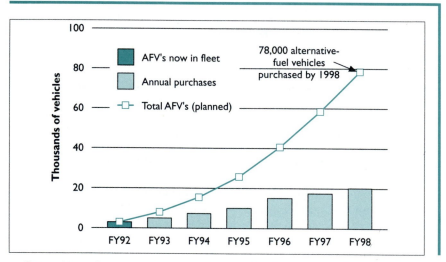

Source: *National Energy Strategy. Powerful Ideas for America. One Year Later.* National Technical Information Service, US Department of Commerce, February 1992.

FIGURE 12–11
Combination Bar/Line Graph

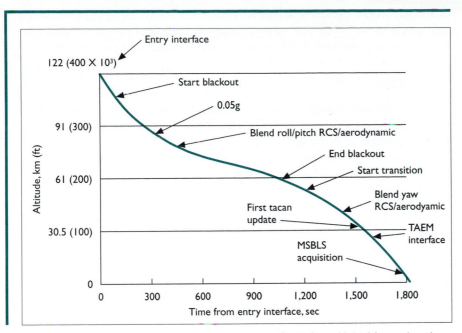

Source: John F. Hanaway and Robert W. Moorehead, *Space Shuttle Avionics System,* National Aeronautics and Space Administration, Office of Management, Scientific and Technical Information, NASA SP–504, 1989.

FIGURE 12–12
Complex Line Graph

- Mark individual data points on each line so the audience can see how you used the data to determine the slope.
- Place only three or four lines on each graph.
- Use colors or symbols to distinguish your lines.
- Label each line or use a legend if necessary to avoid clutter.
- When showing positive and negative values, use the X axis as the mean.

Look back at Figure 12-1 for an example of a simple line graph. Rather than comparing the numbers of birds trapped in each year, the lines in this figure show increases and decreases as well as trends and projections. For example, a reader could be fairly certain that the number of grackles trapped would not increase in 1996, but would not be surprised if the number of starlings increased, given the trend from 1970 to 1990.

A variation on the line graph (Figure 12-12) depicts events during movement or action. The points marked on the line indicate each stage of the entry phase of the Space Shuttle Avionics System measured by a decrease in altitude over a period of seconds. An audience with technical knowledge of the system would be able to understand not only the events but also why each event appropriately follows the preceding one at that time and altitude.

AREA GRAPHS

You can use area graphs (also called multiple band graphs) to show both trends in data and the sum of those data. The area below each line is filled in by a pattern or color and represents a unique value. The topmost line represents the total or sum of all the data. Usually the largest area or the line with the least slope appears closest to the X axis.

Because area graphs display two aspects of data, they are difficult to interpret, and you should use them for experienced and educated audiences. An audience seeing Figure 12-13, for example, might need to be reminded that the top line represents the total number of birds trapped in a given year, while the areas between each line emphasize the trends for that kind of bird only. In other words, the number of blackbirds trapped in 1970 is about 1500, not over 4000. You can add **grid lines** like those in Figure 12-13 to help an audience read the data for more recent years; however, remember that area graphs are abstract and therefore are best used to demonstrate trends, not exact data.

grid lines

can be used on a graph to help readers pick out data from given categories

100 PERCENT AREA GRAPHS

You would use a 100 percent area graph to show trends in the components of a whole. For example, an audience looking at Figure 12-14 could see how the amount of organic material has increased since 1900, while the number of forestry products has gone down. The Y axis indicates the part of 100 percent that each component makes up. Although stacked bar graphs and pie graphs show proportions of a whole unit, the 100 percent area graph is more difficult to read because it shows proportions of a whole trend.

PIE GRAPHS

coordinate system

the X and Y axes on a graph

Pie graphs represent proportions or percentages of a whole, rather than figures plotted on a **coordinate system.** You will find that they work well with inexperienced audiences who have little technical knowledge or who read with difficulty. However, although pie graphs gain attention and convey impressions, they offer little help in close comparisons. Audiences have difficulty assessing the differences between wedges similar in proportion. Also, pie graphs cannot be used to compare component parts of two or more wholes. Although exact numbers can be added to the pie, experienced audiences prefer line, area, or bar graphs.

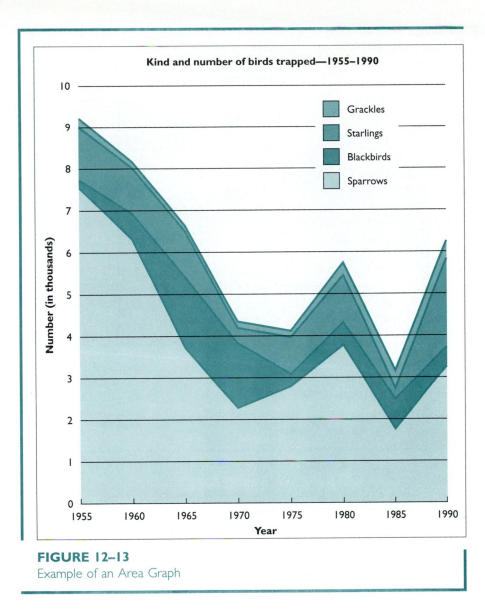

FIGURE 12–13
Example of an Area Graph

When creating pie graphs, keep in mind the following guidelines:

■ Start the pie graph at the 12 o'clock position with the largest wedge.
■ Make sure that wedges add up to 100 percent with each 3.6 degree segment of the pie equaling 1 percent.

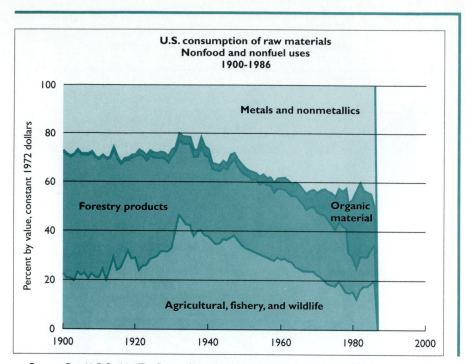

Source: Donald G. Rogich. "The Future of Materials: Plastic Components Are Growing," *Minerals Today,* US Department of the Interior, Bureau of Mines, MS2501, June 1991.

FIGURE 12–14
Example of a 100 Percent Area Graph

- Keep wedges at least 5 percent or 18 degrees.
- Limit your pie graph to eight wedges.
- Combine small segments under "Other."
- Separate only one wedge for emphasis.
- Place labels horizontally outside the pie. Avoid lines and arrows.
- Indicate exact percentages in labels.
- Avoid 3-D pies, because the closest wedge looks larger than it actually is.

The pie graph in Figure 12–15 would be best used in an oral presentation to gain the attention of an audience interested in the high percentage of sparrows trapped in 1990. By pulling out the sparrow wedge and adding the exact percentage, the audience can see immediately that sparrows constituted over half of the number of birds trapped.

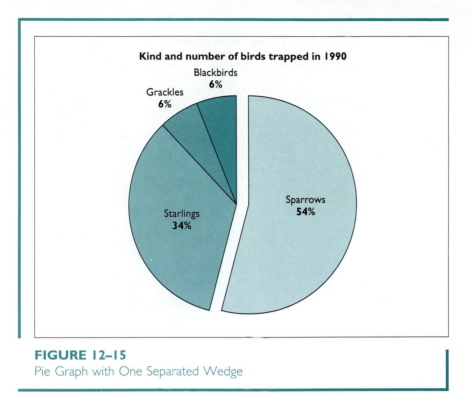

FIGURE 12–15
Pie Graph with One Separated Wedge

PICTORIAL GRAPHS

You can use pictorial graphs (or pictographs) to communicate quickly to most audiences abstract ideas or concepts by using symbols or **icons.** These symbols can be used as numerical counting units, with each representing a specific number. Perhaps the most effective pictorial graph is a bar graph with the bars replaced by stacks of symbols. A pictorial graph is visually interesting and memorable but, designed poorly, can distort or misrepresent data. For that reason, keep in mind the following guidelines when creating a pictorial graph:

icons
symbols conveying abstract ideas or concepts

- Make sure that symbols have a strong association with your topic, value, or idea.
- Make sure that symbols are readily understood, are simple and recognizable when reduced or enlarged, and are easily divisible (if necessary). If more than one symbol is used, symbols should contrast well in shape.
- Design all symbols so that they are the same value and size.
- Space all symbols equally.

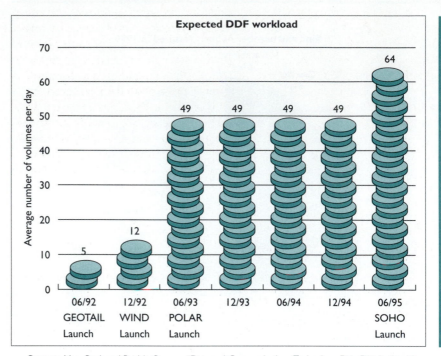

Source: Mary Reph and Patricia Carreon, "Data and Communications Technology. Data Distribution Media Study," *Research & Technology*, Goddard Space Flight Center, NASA, 1990.

FIGURE 12–16
Pictorial Graph or Pictograph

- When displaying different quantities, change the number of symbols, not the size or height.
- If possible, round off numbers to eliminate fractions (and therefore divided symbols).

In Figure 12–16, the daily workload of a Data Distribution Facility (DDF) for distributing spacecraft data from various satellites to science centers is depicted in a pictorial graph. This graph was shown to vendors who must demonstrate that they can provide the needed number of rewritable optical disks of the anticipated production workload. The symbols used in the graph take the place of traditional bars and represent the optical disks. The audience can grasp immediately the anticipated increase in the number of disks needed as more satellites are launched, and the audience can later study exact numbers that appear at the top of the stacked symbols.

Before we discuss other types of visual display, we need to remind you about accurate representation of data. Most of the common problems that might lead to **distortion** of your data presentation occur in bar, line, and pictorial graphs. Of course, you should make sure that your sample of data is randomly collected, large enough to represent your topic, and gathered accurately, as discussed in Chapters 6, 7, and 8. You must present all data gathered, even the unfavorable or atypical. You must be sure that your data measure what you claim and that the causal relationships you describe are really supported by your data. If your data are taken from another source, you should check the accuracy of that study carefully and credit that source for your audience.

Also, you need to avoid the following five common pitfalls in displaying data in graphs:

- Distortion of the X or Y axis in graphs;
- Distortion by **suppressing the zero** in graphs;
- Distortion by increasing the width or height of bars in graphs;
- Distortion in 3-D pie graphs; and
- Distortion through improper use of icons or isotypes in pictorial graphs.

DISTORTION OF THE X OR Y AXIS

The values on the X and Y axes of graphs should be spaced equally to portray relationships and trends accurately. This rule is particularly important in line graphs. You can see that the information display in Figure 12–1 is distorted in Figures 12–17 and 12–18. In Figure 12–17 the increases and decreases in the number of birds trapped appear much greater than they actually are because the tick marks on the Y axis are spaced farther apart than those on the X axis. The same distortion would take place if the values on the Y axis were measured at more frequent intervals or at fewer or more intervals midway through the axis; for example, marking values after 3,000 at 3,500, 4,000, 4,500, 5,000, 5,500, 6,000, and 6,500 would cause the drop in the number of sparrows trapped to appear quite dramatic. In turn, there appear to be few significant increases or decreases in the trends displayed in Figure 12–18 because the X axis has been distorted.

DISTORTION BY SUPPRESSED ZERO

As we discussed previously, graphs should always begin at zero. You can see the distortion in Figure 12–19 because the values begin at 30,000, and therefore, the salaries of the vice presidents look rather modest. You might be tempted to suppress the zero if none of your values come close to zero; for example, every vice president gets more than $30,000 per year. However,

Accurate Representation of Data

distortion
misrepresentation of data in a visual display

suppressing the zero
a distortion of data by starting graphs at a number other than zero

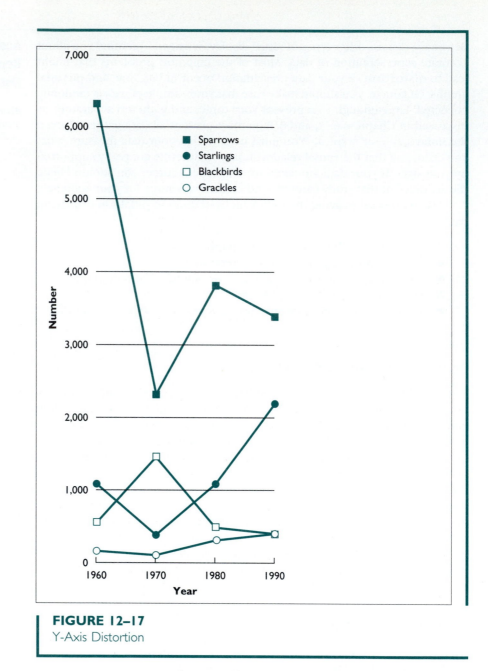

FIGURE 12–17
Y-Axis Distortion

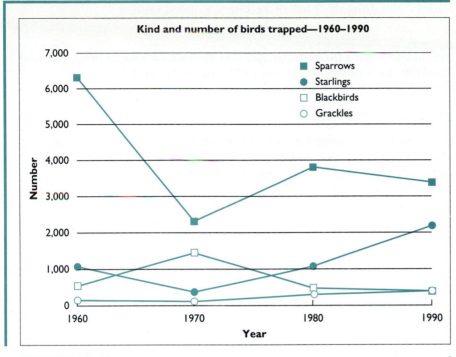

FIGURE 12–18
X-Axis Distortion

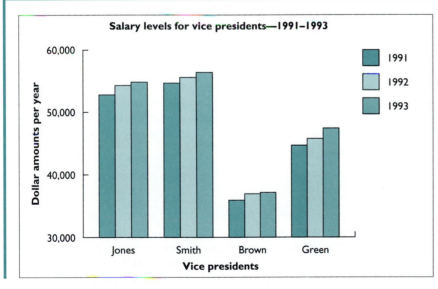

FIGURE 12–19
Data Distortion by Suppressed Zero

373

you need to begin the graph at zero and indicate by a visual break that you have left out a portion of the bar or line (as in Figure 12-20).

DISTORTION IN WIDTH OR HEIGHT OR BARS

If you make any changes in one dimension of your graph without making corresponding changes throughout the graph, you will introduce distortions into your visual display. Increasing the width of a bar in a graph, without making a change in the height, and vice versa, will distort the visual appearance of the data. For example, the bars in Figure 12-21 are so wide that the audience will perceive the number of birds trapped to be much greater than they are. If you increase the width of your bar, you need to increase the height in proportion, and vice versa.

DISTORTION THROUGH 3-D

Particularly when creating a pie graph, you may be tempted to make your presentation more dramatic by using 3-D. However, the data "nearest" the audience in the 3-D version will appear greater than it is. The data from Figure 12-15 have been displayed in Figure 12-22 in a way that makes the number of blackbirds and grackles appear much greater than they are (particularly because we have also pulled out these two wedges from the pie).

DISTORTION THROUGH IMPROPER USE OF ISOTYPES OR ICONS

isotypes
another term for icons used as numerical counting units in visual displays

When you use a pictorial graph, remember that **isotypes** or icons are counting units. If you attempt to display your data by enlarging or shrinking one icon (as in Figure 12-23), you will misrepresent your data. The audience cannot gauge how many sparrows were actually trapped because the contrast between the smallest and largest icon is so dramatic. The same data are accurately displayed in Figure 12-24.

Creating Photographs

You can use a photograph to depict a subject in detail and in a realistic setting, particularly if you use color. Photographs can be reduced, cropped, and enlarged (although enlargement sometimes decreases sharpness) to enhance a subject; the angle of view can add emphasis to a particular part or approach. For audiences unfamiliar with your subject, a photograph can bring instant recognition and orientation as photographs are concrete, rather than abstract. However, so much detail may appear in a photograph that your text must direct your audience as to what to look for in that photograph.

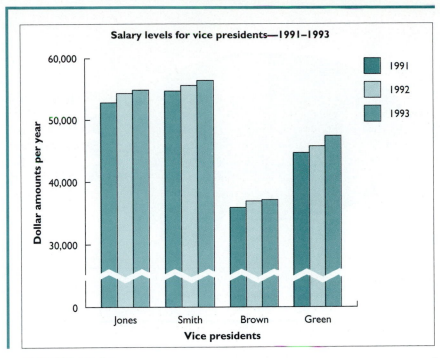

FIGURE 12–20
Accurate Way to Indicate Break in Axis

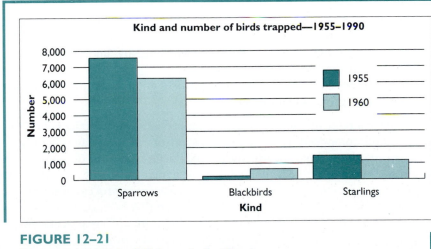

FIGURE 12–21
Distortion in Width of Column in Bar Graph

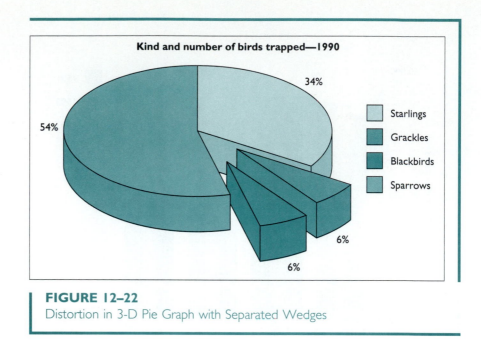

Kind and number of birds trapped—1990

34%

54%

6%

6%

Starlings

Grackles

Blackbirds

Sparrows

FIGURE 12–22
Distortion in 3-D Pie Graph with Separated Wedges

When taking a photograph of your subject, consider the following suggestions:

- Avoid too much detail in the photograph; for example, be certain that what surrounds and provides background for your subject is necessary and not distracting.
- Consider adding callouts or small arrows to the photo to point to the main features.
- Consider adding a caption to help the audience understand what they are seeing.
- Consider adding a familiar object to the photograph to help the audience gauge size and dimensions.

A photograph of the "L" blast furnace complex of Bethlehem Steel Corporation comprised part of a press packet on opening day (Figure 12–25). Because the photo clearly shows the facility (and its dimensions in contrast to the two workers on the right), the general audience reading about the furnace opening in a newspaper could visualize the process described in the caption to the photograph. However, notice that because there is such detail in the photograph, the caption must help the audience decide what to look at (e.g., the facility rather than the railroad tracks):

Large facilities designed to clean 360,000 cubic feet per minute of blast furnace gas occupy this end of the "L" blast furnace complex at Bethlehem

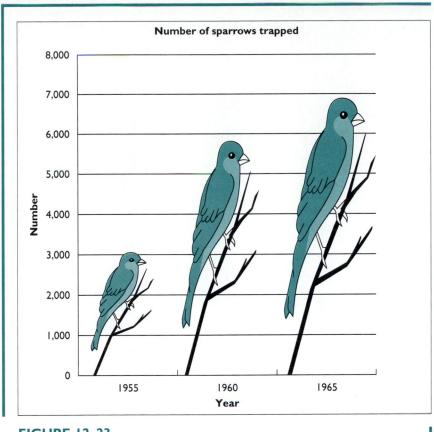

FIGURE 12–23
Misuse of Isotypes or Icons in Pictorial Graph

Steel Corporation's Sparrows Point, MD, plant. High-energy scrubbers, using about 10,000 gallons per minute of industrial water, help clean the gas. The water is clarified in two 90-foot-diameter thickeners and then cooled in the towers shown at left. It is then chemically treated before being recirculated. The iron-bearing slurry recovered from the thickeners is processed and recycled through the sinter plant to the blast furnaces. Expenditures for environmental control facilities on the furnace, the largest in the Western Hemisphere, totaled $22 million.

The angle of view, a wide shot of the furnace looking slightly upward, emphasizes the size of the furnace. Again, because the photograph captures the exact image of the furnace, this visual display is most appropriate for a general audience reading a newspaper or an annual report, an audience that might never have seen such a facility before and therefore could not understand a more abstract depiction.

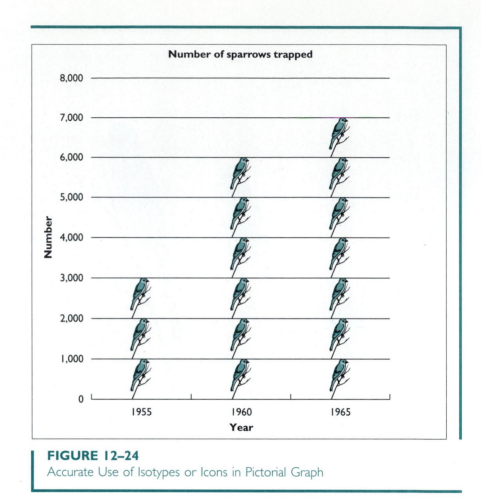

FIGURE 12–24
Accurate Use of Isotypes or Icons in Pictorial Graph

Creating Drawings and Diagrams

Rather than depicting the actual appearance of a subject, drawings and diagrams usually convey structure or organization symbolically. Therefore, audiences need experience or education in decoding drawings and diagrams. Because they are more abstract than photographs, drawings and diagrams emphasize important points, focus on specific components, or convey through symbols specific components of an object, system, or process.

For example, the drawing of the Bethlehem Steel "L" blast furnace (Figure 12-26) shows the main parts of the furnace (notice that the local audience can still appreciate the size of the furnace in contrast to the sketch of Baltimore's Washington Monument or read the exact dimensions in the drawing). However, extraneous detail has been eliminated, and the drawing comes close to that of an engineering blueprint.

Source: Courtesy of Bethlehem Steel.

FIGURE 12–25
Photograph of "L" Blast Furnace

The type of drawing or diagram you create depends on the education and experience of your audience and on what aspects of your subject you want to depict. In general, keep in mind the following guidelines:

- Select the amount of detail your audience will find useful and manageable.
- Label clearly all parts, steps, or stages by horizontal, parallel labels and arrows. Use a key only when absolutely necessary.
- Select symbols that are familiar to your audience.
- Give dimensions, orientation, and point of view information about your object. (For example, you might specify that the object is "seen from the top right, which is 11 feet off the ground.")
- If useful, show how parts relate or are connected or attached to the whole and to each other.

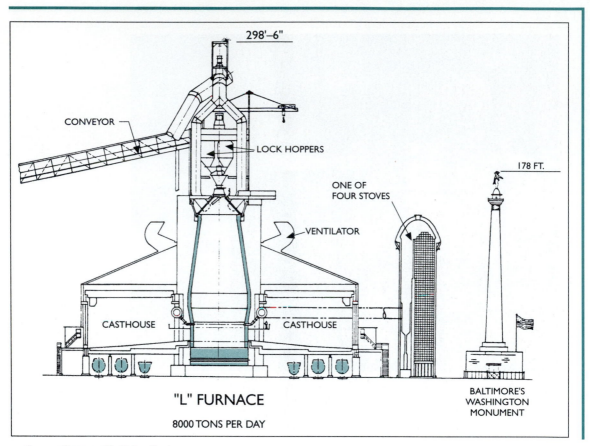

298'-6"

CONVEYOR

LOCK HOPPERS

ONE OF
FOUR STOVES

178 FT.

VENTILATOR

CASTHOUSE

CASTHOUSE

"L" FURNACE

8000 TONS PER DAY

BALTIMORE'S
WASHINGTON
MONUMENT

Source: Courtesy of Bethlehem Steel.

FIGURE 12–26
Drawing of "L" Blast Furnace

ACTION VIEW DRAWINGS

You can use an action view drawing to show how components interact or how one step in the process leads to another. For example, Figure 12–27 displays the effect of a passenger thrown forward during a train crash. The display was created by a computer graphics program developed at the Department of Energy's Lawrence Livermore National Laboratory and is used to measure the effect of stress traveling through structures. Even an inexperienced audience would have no difficulty understanding the movement of the human body upon impact.

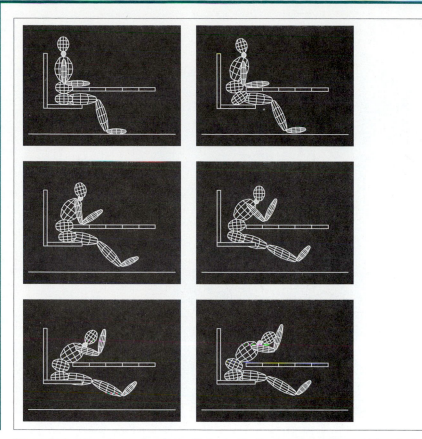

Source: *National Energy Strategy. Powerful Ideas for America. One Year Later.* National Technical Information Service, US Department of Commerce, February 1992.

FIGURE 12–27
Action View Drawing

TRANSLUCENT AND PHANTOM VIEW DRAWINGS

Translucent or phantom view drawings show the inside structure and parts of objects or organisms in relationship to the outside frame or covering. A centrifugal grinder pump, a part of a sewer system, shown in Figure 12–28, depicts the inside components in relation to the outside shell of the system. The audience gets an impression of the inside without being overwhelmed by details and labels.

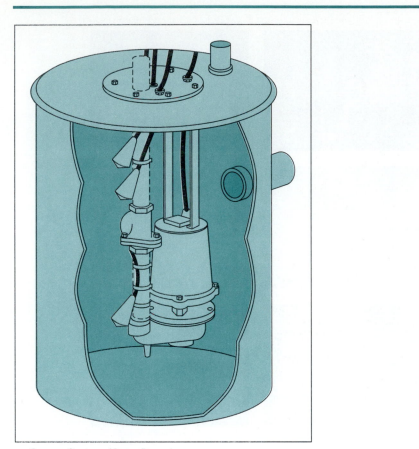

Source: Courtesy of Barnes Pumps, Inc.

FIGURE 12–28
Translucent View Drawing

CUTAWAY VIEW DRAWINGS

For an experienced or technical audience, cutaway view drawings depict the structure and parts of a cross section of a device or organism. Figure 12-29, a cutaway view of the grinder pump package, depicts the object as if it were sliced down the middle and opened up to show the inside.

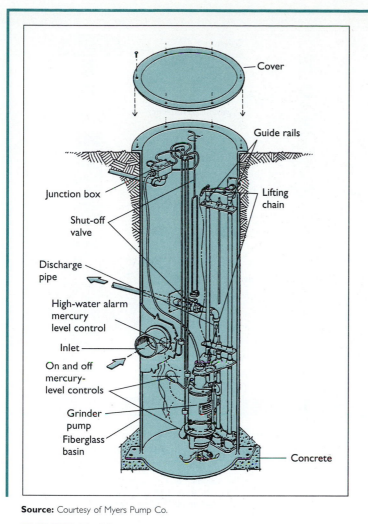

Source: Courtesy of Myers Pump Co.

FIGURE 12–29
Cutaway View Drawing

EXPLODED VIEW DRAWINGS

Even more abstract than cutaway view drawings are exploded view drawings, useful when your technically experienced audience must construct or assemble an object. In Figure 12-30, the parts of the external breather dial of a wastewater treatment system are blown apart or separated so that the audience can see how the parts connect.

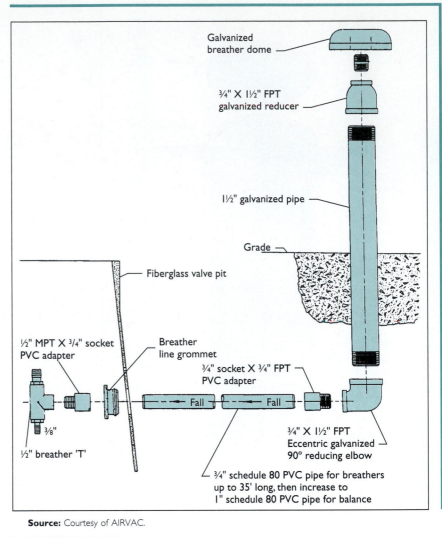

Source: Courtesy of AIRVAC.

FIGURE 12–30
Exploded View Drawing

SCHEMATIC OR SYMBOLIC DIAGRAMS

Only technical audiences experienced in the specific subject can read schematic or symbolic diagrams. These diagrams rely on specific symbols understood by those in the field and are highly abstract. Only audiences trained

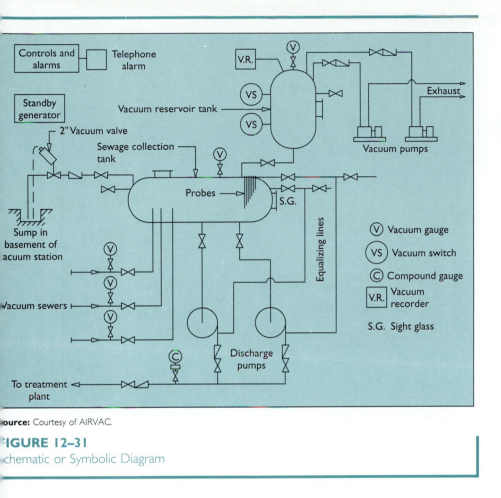

Source: Courtesy of AIRVAC.

FIGURE 12–31
Schematic or Symbolic Diagram

in reading schematics and who are familiar with a vacuum station could understand Figure 12–31.

Creating Maps

Maps help an audience locate geographical information by showing features of a landscape, locating the site of an event, or displaying data as they pertain to a location. For example, to understand the progression of the oil spill from the *Exxon Valdez,* an audience could study the map presented in Figure 12–32. The dates and mileage imposed upon the map, as well as the enlarged version of Prince William Sound and the small map of Alaska, help the audience visualize what might be an unfamiliar location and how that location was affected over time.

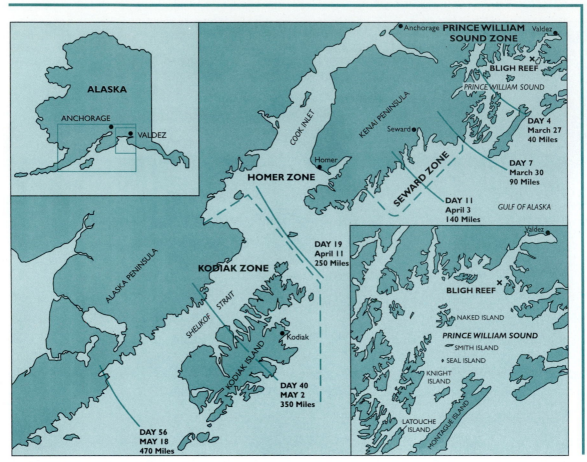

Source: Courtesy of Exxon USA.

FIGURE 12–32
Map of the *Exxon Valdez* Oil Spill Location and Movement

Creating Charts   When you want to display the components of an organization or an object or the steps in a process, you can use a chart. Much as do exploded drawings, charts separate the parts of a whole or demonstrate how parts interact. In creating a chart, it's helpful to keep in mind the following:

- Label components, steps, and subdivisions with meaningful labels within blocks.
- Use arrows or lines to show relationships within the system or over time.

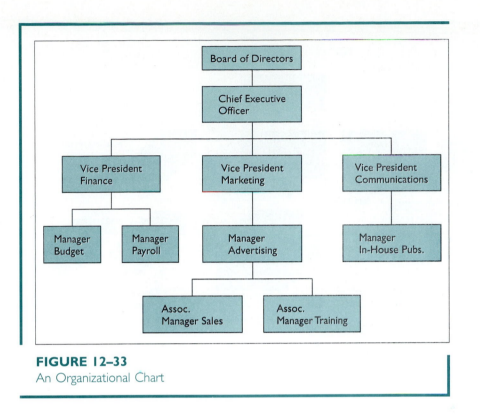

FIGURE 12–33
An Organizational Chart

ORGANIZATIONAL CHARTS

Organizational charts (such as in Figure 12-33) display the **hierarchy** (chain of command as well as formal communication links) of the positions within an organization through a series of interconnected blocks.

hierarchy
the arrangement of positions and power within an organizational structure

BLOCK COMPONENTS CHARTS

You can use block components charts to represent the components, parts, or subdivisions of a whole system. For example, a space shuttle avionics system, its components, and the relationship between components are shown in Figure 12-34.

FLOWCHARTS

Flowcharts display the steps within a process. Often the time that each step takes or how one step causes another will be indicated in the flowchart. Figure 12-35 demonstrates how product design affects materials flow in a "Green Design." The flowchart shows that the impact on the environment

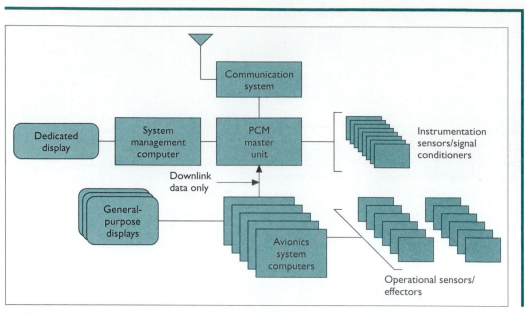

Source: John F. Hanaway and Robert W. Moorehead, *Space Shuttle Avionics System,* National Aeronautics and Space Administration, Office of Management, Scientific and Technical Information, NASA SP-504, 1989.

FIGURE 12–34
A Block Components Chart

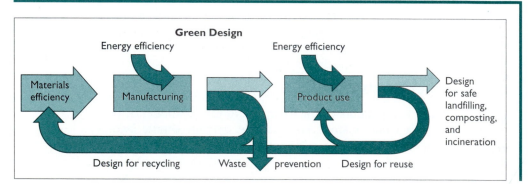

Source: *Green Products by Design: Choices for a Cleaner Environment,* US Congress, Office of Technology Assessment, US Government Printing Office, 1992.

FIGURE 12–35
A Flowchart

is reduced through recycling, reduction of waste, and efficient use of material and energy.

Increasingly, technical communicators are finding computers to be useful—and, in some cases, essential—tools for the creation, manipulation, and production of visual presentations. In the sections that follow, we have listed a few of these tools. Some of them allow for graphic representations in print environments; others support computer-based presentations that combine text and graphics in multimedia environments.

Computer tools for communicators are being developed and improved so rapidly that this particular list will soon be outdated. You'll need to stay on top of new developments in the field of computer graphics by consulting graphic designers who use computers and computer-supported graphics and by reading trade publications on desktop publishing.

Using Computer Software and Hardware to Create Visual Displays

computer tools
computer applications designed to create a variety of visual displays based on information stored in data files

IMAGE/TEXT SCANNING PACKAGES

Image and text scanning packages allow you to scan images electronically from books, magazines, and photographs, and to import these images in a digital environment. When these images are **digital**—that is, converted into the electronic impulses computers use—you can manipulate them according to the rhetorical contexts for which they will be used, the audiences for whom they are designed, and the information that they are intended to convey.

A representative image scanning product, for example, would allow you to import and export images in a variety of industry standard formats, correct and balance the color or tone of images, increase or decrease brightness and contrast, capture computer screens, crop and scale images, and compress images so that they take up less memory in a computer.

digital images
electronic form of data storage; increases the flexibility with which images can be modified to suit a variety of rhetorical contexts

SPREADSHEETS

Spreadsheets are computer programs that allow you to store, manipulate, and represent data in columns and rows. Most full-featured spreadsheets also provide tools for the visual representation of data in print contexts and can help you create area charts, bar charts, column charts, stacked column charts, line graphs, pie charts, scatter plots, 3-D charts, line/column charts, and volume charts. Remember that we used Microsoft Excel™, a widely used spreadsheet, to create the bird trapping illustrations in this chapter.

MULTIMEDIA PACKAGES

Multimedia packages will help you create computer-based documents that can incorporate video clips, animation, sound effects, music, still images, and text. The multimedia presentations that you create using these packages require computer support when they are displayed to an audience. Such packages can help you show the operation of a mechanism, the change associated with a phenomenon over time, the aging or growth of an organism, or the unfolding of a technical process or procedure, among many other temporal representations. Generally, multimedia presentations are meant to be viewed on a computer or a television screen rather than in print, but crude animation sequences can be depicted in print. Figure 12–36 illustrates an animation sequence and score in a multimedia program called MacroMind Director™. Figure 12–37 shows the more complex digital environments in which multimedia presentations are "scripted."

COMPUTER-AIDED DESIGN PACKAGES (CAD)

You can use computer-aided design packages (CAD) to support the creation and manipulation of 3-D images. These packages commonly give you such features as coloring and shading capabilities, rotation options, and cross-sectioning and exploding views. CAD is used in circuit design, architecture, geology, engineering, manufacturing, and medicine, among other fields. Images created with CAD programs can usually be imported into print documents.

VIDEO EDITING PACKAGES

Increasingly, technical communicators are called upon to be familiar with video used for training, sales, and public relations. Video editing packages support direct video input from cameras into computers. When video sequences are digitized, these programs allow users to select video clips, zoom in on parts of sequences, slow or speed-up sequences, create various fades and cuts, edit sound tracks, and overlay text on video images. Figure 12–38 shows how input from video editing programs—in this case QuickTime™—can be imported in multimedia presentations through digital linking.

CLIP ART

If you are involved in any of your company's communication activities—flyers, newsletters, posters—you will soon become accustomed to turning to clip-art packages for ready-made artwork. Clip-art packages contain a range of preselected images, often accessible by topic (e.g., celebrations, maps, animals, objects, business images, sports, borders, food, or transportation)

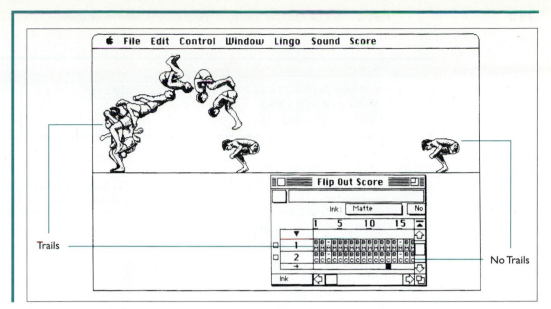

Trails

No Trails

Source: MacroMind Director, Version 3.0, *Studio Manual.* 2nd ed. Reprinted with permission of Macromedia, Inc. © 1992.

FIGURE 12–36
Animation Sequence

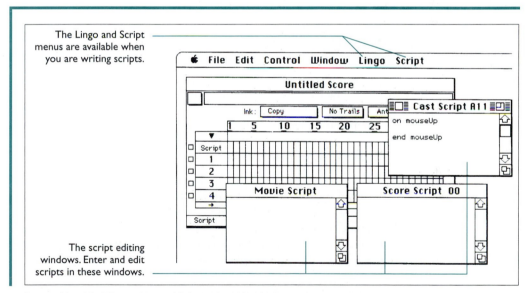

The Lingo and Script menus are available when you are writing scripts.

The script editing windows. Enter and edit scripts in these windows.

Source: MacroMind Director, Version 3.0. *Studio Manual.* 2nd ed. Reprinted with permission of Macromedia, Inc. © 1992.

FIGURE 12–37
Digital Environment

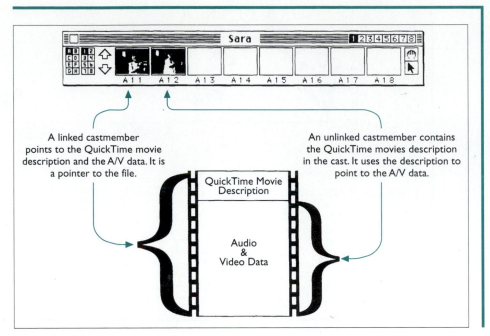

Source: MicroMind Director, *Getting Started and New Features Guide.* Version 3.1. Reprinted with permission of Macromedia, Inc. © 1992.

FIGURE 12–38
Using Video Input in Multimedia Presentations

that can be selected, cut, and pasted into documents to illustrate or enhance text. If you don't have access to clip-art software, you can purchase the hard copy and scan the images in.

PAINT PROGRAMS

Paint programs allow you to do on the computer many of the routine tasks that formerly had to be sent out to graphic artists. These packages provide you with access to tools similar to those used by artists: brushes, compasses, lines, erasers, spray paints, texturizers, dots, dashes, and colors, among others. These tools enable you to design original images or modify images imported through scanning programs.

DRAW PROGRAMS

With draw programs you can create and manipulate "freehand" art on the computer. Among the common tools available on such packages are rectangles, lines, ovals, arcs, and freehand polygons. These items can be colored,

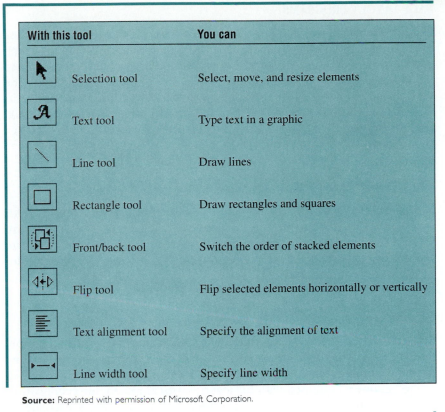

With this tool		You can
	Selection tool	Select, move, and resize elements
	Text tool	Type text in a graphic
	Line tool	Draw lines
	Rectangle tool	Draw rectangles and squares
	Front/back tool	Switch the order of stacked elements
	Flip tool	Flip selected elements horizontally or vertically
	Text alignment tool	Specify the alignment of text
	Line width tool	Specify line width

Source: Reprinted with permission of Microsoft Corporation.

FIGURE 12–39
Graphics Tools from a Word Processing Program

filled with gray scale tones, grouped and ungrouped, shadowed, or moved. Often, draw programs are used on systems that include additional support hardware such as scanning devices, digitizing tablets and pens, and color printers.

WORD PROCESSING PROGRAMS

Even if you do not have access to more software than your word processing program, you will find that such programs increasingly support a number of graphic features including charts, graphs, lines and rules, borders, fonts, and columns. Version 5.0 of Microsoft's Word™, for example, provides graphics tools, a portion of which is portrayed in Figure 12-39.

PAGE LAYOUT PROGRAMS

Page layout programs will help you to design pages that incorporate sophisticated combinations of text and images. Using cut-and-paste metaphors, such programs generally support indexing, importing of graphics, cropping and scaling of gray-scale images, and using columns and wraparound text. The key to such packages is that they allow page designers to see various effects in real time. Because they show onscreen pretty much what readers will see on the page, they are called what-you-see-is-what-you-get (WYSIWYG, pronounced "wizzy wig") programs.

DIGITAL PRESENTATION SOFTWARE

Sometimes you will want to prepare overhead projection sheets or color slides for a presentation. Many companies have digital presentation software available for employees to use in preparing their presentations. This software helps writers create overhead projection sheets or 35 mm slides with color or black and white images (e.g., bar charts, graphs, still images).

SCANNERS

optical character recognition (OCR)
a feature of some scanners that allows them to scan text and import it as text rather than as an image of a page; text scanned this way does not have to be retyped

Scanners are devices that allow you to take a still image (a page, a photograph, line art, cartoons), digitize, manipulate it on the computer (e.g., changing shapes, outlines, size, color), and print it in color, gray scale, or black-and-white. Scanners equipped with full-feature **optical character recognition** (OCR) can import text and recognize diacritical marks. This is a wonderful time saver because you don't have to retype pages into the computer. Scanners come in all sizes, from flatbed scanners, large devices that resemble a photocopy machine, to smaller, handheld, portable devices. These tools are especially helpful if you want to use both a photograph and a related drawing in a document. After scanning in the photograph, you can use a software program to manipulate the photograph, eliminating extraneous detail to create a drawing.

SUMMARY

In this chapter, you have learned that visual display of data can be used for everything from making complex information accessible to symbolizing abstract structures. However, all visual displays of data must be introduced, interpreted, and explained in your text for the data to support an argument or inform an audience. Generally you select and create a visual display of data according to your purpose and audience of your document, particularly according to the experience, technical understanding, and language and culture of that audience. However, we also know that certain types of visual displays of data lend themselves to specific purposes. For example, line drawings show selective features of objects while photographs show complete detail.

Before creating a visual display of data, you need to gather all pertinent data from a large enough sample, indicate all possible interpretations of those data, make sure your data really support what you are saying, indicate units of measurement, and condense the data into a manageable form.

You might select a table if you need to present complete and specific data to an audience that understands your topic. Graphs, on the other hand, translate numbers into pictures—bars, lines, areas, or pictorials. More abstract than tables, graphs illustrate changes in one amount relative to another. Bar graphs, whether simple, multiple, stacked, 100 percent, or deviation, demonstrate magnitude or size of several items or emphasize difference in one item at equal time intervals. Line graphs show trends or movement in series of data. Simple area graphs show not only trends in data but also the sum of those data, whereas 100 percent area graphs show trends in the components of a whole. Pie graphs represent proportions or percentages of a whole, and pictorial graphs use icons or symbols to represent data.

When creating bar, line, and pie graphs, you have to be particularly careful not to misrepresent your data by distortion of the X or Y axis, distortion by suppressing the zero, distortion by increasing the width or height of bars, distortion in 3-D, and distortion through improper use of icons.

Photographs depict subjects in detail and in a realistic setting. Drawings or diagrams, on the other hand, are more abstract and symbolically convey structure or organization. Maps helps audiences locate geographical information, and charts display the components of an organization or object or the steps in a process.

Finally, an ever-growing number of computer software programs and computer hardware are available to help you create your visual displays of data. These scanning programs, spreadsheets, multimedia packages, paint/draw programs, and so forth are constantly being updated and improved.

---

1. Analyze the visual displays presented in Figures 12–40 through 12–43 and answer the following questions:

    a. Who might be the audience for the visual display? What can you tell about their level of education, technical knowledge, ability to read English, and so on?

    b. What did the creator of the visual display want to show? How might that person interpret for the audience the data in the display? In what type of publication might you find such a display?

    c. What type of visual display do you see? Is the visual display successful according to the guidelines of displays in general and the type of display specifically? In what ways? In what ways is it unsuccessful?

    d. How would you revise the display to improve its effectiveness?

    Be prepared to discuss your impressions with your classmates.

2. A small-diameter gravity sewer is displayed in Figure 12–44. It has the following features:

    ■ Collector mains do not carry solids;
    ■ Primary treatment occurs at each connection;
    ■ Only settled wastewater is collected;

ACTIVITIES AND EXERCISES

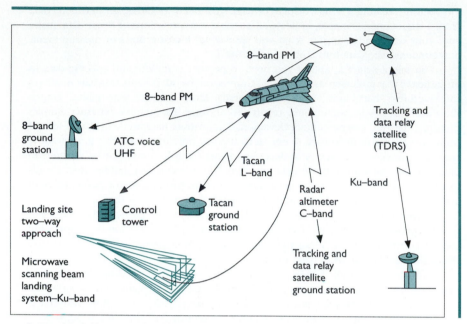

Source: John F. Hanaway and Robert W. Moorehead, *Space Shuttle Avionics System*, National Aeronautics and Space Administration, Office of Management, Scientific and Technical Information, NASA SP-504, 1989.

FIGURE 12–40

- Construction cost is low;
- Grit, grease, and similar contaminants are separated from wastewater;
- Fewer obstructions occur in collector mains;
- Interceptor tanks appear upstream of each connection;
- Grit, grease, and similar contaminants are retained in interceptor tanks.

Write a paragraph in which you introduce the visual display, interpret it, explain the main features by referring to the display, and recommend it for consideration over conventional sewer systems.

3. Present the following information in a visual display or displays that would be meaningful to graduating science and engineering students at your university or college.

In 1989 the average salaries of doctoral scientists and engineers were gathered by field, sex, and years in the profession. In environmental science men earned $55,600 and women $43,600 (women's salaries were 78.4 percent of men's); in physics men earned $59,100 and women $48,700 (women's salaries were 82.4 percent of men's); in computer science men earned $60,100 and women $50,000 (women's salaries were 83.2 percent of men's); in math men earned $52,400 and women $43,800 (women's salaries were 83.6 percent of men's); in psychology men earned $51,300 and women $44,300 (women's salaries were 86.4 percent of men's); in

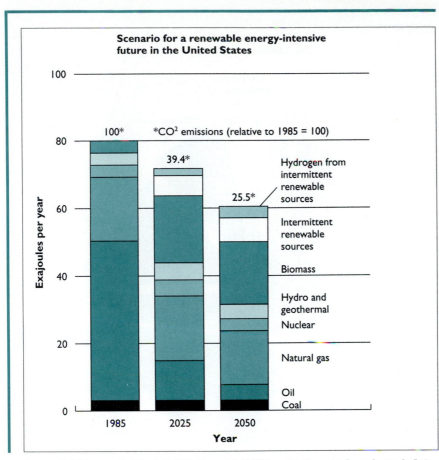

Source: T. B. Johansson, H. Kelly, A. K. N. Reddy, and R. H. Williams, Eds. *Renewable Energy Sources for Fuels and Electricity,* Washington, D.C. and Covelo, CA: Island Press, 1992. Granted with permission of Island Press © 1992.

FIGURE 12–41

chemistry men earned $55,900 and women $46,900 (women's salaries were 83.9 percent of men's); in engineering men earned $62,900 and women $53,400 (women's salaries were 84.9 percent of men's); in social science men earned $52,000 and women $44,200 (women's salaries were 85 percent of men's); in life sciences men earned $53,200 and women $43,100 (women's salaries were 81 percent of men's). In all fields men earned $56,000 and women $44,800 (women's salaries were 80% of men's).

In terms of years in these professions, for less than five years, men earned an average of $42,500 and women $36,900 (women's salaries were 86.8 percent of men's); for between five and nine years men earned $48,900 and women $42,700 (women's salaries were 87.3 percent of men's); for

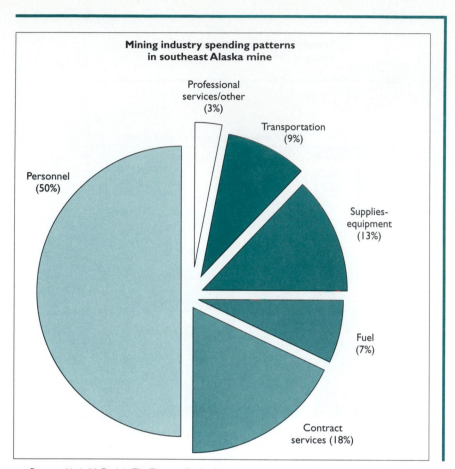

Source: Linda M. Daniel, "The Tongass: Getting Beneath the Surface," *Minerals Today,* US Department of the Interior, Bureau of Mines. MS2501, August 1992.

FIGURE 12–42

between 10 and 14 years men earned $54,800 and women $48,000 (women's salaries were 87.6 percent of men's); for between 15 and 19 years men earned $60,400 and women $51,200 (women's salaries were 85.4 percent of men's); for between 20 and 24 years men earned $63,700 and women $55,200 (women's salaries were 86.7 percent of men's); for between 25 and 29 years men earned $67,400 and women $58,300 (women's salaries were 86.5 percent of men's); for between 30 and 34 years men earned $70,100 and women $63,400 (women's salaries were 90.4 percent of men's); for over 34 years of experience men earned $74,500 and women $62,300 (women's salaries were 83.6 percent of men's).

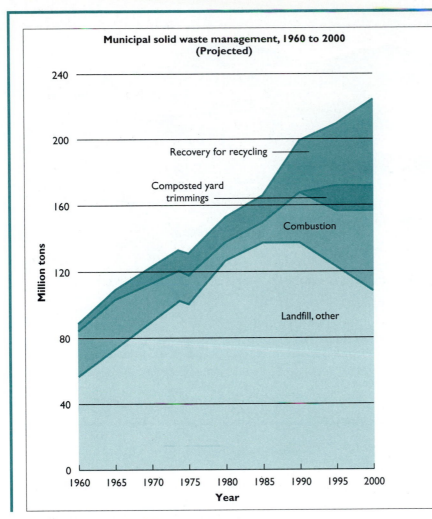

Municipal solid waste management, 1960 to 2000 (Projected)

Recovery for recycling

Composted yard trimmings

Combustion

Landfill, other

Source: Courtesy of Franklin Associates, Prairie Village, Kansas.

FIGURE 12–43

Source: Betty Vetter, "Ferment: Yes/Progress: Maybe/Change: Slow," *MOSAIC* 23.3 (Fall 1992), p. 35. National Science Foundation, Washington, DC

4. Survey the visual presentation software and hardware available at your college or university. What is available in the computer labs? In the administrative offices? In the printing or communication department? Write a memo to your teacher and to your classmates describing available resources and including, where possible, examples of visual displays produced on these systems.

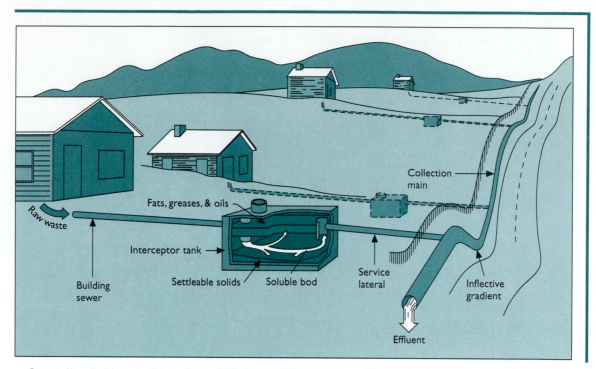

Source: *Alternative Wastewater Collection Systems*, US Environmental Protection Agency. 625/1-91/024. Office of Research and Development and Office of Water, Washington, DC, October 1991.

FIGURE 12–44

5. Study the data displayed in the tables in the "Minnesota Department of Health AIDS Epidemiology Unit/Pediatric/Adolescent HIV/AIDS Monthly Surveillance Report, November 1, 1992" (Appendix A, Case Documents 1). Create alternative displays of these data to distribute to the general public that would emphasize aspects of the data that you find important.

6. In the "Fact Sheet" on AZT (Appendix A, Case Documents 1), find data within sentences and paragraphs that might be displayed better visually. Create these visual displays and explain how they would be integrated into the fact sheet.

7. The following information pertains to the bird trapping at the University of the Midwest Agricultural Experiment Station (Appendix A, Case Documents 3). Take a stand on one side or the other of the controversy and display the data below in a way that would support your points at the public hearing on the issue. The following data represent the total number of birds (all species) trapped for the year (Notice that for some years data were not gathered.): 1971—4307; 1972—5114; 1973—3407; 1974—6145; 1975—no data; 1976—4752; 1977—9846;

1978—no data;1979—4989; 1980—6488; 1981—6014; 1982—773; 1983—735; 1984—no data; 1985—3648; 1986—6322; 1987—2205; 1988—6187; 1989—10,380; 1990—5951.

Cochran, Jeffrey K.; Sheri A. Albrecht; and Yvonne A. Green. "Guidelines for Evaluating Graphical Designs: A Framework Based on Human Perception Skills." *Technical Communication* 36.1 (1989), pp. 25-32.

WORKS CITED

Creating Effective Documents

Definitions and Descriptions

Creating Definitions ▌ Including Contexts ▌ Deciding When and
Where to Include Definitions ▌ Creating Descriptions

Sharing a language means sharing a conceptual universe.
Evelyn Fox Keller, *Secrets of Life, Secrets of Death:
Essays on Language, Gender, and Science.* New York: Routledge, 1992, p. 27.

Since the meaning of a word helps organize the way we encounter an entire
situation, misusing a word encourages false perceptions, occludes possibilities, and
gives us bad habits.
David Dobrin, *Writing and Technique.* Urbana, IL:
National Council of Teachers of English, 1989, p. 310.

405

Introduction

I n Chapter 5, we stated that the most powerful kind of good reason to support an idea is an argument from definition. And, in Chapter 12, we discussed how visual displays, such as photographs, drawings, and diagrams, could help an audience see an object or organism that might be described in a document. In this chapter, we cover in detail definitions and descriptions—two rhetorical patterns that drive arguments and convey information in a great many technical communications. Whether you're giving a progress report or deciding on the feasibility of a recommendation, writing a resume or delivering a sales pitch, you'll need to define and describe.

In the past, technical communication—perhaps because it has been associated so closely with the concerns of scientists, engineers, and technology developers—has been portrayed as a straightforward process of representing objective reality as accurately as possible through language. Certainly, the project of *defining* words and *describing* objects or processes grows out of such assumptions.

We now understand, however, that techniques of definition and description are neither neutral nor objective, and that even those writers who use definitions with precision and care cannot always define things in a way that will avoid all confusion or eliminate the need for interpretation on the part of audiences. In turn, audiences interpret definitions and descriptions according to the contexts within which they find themselves, their experiences and education, their belief systems and values, the goals of the companies and organizations within which they work, and their understanding of the subject matter.

By defining and describing, you can help readers approach a topic, avoid error, reduce confusion, or reach a goal, but you can never be sure that you have conveyed an *absolutely* precise meaning or understanding. You *can* work to provide your audience with the clearest possible understanding of a concept, object, process, or phenomenon, however. In this chapter, we provide some techniques that help you describe and define.

Creating Definitions

categorical proposition
a subject, a verb, and a predicate

Chapter 5 addressed writing information definitions or categorical propositions. A **categorical proposition** consists of a subject (the word to be defined), a linking verb (such as *is* or *was*), and a predicate (something to be said about the word being defined). The following examples are information definitions or categorical propositions. As practice in recognizing the elements of categorical propositions, identify the subject, linking verb, and predicate in each example below:

Science and engineering curricula are usually rather demanding.
A recycling program is something that this campus badly needs.
Ms. Protevi was an ideal boss.
Nursing is a more interesting profession than most people imagine.
Certain climatic changes now in evidence are truly disturbing.

Although information definitions are useful when your audience is familiar with your terms, you need a more **formal definition** if you are defining unfamiliar terms for your audience or if your purpose depends on the clearest possible understanding of your terminology. Throughout this chapter, when we mention defining a term, remember that a term may convey information about a concept, process, object, or procedure.

The classical format for constructing definitions involves the following formula:

name (term) + class/family/species/group + special characteristics = definition

Especially important in formal definitions of objects is the task of characterizing the **special features** that distinguish a particular item from other items in a related classification group, such as in the following examples:

light pen—A light-sensitive input device used to select an entry or indicate position. (Szymanski et al., p. 738)

liquid-based solar heating system—A solar heating system in which liquid, either water or an antifreeze solution, is heated in solar collectors. (Beckman et al., p. 172)

In defining processes, procedures, or phenomena that occur over time, you can modify this format, placing the focus on function or operation:

name (term) + function or operation + special uses or operations = definition

These definitions are often called **functional or operational definitions,** as in the following examples:

selecting—The process of retrieving only certain records in a table in a relational database management system. (Szymanski et al., p. 745)

kerning—In text composition or typesetting, selectively nudging letters together to improve spacing gives the illusion of equal spacing between letters.

In both patterns, you indicate through language—as accurately, precisely, and clearly as possible—what a term *is* by indicating what differentiates it from other similar terms.

This approach to definition takes two forms: **differentiation by precision** and **differentiation by efficiency.** To be absolutely precise, you would have to differentiate a term *from all other similar terms.* For example, to define accurately and precisely a ¾-inch, plated, hex-head lag screw, you would have to differentiate it from every other type of screw: from bolt screws; from slotted-head screws and Phillips-head screws; from decking screws and brass furniture screws; and from ½-inch hex-head lag screws and 1½-inch hex-head lag screws. Fortunately, most readers bring a great

formal definition
definition consisting of the term to be defined, the class/family/species to which the term belongs, and the special characteristics that distinguish the term from other terms in the same class/family/species

special features
characteristics that distinguish a particular item from other items in a related classification group

functional or operational definitions
definitions used for processes, procedures, or phenomena that occur over time; they are constructed by using the formula:
name + the function or operation = special uses

differentiation by precision
differentiating a term from all other similar terms; balanced against efficiency

differentiation by efficiency
differentiating a term from similar terms by mentioning only the key characteristics that differentiate it from similar terms

deal of contextual knowledge to situations that require definitions. A customer putting together a set of bookshelves, for example, probably knows that 6-inch, galvanized decking screws are not appropriate. She also knows it is customary for manufacturers to include the appropriate hardware in a little plastic bag that comes with shelves requiring assembly, and she knows the approximate size of the hardware from observing the size of the pre-drilled holes. As a result, such customers need only a picture or diagram of the appropriate screw to identify a hex-head lag screw, not a fully fleshed-out description. Therefore, you often need to mention only the *key* characteristics that differentiate a term from other, similar terms customers or users might encounter in the same context.

Thus, you need to analyze your audience's expertise and the context in which your definition appears. For example, see the various definitions of the word *gas* listed below. Some of the definitions are more appropriate to a general audience and some to a specialized audience. Also, the particular meaning of the word *gas* depends on the context in which it is used and defined.

- *Gas*—A shortened term for *gasoline.* The fuel used to power most internal combustion engines.
- *Gas*—In the intestinal tract, oxygen, nitrogen, hydrogen, carbon dioxide, methane, and in decomposition of proteins, hydrogen sulfide, indole, skatole, ammonia, and such.
- *Gas*—The action of treating chemically with gas. Because of its expansion capabilities, gas immediately penetrates large areas. For example, during World War I many soldiers suffered the ill effects of poisonous gas spread over the trenches and battlegrounds. Or, certain insects are best destroyed or controlled by spreading gaseous pesticides over the area in question.
- *Gas*—The state of matter, as opposed to solid or liquid, characterized by low viscosity and density. Gases respond with relatively great expansion and contraction to changes in pressure and temperature. Gases diffuse easily and distribute uniformly in a container.
- *Gas*—The control device for an engine powered by gas. For example, to move the car forward one must step on the *gas.*
- *Gas*—The occult principle thought to be present in all bodies. From the Dutch. Named by Belgium chemist J. B. van Helmont, 1577-1644, and taken from the Greek word *khaos,* meaning chaos or chasm. The gases within dead bodies cause swelling and decomposition.

The level of specificity required by a definition and the nature of the definition itself depend on your analysis of the audience's knowledge, experience, and interest. Definitions can be incorporated within the text through modification at the word, phrase, clause, sentence, or paragraph level; incorporated within the text through example, metaphor, analogy, negation, or comparison–contrast; covered in informational footnotes, endnotes, or captions; illustrated in figures or tables; or included in an appendix or glossary.

DEFINITION BY MODIFICATION

Depending on the amount and extent of definition that you think your audience may need, you can define your terms by using a technique known as **modification.**

modification
process of adding words, phrases, clauses, or sentences to clarify or hone a definition

■ *Modifying Individual Words*

The *spring-operated* closing mechanism is located on the inner surface of the oven door.

■ *Modifying Phrases*

Double poling—*used primarily by skiers in racing and skating situations*—is characterized by identical poling motions on both sides of the body.

■ *Modifying Clauses*

The antenna, *which is the device receiving the signal,* should be repositioned to avoid interference.

■ *Modifying Sentences*

Cumulonimbus clouds produce what people call ''thunderstorms.'' *To be termed a ''thunderstorm,'' a storm must include visible lightning, audible thunder, and precipitation of some kind.*

Extended definitions, as in Example 13–1, may involve a paragraph, several paragraphs, or even an entire document.

DEFINITION BY SYNONYMS AND ANTONYMS

Definitions often include or use **synonyms** (words that mean the same thing) for key terms or **antonyms** (words that mean the opposite). Both synonyms and antonyms help your audience by providing terms that may be more familiar to them or that expand their frame of reference for a concept, as in the following examples:

synonyms
words that mean the same (or almost the same) thing

antonyms
words that mean the opposite thing

Assembler—A computer program that converts (i.e., translates) programmer-written symbolic instructions, usually in mnemonic form, into machine-executable (computer or binary-coded) instructions. This conversion is typically one-to-one (one symbolic instruction converts to one machine-executable instruction). (Giesecke, p. 635)

. . . in this molecular orbital, the likelihood that an electron will be found in the region between the two nuclei is high. Electrons in such a molecular orbital tend to be in the region where they can hold the nuclei together. This molecular orbital is called a *bonding orbital*. . . . Another molecular orbital is obtained by subtracting the $1s$ orbital on one atom from the $1s$ orbital on the other . . . the resulting values in the region of the overlap are close to zero. This means that in this molecular orbital, the electrons spend little time between the nuclei. We call it an *antibonding orbital*. . . . (Ebbing and Wrighton, pp. 258–59)

DEFINITION BY EXAMPLE AND ENUMERATION

examples or enumeration

way of extending definitions by including items that fall within the scope of the word being defined

You can frequently extend definitions by including **examples** (sometimes indicated by e.g., which stands for the Latin words *exempli gratia*) or by **enumerating** items that fall within the scope of the word being defined:

> *Noise* has been defined as unproductive sound. Sometimes the noise level may be so high that permanent aural damage may result, e.g., from continued

---

EXAMPLE 13–1   Extended Definition through Modification

1.   Nature of radiant heat transfer. In the preceding sections of this chapter we have studied conduction and convection heat transfer. In conduction heat is transferred from one part of a body to another, and the intervening material is heated. In convection the heat is transferred by the actual mixing of materials and by conduction. In radiant heat transfer the medium through which the heat is transferred usually is not heated. Radiation heat transfer is the transfer of heat by electromagnetic radiation.

Thermal radiation is a form of electromagnetic radiation similar to x-rays, light waves, gamma rays, and so on, differing only in wavelength. It obeys the same laws as light: travels in straight lines, can be transmitted through space and vacuum, and so on. It is an important mode of heat transfer and is especially important where large temperature differences occur, as, for example, in a furnace with boiler tubes, in radiant dryers, and in an oven baking food. Radiation often occurs in combination with conduction and convection. . .

In an elementary sense the mechanism of radiant heat transfer is composed of three distinct steps or phases:

1.   The thermal energy of a hot source, such as the wall of a furnace at T_1, is converted into the energy of electromagnetic radiation waves.
2.   These waves travel through the intervening space in straight lines and strike a cold object at T_2 such as a furnace tube containing water to be heated.
3.   The electromagnetic waves that strike the body are absorbed by the body and converted back to thermal energy or heat.

Source: Christie J. Geankopolis, *Transport Processes and Unit Operations,* 2nd ed. © 1983, p. 258. Reprinted by permission of Prentice Hall, Inc., Englewood Cliffs, NJ.

use of a chainsaw or considerable exposure to a musical group with amplifiers. (Hook, p. 57)

cathode ray tube (CRT)—The principal component in a CAD display device. A CRT displays graphic representations of geometric entities and designs and can be of various types: storage tube, raster scan, or refresh. These tubes create images by means of a controllable beam of electrons striking a screen. (Giesecke, p. 635)

dedicated—Designed or intended for a single function or use. For example, a dedicated work station might be used exclusively for engineering calculations or plotting. (Giesecke, p. 637)

A *computist* is a mathematician who specializes in computing such things as time of high and low tides, eclipses, dates, business accounts, and the like in which specific answers useful in practical applications are the goal. (Hook, p. 243)

DEFINITION BY METAPHOR AND ANALOGY

You can use a **metaphor** in a definition to explain an unfamiliar item by implying comparison to a familiar item (recall our discussion of metaphor and analogy as good reasons in Chapter 5). The effective use of metaphor depends on cultural knowledge shared by you and your audience, as the following example illustrates:

> Inside the CRT is a "gun" that "shoots" electrons out to a screen. The electrons hit a phosphorescent coating on the inside of the picture tube. When the phosphorous on the screen gets "hit," it glows. (Brecher, p. 27)

metaphor
comparison without the use of the words *like* or *as;* e.g., "The wind was a bulldozer in the forest"

An **analogy** also works by comparing two things to illustrate elements of similarity. As you can imagine, the effective use of analogy also depends on your having a sense of the wide variety of experience audiences bring to reading tasks. In the following example, for instance, the authors of a bicycle maintenance and repair manual count on their readers to have a broad knowledge of other sports activities and the equipment that goes along with these activities:

> *clipless pedals*—Road bike pedals that use a releasable mechanism like that of a ski binding to lock onto cleated shoes. . . . (*Bicycling Magazine's Complete Guide,* p. 298)

analogy
comparing two items to illustrate elements of similarity; one of the items must be familiar to the audience

DEFINITION BY NEGATION

Defining by **negation** involves describing an object, concept, or term by what it is *not* or by identifying characteristics it does *not* have:

> *Diffuse radiation*—Solar radiation that is scattered by air molecules, dust, or water droplets before reaching the ground and not capable of being focused. (Beckman et al., p. 171)

negation
defining an object, concept, or term by stating what it is not

Collateral publications—Ad agency term for printed pieces, such as brochures and annual reports, that are not directly involved in advertising. (Beach et al., p. 197)

DEFINITION BY COMPARISON–CONTRAST

Definition by comparison can extend the range of your audience's understanding by connecting several objects or concepts that are related in some way, either by similarity or difference. The following examples involve definition by **comparison or contrast:**

comparison or contrast
connecting several objects or concepts by similarity (comparison) or by difference (contrast)

stride rate—The number of times the basic movement or cycle is completed in a given time period. Average race pace is 2.2 strides per second for hill running and 1.6 strides per second for diagonal. (Hall, p. 234)

direct-view storage tube (DVST)—One of the most widely used graphics display devices, DVST generates a long-lasting, flicker-free image with high resolution and no refreshing. It handles an almost unlimited amount of data. However, display dynamics are limited since DVSTs do not permit selective erase. The image is not as bright as with *refresh* or *raster*. (Giesecke, p. 637).

digital computer—A high-speed programmable electronic device that stores processes and retrieves data by *counting* discrete signals, as opposed to measuring a continuous signal, as is done in an analog computer. (Szymanski, et al., p. 732)

What is damp haze? Damp haze lies somewhere in the scale between dry haze and light fog. It consists of microscopically small water droplets or particles that attract water (hygroscopic nuclei) suspended in the air. It differs from dry haze in that it has a grayish appearance and occurs with high relative humidities. It differs from light fog since its droplets are most widely dispersed and smaller in size. (Forrester, p. 61)

DEFINITION BY VISUAL DISPLAY AND PRESENTATION

There are times when a visual presentation, such as a photograph or a drawing, can do a much better job of defining an object or concept than can words, given the slippery nature of language and the economy of visual images. In Example 13–2, various kinds of book bindings are defined through illustrations. Although the writer could have tried to convey the differences between saddle stitching and loop stitching through words, the illustrations are much more economical and informative.

As we discussed in Chapter 12, you need to understand your audience's experience, technical expertise, language, and culture as well as your purpose when creating a visual presentation. In Example 13–3, the writers addressed an audience of engineering students. Because the writers were defining a device somewhat unfamiliar to the students but could rely on the students' experience reading technical diagrams, they chose to define the liquid propellant rocket with a line drawing. The drawing is abstract in that

EXAMPLE 13–2 Definition through Visual Presentation

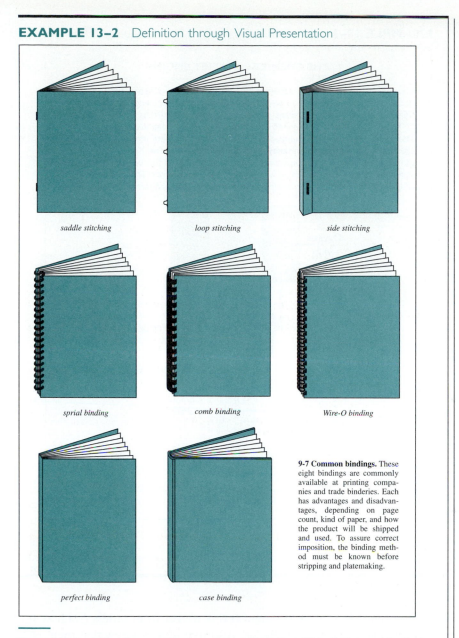

saddle stitching · loop stitching · side stitching

sprial binding · comb binding · Wire-O binding

perfect binding · case binding

9-7 Common bindings. These eight bindings are commonly available at printing companies and trade binderies. Each has advantages and disadvantages, depending on page count, kind of paper, and how the product will be shipped and used. To assure correct imposition, the binding method must be known before stripping and platemaking.

Source: Mark Beach; Steve Shepro; and Ken Russo, *Getting It Printed: How to Work with Printers and Graphics Arts Services to Assure Quality, Stay on Schedule, and Control Costs.* Portland, Oregon: Coast to Coast Books, 1986. p. 120.

EXAMPLE 13-3   Definition through Visual Presentation and Text

THE CHEMICAL ROCKET ENGINE

The advent of missiles and satellites has brought to prominence the use of the rocket engine as a propulsion power plant. Chemical rocket engines may be classified as either liquid propellant or solid propellant, according to the fuel used.

Figure 1.13 shows a simplified schematic diagram of a liquid propellant rocket. The oxidizer and fuel are pumped through the injector plate into the combustion chamber where combustion takes place at high pressure. The high-pressure, high-temperature products of combustion expand as they flow through the nozzle, and as a result they leave the nozzle with a high velocity. The momentum change associated with this increase in velocity gives rise to the forward thrust on the vehicle.

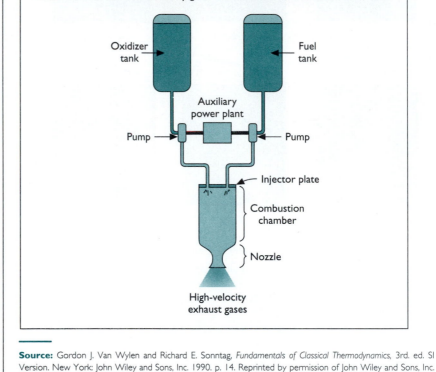

Source: Gordon J. Van Wylen and Richard E. Sonntag, *Fundamentals of Classical Thermodynamics*, 3rd. ed. SI Version. New York: John Wiley and Sons, Inc. 1990. p. 14. Reprinted by permission of John Wiley and Sons, Inc.

only important details that distinguish this rocket from others (such as the solid propellant rocket) are included. Also, the writers know that their audience will be familiar with the technical terms that help define the rocket, such as oxidizer and combustion chamber. Finally, because the drawing defined the liquid propellant rocket according to its parts, the writers could reserve their text for a description of the rocket's functioning.

To create useful and clear definitions you need to consider *context.* Adding contextual information—information about the social, cultural, environmental, systemic, ideological, or historical contexts within which terms are used or understood and within which technology is set—will enrich your definitions.

SOCIAL AND CULTURAL CONTEXTS

Consideration of the social and cultural settings within which technological terms exist can provide your audience with a way of understanding the terms in connection with human beings and the institutions and organizations human beings create. These **social and cultural contexts** are particularly important when the purpose of a document is to solve problems for people. Definitions that include some reference to social contexts often focus on human agency, social institutions, or cultural practices.

social and cultural contexts
the human settings in which technological terms exist

The following definitions of *data entry* and *pathnames,* for instance, illustrate definitions placed in social and cultural contexts. The definitions convey information about the culture that makes and uses technology, and the relationship that exists between humans and the technology:

> *Data entry*—Entering data into a computer in a timely manner, at a reasonable cost, and with minimal error. (Szymanski, p. 731)
>
> *Stack, document, and education pathnames*—In a Macintosh desktop world of folders and icons, the three pathname cards in the HomeStack are as mysterious as an MS–DOS *c*⟩ command prompt. If you never let the user get within sight of these cards while setting up your stack, all the better. (Goodman, p. 28)
>
> . . . a *stack developer* is anyone who designs a HyperCard stack that one or more people will be using. That includes corporate stacks developed for in-house use, perhaps as training vehicles or as the basis for departmental information systems. . . A stack developer is a computer consultant whose charter is to create information tools for clients, whether the tools are for time and money management or a freestanding kiosk of trade show exhibitors and products. (Goodman, p. 3)

Defining terms in their social and cultural contexts differs in a fundamental way from defining technical terms as existing apart from the social norms that shape them. Defining objects and concepts in terms of the ways in which human beings make, design, and use them can help your readers envision the ways in which they and other human beings relate to technology. This practice calls upon you to focus on the human uses of technology rather than on technology itself.

ENVIRONMENTAL OR SYSTEMIC CONTEXTS

environmental or systemic contexts
the environments or systems in which technical terms can be placed

Consideration of the **environmental or systemic contexts** within which technological terms are used can expand your audience's understanding. Such definitions assist people in troubleshooting and problem solving when mechanisms fail to work as they should. In a manual of bicycle maintenance, for example, the definitions of *spoke damage* and *low gear shifting* include the following information:

> Spokes are regularly damaged in two ways: something gets caught in them or the chain overshifts low gear and lands on them. Common sense can minimize the chances of either problem occurring. . . .

> The rule with *low-gear shifting* is conservatism. Always shift gently into low gear when riding after a wheel change or when riding a bike whose derailleurs may be improperly adjusted. Test the gear before you reach a steep section of the road where your shift will be rapid and pedaling force extreme. And when adjusting the derailleur, do not allow it to move closer to the spokes than necessary. . . Once scratched, drive-side spokes are much more likely to break when they are subjected to a lot of pressure. (*Bicycling Magazine's Complete Guide,* p. 36)

ECONOMIC CONTEXTS

In definitions, considerations of the ways in which terms are connected to factors of cost, supply and demand, or systems of exchange can expand an audience's understanding. The following examples of *VHS* and *compatibility,* for instance, illustrate information about **economic contexts** that may be useful to audiences in specific technical environments:

economic contexts
the environments of cost, supply and demand, and systems of exchange in which terms can be placed

> *VHS (S-VHS)*—A trademarked name for the first popular videotape cassette recording, storage, and playback format whose hallmark is consumer acceptance and the resulting economies of scale. (*MacroMind Director,* p. 373)

> *compatibility*—The ability of a particular hardware module or software program, code, or language to be used in a CAD/CAM system without prior modification or special interfaces. *Upward compatible* denotes the ability of a system to interface with new hardware or software modules or enhancements (i.e., the system vendor provides with each new module a reasonable means of transferring data, programs, and operator skills from the user's present system to the new enhancements). (Giesecke, p. 636)

IDEOLOGICAL CONTEXTS

ideological contexts
the environments of human beliefs and values in which terms can be placed

Sometimes, definitions can be enhanced, clarified, or focused by making implicit **ideological contexts** explicit. Such definitions may focus on the relationships between objects, phenomena, and concepts and the human beliefs that shape or influence them. The following passage, for example, has been taken from a text that attempts to define *human information processing* within a framework of belief that assumes computers can imitate:

> [Computer] scientists seem to work under two assumptions that lead to a fallacy par excellence: that *human information processing* is optimal, and that programs fare better by imitating and emulating human intelligence. It is important to note that this contention leads to a fallacy of "faulty hypostatization." The use of human behavior as a criterion for success or failure of emulation procedures should always be regarded in relation to the qualitative differences between man [sic] and machine. (Obermeier, p. 158)

HISTORICAL CONTEXTS

Some definitions include historical information that expands an audience's understanding of the **historical contexts** within which technology is necessarily designed, used, and modified. Historical information helps instruct or train an audience, shows change or timing, or illustrates a sequence. You might use historical information to provide the **etymology,** or word history, of a term. The following examples of definitions contain historical information:

historical contexts
the temporal environments in which terms can be placed

etymology
history of a word and its formation

> *database query language*—A fourth-generation language used in conjunction with a relational database; acts as an interface between a user and a relational database management system to facilitate easy access without use of complex programming code. (Syzmanski et al., p. 731)

> *Gears,* defined as toothed members transmitting rotary motion from one shaft to another, are among the oldest devices and inventions of man [sic]. In about 2600 BC the Chinese are known to have used a chariot incorporating a complex series of gears. . . Aristotle, in the fourth century BC, wrote of gears as though they were commonplace. In the 15th century AD, Leonardo da Vinci designed a multitude of devices incorporating many kinds of gears. (Juvinall and Kurt, p. 550)

The placement of definitions differs according to the purpose of the document, its layout and design specifications, the ways in which the audience will use the text, and the conventions of a given discipline.

Deciding When and Where to Include Definitions

IN-TEXT DEFINITIONS

Some texts identify key words with **highlighting** (bold or italics) and provide a definition immediately after the term, generally in an **appositive, noun phrase,** or **noun complement:**

highlighting
identification of key words by the use of bold or italic typefaces

> *Gravure,* a process of photomechanical printing [appositive definition], used in newspapers to provide visual references for readers.

appositive, noun phrase, or noun complement
grammatical forms that usually follow the statement of the key word in a definition

An *acid* is any substance other than a salt that increases the concentration of hydrogen ions in aqueous solution [noun complement]. (Ebbing and Wrighton, p. 283)

You might prefer, especially if a definition is rather long or complicated, to devote an entire sentence or more to the definition following the key term, as shown below:

Warning!
Kickback may occur when the nose or tip of the guide bar touches an object, or when the wood closes in and pinches the saw chain in the cut. This contact may abruptly top the chainsaw and in some cases may cause a lightning fast reverse reaction, kicking the guidebar up and back toward the user. . . Kickback may cause you to lose control of the saw. (inside cover, *Instruction Manual/Owner's Manual*, Stihl, p. 028)

footnotes
placement of annotations at the bottom of a page, in which definitions may be included

endnotes
placement of annotations at the end of a chapter or section of text, in which definitions may be included

FOOTNOTE/ENDNOTE DEFINITIONS

In some texts, and for some disciplines, accepted placement of definitions involves **footnotes** (located at the bottom of the page) or **endnotes** (located at the end of a chapter or section). The following definition of the computer program ELIZA, for example, was presented in an endnote form:

[61]ELIZA mimics a Rogerian psychotherapist, whose technique consists largely of echoing utterances of the patient; it therefore uses very little memory, and arrives at its "answers" by combining transformations of the "input" sentences with phrases stored under key words. (Kurzweil, p. 491)

EXAMPLE 13–4 Definition within a Sidebar

Group IVA Elements: Valence-Shell Configuration ns^2np^2

As mentioned earlier, Group IVA elements show a distinct trend from non-metal (carbon C) to metalloid (silicon, Si. and germanium, Ge) to metal (tin, Sn, and lead, Pb). Carbon exists in two well-known allotropic forms: graphite, a soft, black substance used in pencil leads, and diamond, a very hard, clear, crystalline substance. Both tin and lead are metals that were known to the ancients. Tin is mixed, or alloyed, with copper to make bronze.■

These elements form oxides of general formula RO_2. Examples are CO_2, a gas used to make Dry Ice (solid CO_2) and carbonated beverages; SiO_2, a solid that exists as quartz and white sand (particles of quartz); and SnO_2, the mineral cassiterite, principal ore of tin. Lead forms the dioxide, PbO_2, but the monoxide, PbO, is more stable. Carbon also forms a stable monoxide, CO.

■ Bronze, one of the first alloys (metallic mixtures) used in history, contains about 90% copper and 10% tin. It melts more easily than copper, but is much harder.

Source: Darrell D. Ebbings and Mark S. Wrighton, *General Chemistry*, 1st ed. Copyright © 1984 by Houghton Mifflin Company. Used with permission. p. 189.

EXAMPLE 13–5 Definition within a Glossary

bit   The smallest unit of information that can be stored and processed by a digital computer. A bit may assume only one of two values: 0 or 1 (i.e., ON/OFF or YES/NO). Bits are organized into larger units called *words* for access by computer instructions.

Computers are often categorized by word size in bits, i.e., the maximum word size that can be processed as a unit during an instruction cycle (e.g., 16-bit computers or 32-bit computers). The number of bits in a word is an indication of the processing power of the system, especially for calculations or for high-precision data.

bit rate   The speed at which bits are transmitted, usually expressed in bits per second.

Source: Frederick E. Giesecke; Alva Mitchell; Henry Cecil Spender; Ivan Leroy Hill; Robert Olin Loving; and John Thomas Dygdon, *Principles of Engineering Graphics*. New York: MacMillan Publishing Co., Inc., 1990, p. 635. Reprinted with the permission of Macmillan College Publishing.

VISUAL DISPLAY/SIDEBAR DEFINITIONS

In some documents, especially those that support immediate reference to technical materials or processes, you might define key terms or words in an accompanying visual display or in a **sidebar** of text. Example 13–4 shows a portion of a page from a chemistry text. On this page, the alloy bronze is cued by a text flag (a small square) and defined in a nearby sidebar marked with the same flag.

sidebar
placement of annotations in the margins of a text, in which definitions may be included

GLOSSARY DEFINITIONS

Glossaries are collections of terms located either at the beginning of a text (if readers need to know the information contained in the descriptions before proceeding through the text) or at the end of the document (if readers can refer to definitions as they work through a text). Example 13–5 shows the glossary from a computer science book.

To help you create effective definitions, the Writing Strategies 13–1 contains questions you can ask in preparing your definition.

glossaries
collections of terms located either at the beginning or end of a document, depending on the needs of the readers

Creating Descriptions

Technical descriptions can be found in all sorts of communication settings. Frequently included in reports and proposals, descriptions help audiences make decisions about designs, manufacturing, or purchasing. Descriptions of processes—such as chemical processes, organizational procedures, or safety measures—are found in manuals that accompany a specific product

---

WRITING STRATEGIES 13–1   Creating Effective Definitions

Ask and answer the following questions:

1. What is the purpose of this definition or set of definitions? Are there several purposes? If so, how are they related?
2. For what audience(s) is this definition intended? What is the audience's expertise in this area? Experience? What do they need to know or do?
3. What is the term being defined? To what class/group/family does it belong? What special characteristics differentiate it from other things in this group?
4. What conventional definition strategies best lend themselves to this task? Among those frequently used include the following:

   - Modification
   - Synonyms and antonyms
   - Example and enumeration
   - Metaphor and analogy
   - Negation
   - Comparison–contrast
   - Visual display and presentation

5. What special contextual information will help meet audience's needs? Information about:

   - Social or cultural contexts?
   - Environmental or systemic contexts?
   - Economic contexts?
   - Historical contexts?

6. Where should the definition be placed?

   - In text?
   - In a footnote or endnote?
   - In a visual display or a sidebar?
   - In a glossary?

---

or that direct activities. Literature written to sell or explain products includes technical descriptions of products or processes, as do documents written for technical training.

As you might suspect, like definitions, descriptions written for differing purposes and audiences differ in length, content, detail, tone, vocabulary, and format. Some technical descriptions of simple mechanisms, for example, are limited to a single visual display with accompanying text labels. Other technical descriptions may continue for pages. Depending on the purpose of a document, a technical description, much like a definition, may include contextual information about social and cultural contexts; about environmental or systemic contexts; about historical background and economic factors; and about ideological assumptions. Generally, technical descriptions fall into

EXAMPLE 13–6 Description of Boiling for a Nonspecialized Audience

Boil—To cook in liquid at boiling temperature (212° at sea level) where bubbles rise to the surface and break. For a full rolling boil, bubbles form rapidly throughout the mixture.

Source: *Better Homes and Gardens New Cook Book,* Des Moines, IA: Meredith Corporation, 1976, p. 408.

two main categories: (1) descriptions of objects/mechanisms/phenomena, and (2) descriptions of processes/procedures/activities.

The exact nature of technical descriptions, as well as the scope and depth of these answers, depends on your audience(s)—their education, experience, attitude, responsibilities, among other factors—and the purpose(s) for which the description is written, distributed, and read. Your description also depends on the nature of the information involved in the description itself and the contexts that surround your creation of the description: the time allocated for writing, the corporate contexts and purposes for the writing, the historical factors influencing the descriptive act, and so forth.

Examples 13–6 and 13–7 show two different technical descriptions of the same process, boiling. One comes from a chemistry textbook and the other from a cookbook; one addresses specialists—chemical engineers with an interest in the processes of steady state heat transfer, evaporation, and distillation—and the other is written primarily for nonspecialists with a general interest in those chemical processes associated with cooking. Notice that the description for specialists is supported by a visual display. Look at Writing Strategies 13–3. What other characteristics does the description in Example 13–7 contain?

In the rest of this chapter, we offer you suggestions on how to create the two main types of technical descriptions (descriptions of objects/ mechanisms/phenomena and descriptions of processes/procedures/activities).

DESCRIPTIONS OF OBJECTS, MECHANISMS, OR PHENOMENA: PARTS, WHOLES, AND SYSTEMS

You can describe an object, mechanism, or phenomenon by focusing on the parts or elements that make up the whole—sometimes an entire system. The technical descriptions in Examples 13–8 and 13–9 illustrate two ways of representing parts, wholes, and systems. Example 13–8 emphasizes the structural parts of a whole, and Example 13–10 emphasizes the function of parts within a system.

EXAMPLE 13–7   Description of Boiling for a Specialized Audience

4.8A Boiling

1. Mechanisms of boiling.   Heat transfer to a boiling liquid is very important in evaporation and distillation and also in other kinds of chemical and biological processing, such as petroleum processing, control of the temperature of chemical reactions, evaporation of liquid foods, and so on. The boiling liquid is usually contained in a vessel with a heating surface of tubes or vertical or horizontal plates which supply the heat for boiling. The heating surfaces can be heated electrically or by a hot or condensing fluid on the other side of the heated surface.

In boiling the temperature of the liquid is the boiling point of this liquid at the pressure in the equipment. The heated surface is, of course, at a temperature above the boiling point. Bubbles of vapor are generated at the heated surface and rise through the mass of liquid. The vapor accumulates in a vapor space above the liquid level and is withdrawn.

Boiling is a complex phenomenon. Suppose we consider a small heated horizontal tube or wire immersed in a vessel containing water boiling at 373.2 K (100°C). The heat flux is q/A W/m^2, $\Delta T = T_w - 373.2$ K, where T_w is the tube or wire wall temperature and h is the heat-transfer coefficient in W/m$^2 \cdot$ K. Starting with a low ΔT, the q/A and h values are measured. This is repeated at higher values of ΔT and the data obtained are shown in Figure 4.8–1 plotted as q/A versus ΔT.

In the first region A of the plot in Fig. 4.8–1, at low temperature drops, the mechanism of boiling is essentially that of heat transfer to a liquid in natural convection. The variation of h with $\Delta T^{0.25}$ is approximately the same as that for natural convection of horizontal plates or cylinders. The very few bubbles formed are released from the surface of the metal and rise and do not disturb appreciably the normal natural convection.

In the region B of nucleate boiling for a ΔT of about $5 - 25$ K ($9 - 45$°F), the rate of bubble production increases so that the velocity of circulation of the liquid increases. The heat-transfer coefficient h increases rapidly and is proportional to ΔT^2 to ΔT^3 in this region.

In the region C of transition boiling, many bubbles are formed so quickly that they tend to coalesce and form a layer of insulating vapor. Increasing the ΔT increases the thickness of this layer and the heat flux and h drop as ΔT is increased. In region D or film boiling, bubbles detach themselves regularly and rise upward. At higher ΔT values radiation through the vapor layer next to the surface helps increase the q/A and h.

The curve of h versus ΔT has approximately the same shape as Fig. 4.8–1. The values of h are quite large. At the beginning of region B in Fig. 4.8–1 for nucleate boiling, h has a value of about 5700–11400 W/m$^2 \cdot$ K, or 1000–2000 btu/h $\cdot$ ft$^2 \cdot$ °F, and at the end of this region h has a peak value of almost 57 000 W/m$^2 \cdot$ K, or 10 000 btu/hr $\cdot$ ft$^2 \cdot$ °F. These values are quite high, and in most cases the percent resistance of the boiling film is only a few percent of the overall resistance to heat transfer.

The regions of commercial interest are the nucleate and film-boiling regions (P1). Nucleate boiling occurs in kettle-type and natural-circulation reboilers.

EXAMPLE 13–7   Concluded

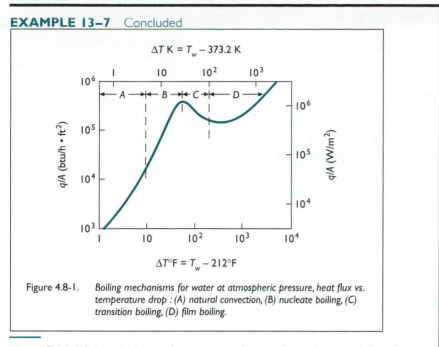

$$\Delta T \text{ K} = T_w - 373.2 \text{ K}$$

$$\Delta T°\text{F} = T_w - 212°\text{F}$$

Figure 4.8-1.   *Boiling mechanisms for water at atmospheric pressure, heat flux vs. temperature drop : (A) natural convection, (B) nucleate boiling, (C) transition boiling, (D) film boiling.*

In the technical description of the AIWA compact stereo (Example 13–8), the graphics and text identify various parts (e.g., switches, jacks, controls, knobs, buttons) of the stereo that users should know how to locate. Except for hints provided by the names of the individual elements, however, these labels describe the components as *structural* elements rather than detailing their *function*. The description of the fractionating column (Example 13–9), on the other hand, gives information not only about the various structural parts of the apparatus, but also about the function or purpose of these component parts within the system of the entire fractionation column. Thus, we learn that the glass beads in this system "provide a surface on which the less volatile part of the vapor condenses." Often technical descriptions of objects, mechanisms, or phenomena include both structural and functional information to provide readers with as much data as possible.

To create useful descriptions of objects, mechanisms, or phenomena, you need to begin with three steps:

EXAMPLE 13–8 Technical Description Representing Parts as Structural Elements

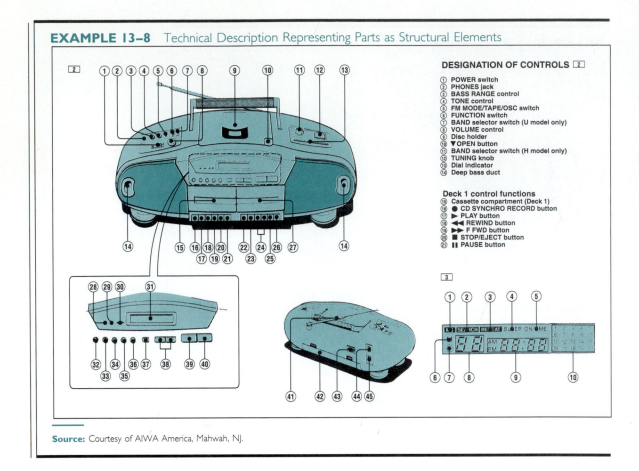

DESIGNATION OF CONTROLS 2

① POWER switch
② PHONES jack
③ BASS RANGE control
④ TONE control
⑤ FM MODE/TAPE/OSC switch
⑥ FUNCTION switch
⑦ BAND selector switch (U model only)
⑧ VOLUME control
⑨ Disc holder
⑩ ▼OPEN button
⑪ BAND selector switch (H model only)
⑫ TUNING knob
⑬ Dial indicator
⑭ Deep bass duct

Deck 1 control functions
⑮ Cassette compartment (Deck 1)
⑯ ● CD SYNCHRO RECORD button
⑰ ▶ PLAY button
⑱ ◀◀ REWIND button
⑲ ▶▶ F FWD button
⑳ ■ STOP/EJECT button
㉑ ‖ PAUSE button

Source: Courtesy of AIWA America, Mahwah, NJ.

1. **Identify the rhetorical context of the description.** Begin by thinking about the *purpose* of the description and the *audience(s)* who will read it. What do these people need to know about the complex whole you are describing or the parts that comprise this whole? Why do they need to know these things? This rhetorical information will help you make subsequent decisions about how best to represent your topic.

   Example 13-10 (page 426) demonstrates three ways of describing a dishwasher based on different purposes—operating, troubleshooting, and repairing. Imagine that these lists represent topics to be covered in the print documentation that accompanies the dishwasher when it is sold to a customer. The first list, detailing the various cycles the washer offers, would help individuals interested in understanding the machine's

EXAMPLE 13–9 Technical Description Representing Parts as Functional Elements

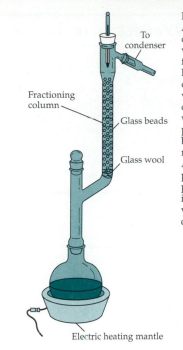

To condenser

Fractioning column

Glass beads

Glass wool

Electric heating mantle

Figure 2.11
An apparatus for fractional distillation. The temperature varies up the length of the fractioning column, being hotter at the lower end and cooler at the top. The most volatile component passes over into the condenser, whereas less volatile components condense on the beads in the column or remain in the distillation flask. As the most volatile component distills over, the temperature of the boiling liquid increases until the next most volatile component distills over.

Fractional distillation is useful when the solution consists of two or more volatile components. The temperature at which a liquid boils is known as its *boiling point*. When a solution containing liquid substances of different boiling points is distilled, the substance with the lowest boiling point normally distills over first. ■An apparatus for fractional distillation is shown in Figure 2.11. A *fractioning column*, containing glass beads, is placed at the top of the distillation flask. The glass beads provide a surface on which the less volatile part of the vapor condenses. The temperature varies along the length of the column, being hotter at the bottom and cooler at the top. The substance with the lowest boiling point passes over into the condenser from the top of the column, whereas substances with higher boiling points condense on the beads in the column or remain in the distillation flask. As the substance of lowest boiling point is removed from the flask, the temperature of the boiling liquid increases and the next substance, now the one with the lowest boiling point, begins to distill.

■ Although solutions of volatile liquids usually increase in boiling point as the more volatile components distill off, some solutions exhibit a constant boiling point. For example, a solution of 95.6% ethanol (grain alcohol) and 4.4% water by mass boils at the constant temperature of 78.15°C; pure ethanol boils at 78.5°C. Such a constant-boiling mixture is called an *azeotrope*. It is impossible to separate ethanol and water completely by fractional distillation of simple ethanol-water mixtures.

Fractional distillation is employed commercially to separate crude oil or petroleum into useful products. For this, the distillation is operated continuously, various fractions being taken off at different heights up the fractionating column. Thus, the lowest boiling fraction (20°–60°C), called petroleum ether, is obtained near the top of the column. Light naphtha or ligroin is a fraction with a slightly higher boiling point (60°–100°C), and it is taken off the column just below the petroleum ether. Gasoline boils between 50°C and 200°C, and kerosene between 175°C and 275°C; these come off even lower on the column. Furnace and diesel fuels boil at still higher temperatures.

Source: Darrell D. Ebbing and Mark S. Wrighton, *General Chemistry*, 1st ed. Copyright © 1984 by Houghton Mifflin Company. pp. 43–44. Reprinted with permission.

EXAMPLE 13–10   Description Topics Listed according to Purpose

Operating	Troubleshooting	Repairing
Light Soil Cycle	Spots and Filming	Motor
Normal Wash Cycle	Cloudiness	Pump
Pot-Scrubber Cycle	Chipping of China	Hub Connection
Delay Start Cycle	Suds in the Tub	Soap Cup
Energy Saver Cycle	Water Won't Pump	Washing Tower
Rinse and Hold Cycle	Leaking	Spray Arm

---

features or how the machine functions. The second list, detailing the problems that might occur in connection with the machine's use, would help people trying to solve a problem with the machine's performance. Such people may be everyday users of the machine trying to cope with a minor problem or repair people called in to cope with a more serious malfunction. The third list, detailing the various parts of the dishwasher, would help a repair person or an owner who needed to understand the structural components of the larger mechanism.

2. **Decide on a useful scheme for identifying important parts of the whole.** Your decisions about what the "important" parts are will depend on your knowledge of the audience(s) and their needs. In Example 13–10, note that the first list is an identification of dishwasher cycles by function. These cycles are important "parts" of the dishwasher for customers who will have to know how to use it properly. Similarly, consider the list of potential problems—these items represent symptoms that are important "parts" of a whole troubleshooting diagnosis.

3. **Present the parts of the description in a way that is most congruent with your information about audience, purpose, and context and the "whole" you are presenting.** The ways in which you present parts of a whole, once these parts have been identified, will differ according to your understanding of audience(s) and their needs. The order of parts, the grouping of like items, the presentation of parts, and the use of visual display, for example, will change according to rhetorical context.

The "repairing" list, for instance, in Example 13–10 could be composed in two different ways depending on whether you see your audience mainly as experienced mechanics or weekend troubleshooters. For the experienced mechanics, the various structural components of the dishwasher could be presented in an order determined by their listing in the parts manual. (Because motor parts and subassemblies within the motor are listed first in the technical repair and parts manual, they could be listed first in the owner documentation, along with references to parts numbers in the technical repair and parts manual.) These parts could also be represented by a technical diagram to indicate their structural relationship to one another.

For less experienced, weekend problem solvers, the list of components might be presented in a reverse order—perhaps by the component's frequency of repair record—in an effort to facilitate locating the most likely problem areas (motors require repair less frequently than do spray arms). A visual display might help here as well, but parts could be identified with more common—and less technical—names, perhaps linked to troubleshooting directions

You can use the questions in Writing Strategies 13–2 to develop descriptions of objects, mechanisms, or phenomena. The answers to these questions would appear in your description.

DESCRIPTION OF PROCESSES, PROCEDURES, OR ACTIVITIES: STEPS AND OVERVIEWS

Technical descriptions of processes or procedures also generally break down complex wholes (in this case, processes that happen over some period of time) into parts (designated as events, steps, activities, or sequences). Examples 13–11 through 13–13 provide three different types of technical descriptions that focus on processes, procedures, or activities.

The technical description of food digestion (Example 13–11) begins with an *overview of the process* ("Digestion breaks down the complex molecules in foods to the simple molecules that can be used in metabolism for the generation of energy and the biosynthesis of cellular components"). This overview is followed by a narrative of sorts—a record of digestive events that we recognize as smaller successive *events within the larger process,* as these events occur in chronological order.

The description of the Norton Utilities "ASK command" (Example 13–12) starts with an *overview of the procedure* invoked by the command and the function of this particular procedure ("The Norton ASK Command provides an easy way to create batch files that execute different commands . . ."). The description then proceeds with a description of the various *subprocesses within this procedure* (e.g., user invokes the command, command invokes the prompt list, user responds to list, ASK returns control to the batch file, batch program branches to "different labels"). The description is marked—for ease of reading and comprehension—with key words that signal the temporal order in which the various steps happen (e.g., "when," "then," and "after").

WRITING STRATEGIES 13–2 Questions Answered by Descriptions of Objects, Mechanisms, and Phenomena

What is it?

What family or group of things does it belong to?
What is it called?
What is its model or part number?
Does the object/mechanism/phenomenon have a historical background that helps people understand it? A cultural, social, or economic context that would help people understand it?

Where is it located?

What are its surroundings, setting, environment? Why are these important to an understanding of it?
How does one get to it within that location?

What are its sensory characteristics?

What does it look like? What is its size? Shape? Color?
What does it smell like?
What does it feel like? Weight? Texture? Density?
What does it taste like?
What does it sound like?

How is it put together or constituted?

Of what materials is it made?
What are its characteristics? State (gas, liquid, solid)? Stage (immature, mature, elderly)?
Of what parts or elements is it constituted? What are the distinctive characteristics of each part? How are these related?
How does one put it together or take it apart? Does one?
Is this object/mechanism/phenomenon part of a larger system? An environment?

How does it change/act/function over time?

How long does it last or live?
What special characteristics (dangers, precautions, benefits) are associated with its use, care, or maintenance?

How do people relate to it?

Who uses/observes/attends to this object/phenomenon/mechanism?
Who does not use/observe/attend to it? Why?
What is its function? Within what settings/environments?
What assumptions do people have about this object/mechanism/phenomenon that might help people understand it?
How can people use this product/follow this mechanism safely?

EXAMPLE 13–11   Description of Food Digestion

Digestion

A biochemical process that all people experience is the digestion of foods. Certainly, a failure in digestion is a memorable experience for everyone. Digestion breaks down the complex molecules in foods to the simple molecules that can be used in metabolism for the generation of energy and the biosynthesis of cellular components. Digestion of food is moderately well understood in biochemical terms for human beings and animals. Digestion in biochemistry extends to the breakdown *within cells* of complex compounds to simple molecules that can be either excreted or used in cellular metabolism. Both of these digestive processes are essential to the well being of complex animals.

Digestion of Food

The major nutrients in food that require digestion are proteins, starches, and fats, all of which are large molecules that are digested by hydrolysis into smaller molecules. Proteins are long-chain polymers of *x*-amino acids, and starches are polymers of sugars, whereas fats are *triglycerides* or *phospholipids,* which are glycerol esters of fatty acids. (The chemical structures of triglycerides and phospholipids are given in Chapter 11 and those of sugars and starches are in Chapter 15.) The digestive reactions are catalyzed by enzymes in the digestive tract. Enzymes are themselves proteins and are often defined as biological catalysts.

The first enzyme encountered by food during ingestion into the mouth is *salivary amylase.* This enzyme initiates digestion by catalyzing the partial hydrolysis of starches into shorter oligomers. Its main importance may be to protect teeth from decay by digesting adhering food particles into soluble molecules that flow easily into the stomach, thereby removing potential nutrients for bacteria. Food quickly reaches the stomach. . . .

Source: Robert H. Abeles; Perry A. Frey; and William P. Jencks, *Biochemistry.* Boston: Jones and Bartlett Publishers, 1992. p. 16. Reprinted with permission.

The technical description of mouse cleaning from the *Macintosh Reference* manual (Example 13-14) begins with an exceptionally brief *overview of the procedure* ("To clean the mouse") and then breaks the process down into discrete *steps within the procedure* that are labeled so that the audience can follow them easily. Visual displays provide additional information.

Let's examine more closely the stylistic techniques that the authors of the preceding examples have employed to describe processes. These techniques can best be explained in relation to the **concept of performance.**

Look at the description of digestion (Example 13-11). This process is obviously not meant to be performed (at least in a conscious way) by a user.

concept of performance the use of the product or process being defined or described; often includes the tasks a user must perform

---

EXAMPLE 13–12   Description of Norton Utilities "ASK Command"

Description

The Norton ASK command provides an easy way to create batch files that execute different commands, depending on user input at the time the batch file is run.

When the ASK command is invoked, it displays the prompt text, then awaits a response. The prompt will generally list the keys the user may type in response to the ASK command (any of the keys in the key-list) and explain the results of choosing each.

The user responds to ASK by typing any one of the keys in the key-list. The key-list itself is not displayed on screen; only the prompt is displayed.

The key-list is optional. When no key-list is given, any key the user presses is accepted.

After the user chooses one of the keys in the key-list (or any key, if no key-list is specified), ASK returns control to the batch file. ASK passes the key that was chosen by the user as an ERRORLEVEL code. The first choice (key) in the key-list corresponds to ERRORLEVEL 1, the second to ERRORLEVEL 2, the third to 3 and so on (if no key-list is supplied, ASK returns ERRORLEVEL 0 in response to any key being pressed). The batch program can then branch to different labels in the batch file, in accordance with the ERRORLEVEL.

Source: *The Norton Utilities Reference Manual.* Advanced Edition, Version 4.5. Santa Monica, CA: Peter Norton Computing, Inc, 1988. p. 41.

Note that the sentences focus on the process of digestion, and that no human agent, or actor, is represented as *doing* the process. As a result, the purpose of this passage is not to explain *how* to accomplish some task, but, rather, *what happens and when.* The words the writer chose support this purpose. In the first paragraph, for instance, there are six sentences; the writer has made sure that the first **noun cluster** in each of these six sentences ("A biochemical process," "a failure in digestion," "Digestion," "Digestions," "Digestion," "Both of these digestive processes") focuses readers' attention on the process of digestion.

In contrast, look at the stylistic features of the mouse cleaning passage (Example 13–13). The purpose here is to describe a procedure that manufacturers of the product expect customers to perform: cleaning a mouse. As a result, the text and graphics focus on the users' actions; the text, then, constitutes directions for performing a task as well as a description of a procedure. The **imperative form of verbs** in many sentences ("turn off the Macintosh," "turn the mouse over and open the plastic ring," "remove the ring") supports this focus. The subject of these phrases—understood, by readers of English, to be "you"—is clearly intended to perform the actions described by the text and graphics.

noun cluster

one or more nouns used at the beginning of a sentence

imperative form of verbs

form of the verb used when giving commands; e.g., "Unplug the machine before servicing"

EXAMPLE 13–13 Description of Mouse Cleaning

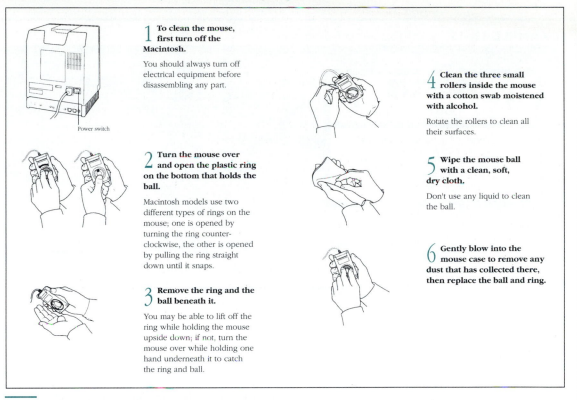

1 **To clean the mouse, first turn off the Macintosh.**

You should always turn off electrical equipment before disassembling any part.

Power switch

2 **Turn the mouse over and open the plastic ring on the bottom that holds the ball.**

Macintosh models use two different types of rings on the mouse; one is opened by turning the ring counter-clockwise, the other is opened by pulling the ring straight down until it snaps.

3 **Remove the ring and the ball beneath it.**

You may be able to lift off the ring while holding the mouse upside down; if not, turn the mouse over while holding one hand underneath it to catch the ring and ball.

4 **Clean the three small rollers inside the mouse with a cotton swab moistened with alcohol.**

Rotate the rollers to clean all their surfaces.

5 **Wipe the mouse ball with a clean, soft, dry cloth.**

Don't use any liquid to clean the ball.

6 **Gently blow into the mouse case to remove any dust that has collected there, then replace the ball and ring.**

Source: *Macintosh Reference,* Cupertino, CA: Apple Computer, Inc. pp. 295–96. Used with permission.

EXAMPLE 13–14 Description of Paper Making

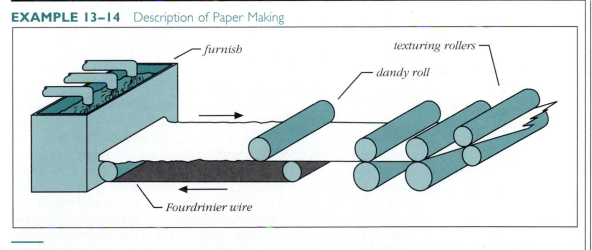

furnish

texturing rollers

dandy roll

Fourdrinier wire

Source: Mark Beach; Steve Shepro; and Ken Russo, *Getting It Printed: How to Work with Printers and Graphics Arts Services to Assure Quality, Stay on Schedule, and Control Costs.* Portland, Ore: Coast to Coast Books, 1986. pp. 95–96.

EXAMPLE 13–15   Safety Features in Descriptions

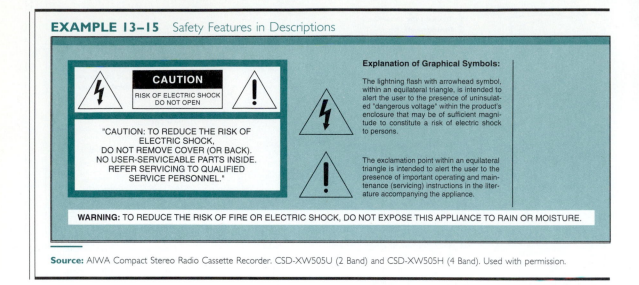

Source: AIWA Compact Stereo Radio Cassette Recorder. CSD-XW505U (2 Band) and CSD-XW505H (4 Band). Used with permission.

To create effective technical descriptions of processes, procedures, and actions, begin with the following four considerations:

1. **Identify the rhetorical context of the process or procedural description.** Begin by thinking about the *purpose* of the description and the *audience(s)* who will read it. What do these people need to know about the process or procedure you are describing or the steps and activities that comprise this process? Why do they need to know these things? What parts of the process are they familiar/unfamiliar with? This rhetorical information will help you decide how best to represent the steps and subprocesses you identify.

2. **Working with a chronological framework, identify the steps/ activities/subprocesses that make up the entire process and the appropriate order of these subparts.** One good way to accomplish this task is to observe a situation that closely parallels the context in which your audience will be operating. If you are describing a complex procedure that must be performed safely and precisely, for example, observe an expert performing it and take careful notes or videotape the process. If your intended audience is a novice in a field with which you are not familiar, you may simulate the conditions by performing the task yourself. Given your knowledge of the task and the audience's needs, you should also determine the level of specificity and completeness that your audience must achieve. Novices, for example, may require description that identifies even the smallest event or step. An audience

WRITING STRATEGIES 13–3 Questions Answered by Descriptions of Processes/Procedures/Activities

What is it?

What kind of action or process is involved?

What is the name of the process/action/procedure?

What is the focus? People? Mechanical process? Biological process? Natural phenomenon? Chemical process?

Does the process/procedure/activity have an historical background that helps people understand it? A cultural, social, or economic context that would help people understand it?

Where and how does it happen?

What are the surroundings, settings, or environments for the action?

How does one observe/take part in/do these actions? Why?

Who does *not* observe/take part in/do these actions? Why?

What are the sensory characteristics of the action/procedure/process?

What does it look like when it happens? Before? After?

What does it smell like when it happens? Before? After?

What does it feel like as it is happening? Before? After?

What does it taste like as it is happening? Before? After?

What does it sound like as it is happening? Before? After?

How is the process/procedure/activity put together?

What steps or stages can be identified? Would different people identify different steps or stages? If so, why?

What are the distinctive characteristics of each? How are these related?

Can steps be taken out or added? If so, by whom and why?

Who determines the steps of this process/procedure/activity? Why and how? Who does not? Why not?

Is this process/procedure/activity part of a larger system? An environment? A context?

How does the process/procedure/activity change over time?

What is the function?

What characterizes the beginning state?

What characterizes the ending state?

What is the input? What is the output?

What special characteristics (dangers, cautions, benefits) are associated with the action/procedure/process?

How do people relate to it?

Who does/observes/attends to this process/procedure/activity? Why? Who does not do/observe/attend to it? Why not?

What assumptions do people have about this process/procedure/activity that might help people understand it?

How can people use this process/follow this procedure safely?

with experience may be able to complete subprocesses without detailed explanations and may need only safety precautions.

3.   **Group the steps/activities/subprocesses in useful chunks so the audience can use them efficiently. Relate these chunks one to another for the same purpose.** Simply listing the steps of a procedure or process may not help an audience conceptualize the entire process or its major stages. Often audiences need to see steps chunked into meaningful groupings to understand the steps as a coherent whole. Example 13-14 describes a coherent procedure of making paper by chunking steps of the process into logical paragraphs. Each paragraph represents a related group of subprocesses (e.g., "cleaning," "beating" and "sizing" the pulp, mixing the "furnish" and draining the furnish, applying a "finish," trimming, drying, pressing, etc.). Note also that the writer demonstrates how the various steps in this process are related to one another with phrases such as, "After the pulp is beaten, cleaned, and sized, it is mixed at the rate of one pound of pulp to 100 pounds of water," and, "When the furnish reaches the end of the wire. . ." .

4.   **Describe safety factors.** Increasingly, companies and manufacturers are being held legally responsible for safe use of their products (Helyar). As a result, warnings and risk descriptions such as those represented in Example 13-15 are becoming common.

Pamela S. Helyar, in an article about writing "adequate instructions" that meet legal standards, notes that adequate instructions must:

- Contain clear, correct, and tested instructions.
- Use words and graphics that suit the intended audience.
- Appropriately warn of product hazards.
- Offset claims of product safety in advertising or other materials with intense warnings.
- Present important directions or warnings so consumers can spot and follow them.
- Meet government, industry, and company standards.
- Reach product users.
- Inform consumers of defects discovered after marketing in a timely fashion (p. 142).

You can use the questions within Writing Strategies 13-3 to develop descriptions of processes, procedures, or activities. The answers to these questions would appear in your description.

SUMMARY   This chapter began with a discussion of formal definitions, as opposed to the informal definitions or categorical propositions we discussed in Chapter 5. A formal definition usually contains the term to be defined; the class, family, species, or group in which the term belongs; and the special characteristics that distinguish the term from other terms in the same class, family, species, or group. Functional or operational definitions stress the *use* of the term as the special feature. In differentiating the term you are defining from the others in the same group, you need to consider the precision and efficiency of your definition, both of which depend on your rhetorical situation.

When creating a definition, you can define your term by modification, by synonyms and antonyms, by example and enumeration, by metaphor and analogy, by negation, by comparison or contrast, and by visual display and presentation. You also need to place your definition within an appropriate context for your audience. You might explain the social and cultural, the environmental or systemic, the economic, the ideological, or the historical contexts of the term you are defining. Finally, you must decide where to place your definition in your document. You might put it in your text immediately after you mention your term and signal the definition by a highlighting cue such as boldface or italics. You can place the definition in a footnote or endnote, a glossary, a visual display, or a sidebar.

The second part of this chapter addressed descriptions, which are often developed with the similar techniques as those used in definitions. Descriptions also may be placed in social, cultural, environmental, systemic, historical, economic, or ideological context.

Generally, technical descriptions fall into two main categories: (1) descriptions of objects, mechanisms, or phenomena; and (2) descriptions of processes, procedures, or activities. When creating descriptions of objects, mechanisms, or phenomena, you should consider the parts and the whole of your subject; that whole may be an entire system. To create an effective description of an object, mechanism, or phenomenon, identify your rhetorical context, decide on a useful scheme for identifying important parts of the whole, and finally present the parts of the description in a way that is most congruent with your information about audience, purpose, and context and the whole you are presenting.

To create an effective description of a process, procedure, or activity, you should consider the steps and overviews required in your description. Many of the techniques used in this kind of description pertain to performance—actually doing the process or determining what happens at what time during the procedure or activity. When describing your process, procedure, or activity, identify your rhetorical context and work with a chronological framework to identify the steps, activities, or subprocesses that make up the entire process and the appropriate order of these subparts. Then group the steps, activities, and subprocesses in useful chunks so the audience can use them effectively and relate these chunks one to another for the same purpose. Finally, describe any safety factors required for your audience, purpose, and context.

In the next chapter, we take another look at descriptions—in this case, those that appear as instructions, specifications, and procedures in technical documents.

ACTIVITIES AND EXERCISES

1. Consult the AIDS case documents (Appendix A, Case Documents 1). Using this material and additional information you can get from the library or from your local health services organization, write the copy for a brochure on AIDS that is appropriate for junior high school students enrolled in a sex education class. Make sure the brochure includes a definition of AIDS, a description of how AIDS

is acquired, and instructions about how to avoid HIV infection. Make sure that the material you write is appropriate for a junior high school audience. If possible, get permission to test the brochure copy with several individuals of the appropriate age group.

Write the copy for another brochure about AIDS for the parents of the students described above. This brochure will be distributed during a parent-child AIDS information night. Make sure the brochure includes a definition of AIDS that is appropriate for parents of children who might soon be sexually active, a description of how AIDS is acquired within this population, and instructions about how to help children avoid HIV infection. Make sure that the material you write is appropriate for parents. If possible, get permission to test the brochure copy with several individuals of the appropriate age group.

2.   Working collaboratively in teams of four, write a description of the registration process on your campus that can be sent to enrolling first-year students so that they get an accurate picture of the complex procedures involved and the role they must play in these procedures. In this description, make sure to identify all of the major stages within the registration process, and the subprocesses within these (e.g., planning a schedule, meeting with your advisor, choosing classes, dealing with time conflicts, paying the bill). Also, make sure to define the key terms associated with the registration process (those specialized terms that students learn only after they have a semester or two of experience— registrar, bursar, discussion or recitation versus lecture section, teaching assistant, advisor). You may also want to include warnings of registration dangers, shortcuts for the adventurous, and ideas for preparation. Be sure to check with the registrar's office about the accuracy of your document, but send only one member of your group to the registrar so you don't overwhelm the staff.

3.   Find an example of a poorly written technical description of a process/procedure/activity or object/mechanism/phenomenon. Rewrite and revise this description using the strategies we have identified in this chapter. Be sure to identify the audience(s) for the revised description and the rhetorical context within which it will be used. Submit a copy of both the original and the revised description to your teacher for review.

4.   Find an example of a poorly written definition. Rewrite and revise this definition using the strategies we have identified in this chapter. You might want to try rewriting the definition for a particular audience and situational context. Submit a copy of both the original and a copy of the revised definition to your teacher for review.

5.   Choose a small device or machine with which you are familiar. It is best to choose something simple—such as a manual pencil sharpener, an egg beater, a manual can opener, or manual grass clippers—something with moving parts, and something that allows you to see all the parts of the device. Study the device by taking measurements, learning what materials make up the parts, and discovering how the parts are attached to each other. Observe how the parts move and how that movement might cause other parts to move in turn.

Once you have completed your observations, imagine that you are writing about the device for an audience who needs to assemble the device. First, write a definition of the device. Then, describe the device and how the parts relate. Include the measurements and materials of the parts and their method and location of attachment. Use Writing Strategies 13–1 to help you write the most effective description.

6. Find a term that has taken on a new meaning for a particular audience or is used to describe a new technology within your own major field of study. For example, in the computer field, words such as *disk, network,* and *mouse* have come to mean different things than they originally did. Using Writing Strategies 13-1 for inspiration, create a definition of the term that fully explains and illustrates a term. Imagine that your definition will be part of a manual that accompanies a device or a textbook for high school students. You may include a visual illustration if you think it appropriate.

7. Assume that you and other members of a collaborative team must present a brief oral presentation on the bioremediation process used to clean up Prince William Sound in Alaska. Read over the case documents in Appendix A (Case Documents 2) that pertain to this issue and create a definition and description of bioremediation. You may use visual displays if you like. Choose one of the following audiences:

- The Environmental Protection Agency.
- The citizens of Prince William Sound.
- The stockholders of the Exxon Corporation.
- The viewers of a major network's evening news broadcast.

8. Gather warning, caution, and danger labels from appliances and products you have around your home. Try to determine from the labels you find when a caution, warning, or danger label is appropriate. Using the guidelines in the chapter and your own observations, write a memo to your teacher and classmates indicating what kind of labeling and precautions would have been appropriate in the DuPont accidental poisoning case described in Appendix A, Case Documents 4.

WORKS CITED

Beach, Mark; Steve Shepro; and Ken Russo. *Getting It Printed: How to Work with Printers and Graphic Arts Services to Assure Quality, Stay on Schedule, and Control Costs.* Portland, OR: Coast to Coast Books, 1986.

Beckman, William A.; Sanford A. Klein; and John A. Duffie. *Solar Heating Design: By the f-Chart Method.* New York: John Wiley and Sons, 1977.

Bicycling Magazine's Complete Guide to Bicycle Maintenance and Repair. Emmaus, PA: Rodale Press, 1990.

Brecher, Debra L. *The Women's Computer Literacy Handbook.* New York: New American Library, 1985.

Ebbing, Darrell D., and Mark S. Wrighton. *General Chemistry.* Boston, MA: Houghton Mifflin, 1984.

Forrester, Frank H. *1001 Questions Answered About the Weather.* New York: Dover Publications, Inc., 1981.

Geankopolis, Christie J. *Transport Processes and Unit Operations*. 2nd ed. Englewood Cliffs, NJ: Prentice Hall, 1983.

Giesecke, Frederick E.; Alva Mitchell; Henry C. Spender; Ivan L. Hill; Robert O. Loving; and John T. Dygdon. *Principles of Engineering Graphics*. New York: Macmillian, 1990.

Goodman, Daniel. *HyperCard Developer's Guide*. New York: Bantam Books, 1990.

Hall, Marty. *One Stride Ahead: An Expert's Guide to Cross-Country Skiing*. Tulsa, OK: Winchester Press, 1981.

Helyar, Pamela S. "Product Liability: Meeting Legal Standards for Adequate Instructions." *Journal of Technical Writing and Communication* 22.2 (1992), 125–47.

Hook, Julius Nicholas. *The Grand Panjandrum: And 1,999 Other Rare, Useful, and Delightful Words and Expressions*. New York: Macmillan, 1980.

Instruction Manual/Owner's Manual, Stihl 028. Waiblingen, West Germany: Andreas Stihl, nd.

Juvinall, Robert C., and Kurt M. Marshek. *Fundamentals of Machine Component Design*. 2nd ed. New York: John Wiley and Sons, 1991.

Kurzweil, Ray. *The Age of Intelligent Machines*. Cambridge, MA: MIT Press, 1990.

MacroMind Director, Version 3.0: Interactivity Manual. San Francisco, CA: MacroMind, 1991.

Obermeier, Klaus K. "Computers and Their Frame of Mind—The Linguistic Component." *Humans and Machines*. Proceedings at the 4th Delaware Symposium on Language Studies, October 1992, University of Delaware. Ed. Stephanie Williams. Norwood, NJ: Ablex Publishing, 1985. pp. 154–65.

Szymanski, Robert A.; D. P. Szymanski; N. A. Morris; and D. M. Pulschen. *Introduction to Computers and Information Systems*. New York: Macmillan, 1991.

Instructions, Specifications, and Procedures

Walter and Corrie contributed significantly to this process in several ways. What was unique about this particular process was that they helped programmers write both programming documentation and certain types of programming reports that were sent to customers. Furthermore, Walter and Corrie participated in writing the product design specification. Finally, they wrote some of the user documentation and consulted with other technical writers—writers who had not participated in the team project—for help in writing the remaining user documentation for the product. In this capacity, they served as what they called scout writers—writers who went off into unknown territory and brought back technical knowledge that the other writers could tap into.

Stephen Doheny-Farina, *Rhetoric, Innovation, Technology: Case Studies of Technical Communication in Technology Transfers.* Cambridge, MA: MIT Press, 1992, p. 184.

Introduction

performance
the focus in instructions, specifications, and procedures on the performing of a task; usually involves a user's activities

procedures
descriptions of how to coordinate activities in which many people are involved

instructions
details of the specific steps that an individual must take to complete a task

specifications
standards that must be followed for a task to be completed safely and according to applicable codes

In Chapter 13, you learned how to describe objects, mechanisms, and phenomena as well as processes, procedures, and activities. In this chapter, we focus more on the **performance** aspects of processes, procedures, and activities, in particular users' actions. You may remember that the directions on how to clean a computer mouse (Example 13–14) centered on users' actions. We now look at instructions, procedures, and specifications—three related documents or parts of documents that focus on performance and frequently *support* technology. However, in the last half of this chapter we focus exclusively on the most common type of support document created by technical communicators—instructions.

Instructions, procedures, and specifications serve as important interfaces between people and technology, as these documents provide the information necessary for people to use products, complete procedures, and follow professional codes and standards. **Procedures** describe how to coordinate an activity in which many people may be involved. For example, procedures might be used to describe how operators of a printing press would coordinate their activities while the press is running: some people would apply the appropriate ink, others feed the blank paper into the press, and still others monitor the final folding and cutting of the printed material.

Instructions detail the specific steps that an individual would follow to complete a task. For example, the printing press operator responsible for folding and cutting would adjust certain mechanical parts of the press for each job. **Specifications** state what standards must be followed for a task to be completed safely and according to professional codes. For example, the press operators along with the customer would specify the paper weight and quality of ink for each printing press run, while unions or professional groups would specify the safety measures that operators must follow to avoid injury.

To see the importance of such documents, let's look at a very early example. In 1755, Benjamin Franklin recorded his directions for conducting experiments with electricity (Example 14–1). He wrote instructions for one of his most famous experiments—the one designed to prove that lightning and electricity were the same. In this experiment, Franklin flew a kite in a lightning storm and had a metal key attached to the kite string to attract a lightning strike. Franklin was able to prove his hypothesis that lightning and electricity are related, but others who sought to duplicate his experiment by following his directions were electrocuted when the lightning strike passed down the damp string to their bodies. Franklin just may have been lucky, or he may have omitted some important information from his instructions that he may have considered self-evident. If Franklin had published his instructions today, he certainly would have had to include cautions and warnings as well as some illustrations indicating how to shield one's hands from the damp string along which the electric current moved.

EXAMPLE 14–1 Benjamin Franklin's Instructions

Preparation

Fix a tassel or fifteen or twenty threads, three inches long, at one end of a tin prime conductor (mine is about five feet long, and four inches in diameter) supported by silk lines.

Let the threads be a little damp, but not wet.

Experiment I

Pass an excited glass tube near the other end of the prime-conductor so as to give it some sparks, and the threads will diverge.

Because each thread, as well as the prime-conductor, has acquired an electric atmosphere, which repels and is repelled by the atmospheres of the other threads: if those several atmospheres would readily mix, the threads might unite, and hang in the middle of one atmosphere, common to them all.

Rub the tub afresh, and approach the prime-conductor therewith, crossways, near that end, but not nigh enough to give sparks, and the threads will diverge a little more.

Because the atmosphere of the prime-conductor is pressed by the atmosphere of the excited tube, and driven towards the end where the threads are, by which each thread acquires more atmosphere.

Withdraw the tube, and they will close as much. . .

Source: Benjamin Franklin, "Electrical Experiments, dated 1753," *Benjamin Franklin's Experiments*, ed. Bernard Cohen. Cambridge: Harvard UP, 1941, p. 303. Used with permission of the editor.

To create safe, accurate, and complete instructions, procedures, and specifications, you need to begin by creating **product and process information.**

product and process information
information from which to base instructions, gathered during the early stages of design and development

Creating Product and Process Information

Ideally, whenever you create instructions, procedures, or specifications, you will be involved in the early stages of design and development of the technology that your document supports. For example, if you must write instructions on how to use a particular computer graphics package, you should be part of a team that creates the computer code for the package. In that role, you can follow and even influence the design and development stages. Often the person writing such instructions becomes the **user's advocate.** For example, you might suggest a 3-D option be added or the spreadsheet be simplified. Or you might maintain a list of requirements that must be followed in all of the next stages of design and development to maintain quality control. You are likely then to find yourself part of a design team

user's advocate
a representative of the user's needs and interests

that includes technical experts, human factors people, marketing experts, writers, and graphic artists. Finally, whatever documents you create must be ready when the product or process they support is ready for the market or for implementation.

To ensure that their products and processes are well documented, some companies carefully map out the stages of development. Example 14-2 shows one company's processes roadmap. In this company, **product documentation** starts about the same time as **product development,** about the same time as the first marketing review within the project management plan (PM). Stages in this roadmap include gathering information, scheduling development and testing, writing, and distributing.

GATHERING INFORMATION

Before you can begin your instructions, specifications, or procedures, you will need to know a lot about the product or process at hand. Often, the product's **functional specifications** will provide you with essential information. Functional specifications describe in detail exactly what the product will do and how it will perform. In some cases you might be asked to write these specifications; in other cases, you'll study these specifications before writing instructions or procedures for users. Proposals, diagrams and schematics, marketing information, and interviews and surveys are also important sources of information. You must study what has been written already about the product or process before you begin your own writing.

Once the design is relatively firm, you can complete a **task analysis** of the product or process. A task analysis systematically analyzes performance. And, at this stage, task analysis might identify what the product or process will do, not necessarily what tasks a user might be doing when using the product or completing the process. Example 14-3 lists some questions that one writer answered about task analysis of a computer graphics program.

To get a full picture of the product or process you are documenting, you need to ask questions not only about the specific tasks involved but also about the environment in which the tasks will be performed, the equipment or materials needed to carry out the task, the skills or knowledge needed to complete the task, and the time and cost constraints on the tasks.

If you develop your information as the product or process itself is being developed, you will constantly update your information as the product or process changes. If you become involved after design is well along, you will have to work harder to get information because many of the people involved in development will have moved on to their next project and may be less enthusiastic about discussing an "old" project with you. In that case, you'll need to schedule extra time for hunting down information.

---

product documentation
process of creating the instructions and other documents that will accompany a new product release; in the best circumstances, it is cotemporal with product development

product development
process of creating and testing a new product

functional specifications
detailed descriptions of exactly what a product will do and how it will perform

task analysis
systematic analysis of product performance

EXAMPLE 14–2 Processes Roadmap

Processes Roadmap

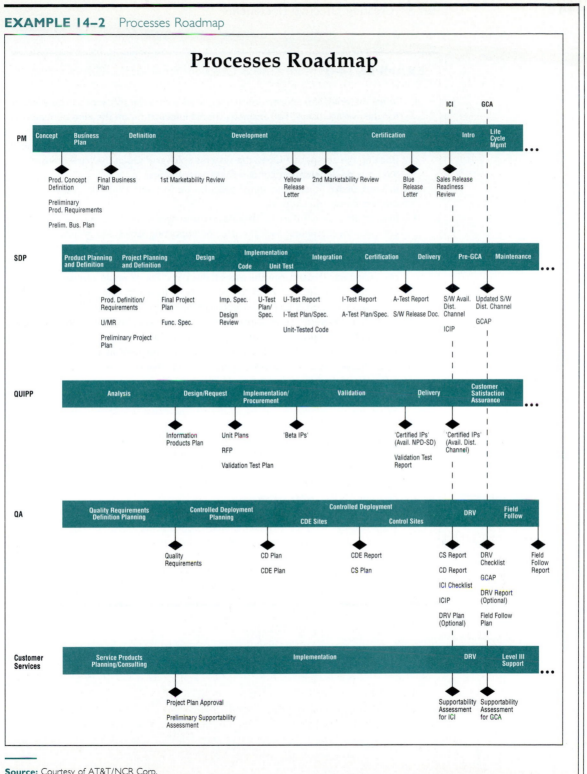

Source: Courtesy of AT&T/NCR Corp.

EXAMPLE 14–3   Task Analysis Questions for a Specific Graphics Program

1. What materials and information are required before the software can be effective for the average users? For example, should information on the kind of numerical data needed for a spreadsheet be indicated in the documentation?

2. What tasks must be completed before this program will become operational? Boot up the computer? Open the systems software? Open the hard drive menu? Will our documentation contain these instructions?

3. In what order must these initial tasks be performed? Will our documentation contain these instructions?

4. In what situation will the user depend on the program? Will the user be maintaining complex spreadsheets? Will the user be importing tables and graphs into text? Will the user be creating visual displays for oral presentations?

5. What specific skills will the user need to complete the task? Will other aspects of our software have taught the users these skills? For example, mouse operation? Menu use? Will the user need to learn how to format a spreadsheet within our documentation, or is that knowledge assumed?

6. What knowledge of visual display will be needed, e.g., how to choose between tables, charts, and graphs? Will our documentation contain these principles?

7. What steps are needed in the creation of a spreadsheet? In the creation of the visual display? Are these steps separate or do they overlap? Will the design of a specific spreadsheet determine which visual display is best for the data?

8. What will the final product look like? Will the program include color options? What textual elements can be added to the visual display—titles, axis labels, arrow labels?

9. What warnings and cautions should be added? For example, will the user's personal computer need a certain amount of memory to run the program? Will the user be able to run the graphics program and a word processing program at the same time? How often should the user save the spreadsheet while choosing between visual display options?

10. How much time will be required to complete both the spreadsheet and the visual display? Should the documentation contain information on how much time will be needed to create the spreadsheet so the user need not stop in the middle of the task?

11. With the suggested options for the program, how much cost will be involved in developing the documentation? How much will the final product cost the user?

IDENTIFYING YOUR AUDIENCE AND PURPOSE

Identifying the *audience* and *purpose* of instructions, specifications, and procedures is essential. If you provide instructions that users of a product cannot understand, the users may damage the equipment or endanger their lives. If you leave out important specifications on how to maintain a product, parts might need to be replaced too frequently. If you fail to envision everyone who needs to follow a procedure, you might address only one group of readers, leaving another group frustrated or confused. Your documents must be tailored to your audience's level of technical knowledge and expertise, their situation and environment, and their attitudes and education.

You will need to address at least the following general audience needs when creating your documents:

- Prior knowledge about the process and product, if needed.
- Prior experience with the process and product, if needed.
- Amount of detail.
- Number of examples.
- Attention-getting devices.
- Justification of individual steps.
- Level of vocabulary.
- Placement and type of definitions and descriptions.
- Number and type of visual displays.
- Document design features.

Instructions, specifications, and procedures all assume that the audience will be involved directly with the process or product.

Audiences and Purposes of Procedures

Procedures frequently describe how a group of people are to cooperate in various situations or operations. If you think about how firefighters handle large residential blazes, how companies deal with the removal of asbestos from buildings, or how kidney transplants are coordinated, you can imagine what kind of audiences and activities you address in procedures. Often your audience consists of individuals who have been trained to do specific jobs, all of which are an important part of a procedure. The procedures you create tell them how to accomplish the overall goal of the activity. Your document may be used to clarify, standardize, or modify existing procedures, or your document may be used to train new employees who must see how their work affects others.

Often a procedure outlines who is responsible for what so that a larger goal can be reached. Example 14–4 describes how a plan administrator, an insured person, an attending physician, and a specialist all participate in a common goal—making sure that the cost of a second opinion is justified and therefore covered by an insurance company. To create a procedure that

---

EXAMPLE 14–4 Second Opinion Procedure

When a program is established, it is not necessary to publish a list of the procedures on which a second opinion will be required. Instructions to the insured simply state that when he or she is told by the attending physician that surgery of a nonemergency nature is required, the insured should promptly contact the plan administrator in person or by phone. The insured will be asked to furnish the administrator with the name of the attending physician, his or her own version of the medical problem involved, the date surgery is scheduled, the insured's address and phone number, and a choice of two preferred appointment dates and most convenient times.

On the basis of this information, the plan administrator decides whether a second opinion is necessary. If the administrator feels that it is, he or she consults the directory of specialists, picks the appropriate specialist, and makes an appointment for the examination. The administrator then notifies the claimant of the appointment time and sends him or her a presurgical screening form. The presurgical screening form must be completed promptly by the attending physician and the insured. The insured is required to authorize release of necessary information to the specialist. The attending physician is asked to indicate the surgical procedure, diagnosis, pertinent history, X-rays, and laboratory findings, where the procedure will be performed, expected date of surgery, date of hospital admission, expected length of hospital stay, and total surgical care, including aftercare.

This form is submitted to the specialist who will make the second examination. He or she will consider not only the need for the surgery but also the qualifications of the specialist who plans to do it. If hospitalization is planned, the specialist will also determine whether such hospitalization and expected duration are necessary. The specialist must have access to the X-rays and lab work done by the attending physician. Prior approval of the plan administrator is needed before additional tests can be made.

Source: Excerpts adapted from Mary M. Lay, *Strategies for Technical Writing: A Rhetoric with Readings.* Copyright © 1982. New York: Holt, Rinehart and Winston, Inc. p. 201. Reprinted with permission of the publisher.

---

all readers could understand, the writer of Example 14–4 had to describe the interaction between each person involved chronologically.

Audiences and Purposes of Instructions

Unlike the audience of procedures that needs to know its special role in a larger operation, the audience of instructions may be a single person completing an entire task. Your audience for instructions may be the general consumer or the expert who wants to use your documents to operate, maintain, assemble, install, and repair devices. In Example 14–5, the writer instructing someone how to use a cellular telephone imagined that the audience had little knowledge of the device but would understand certain terms, such as memory system and storage methods.

EXAMPLE 14–5 Instructions for a Cellular Telephone

Memory Dialing

Your telephone provides an alphabetic and numeric memory system capable of storing up to 99 frequently called names and phone numbers. Before setting up your memory, we suggest you read the rest of this section to become familiar with the ways in which you will access and use it.

Enter Memory (Alpha Entry Mode)   Telephone numbers can be entered into the memory three different ways through the Alpha Entry Mode. Choose one of the following storage methods and use the directions given to enter the information you wish to store.

Source: Motorola Inc., Pan American Cellular Subscription Group.

We look at instructions in greater detail in the second half of this chapter.

Audiences and Purposes of Specifications

Specifications are documents that designate the required methods and materials for a projected work, such as plans for creating a landfill or constructing a synagogue. Specifications are often used to provide information to independent firms when bids are being solicited for construction or reclamation projects. Audiences for specifications are almost always experts, and when you create specifications for a new project or product, you must follow standard specifications formats because specifications are legally binding. Example 14–6 shows specifications used for state highway construction. You can see that the audience must be familiar with the construction materials and methods.

PLANNING YOUR DOCUMENTS

Although you know a great deal about planning already, when planning instructions, procedures, and specifications, you will find that design, organization, and structural decisions will often be made for you. A number of companies and funding agencies (such as the US Government) have style manuals or specification guidelines that dictate what your information must look like. Example 14–7 shows military specifications for the general style and format of technical manuals. These specifications dictate everything from page sizes to acceptable symbols.

Many companies have a style manual to which all documents written for the company must conform. We have already mentioned the ISO-9000

EXAMPLE 14–6   The Audience of Specifications

The concrete in areas to be refinished shall be removed to a depth of approximately ¾ inch below the adjoining surface. A cut ¾ inch deep shall be made with an abrasive saw along the perimeter areas prior to removing concrete.

   Concrete removal shall be done by abrasive blast cutting, abrasive sawing, impact tool cutting, machine rotary abrading, or other methods. The method used to remove concrete shall be approved by the engineer. Cut areas shall be cleaned free of dust and other loose and deleterious materials by brooming and high pressure air jets.

   Epoxy adhesive confirming to the adhesive required for bonding dowels shall be applied to the surfaces to be refinished before placing Portland cement concrete filling.

Source: State Highway Construction, Department of Transportation, State of California.

Series of International Standards, which have an impact not only on quality and consistency in the processes used to develop products and services but also on their supporting documents. When style manuals do not exist, you may need to create one of your own. Your style guide may address language questions such as the spelling of *data base* (one word or two?) and hyphenation of such words as *four-wheel drive.* Follow the suggestions in Chapters 10 and 11 of this textbook when developing your style manual.

EXAMPLE 14–7   Military Specifications for Design

3.1.4 Manual outline. When specified in the contract or order, a manual outline shall be provided and shall contain the following:

   a.   A text plan that shall be in accordance with the requirements of the technical content specification, showing paragraph titles to indicate the intended coverage of the various aspects of the equipment or system. Each paragraph title or notation shall be followed by a brief statement outlining the information to be presented.
   b.   An illustration plan and a table plan that shall be keyed to the text plan. Each illustration and table listed in the plans shall be described. The illustration plan shall contain figure numbers, title, information, intent, approximate size and nature of illustration (halftone, schematic, line drawing). The table plan shall describe the tables by table number and information content . . .

Source: Military Specifications M-M-38784A Technical Manuals.

Because your audience may actually carry out the task while reading your instructions, you will need to address them directly when writing instructions. Your instructions must be simple and clear as well as accurate, complete, and safe. In the past 20 years or so, the concept of **user-friendly** manuals has demanded that instruction manuals be oriented according to what a user does with the product; instructions, then, are **task-oriented.**

Suppose you are writing a user guide for a word-processing program. In your task analysis you discover that one essential task is saving a file. You need to anticipate the steps related to this task by answering the following questions:

- How is a file different from a folder?
- Is there more than one way to save a file?
- How often should a file be saved?
- How many copies can the user save, including backup copies?
- Where (in what folder) will the file be saved? Can this location be changed?
- What happens if a user forgets to save the file before turning off the computer?
- What's the difference between saving on the hard drive and saving on a disk?
- When in the process of saving a file should the file be named?

Generally, because instructions guide a person through a task, instructions are arranged **chronologically,** as in Example 14-8. However, in troubleshooting instructions, you might direct a user back and forth through a set of alternative steps, as shown in Example 14-9.

The amount of detail needed in your instructions depends on your audience's experience with your product or process. When you write for the general consumer, you must make it possible for your reader to carry out the task he or she wishes to perform efficiently and safely. The audience will often need to know why each step is important—and even why using the device properly or completing the process exactly is essential.

You should start your instructions with the earliest possible action and include every action, no matter how obvious it might seem to you. Place in the beginning of the instructions all parts and materials needed to complete the operation. Notice the simple, clear illustrations that accompany the Macintosh Powerbook™ (Example 14-10). The reader knows why each step is important and what could damage the computer. Action verbs appear in the imperative voice, and the visual displays illustrate not only the device but also the user's actions.

You can begin your instructions with definitions and other contextual information. The American Lung Association begins its instructions on installing a carbon monoxide detector with a definition of CO, the conditions in

Creating Effective Instructions

user-friendly
concept of creating instruction manuals (and other documents) oriented toward the user's needs

task-oriented instructions
user-friendly instructions designed to help users perform the tasks they wish

chronologically arranged instructions
instructions arranged in the order a user will perform a task

---

EXAMPLE 14–8 Chronological Organization in Instructions

Cleaning Fuel Tank Filter

If the fuel tank filter is obstructed and requires cleaning, proceed as follows: To remove filter:

- Empty fuel tank.
- Remove fuel line tie strap (cut strap) and disconnect fuel line.
- Unscrew shut-off valve and filter assembly from tank.
- Wash filter element with clean solvent and a brush.

Source: *Evinrude Owner's Operator's Manual 1994. Outboard Marine Corp., Waukegan, IL.*

---

which the instructions are needed, the justification for completing the whole process, the locations affected, and the timing (Example 14–11).

When writing instructions for expert audiences, you may ask your audience to judge whether an action has been successful before moving on to the next step. Your instructions might also incorporate actions that the audience has already been trained to perform elsewhere. Example 14–12 addresses physicians who would treat patients poisoned by a certain pesticide. The instructions assume that the physician knows how to clear an

---

EXAMPLE 14–9 Troubleshooting Instructions for a Vacuum Cleaner

Problem	Possible Cause	Possible Solution
Cleaner won't run.	1. Not firmly plugged in. 2. No voltage in wall plug. 3. Blown fuse/tripped breaker.	1. Plug unit in firmly. 2. Check fuse or breaker. 3. Replace fuse/reset breaker.
Cleaner suction low	1. Bag full 2. Obstruction in nozzle connector, hose, or wand. 3. Secondary filter dirty. 4. Suction regulator improperly set. 5. Hose not properly connected to cleaner.	1. Replace paper bag. 2. Remove obstruction. 3. Clean filter 4. Reset suction regulator. 5. Ensure hose is properly connected. . .

Source: *Hoover Futura Cleaner Owner's Manual, p. 19. Used with permission of the Hoover Company.*

EXAMPLE 14–10 Simple, Detailed Instructions for the General Consumer

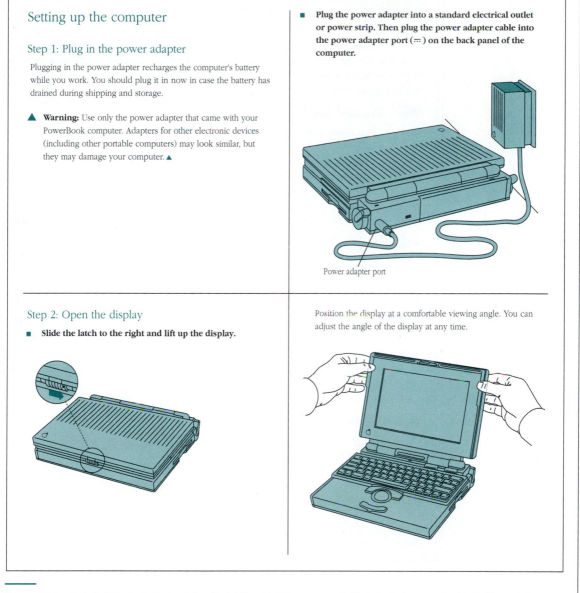

Setting up the computer

Step 1: Plug in the power adapter

Plugging in the power adapter recharges the computer's battery while you work. You should plug it in now in case the battery has drained during shipping and storage.

▲ **Warning:** Use only the power adapter that came with your PowerBook computer. Adapters for other electronic devices (including other portable computers) may look similar, but they may damage your computer. ▲

■ **Plug the power adapter into a standard electrical outlet or power strip. Then plug the power adapter cable into the power adapter port (⎓) on the back panel of the computer.**

Power adapter port

Step 2: Open the display

■ **Slide the latch to the right and lift up the display.**

Position the display at a comfortable viewing angle. You can adjust the angle of the display at any time.

Source: *Macintosh User's Guide for the Macintosh PowerBook 160 and 180 Computers*, pp. 2–30, © Apple Computer, Inc. Used with permission.

EXAMPLE 14–11   Project CO

What Is Carbon Monoxide (CO)?

Toxic, odorless, tasteless, colorless, gas produced by incomplete combustion of gasoline, kerosene, natural gas, butane, propane, fuel oil, wood, coal, or anywhere fossil fuels are burned. Dangerous levels of CO come from internal combustion engines such as cars and lawn mowers or from improperly adjusted or poorly ventilated fuel-burning appliances.

What Should I Test?

Flu or virus is often blamed when in fact CO poisoning is the case. Even small amounts of CO can cause headaches, dizziness, shortness of breath, anxiety, and irritability. Severe heart and brain damage and even death can result from prolonged exposure to CO. Infants and small children are at increased risk due to lower body mass. Pregnant women should avoid exposure. CO can cause premature birth, low birth weight, and increased risk of miscarrige. CO has been linked to learning and memory defects in a developing fetus.

Where Should I Test?

Home, kitchen, furnace area, garage, car, RV, camper, boat, airplane, cottages, tents, wood stoves, fireplaces, spaceheaters, water heaters.

How Do I Test?

- Remove from sealed packet only when ready to use.
- If CO is present, the spot will turn gray or black.
- Spot will return to original color when air freshens.
- If spot darkens after several months, replace.
- Replace at least once per heating season.
- Shelf life in unopened pack is three years.

How Fast Does It Work?

Concentration	*Reaction Time*
100 ppm (.010%)	15–45 minutes
200 ppm (.020%)	4–15 Minutes
400 ppm (.040%)	2–4 Minutes
600 PPM (.060%)	1–2 Minutes

Source: American Lung Association of Hennepin County and Hennepin County Medical Center.

EXAMPLE 14–12   Instructions for an Expert Audience

1. Establish CLEAR AIRWAY and TISSUE OXYGENATION by aspiration of secretions, and if necessary, by assisted pulmonary ventilation by oxygen.
2. CONTROL CONVULSIONS. The anticonvulsant of choice is DIAZEPAM (VALIUM). Adult dosage, including children over 6 years of age or 23 kg in weight: inject 5–10 mgm (1–2 ml) slowly intravenously (no faster than one ml per minute), or give total dose intravenously (deep). Repeat in 2–4 hours if needed. Dosage for children under 6 years or 23 kg in weight: inject 0.1–0.2 mgm/kg (0.02–0.04 ml/kg) slowly intravenously (no faster than one-half total dose/minute), or give total dose intramuscularly (deep). Repeat in 2–4 hours if needed.

CAUTION: Administer intravenous injection slowly to avoid irritation of the vein, occasional hypertension, and respiratory depression. Because of a greater tendency to cause respiratory depression, BARBITURATES are probably of less value than DIAZEPAM. One used successfully in the past is PENTOBARBITAL (NEMBUTAL). Maximum safe dose: 5 mgm/kg body weight, or 0.20 ml/kg body weight, using the usual 2.5% solution.

Source: Donald P. Morgan, *Recognition and Management of Pesticide Poisoning.* 2nd ed. Washington, D.C.: US Environmental Protection Agency, August 1977, p. 6.

airway, will recognize the abbreviations and medications, and can make choices as to what steps or actions pertain to any particular case.

SPECIAL TYPES OF INSTRUCTIONS

Instructions for many products include **templates** that fit over keyboards or on control panels of equipment to remind users about functions of a device. Most computer users are familiar with the templates that fit over the computer keyboard for many word-processing programs. Example 14-13 illustrates a template explaining the use of voice mail that fits right over the face of a touch-tone telephone.

Quick reference cards provide enough information to get users started and to keep users performing basic tasks. Look back at Example 11-7 to see a quick reference card. **Tutorials,** on the other hand, are extended instructions that help beginning users understand a product's functions. When online, tutorials are sometimes called **computer-based training** (CBT). The tutorial shown in Example 14-14 takes users step-by-step through an exercise that teaches them how to insert text while they are editing documents. Notice that the tutorial not only directs users' actions—"position," "type," and "add"—but also creates sample documents for users to

templates
overlays that fit on keyboards or control panels to remind users about the functions of a device

quick reference cards
single-page reference documents to get users started and to provide basic information about a product

tutorials
extended instructions that help beginning users understand a product's functions

computer-based training (CBT)
step-by-step training exercises designed for the user to perform on the computer

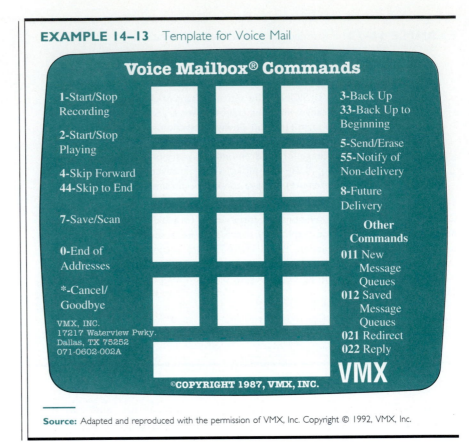

EXAMPLE 14–13 Template for Voice Mail

Voice Mailbox® Commands

1-Start/Stop Recording

2-Start/Stop Playing

4-Skip Forward
44-Skip to End

7-Save/Scan

0-End of Addresses

*-Cancel/ Goodbye

VMX, INC.
17217 Waterview Pkwy.
Dallas, TX 75252
071-0602-002A

3-Back Up
33-Back Up to Beginning

5-Send/Erase
55-Notify of Non-delivery

8-Future Delivery

Other Commands

011 New Message Queues

012 Saved Message Queues

021 Redirect
022 Reply

VMX

©COPYRIGHT 1987, VMX, INC.

Source: Adapted and reproduced with the permission of VMX, Inc. Copyright © 1992, VMX, Inc.

---

practice on. Finally, users are shown how the document should appear if the users have followed the tutorial successfully.

User guides provide step-by-step instructions for completing all the tasks associated with a product or process. Although the tutorial shown in Example 14–14 helped the *beginning* user insert some *sample* text into a document, the passage from the user guide shown in Example 14–15 tells *any* user how to add *any* text to any document.

Reference guides are a kind of glossary. They list in alphabetical order the information users will need to know to continue using the product. Usually a reference guide describes functions of the product rather than tasks users will perform. Although most reference guides include some step-by-step instructions, these instructions are usually listed by product function rather than user task. Example 14–16 shows a page from a reference guide that explains the autohyphenation checker in a word-processing program. Reference guides generally contain much more technical information than

user guides
step-by-step instructions for completing all tasks associated with a product or process

reference guides
glossaries containing information arranged alphabetically that users will need to know to use a product

EXAMPLE 14–14 Tutorial

Your document should look like the following illustration.

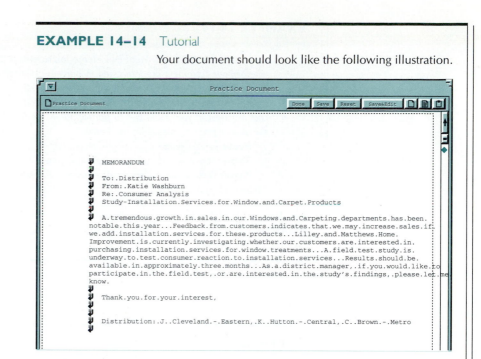

Inserting Text

You can insert additional text at any time.

1. **Position** the pointer on the O in the word MEMO and **click** the Select mouse button.

   The caret appears to the right of the point where you clicked the mouse button.

2. **Type:** *RANDUM* and **press** RETURN.

   The word "MEMORANDUM" is now completely spelled out.

   Characters are always added to the right of the point where you click.

   A non-printing paragraph character is also added, and the caret moves down to appear after the paragraph character.

3. **Position** the pointer on the Paragraph character after the word "Distribution" and **click** the Select mouse button.

4. **Add** the following text to complete the memo. Press RETURN only where indicated.

   Type: *From: (Your name)* and **press** RETURN.

   Type: *Re: Consumer Analysis Study—Installation Services for Window and Carpet Products* and **press** RETURN twice.

   Type (without pressing RETURN): *A tremendous growth in sales in our Windows and Carpeting departments has been notable this year. Feedback from customers indicates that we may increase sales if we add installation services for*

EXAMPLE 14-14   *(continued)*

these products. Lilley and Matthews Home Improvement is currently investigating whether our customers are interested in purchasing installation services for window treatments. A field test study is underway to test consumer reaction to installation services. Results should be available in approximately three months. As a district manager, if you would like to participate in the field test, or are interested in the study's findings, please let me know.

Press RETURN twice.

Type: *Thank you for your interest.* and **press** RETURN three times.

Type: *Distribution: J. Cleveland—Eastern, K. Hutton—Central, C. Brown—Metro* and **press** RETURN.

Source: Examples of instructions from XSoft software manuals are copyright of Xerox Corporation and are reprinted here with permission.

EXAMPLE 14–15   User Guide

To add text to existing text:

1. Move the mouse pointer to the position in the text where you want to enter new text.
2. Click the Select mouse button to place the caret.
3. Type the new text.

You can use the AGAIN key to repeat text entry operations, including inserting paragraph, new line, and tab characters, and backspacing over characters. The system "remembers" up to 100 keystrokes entered since the last time you made a selection. This feature is useful for repetitive text entry in tables or forms. It works within a document and between documents.

For example, if you want to enter a name and address in several different places on a form, or on different forms, you can type the name and address once and use the AGAIN key to enter it in the other places.

To repeat a text entry operation:

1. Type the text or characters.
2. Move the pointer to a new location for the text or characters and click the Select mouse button.
3. Press AGAIN.

Source: Examples of instructions from XSoft software manuals are copyright of Xerox Corporation and are reprinted here with permission.

EXAMPLE 14–16 Reference Guide

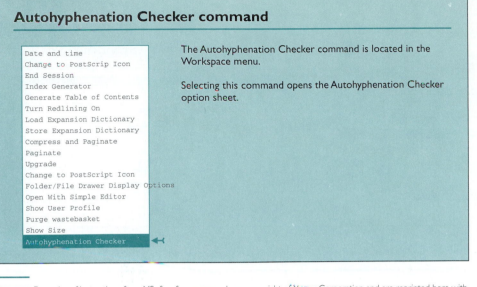

Autohyphenation Checker command

Date and time
Change to PostScrip Icon
End Session
Index Generator
Generate Table of Contents
Turn Redlining On
Load Expansion Dictionary
Store Expansion Dictionary
Compress and Paginate
Paginate
Upgrade
Change to PostScript Icon
Folder/File Drawer Display Options
Open With Simple Editor
Show User Profile
Purge wastebasket
Show Size
Autohyphenation Checker

The Autohyphenation Checker command is located in the Workspace menu.

Selecting this command opens the Autohyphenation Checker option sheet.

Source: Examples of instructions from XSoft software manuals are copyright of Xerox Corporation and are reprinted here with permission.

do user guides. For example, they may describe the internal operation of a product and the technical information needed for its maintenance and repair.

Online help provides on the computer screen the information found in other types of instructions. Some online systems provide **context-sensitive help;** that is, information presented on the screen will be related to the procedure you are performing at the time of calling on the help screens. Other help systems are linear and are simply a computerized version of what you would find in a paper manual. These linear online instructions are often not very effective because users lose their place in the online document easily in the absence of the types of visual cues (table of contents, headers and footers, etc.) found in a paper manual. Newer **hypertext** helps are not linear in nature and allow a user to work through a web of interlinked information.

Generally speaking, we have not yet developed hypertext or online instructions sufficiently for them to be more effective than paper manuals in all contexts. So, if you are preparing support documents for a complex product, you may have to develop several different types of instructional material. Example 14–17 shows a roadmap to all the instructional material prepared for the *GLOBALVIEW for X Windows* by Xerox Corporation. This

online help
information on the computer screen; can be context sensitive

context-sensitive help
online help systems that present information on screen that is related to the procedure being performed at the time when the help is called up

hypertext
nonlinear arrangements of information on a computer; information is interlinked in a weblike structure

---

EXAMPLE 14–17 Roadmap

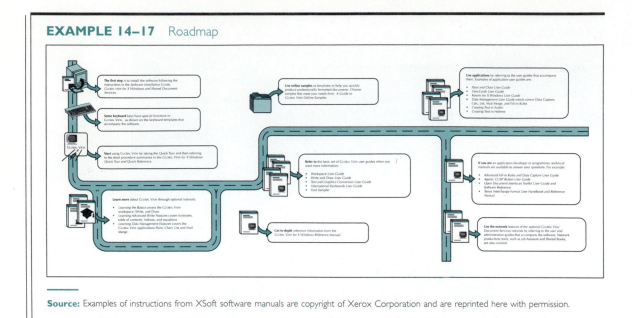

Source: Examples of instructions from XSoft software manuals are copyright of Xerox Corporation and are reprinted here with permission.

---

pamphlet gives a tour of all the information products so that users can choose which one best suits their needs.

ORGANIZATION OF INSTRUCTIONS

Creating an outline of the information you will include in your instructions helps you decide what information goes where. You can also keep in mind two guiding principles:

- Users bring with them expectations about what they will find in instructions, and
- Users rarely read instructional materials from the beginning to the end.

In fact, typical users, when confronted with a new product, attempt to use it without referring to any instructional help. They may find someone else to help them operate the product or answer questions, perhaps a colleague who is more familiar with the product. Users may call the technical assistance number provided by manufacturers to get help with a particular problem. As a last resort, they refer to the instructional material, looking for an answer to the particular problem they need solved.

   Make your information easy to find for this user by **chunking** your information in a logical way; that is, cluster information according to tasks,

chunking
clustering information by task, problem, or situation so that a user can navigate it more easily

problems, or situations. Pay attention to the headings you give these chunks of information. Although your headings may change as you write and revise your instructions, taken together they represent the first draft of your table of contents, an important point of access used by readers. It is at this level of planning that structural decisions can be switched around easily, and organizational flaws, such as omission or slighting of significant information, can become glaringly obvious.

Users need to find certain kinds of information in instructions. Therefore, your outline for complete instructions will generally include the following components:

- *Introduction*—Your introduction provides an overview of the instructions for the user. Sometimes these introductions are given such titles as "How to Use This Manual."

    If applicable, your introduction should identify the specific product by the product name, model number, and trademark. Just as you identify the product with which you are dealing, your introduction should define the assumed audience of the instructions, the background knowledge the user is expected to have, and any requirements the user needs to have (such as licenses, certifications, and so on).

    Use the introduction to provide the reader with a roadmap for navigating the rest of the manual. Explain the conventions you are following and how any special information such as warnings and cautions will be displayed in the manual itself. List what materials and equipment the reader needs to complete the instructions, and, if the reader is constructing something, show an illustration or give a description of that final product.

- *Description of the mechanism*—Your manual may require a section that tells your readers the names of the parts they will be using to perform their tasks. This section then defines as it describes, emphasizing not what these parts do, but instead what they are. Chapter 13 discusses in detail how to write definitions and descriptions, and you will want to review its recommendations. When writing instructions, chunk this kind of information so readers can find the definition of a particular part they may be looking for and a description of where to find it and what it looks like.

- *Theory of operation*—In instructions, most readers expect to find out how a product works. Most nontechnical readers will be satisfied with a description of the mechanism in process. Example 14-18 contains a sample paragraph from an explanation to a nontechnical audience about how a diamond rocksaw works. Although the paragraph uses technical information, a novice user would be able to get a general idea of what to expect. More technical audiences, of course, would be impatient with such information and prefer to hear more about the theoretical principles

---

EXAMPLE 14–18 Theory of Operation for a Rocksaw

Diamonds are the hardest substance on earth and can cut anything. Each rocksaw blade has many tiny, industrial-grade diamonds embedded in its edge. A sliding carriage, mounted beside the sawblade, holds the rock securely. The carriage is part of a passive counterbalance system. A 7-lb. counterweight pulls the carriage and rock past the blade. The friction of the blade against the rock produces tremendous heat. Therefore, a nonvolatile cutting oil bathes the blade to remove the heat. The oil also carries the debris away from the cut.

---

of the product's internal operations. For electronic equipment, for example, expert audiences would prefer to see schematics of circuitry and formulas or algorithms upon which its operation is based.

■ *Step-by-step actions*—These are at the heart of all instructions. These step-by-step instructions are generally based on user task analysis, as discussed earlier.

Your instruction manual is likely to have other sections too, such as a troubleshooting guide and suggestions for maintenance and care. Navigational aids, such as tables of contents and indexes, help users find what they are looking for. If your instructions contain terms that may be unfamiliar, your manual should also have a glossary.

VISUAL PRESENTATION

Visual displays in instructions help readers orient themselves with the product before them. You can use callouts to label each part of a product, as in Example 14–19, the illustration of the cleaner referred to in Example 14–9. If you use this technique, you will want to call the reader's attention to it by referring to the numbering system in your prose. You would, for example, tell your reader when describing the power nozzle, "The power nozzle (labeled *1* in Figure 13–24) . . ." Or, if you have already told the reader that you have labeled parts by numbers, you can then just refer to the number itself: "The power nozzle (1) . . ."

Plan the visual displays for your instructions as you plan your content. Depending on the level of your audience, you will be using visual displays to augment your writing at ratios as high as 60 percent prose to 40 percent visual displays. Some instructions, especially those aimed at international audiences, are 100 percent visual displays. Most airlines, for example, present their safety and evacuation procedures visually, making them understandable to a nonreading audience. When planning pictures of people, remember to be inclusive, representing both genders as well as ethnic diversity.

EXAMPLE 14–19   Callouts

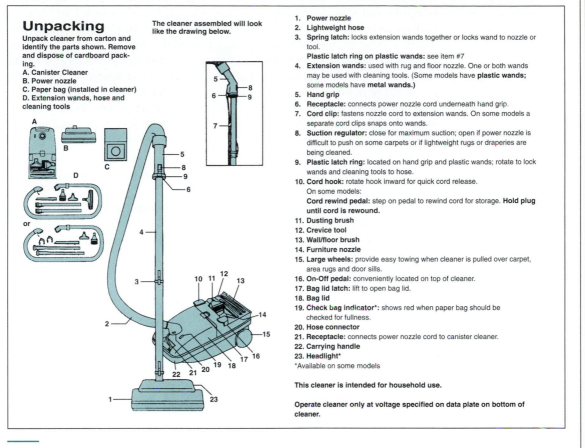

Unpacking

Unpack cleaner from carton and identify the parts shown. Remove and dispose of cardboard packing.

A. Canister Cleaner
B. Power nozzle
C. Paper bag (installed in cleaner)
D. Extension wands, hose and cleaning tools

The cleaner assembled will look like the drawing below.

1. **Power nozzle**
2. **Lightweight hose**
3. **Spring latch:** locks extension wands together or locks wand to nozzle or tool.
   Plastic latch ring on plastic wands: see item #7
4. **Extension wands:** used with rug and floor nozzle. One or both wands may be used with cleaning tools. (Some models have **plastic wands;** some models have **metal wands.**)
5. **Hand grip**
6. **Receptacle:** connects power nozzle cord underneath hand grip.
7. **Cord clip:** fastens nozzle cord to extension wands. On some models a separate cord clips snaps onto wands.
8. **Suction regulator:** close for maximum suction; open if power nozzle is difficult to push on some carpets or if lightweight rugs or draperies are being cleaned.
9. **Plastic latch ring:** located on hand grip and plastic wands; rotate to lock wands and cleaning tools to hose.
10. **Cord hook:** rotate hook inward for quick cord release.
    On some models:
    Cord rewind pedal: step on pedal to rewind cord for storage. **Hold plug until cord is rewound.**
11. **Dusting brush**
12. **Crevice tool**
13. **Wall/floor brush**
14. **Furniture nozzle**
15. **Large wheels:** provide easy towing when cleaner is pulled over carpet, area rugs and door sills.
16. **On-Off pedal:** conveniently located on top of cleaner.
17. **Bag lid latch:** lift to open bag lid.
18. **Bag lid**
19. **Check bag indicator*:** shows red when paper bag should be checked for fullness.
20. **Hose connector**
21. **Receptacle:** connects power nozzle cord to canister cleaner.
22. **Carrying handle**
23. **Headlight***
*Available on some models

This cleaner is intended for household use.

Operate cleaner only at voltage specified on data plate on bottom of cleaner.

Source: *Hoover Futura Cleaner Owner's Manual.* Used with permission of the Hoover Company.

STYLE IN INSTRUCTIONS

Alert your readers frequently to what your instruction manuals are doing. Introductions should provide overviews of your manual's content as well as the contents of each section. Diagrams help readers find the information they are looking for. A table of contents at the beginning of each section as well as the one at the beginning of the manual make information easy to find. Frequently these sectional tables of contents can be more detailed than the full table of contents, including all headings and subheadings in the manual.

Use headers and footers for all pages—they are effective navigational tools for readers to use as they move through the manual. Write meaningful headings and subheadings that chunk the content for readers. Use verbs or verbals rather than nouns in your headings and subheadings to emphasize action. Sometimes writers worry about insulting the intelligence of their readers by providing too much assistance. Don't worry. Remember that no one uses a manual who already knows the answers or where to find them.

Also, remember that perceptual aids reinforce learning. That is, people learn by doing, touching, acting, viewing, and hearing. The more you can engage all the senses, the more your audience will understand and retain:

- Address the reader by "you" or the imperative voice: "You must unplug the radio before servicing," or "Unplug the radio before servicing."
- Use active voice. Rather than "The bolt is placed on top," say, "Place the bolt on top."
- Select precise, action verbs (e.g., "place," "connect," "plug," "type").
- Number and indent steps.
- Use simple, direct statements.
- Write clearly so no requirement is ambiguous.
- Use standard symbols and abbreviations consistently.
- Use parallel structure.

WARNINGS AND CAUTIONS IN INSTRUCTIONS

Legally, companies are responsible for the safety of users of their products. Most companies require a legal review of their manuals to ensure the safety of users of the product, and, of course, to protect themselves from legal action. You should take special care to warn users of any dangers possible from the product or process. For example, Westinghouse's product safety label program uses color (red for danger, orange for warning, yellow for caution, and blue for notes) to classify the seriousness of the hazard. These colors are compatible with American National Standards Institute (ANSI) standards for warnings and cautions.

When writing safety instructions, you must indicate the seriousness of the potential hazard, the consequences of the hazard, and provide information about how to avoid the hazard. The terms *danger, warning,* and *caution* are not interchangeable and must be used accurately in your work. ANSI has developed standards for creating effective hazard alert messages, which we suggest you always follow (see Example 14-20).

ANSI specifies that all alert messages must contain the following seven basic elements to be effective:

EXAMPLE 14–20   ANSI Standards for Product Signs and Labels

DANGER indicates an imminently hazardous situation which, if not avoided, will result in death or serious injury.

WARNING indicates a potentially hazardous situation which, if not avoided, could result in death or serious injury.

CAUTION indicates a hazardous situation which, if not avoided, may result in minor or moderate injury.

Source: "American National Standard for Product Safety Signs and Labels," Z535.4, American National Standard Institute, 1989.

---

1. A hazard alert symbol.
2. A signal word.
3. The appropriate color: red, orange, or yellow.
4. The pictograph or icon.
5. The hazard identification.
6. The consequences of ignoring the hazard.
7. The description of how to avoid the hazard (Kemnitz 71).

Many manuals include sections that deal exclusively with safety. Usually these are set off graphically to call attention to them—surrounded by black or colored borders, for example. Warnings within the text are usually set off from surrounding material by placing them inside or outside the margins of the text. Placing the warnings in borders alerts readers to their presence. Be sure to remind your audience to heed these warnings by putting a cautionary statement to that effect on the cover of your instructions. Example 14-21 shows a warning and a caution developed to meet ANSI Z535 standards.

Common sense tells us to place warnings *before* telling users to perform an action that may cause damage or harm. Users following procedures take steps one at a time; thus, telling them to take care *after* you've told them to do something will not protect them from a potentially hazardous situation.

TESTING INSTRUCTIONS

Testing your instructions on actual users is an essential part of developing an instructional document. The first stage of testing comes in the design or development stage when you are working with a team of people, production experts, graphic artists, other writers, and subject matter experts. Use this team to provide feedback on what you write and ask them and others to review and comment on your drafts.

EXAMPLE 14–21   Sample Warning and Caution Meeting ANSI Z535
Standards

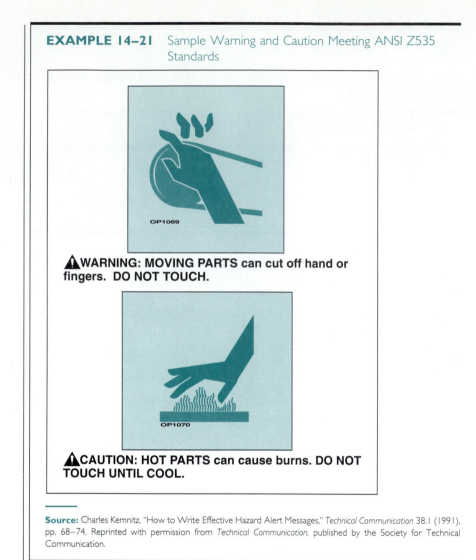

**⚠WARNING: MOVING PARTS can cut off hand or
fingers. DO NOT TOUCH.**

**⚠CAUTION: HOT PARTS can cause burns. DO NOT
TOUCH UNTIL COOL.**

Source: Charles Kemnitz, "How to Write Effective Hazard Alert Messages," *Technical Communication* 38.1 (1991), pp. 68–74. Reprinted with permission from *Technical Communication*, published by the Society for Technical Communication.

Finally, test your document, using one or more of the following procedures:

User Edits

user edits
an edit in which a "typical" user reads document aloud while attempting to follow the steps or points to see if it is possible to follow the instructions

In a **user edit,** a "typical" user of a document reads aloud the document while attempting to follow the steps or points. You can see if your user can follow the instructions step by step, point by point. If the user misreads or stumbles, then you need to revise that step. A user edit works well for short instructions but not for an entire manual in which the user may need to search for information.

Protocol-Aided Revision

In protocol-aided revision or testing, your users perform tasks and talk aloud as they perform each task. Rather than reading the steps of an instruction, these typical users talk about how they are using the instructions and the product. They reveal what they are thinking or how they are interpreting while they are performing the action. This type of testing works best when your users must find as well as follow your instructions to reveal their problems, their misinterpretations, and their questions.

protocol-aided revision
revision or testing in which users perform tasks and talk aloud as they do so to reveal how they are interpreting a text while they are performing an action

Field Trials

The most effective testing occurs when you run a field trial in the actual setting in which your users will need your document. Your users then report back the problems they encounter. Some companies, such as IBM, have information cards that users return with descriptions of problems, which enable the company to receive feedback constantly and revise its manuals accordingly.

field trials
testing of a document in an actual setting to determine what problems users face

Laboratory Testing

Perhaps the most effective way to test a lengthy document is in the laboratory. Users are video- or audiotaped while following the documentation and using the product. Data logging programs are used to record the amount of time it takes a user to read documents and perform actions. Although the number of users involved may be small, the value of data obtained is great, particularly if the users are given meaningful tasks or placed in realistic situations while the testing is conducted (Redish and Schell, pp. 68–71).

laboratory testing
testing lengthy documents in the laboratory by video- or audiotaping users who are following the documentation while using a product

In this chapter, we focused on instructions, procedures, and specifications, three related documents or parts of documents that focus on performance and frequently support technology. Procedures generally describe how to coordinate an activity in which many people are involved. Instructions detail the specific steps that an individual would follow to complete a task, while specifications state what standards must be followed for a task to be completed safely and according to professional code.

SUMMARY

Ideally, within a corporation, product documentation planning should begin at that same time as product development and design, and the writer should function as the user advocate throughout development and planning. Two important sources in the planning stages of product documentation are the product's specifications and your own task analysis. Of course, identifying your audience and purpose are also essential, as the audiences of instructions, procedures, and specifications have different needs and experiences. You should assess the following things about your audience, purpose, and document: prior knowledge about the process and product, if needed; prior experience with the process and product, if needed; amount of detail; number of examples; attention-getting devices; justification of individual steps; level of vocabulary; placement and type of definitions and descriptions; number and type of visual displays; and document design features. Often the style and organization of your support document will be dictated by your company, your professional code, or ISO-9000 standards.

In the second half of this chapter we focused on instructions, perhaps the most common type of document that supports technology. Instructions, like contemporary computer manuals, must be user-friendly and task-oriented. Instructions are arranged chronologically with the exception of troubleshooting instructions, which help a user move back and forth between alternative steps. Instructions start with the earliest possible action and include every action. Parts and materials needed to complete the task are listed at the beginning of instructions, as are definitions and other contextual information. You might also describe any mechanism and explain the theory of operation. Visual displays emphasize not only the devices involved but also the actions taken. Verbs usually appear in the imperative voice, and steps are listed in parallel structure.

Some special types of instructions include templates, quick reference cards, tutorials, user guides, reference guides, and online help. Regardless of the type of instruction you create, consider chunking information in a logical way. Warnings and cautions must be carefully written and carefully placed in instructions, and ANSI guidelines should be followed. Finally, instructions should always be tested, either by user edits, protocol-aided revision, field trials, or laboratory testing.

ACTIVITIES AND EXERCISES

1. Read the instructions that appear in Example 14–22 and compare them with the instructions shown in Example 11–6 (Chapter 11). These instructions appeared in the same document. How do Examples 14–22 and 11–6 relate? What is the purpose of each? When would a user refer to Example 11–6, and when to Example 14–22? What choices in style, organization, and visual display did the writers make in creating these instructions? What experience did they expect their audience to have with the system? What features make these instructions easy or difficult to use? Be prepared to share your responses with your class and instructor.

2. Locate some instructions that have accompanied a product you have recently purchased or acquired. Using the suggestions within this chapter and your own field test, critique those instructions. Describe in your critique how you tested the instructions.

3. Write a set of instructions, procedures, or specifications for a product or process you know well. Follow all the suggestions in this chapter and be prepared to have a classmate test your document.

4. Arrange to interview an employee at a company in your immediate vicinity about the documentation that supports the tasks or products the company engages in or produces. Gather information on what the documents are designed to do:

EXAMPLE 14–22 Instructions for the VMX System

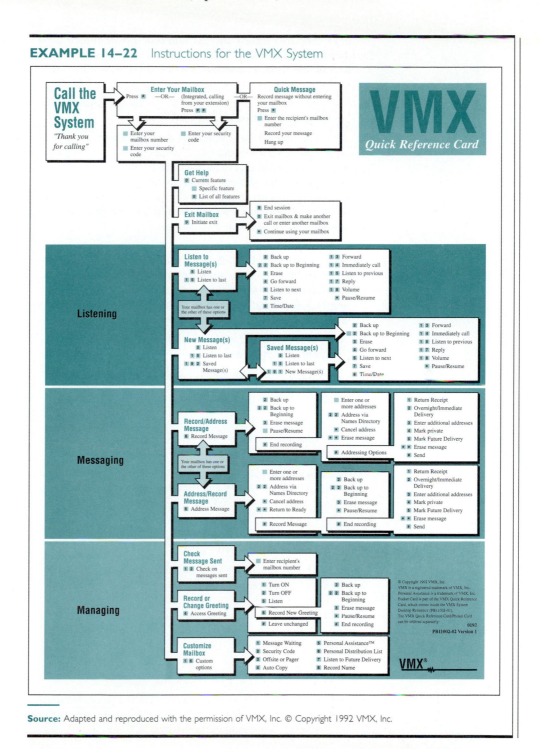

Source: Adapted and reproduced with the permission of VMX, Inc. © Copyright 1992 VMX, Inc.

how they are planned, drafted, and tested; whom their users are; and any other pertinent information. Give the results of your interview, with samples if you can obtain them, in a written or oral presentation to your classmates.

5. Do some primary or secondary research on how ISO–9000 standards are affecting a company you would like to work for in the future. Report your findings in a memo to your instructor and classmates.

6. Read carefully all the documents that pertain to the *Exxon Valdez* oil spill in Appendix A, Case Documents 2. Assume that as a company engineer for Exxon, you need to write a set of instructions for the bioremediation process. You might need to do some library research in order to complete this assignment, but you can get off to a good start with the case documents. Plan, design, and write the instructions so that crews involved in oil spill cleanup efforts in the future can follow your step-by-step instructions.

7. Analyze the "universal precautions" for caring for a person with AIDS (Appendix A, Case Documents 1). Rewrite these instructions for a poster to be displayed in all clinic examining rooms in your local hospital. Be aware that your users will have to follow your instructions while treating patients and therefore have only a few seconds to remind themselves about these steps. You will need to add appropriate danger, warning, or caution labels.

8. Read over *Ziglar* v. *Du Pont Company* in Case Documents 4, Appendix A. Go to the library and obtain a complete copy of the ANSI standards Z535. Design an appropriate label for the pesticide bottle. Justify your design in a memo to your corporate supervisor.

WORKS CITED

Kemnitz, Charles. "How to Write Effective Hazard Alert Messages." *Technical Communication* 38.1 (February 1991), pp. 68–73.

Redish, Janice C., and David C. Schell. "Writing and Testing Instructions for Usability." In *Technical Writing: Theory and Practice,* ed. Bertie E. Fearing and Keats W. Sparrow. New York: MLA, 1989, pp. 63–71.

Reports and Studies

A Process for Planning and Developing Reports ▍ Types of Reports ▍
Conventions of Report Writing

The reader of an industrial report does not possess the writer's knowledge and
has no interest in doing so. In industry, the writer is the expert, not the reader,
even if the reader is the boss. An industrial report should not focus on what the
writer knows, but on what the reader needs to know, what the information means
for the reader, and why it matters to the reader. The industrial writer's task is not
to prove that he or she has learned something, but to produce reports that convey
the significance of information to a reader in a particular context of social interactions.
Peter Hartley, "Writing for Industry: The Presentational Mode versus the Reflective
Mode." *The Technical Writing Teacher* 18.2 (1991), pp. 162–69.

Introduction

Reports accompany and justify or predict all significant activity in the work world, whether the work is research and development, manufacturing, government, law enforcement, medicine, or finance. Along with business letters and memos (which often report activities or decisions), reports are the most common type of writing in the workplace. No matter what career you plan, you will undoubtedly write and read reports.

This chapter provides an overview of reports. In Chapter 16 we'll look more at how reports are used to make decisions. Often those decisions involve solving problems for audiences, as we saw in Chapter 2. Here, we offer a process for planning and developing reports and summarize the contents and purposes of various reports, including literature reviews and progress reports. We also suggest strategies for organization and style to increase the effectiveness of your reports. We describe the formal features of reports, and we offer you some tips to make the task of reporting efficient and even enjoyable.

Reports answer the questions "What happened?" or "What was done?" Report writers, usually observers or participants in what was done, report their observations to someone else who can use them. As writers, they "carry back" (re-port) information. Even though the activities reported have already occurred, the act of reporting assumes that the information can or should be used in the present or future. The various uses to which reports are put result in their classification into subgenres, such as *recommendation reports* or *progress reports.*

Reports may be delivered orally, but they are usually written when the investigation yields too much information for an audience to retain from listening, when some written evidence is required to verify what happened, or when the information will have value beyond the present time and readers. For example, technicians write lab reports so researchers can verify the relationship between the procedures and results. Written records provide evidence that allows researchers to make claims about causes and effects. Or, the secretary of an organization keeps minutes of meetings so that members will know who is to do what and to preserve a record of decisions made. Written records allow people to refer to their decisions so that their follow-up actions can be consistent. Written reports are the focus of this chapter.

Reports all have practical purposes: preserving records and providing information to others who can use it. They have a persuasive purpose as well. Whether you write specifically to persuade, your reports will influence decisions, actions, and policies; you will solve problems. A good starting point in report writing, therefore, is to *think about what should happen as a result of your report and how you can encourage the right things to happen.*

Your reports will also influence the views other people form about your professional competence. As a report writer, you want to persuade your readers that you have performed your work competently. The information you present, your means of gathering it, and the way in which you present it should convey your competence.

Your investment of time and energy in researching and writing a report also suggests that you would like readers to take your ideas seriously. The persuasive purpose is one reason why it is in your best interest to master effective processes of report planning and development.

Readers of reports are generally a select group because reports have a specific focus and purpose. Some reports may have only one reader, such as your immediate supervisor. Often reports are forwarded up the organization to the supervisor's manager or above. They may be filed and used by readers in the future who are trying to reconstruct some decisions or who are building on the information in the report to do a follow-up study. As you consider how your report will be used, you should think not just of how the person who assigned the report will use it but also who else may read it and for what purposes.

A Process for Planning and Developing Reports

Certain conventions of organization and format follow from the varying purposes of reports. However, even within categories of reports, there are variations. Different reporting goals may require different strategies of organization and format. Furthermore, all organizations have their own preferences and conventions for reports. Workplace reporting situations are never exactly like textbook descriptions. Thus, in this chapter we offer guidelines for report writing based on a process of conceptualizing, researching, and managing the reporting process. We do not pretend to offer templates for reports that will work in every situation. Technical communication involves both critical thinking and problem solving. Therefore, your task in learning to write effective reports is not so much to memorize types of reports as it is to learn a process for developing reports that can help you plan effectively for each reporting situation.

The process for writing a report includes the following tasks:

1. **Conceptualization:** conceiving of the purpose in writing and the reader's need for information. Think back on this process as described in Chapter 2.
2. **Research:** researching the problem in ways that produce credible and useful information. Think back on this process as described in Chapters 7 and 8.
3. **Presentation:** choosing the best strategies of organization, format, style, and use of visual displays. Think back on these decisions as described in Chapters 11 and 12.
4. **Management:** managing the research and writing tasks in order to ensure a high-quality report that is finished on time.

conceptualization
process of conceiving of the purpose in writing and the readers' need for information

research
finding credible and useful information

presentation
process of choosing the best strategies of organization, format, style, and visual display

management
dealing with the research and writing tasks in order to produce a high-quality report

The planning and development of reports begins with conceptualization of the uses of reports and the actions that will result from them. You plan by considering the completed report in use and work backwards to plan the steps that will get you to this end. You will write a better report if you can imagine the report used not only in the present but also in the future. You will write your reports more efficiently if you can imagine, from the beginning, the document itself—how it will be organized and formatted, the sections and supplements it will contain, and the visual displays and presentations that will enhance the readers' understanding. Whenever a report is assigned, you should ask for specific expectations for content, format, and length (much as you do when a writing project is assigned in class). You can and should supplement what you are told by examining sample reports written within your organization.

Research often means keeping good records during the course of a project. Many reports in the workplace, including progress reports and lab reports, provide verification of activities and their results. The evidence that forms the substance of the report comes from notes made while the work is in progress; thus anticipating a report assignment helps you determine what kinds of records to keep. Matching research methods with the kind of information that the report will require is also important. The information in Chapters 7 and 8 will help you choose and carry out your research plans. Also, you can use the information in Writing Strategies 15–1 in any reporting situation to help you conceive of and plan your report.

Types of Reports

lab reports
reports that describe methods and results of lab procedures in meticulous detail

trip reports
records of events and observations when people are away from the office

police reports
records of who, what, where, and when information from accidents or crime scenes

Reports are called by different names: *recommendation report, progress report,* or *literature review,* for example. These names reflect different reporting goals. Various types of reports influence action in different ways. For example, **lab reports** describe the methods and results of lab procedures in meticulous detail so that researchers will have credible information to report when they write articles for scholarly journals. These articles, in turn, are kinds of reports that are more comprehensive than lab reports in their details of problems, methods of investigation, and results of experiments. The detail enables other researchers to replicate experiments in order to verify them or enables practitioners to apply the results.

Trip reports present the records of events and observations when people are away from the office to enable a change in a policy or procedure or a decision about reimbursement. **Police reports** record who-what-when-where information from accidents or crime scenes so that police investigators can determine responsibility and degree of damage. **Medical reports** describe symptoms, treatment, responses, and recommended actions so that medical personnel can make sound follow-up decisions about patients. **Personnel reports** evaluate applicants' qualifications or employees' perfor-

Conceptualization: Envisioning the report as it will be used when it is completed.

Purpose

Why is a report needed? (Consider the uses that will be made of it and problems it attempts to solve.)

Readers

Who will read the report?

In addition to the primary reader(s), will the report be forwarded to others within or outside the organization? Will the report be filed and perhaps read by readers in the future? Who will these secondary readers be?

What will the readers do as a result of reading the report?

Content

What questions will the readers have about the report topic?

In what order will the readers be likely to ask the questions?

Research: Planning methods of gathering information.

What are the best ways to get answers to the questions the reader will have? (Consider interviews, surveys, site inspections, literature review, financial analysis, and other suggestions in Chapters 7 and 8. Also plan for the records you will need to keep during the course of the project that is reported.)

Presentation: Imaging the document itself and choosing the best strategies of organization, format, style, and visual displays to make it effective.

How should information be arranged in the report to make the information easy to comprehend and to find? What will be the overall pattern—e.g., comparison, chronology, topics in order of importance? What arrangement may increase its persuasiveness? (Is there some information that should appear early or be delayed?) What information should be displayed in tables, graphs, or illustrations in order to increase comprehension or to persuade?

Management: Identifying research and report preparation tasks, and setting a schedule for completing them.

Research Tasks	Completion Date
a.	
b.	

Writing, Report Preparation Tasks	Completion Date
writing	
introduction	
body	
conclusions and recommendations	
summary or abstract	
revision and editing	
preparation of visual displays	
preparation of preliminary pages	
preparation of supplements	
final proofreading	

medical reports
records of medical symptoms, treatments, responses, and recommended actions so that medical personnel can make sound follow-up decisions about patients

personnel reports
reports used to evaluate applicants' qualifications or employees' performance

financial reports
reports used to evaluate the balance of assets and liabilities in order to determine taxes or investments

feasibility studies
documents designed to analyze the likely outcome of possible future actions

environmental impact statements
documents used to predict possible consequences to land, wildlife, and people of proposed projects

reports used in management
periodic and progress reports, annual reports, minutes, and other reports used to assist in decision making

reports used in research
literature reviews, lab reports, scientific reports, and other reports used in research investigations

sales reports
documents designed to assist managers in deciding about inventory, marketing, and company goals

annual reports
company reports summarizing and evaluating annual activities

mance in order to determine job assignments and pay schedules. **Financial reports** evaluate the balance of assets and liabilities in order to determine taxes or investment options. **Feasibility studies** analyze the likely outcome of possible future actions. **Environmental impact statements** predict possible consequences to land, wildlife, and people of a proposed project to determine whether benefits of the project outweigh possible negative consequences. Thus, the classification of reports by type depends on their intended outcome, on what actions or decisions should result from them. Figure 15–1 classifies some common reports on the basis of action that results from them.

This section describes some common types of reports in order to offer some guidelines for writing and thinking about them. Our discussion is most complete on progress reports and literature reviews, two reports you may be assigned to write for your class. The reports are loosely grouped according to two broad purposes: **reports used in management** (periodic and progress reports, annual reports, minutes, trip reports) and **reports used in research** (literature reviews, lab reports, scientific reports). Managerial reports enable people to make decisions, while research reports present the results of investigations. The purposes of both overlap because all reports require the sharing of information, and all reports invite readers to make decisions, but one purpose usually dominates. Some reports, especially progress reports and annual reports, are important both for management and for research. Some projects require a series of reports; for example, a research project may require a proposal, literature review, several progress reports, and a scientific article.

PERIODIC REPORTS AND COMPLETION REPORTS

Reports are frequently scheduled to appear periodically and at the completion of a project. For example, sales are reported periodically, on a set schedule, so that a company can evaluate how products or services are performing in the market. **Sales reports** help managers make decisions about inventory, marketing, and company goals.

Investment companies send **periodic reports** to investors monthly, quarterly, or semi-annually, as well as annually, summarizing account balances and investments and explaining the performance of the investments. Investors depend on this information to make decisions about maintaining, increasing, or canceling an investment. You receive one version of such a report in your bank statement; you use the information in the statement to guide spending in the next reporting period.

Employees and divisions within an organization are evaluated annually, and an **annual report** indicating achievements and goals for the coming year is often the basis for the evaluation. Likewise, divisions within organizations and the organization as a whole produce annual reports summarizing achievements and goals.

Type of Report	Information Contained	Reader	Action That Results
Lab report	results of a specific procedure	primary investigator	incorporation into a scientific article; interpretation of significance
Scientific article	problem, methods, results, discussion of an experiment	other researchers; practitioners	replication of the experiment; application
Literature review	survey of published work on a topic	researcher; manufacturer; practitioner	continuation or challenge of the work; application to a current project
Trip report	activities, observations, expenses	supervisor; management team	change in policies or procedures; decision about equipment; reimbursement
Medical report	symptoms, treatment, response	treating physician; another physician	follow-up treatment: continuation or change
Personnel report	achievements, evaluation, goals	manager	promotion, salary change, training, dismissal
Financial report	gains, losses; assets, liabilities; investments	investor; banker	continuation or change in investment; credit or denial of credit
Feasibility report	financial, technical, social, and managerial consequences of a possible action	decision maker	decision about an action
Environmental impact statement	consequences to land, water, air, wildlife, people, and economy of a proposed project	executives; government agencies; legislators; concerned citizens	continuation, change, or cancellation of a project that affects land
Progress report	work accomplished during a specific period	manager; project sponsor	continuation, change, or cancellation of the project

Note: M. Jimmie Killingsworth and Michael K. Gilbertson have developed an elaborate theory of technical discourse and the action that results from it in *Signs, Genres, and Communities in Technical Communication.* The book and an article by Carolyn Miller, "Genre as Social Action," have influenced discussion in Chapters 15, 16, and 17 of genres as distinguished by the actions that result from them.

FIGURE 15–1
Reports, Readers, and Resulting Action

completion reports
reports written at the end of a project or service

periodic reports
reports that compare the activities of the current reporting period with that of a previous period

Completion reports are written at the end of a project or service. When you visit the doctor, someone records the reason for the visit, your symptoms, the procedures performed, and the recommended treatments. Likewise, the plumber records services performed after a service call. This written information provides the basis for follow-up treatment or service as well as evidence of services for billing purposes. These reports may be informal and quite brief. Formal completion reports, which occur in research and contracting, indicate the outcome of a project that has ended.

All *periodic reports* compare the activities of the current reporting period with that of the previous period or with goals that were established at the beginning of the project. Achievements are meaningful only in comparison with expectations, and thus the argument of a periodic report—the claim the writer makes—is based on comparison. The argument may also show causes and consequences to explain the results reported.

The basis for the information you use in a report will probably be notes you take throughout the project plus correspondence, financial analysis, or other evidence of the work being done on the project. This type of reporting requires you to keep good records of your work and to ask others who may work with you to keep records. Early in any work assignment, it is a good idea to ask what reports will be required. Examining sample reports will also give you some idea of what kinds of records to keep. In general, when you write any kind of periodic report, you will be able to follow the suggestions for ordering a progress report described in the next section.

Example 15-1 illustrates the first page of a semi-annual report to the shareholders of the Windsor Fund, a mutual fund with investments in the stock market. The letter format personalizes the report, a strategy used to make the shareholder feel important and connected to an important person (the chair of the board) who is in charge of the investments. The first

thesis statement
sentence (or sentences) occurring in an introductory paragraph that states the focus or emphasis of a document

paragraph reports good news and provides a **thesis statement** that shapes the reader's interpretation of the rest of the report. The table emphasizes the good news by drawing the reader's attention. It compares the performance of the Windsor Fund with the performance of the stock market overall. Following the table, however, news that does not look quite so good (the decline in the share value) is buried in paragraph form. The decline is explained to reassure shareholders.

interpretive statements
explanatory statements that occur in the body of a document

Notice how the rest of the letter includes **interpretive statements** as well as financial data. The writer explains why the investment strategy of the fund is sound by making comparisons with other investments and with the fund's previous performance. The third paragraph relates causes and effects: the good performance is the result of a good investment strategy. The last paragraph also explains the investment strategy in terms of market conditions. The writer achieves credibility by referring openly in the right-hand column to past problems with the fund. Referring to the problems also

EXAMPLE 15–1 Periodic Report to Mutual Fund Shareholders, First Page

CHAIRMAN'S LETTER

Fellow Shareholder:

Windsor Fund started its 1992 fiscal year "on the right foot," with a total return of +9.5% for the six months ended April 30. On balance, despite tumbles in November 1991 and March 1992, it was also a favorable period for the stock market.

During the period, the Fund's total return (capital change plus income) outpaced the return of the unmanaged Standard & Poor's 500 Composite Stock Price Index, continuing the trend that began in our 1991 fiscal year. This table presents the figures:

	Total Return
	Six Months Ended April 30, 1992
Windsor Fund	+9.5%
Standard & Poor's 500 Stock Index	+7.3%

The Windsor return is based on net asset values of $12.79 per share on October 31, 1991, and $12.39 on April 30, 1992, with the latter figure adjusted to take into account two semi-annual dividends from net investment income totaling $.57 per share, and a year-end distribution of $.84 per share from net capital gains realized by the Fund during calendar 1991.

The Fund's fine performance during the past six months resulted, to an important degree, from the same factors that shaped our success during fiscal 1991: (1) a concentration in a relatively small number of stocks, several of which provided handsome appreciation; and (2) an emphasis on "value investing" in stocks with generally above-average yields and below-average price-earnings multiples. Particularly during the past four months, such value stocks provided gains in excess of those achieved by their growth stock cousins, which led the overall stock market during 1991.

Windsor's return of 2.2 percentage points above the Standard & Poor's Stock Index for the past six months is satisfying, especially following its 11.3 point advantage for fiscal 1991. What is more, after outpacing the average growth and income mutual fund last year by 11.4 points, our advantage accelerated, as our return exceeded that industry standard by 2.8 points (+9.5% vs. +6.7%) during the past six months.

However satisfying this bounceback in Windsor Fund's relative performance may be, I would emphasize to you that your management is fully cognizant that we still have a great deal to accomplish if we are to recoup our shortfall in the 1990 fiscal year. This period, as you know, was distinctly poor, both on an absolute basis and relative to our competitive benchmarks.

In response to my Chairman's letter in the Annual Report for fiscal 1991, a number of shareholders expressed doubt about the accuracy of the Fund's total return of +44.7% for the fiscal year. Let me assure you that the figure was accurate, even as I note that our return for calendar 1991 was a much more modest +28.6%. The principal difference, simply put, is that the latter period excluded the two very strong months of November and December 1990, when the market jumped upward following its big decline during the crisis in the Persian Gulf. During these two months, Windsor's return was +13.8%.

As John B. Neff suggests in his Investment Adviser's Report on the next page, the stock market, as measured by traditional yardsticks such as dividend yields and price-earnings ratios, cannot be considered "cheap." Reflecting this view, Windsor's common stock position presently represents about 76% of net assets. This conservative stance makes Windsor's results in the rising market of the past six months even more creditable; it also suggests reasonable defensive characteristics should rough weather lie ahead.

Sincerely,

John C. Bogle

John C. Bogle
Chairman of the Board　　　　　May 14, 1992
Note: Mutual fund data from Lipper Analytical Services, Inc.

gives the writer a chance to reassure shareholders that they have made a good investment.

This page is followed by a report from the investment advisor and then by a list of companies in which the fund is invested, the number of shares owned, and the overall values. These companies are grouped according to type of stock, such as utilities and technology, in order to show investors the distribution of investments.

The periodic report is routine and standard in the financial world, and its purpose could be conceived of as information sharing, but it is also persuasive. In addition to providing information, the writer also uses the report to influence readers' actions. The intended outcome is that shareholders will not only maintain their current investments but also invest more.

PROGRESS REPORTS

progress report
report that describes work accomplished during the reporting period and identifies work remaining on a project

Long-term research and development, manufacturing, or contracting projects (lasting six months or a year or more) require progress reports written to managers or project sponsors. The **progress report** describes work accomplished during the reporting period and identifies the work remaining on a project. Progress reports are a type of periodic report specifically written for a project that will ultimately end (unlike the investment that is expected to continue indefinitely), and they demonstrate that work is progressing on schedule. The progress report basically compares the project as it is unfolding with the project as it was proposed. As the starting point for the progress report, the project assignment usually identifies what should be reported as the project progresses, and it specifies where the project should be at the time of each progress report.

The progress report is a good management tool: it enables a manager to maintain productivity and provides an incentive to keep work progressing on schedule. A progress report also allows periodic assessment of project goals and methods so that they may be revised, if necessary, according to the preliminary results of the project. Typically, progress reports are due every three months or every six months on a long-term project.

Three outcomes can result from a progress report:

1. The project may continue as originally scheduled;
2. The project may be expanded or otherwise changed to accommodate new situations; or
3. The project may be canceled.

A project might be canceled because the work seems unsatisfactory, because the benefits are less than predicted, or because of external situations unrelated to the project, such as company reorganization. Because the progress report may shape project directions, you will have a persuasive as well as an informative purpose in writing the report. Indirectly you will make the claim that you are doing competent work that will achieve the established objectives. The best way to do that is to have some progress to report. Excuses for failing to accomplish work will not impress the supervisor or project sponsor. At times, although rarely, you might find yourself in the position of convincing superiors to abandon a project; for example, progress reports pertaining to the *Challenger* space shuttle did recommend such abandonment, although they were ignored or reinterpreted. Although we focus on presenting your progress on a project in the best possible light in the rest of this chapter, keep in mind that you may occasionally find a project just not feasible after all. Your reader will appreciate your honesty and thoroughness in this case.

As for any other periodic report, the data for a progress report come from the records you keep of the work you do. These records may include data from experiments, from regular site inspections, from analysis of figures, or from a log book in which you record activities. Keeping good records makes it easier to write the report.

A progress report has a three-part structure:

1. An introduction that reviews the project and its purpose;
2. A middle section that cites specific accomplishments and work remaining; and
3. A conclusion that assesses the progress.

We look at each part of the progress report separately.

The **introduction** of a progress report should describe the project and problem the project will solve. This description may appear in the same words in each of the progress reports that may be written on a project. Even though the reader of a progress report is supervising the project and presumably is familiar with the project description, it is important to include it in each progress report. In this way, you can reacquaint the supervisor, who may oversee a dozen or more such projects, with your specific project. The introduction also helps to establish your credibility as investigator because your description of the project demonstrates your comprehension of its goals. Including the description also enables the report to stand alone without reference to another report, which makes it more effective if a reader needs to refer to your report at a later time.

introduction
early section of a report that describes the problem the project will solve

You may also preview the rest of the report and its significance by including a statement indicating whether work is on schedule. Although the details of the accomplishments will appear in the body of the report, you might report an accomplishment of particular significance. You should prepare readers for any surprises or changes in the work—something unexpected that might cause delays in completion or recommended changes in the project.

The **body of the report** presents the specific accomplishments in concrete and often numeric statements. This section of the report provides the

body of the report
main section of a report in which specific accomplishments are presented in concrete and numeric statements

evidence that will let you assess the overall progress in the conclusion. Generally avoid evaluative statements in the body of the progress report: instead of saying, "The work is going well," you should say "We completed 14 of the 16 planned trials," or "The site has been excavated and the foundation laid." Readers look for facts more than for evaluation in this section of the report. You should, however, provide interpretive statements that show the significance of the findings, such as, "The level of participation is 17 percent higher than in previous studies." These statements place the facts in a context so the reader will know how to respond to them.

The body of a progress report can be organized chronologically, by task, or by topic. Generally speaking, you will probably use a combination of patterns, with one embedded into another. For example, your main sections may be chronological (work accomplished, work remaining), but subsections will be arranged by task or topic. Alternatively, your main divisions might be ordered according to task or topic, and the subdivisions would be ordered chronologically.

Your instructor may ask you to write a report on your progress on a major assignment. Tasks for such a project include investigation tasks, such as a literature review, interviews, and a site inspection. They also include preparation tasks, such as writing the introduction, designing visual displays, and creating the preliminary pages. Topics for such a project include the particular subjects you are investigating through a literature review, interview, or site inspection. Topics might include cost, size, legal issues, fertility rates, or other objects of your investigation.

Example 15–2 presents the first of three progress reports (here called *status reports*) on a senior design project by electrical engineering students. The team writer has arranged the report by task (hardware development and software development). The report is quite specific about accomplishments. The visual displays aid understanding but are also persuasive because they suggest that the writer's own concept of the task is clear. Work remaining is specified in the conclusion section of the report.

conclusion
section of the report in which the overall assessment of work is presented

The **conclusion** of your progress report provides an overall assessment of the work to date. If everything is going well, you should be able to report that the work is on schedule or ahead of schedule. You will certainly indicate any areas in which you have exceeded the objectives for the project. If you have failed to achieve objectives, you will have to offer a reasonable explanation and state your plans for catching up in the next reporting period. The conclusion is also the place to propose any changes in the project that result from unexpected findings in the work so far. If you do propose changes, ask for some written confirmation that the changes are approved.

Overall, try to maintain an upbeat tone by emphasizing progress rather than problems. If there are legitimate problems, be honest about them, but don't get trapped into complaining about how far behind you are because of all kinds of problems beyond your control. Manage your work with a

EXAMPLE 15–2 Progress Report

FROM: S. David Silk
  TO: Mark Greer
DATE: October 24, 1990
  RE: Status Report on ISU/Motorola ISDN Senior Design Project

This report summarizes all the concurrent tasks performed on the ISU/Motorola Project from September 1 through October 24. The purpose of this project is to build and implement an ISDN TA demonstrating Motorola's ISDN chip set. This status report is being sent prior to the actual proposal for the project so that we may incorporate, in the proposal, some suggestions made by Eric Crane in a memo dated September 12, 1990 that we received the week of October 22, 1990.

The project has been divided into two teams consisting of four students each. One team works on the hardware and one team works on the software.

As you recall, we have discussed hardware quite extensively and the team understands what is required of us. Therefore, this status report treats hardware briefly and software more extensively.

I. HARDWARE DEVELOPMENTS

Further work on the schematic began where the preliminary research left off (the preliminary work on the schematic only included the ISDN chip set). The schematic is currently being designed on a CAD system which when completed will allow us to go straight to fabricating a printed circuit board for the prototypes. Listed below are the areas that have been addressed within the last month pertaining to the schematic.

  1. In addition to the ISDN hardware, the schematic has been refined to incorporate the following components:

     1. A 16 line by 20 character LCD
     2. A dial keypad
     3. The eight function keys
     4. A ringer

  2. Since the preliminary work on the schematic, details on the memory requirements have been refined. Since this is the first attempt at our design, we will not be too concerned with efficiently managing the memory. Time permitting, we will fine tune the memory requirements. So for now, we intend to add additional memory as required while we wire wrap the design for initial testing. Listed below is each type and amount of memory that will be incorporated into the first wire wrap design.

     1. EPROM–1M
     2. RAM–64K
     3. Battery backed RAM–64K
        (Note: This memory has been allocated for storage of the users' personal preferences.)

For your convenience, a printing of the schematic has been enclosed with this report.

2. SOFTWARE DEVELOPMENTS

Developments over the last month have been quite extensive in defining the software for our device. Within the last month, several issues have been resolved. The issues identified relate to how Q.931 and Q.921 will be developed, how the software architecture of our device should be structured, and how code can actually be produced for the MC68302 with the equipment available to us.

EXAMPLE 15–2   *(continued)*

2.1 Implementing Q.931 and Q.921 on Our TA

Within the last month, we have conceptualized the TA's software architecture. Because of the extent of the work involved, it has been decided that we will not write Q.931 or Q.921 code. We have decided to focus our attention solely on writing the software to control our device. To acquire the necessary code, we have solicited the assistance of a group of graduate students who are currently developing Q.931 and Q.921 code. In return for their code, we will provide them with a test bed, our device.

Essentially, the graduate students will be providing us with standard code that is portable and is not designed to operate on any particular hardware. As a result, we will be required to write an interface to their code in order to run it on our TA. To expedite this procedure, a student from our design team has been assigned to work with the graduate students from the start in order to familiarize himself with the protocol and to understand how we will integrate it into our TA.

2.2 system software for our ta

The architecture for our device will be built around the EDX code that will be made available to us by Motorola in the near future. EDX's flexibility to perform as a cooperatively multitasking operating system will lend itself to our application.

The TA's architecture has been designed to be event-driven and object-oriented. The software for our TA will be broken down into self contained modules that will consist of a main routine which does nothing but loop continuously getting messages and calling its appropriate local procedures. No module can call a procedure that exists in another module; instead, a message must be sent to that separate module. Hence, via EDX, messages will be passed around the system to the appropriate modules executing the desired tasks. In a sense, EDX will act as a type of message center responsible for the execution of all TA functions (refer to Figure 1 for a breakdown of the major system software modules).

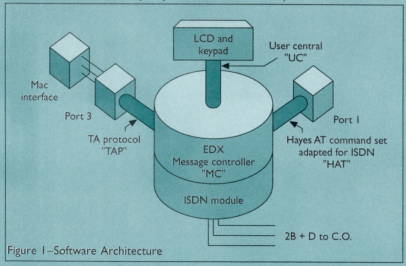

Figure 1–Software Architecture

2.2.1 System Software Modules

Each of the major modules depicted in Figure 1 has been assigned to individual members within the team. Currently, each team member is preparing some preliminary documentation defining his or her modules functionality and how it will interact within the entire system. Within the next week, we will begin generating code.

EXAMPLE 15–2 *(continued)*

2.2.2 A Look at The User Interface Software

From the perspective of the user, the user interface is perhaps the most important aspect of our device. User Central is the module that is responsible for providing an efficient and user friendly means to operate the TA. The preliminary work that has been completed regarding the user interface is shown in the next two figures. Illustrated in Figure 2 is a layout of the procedures within the User Central Module. Also, illustrated in Figure 3 is the graphic layout of the LCD.

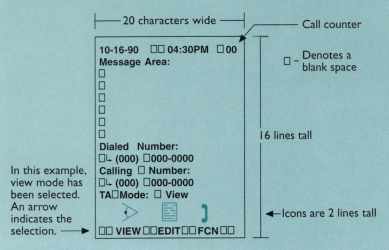

Figure 2–A layout of the internal procedures of User Central

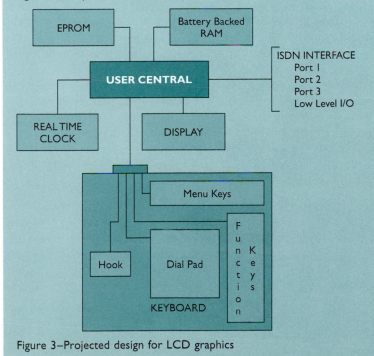

Figure 3–Projected design for LCD graphics

EXAMPLE 15–2   *(concluded)*

2.3 Developing Code For the MC68302

Within the next month, a new cross compiler will be installed on the University's HP System providing us with the proper environment to do our MC68302 development. The package includes a 302 Cross Compiler, Linker, Assembler, and Simulator. The majority of the software development will be done on the HP System and then downloaded to a Mac II, which in turn will be connected to the ADS Board and our prototype.

3. SUMMARY

The following areas have been addressed: refinement of the schematic, acquisition of Q.931 and Q.921 code, and conceptual organization of the framework of the system architecture. Within the next few weeks we will address the following areas: complete the schematic so we can proceed with wire wrapping the design, continue working with the graduate students to evaluate what will be required to get the Q.931 and Q.921 running on our hardware, and continue fleshing out the system software for our device.

A few possible bottlenecks could occur due to delayed arrival of the following equipment: the development system for the University's HP workstations, equipment for the ADS Board, and additional ISDN components. If any one of these setbacks should occur, we will still be able to proceed with the project.

Source: Used with permission of the author, with special thanks to Professor Patricia Goubil-Gambrell, Texas Tech University.

mental eye on the progress report, and accomplish what you have said you would accomplish so that you can report real progress.

Progress reports usually don't have to be very long. You shouldn't have to provide all the data you have collected, but simply give enough detail to show that you have made progress. Report in detail any results of particular interest or importance, and use tables and graphs for important quantitative information. As with every other type of writing you do, you can help judge what is enough by imagining what questions your readers will ask as they read and what format (paragraph, table, or bar graph, for example) will help them read accurately and quickly.

ANNUAL REPORTS

Annual reports, like progress reports, are periodic statements of progress. Common to all organizations, annual reports may be written for internal use, or they may be public. Individual employees and department managers, for example, may be required to report their achievements for the year and to state their goals for the coming year. These reports are used in personnel evaluations as well as to set organization goals.

Internal annual reports may be brief, perhaps no longer than one to three typed pages. They are likely to follow the structure suggested for the progress report:

- An overview that establishes the purpose and specialization of the writer or the writer's department;
- A middle section that reports work accomplished in factual terms; and
- A conclusion that assesses the achievements in comparison with the expectations and sets goals for the next year.

Before you have to complete an annual report as an employee or department manager, ask to see copies of previous reports to get a sense of the organization, format, length, and information expected.

Annual reports distributed to stockholders or contributors are often glossy booklets with pictures and attractive layouts. The audience for these documents are not the organization's executives but rather anyone who has a stake in the organization and a choice about whether to support it or invest in it. Efforts to make these reports attractive reveal their persuasive purpose: the organization wants the readers to maintain their investments, to invest more, or to offer a donation. For example, pictures of appealing children or professionals at work are arguments based on *pathos*. High-quality paper and attractive layout help to create a public image for the organization that invites support. These reports will probably be created by public relations specialists, but the specialists will ask for information from staff members.

Some organizations are required by law to make public their annual financial statements. Publication of these annual reports protects the public by revealing how funds are used. All charitable organizations must report how they spend their funds and make that information available to the public.

MINUTES

Minutes are reports of the business conducted at a meeting. They summarize the discussion of the agenda items as well as the decisions made and actions taken. Minutes are written to provide a history of an organization's activities but also to encourage officers, members, and staff to implement the decisions that the group makes. Good minutes tell not only what was decided but also who will do what and when. They may have some legal significance as well because they establish whether the business was conducted according to organizational guidelines and whether decisions are binding.

Minutes do not have to record every word that was spoken, but they should summarize the gist of the discussion and where necessary provide attribution. If a resolution is passed, the wording must be exact, and names, dates, and financial information should be specific. Minutes must include identifying information, such as the name of the organization, date and time of the meeting, and the presiding officer. Some minutes include the names of all those in attendance and those absent. This information can verify that a quorum of members was present if binding decisions are reached, and it

minutes
reports of business conducted at a meeting

may also encourage attendance as it is embarrassing to see one's name on the list of absentees.

If you are assigned to take and write up minutes for a meeting, get a meeting agenda in advance. Ask for the spellings of names of people who may be introduced so that you can record accurately and without interrupting the meeting. The list of topics on the agenda organizes your minutes. You may find it helpful to format the page on which you take the minutes into columns to encourage complete note taking, as shown in Example 15–3. This example shows a portion of the minutes from a meeting of the advisory committee for a state school. The left column identifies the topic, the middle column summarizes the gist of the discussion and decisions, and the right column says what the committee agreed to do and who will initiate the action.

TRIP REPORTS

Before a company invests in an expensive piece of equipment or a new process, it supplements what it has been told about the equipment or process by gathering firsthand information. Consequently, you may be sent as an organizational representative to a site where the equipment or process is in operation. The trip report you write when you return summarizes what you observed and evaluates the equipment or process in terms of your organization's needs.

The report of such a fact-finding and evaluation trip will probably be structured by topics that will influence the purchase decision, such as the way equipment fits into the overall manufacturing or production process, personnel needed to run the equipment, noise, servicing, or any other topic that will determine whether the equipment will work in its projected setting. These topics reflect the questions the company has about purchase. The topical structure will help decision makers assess the proposed purchase by revealing the issues. On the other hand, your notes may be organized chronologically, according to the order in which you observed things, and this ready-made structure may tempt you to order your report chronologically. You need to decide whether a chronological or topical structure will best help readers to interpret your findings and conclusions.

If the purpose of a trip report is simply to justify reimbursement for expenses, perhaps for attending a professional conference, the information can be structured chronologically. For example, if you attended a conference, your trip report might include the following material:

- Purpose, title, and date of conference
- Your role at the conference (attendee, presenter, session facilitator, etc.)
- Sessions attended (in order throughout the conference)

EXAMPLE 15–3   Minutes from a Meeting, Formatted to
Encourage Action

Regional State School, PRC Committee
Minutes: Meeting of August 11, 1993, 2:30–4:15 p.m. Conference Room.
Presiding: Helena Ramirez

Topic	Discussion and Decisions	Action/Person in Charge
New PRC members	The need for additional committee members was discussed. The new PRC policy allows an unlimited number of members for the facility, but more members will make it more difficult to achieve a quorum at meetings. There are no members to represent the communities outside the city that RSS serves. Recruitment should begin immediately after the holidays.	Begin recruitment after the holidays. Focus on surrounding communities. (Mary and Steve) Develop a means to publicize the membership drive. (Mary and Steve) Publicize the addition of new members to parent organizations, advocacy groups, and the general public. (Mary and Steve)
New PRC poster	After new members have joined, a new poster must be developed and distributed.	Produce a new PRC poster in the spring. (Helena)

- Information gained
- Professional contacts
- Summary or conclusions

This structure argues that you were where you were supposed to be and doing what you were supposed to be doing. On the other hand, if you are sent to evaluate equipment or processes, a topical arrangement will probably help your colleagues in the organization find the information they need to make a decision based on your observations (see Example 15-4).

LITERATURE REVIEWS

Literature reviews summarize and evaluate published material on a topic. In literature reviews you often present the secondary research you have obtained through the processes we described in Chapter 7. The literature review reports the current state of knowledge on a topic by summarizing what you have found as you survey relevant publications. Conducting a literature review is an essential first step in most research. It requires a trip to the library, the study of key reference materials, or an online keyword

literature reviews
summaries and evaluations of published material on a topic

EXAMPLE 15–4   Example of a Trip Report

Ink and Image Professional Communication

TO: Margaret Larson, Director, R & D
FROM: J. Franklin Smith
DATE: 26 August 1994
SUBJECT: Trip Report: Cupertino, California

I went to the three-day executive briefing on hand-held digital devices (or personal digital assistants—PDAs) you asked me to attend on 22–24 August 1994. The briefing was attended by approximately 15 other representatives from industry and education. The three-day briefing covered four aspects of digital devices:

1. Their ability to read handwriting and convert it to type;
2. Their ability to respond to natural language commands;
3. Their ability to send fax, e-mail, and cellular phone messages; and
4. Their connectivity with existing office equipment (fax, photocopy, phone, computer).

Current levels of technology allow for a fairly accurate conversion of handwriting to type, and these devices can easily be used to send faxes. Within the year, they should be able to serve as cellular phones and be used to send e-mail. The technology is not sufficiently developed to have a marketable natural language interface. Also, these devices are primarily stand-alone and are not easily (or inexpensively) integrated into existing office technology.

I met some interesting people who work in development of these devices. They could give us additional information if we need it. Let me know if you would like their names and addresses.

I think that these PDAs have promise for us, but their cost will mean we have to make careful decisions. We should talk over my findings and our plans at the next multimedia group meeting.

---

search. A literature review establishes what is already known about a topic so that new research does not duplicate existing knowledge, and when the literature review turns up an absence of information—a gap in current knowledge—it may establish the need for new research by unveiling a problem that needs attention.

You have already written a version of a literature review whenever you have written term papers. The information for a term paper comes primarily from the library, and you have had to indicate the sources of your information in your list of works cited. Your purpose in writing a term paper, however, was pretty much to demonstrate to your teacher your knowledge of a subject as well as your ability to navigate in the library and cite sources. The purpose for the literature review is not so much to prove your knowledge (though, in fact, citing the literature does strengthen your *ethos* as a writer) but rather to determine how the existing research applies to a current project. Like good term-paper writers, the writers of literature reviews not only summarize published articles but also evaluate and comment on the relevance of the information to the problem at hand.

A literature review may stand alone, or it may be part of another report. Stand-alone literature reviews have the purpose of assessing the current state of knowledge on a topic in order to suggest directions for continuing research. Some disciplines conduct annual reviews of the year's work in the field.

A literature review that is part of another report may have the more limited purpose of identifying knowledge related to a specific research project. For example, an engineer trying to determine whether a particular material will work in a new application will study reports of tests on the material to determine the likelihood that the material will be suitable for the new application. The literature review eliminates the need for duplicate research and helps the engineer assess whether and how to proceed with research in a new situation. Likewise, a committee of a public school system charged with developing a policy on dress codes will read to determine how other school systems have used dress codes and with what results. A consultant charged with recommending computer hardware and software for a company will cite reviews of the components in various consumer and computer magazines.

Literature reviews are common in academic writing. Scientific reports, for example, typically include a literature review as part of the introduction or in a separate section immediately following the introduction. The literature review is one way of showing how the new research is connected to what has been done before, that is, how it extends or refutes the existing research.

A literature review is written in essay or paragraph form. In this way, it differs from an annotated bibliography, which is a list of sources on a topic with brief summaries or descriptions. The literature review may be organized by source or chronologically (early studies to late), but when it is part of another report it is more likely to be organized by topic. A topical pattern best shows the relationship of various sources to one another and to the research problem. Topics in an engineering materials study, for example, might include the ability of the material to withstand stress and temperature and humidity, and the material's strength. The technical communicator putting together the review might cite several sources for each of these topics and might cite the same source in the discussion of more than one topic.

Readers of literature reviews are always asking the question, "What is the relevance of this information to the current problem?" Your evaluation and interpretation answer these questions in ways that mere summaries may not, by connecting existing research to the current problem.

For example, Mark Powell is a mechanical engineer. He wants to build a better remotely operated underwater vehicle (ROUV) equipped with a camera that can take underwater pictures without a human being on board. Why would engineers conduct a literature review before they go into the shop and start building? For one thing, engineers know they will make many mistakes if they work by trial and error or on the basis of their existing knowledge of engineering. Furthermore, if they need to get some money to fund the building project, they will have to write a proposal for funding to a potential sponsor.

Example 15-5 shows a version of the literature review that engineer Mark Powell will ultimately incorporate into his proposal. Through the literature review, Mark accomplishes several objectives:

- He determines what is already known about ROUVs and thus avoids duplicate research;
- He uses the literature to establish the problem that his project will attempt to solve;
- He establishes the "state of the art" on ROUV research and shows its limits in order to show the need for his project; and
- He establishes his own credibility, his *ethos,* with the potential sponsor by showing that he knows what he is talking about.

Mark organizes his review by topic: after an introduction defining the issue and previewing the discussion, he discusses applications of ROUVs (to show that they are important), current technology (to show the state of the art), and problems (to show that his own project is needed). He refers to various sources as they relate to each of these topics and refers to several sources more than once. His own "voice" is obvious in the review, especially in the section on problems. Although he summarizes what others have said, he also states a thesis (that research is moving in the wrong direction) and explains why he believes it. Thus, although the literature review depends on research that others have done, he makes his own statement about it.

Mark will modify his review somewhat to make it part of his proposal for funds. He may be able, for example, to condense the introduction (because the proposal will have an introduction) and place the information there about applications, technology, and problems in the corresponding sections of the review. But writing this review has given him some persuasive arguments for the proposal and even some of the words for it. (We will cover proposals in depth in Chapter 17.)

SCIENTIFIC REPORTS

scientific reports
documents that present the results of experiments or other exact methods of research

Scientific reports present the results of experiments or other exact methods of research. Although no report can be entirely objective, as each report is written within the context of the writer's organization, education, experience, and values, scientific reports attempt to report an accurate description of reality without the interference of bias or personal opinion. They report the test of a hypothesis and discuss whether the test results confirm or deny the hypothesis.

Much of the credibility of a scientific report derives from good research methodology. That is, readers believe the results of experiments when they

EXAMPLE 15–5   Literature Review by Mark Powell

Current Technology of Remotely Operated Underwater Vehicles

Remotely operated underwater vehicles, or ROUVs, are free swimming, remote-controlled submersibles. These submersibles often carry cameras and are used to conduct visual observations at depths too great for human divers. ROUVs, rather than manned submersibles, are used because they allow inspection to be performed with no risk to human life.

ROUVs serve as underwater "eyes" in many situations. One of the most common uses lies in the area of subsea pipeline leak detection and inspection [1]. In this capacity, ROUVs not only inspect the pipeline, but many are equipped for pipeline repair as well. Other ROUVs are used for foundation inspection on offshore oil platforms. Advanced ROUVs, such as JASON and the Scorpio Scout, have been very useful in the areas of underwater archaeology and marine biology [2].

The technology involved in building and operating ROUVs has advanced significantly in the past 20 to 25 years. The greatest advances have come in optical and acoustical sensors, allowing more precise navigation and "vision." Other advances have come in the areas of propulsion and power. Two of the main problems with current ROUV technology are size and cost. Although small when compared to manned submersibles, current ROUVs are actually quite large, requiring weights often in excess of 10,000 pounds for submersion. Costs have skyrocketed with each new step in technology. The cost factor alone has been sufficient to cause the cancellation of many ROUV projects. Yet the cancellations jeopardize the useful applications of ROUVs. If technological problems could be solved, the ROUVs could serve important purposes in research, inspection, and repair.

This report will elaborate on applications for ROUVs, investigate the current state of mechanical technology for ROUV design, and attempt to define the direction for cost-effective application.

Applications of Remotely Operated Underwater Vehicles

The applications of ROUVs are almost as varied as the types of ROUVs themselves. ROUVs are presently used mainly in the oil industry, in support of offshore platforms. Undersea pipeline inspection and repair, cable laying, salvage, and coral harvesting are other uses, as are ocean-bottom geology, marine biological research, underwater archaeology, and environmental missions [5]. *Remotely Operated Vehicles,* published jointly by the United States Department of Commerce (USDoC) and the National Oceanographic and Atmospheric Administration (NOAA), describes ROUV activities as being divided into 7 categories: inspection/determination, monitoring, survey, diver assistance, search/identification, installation/retrieval, and cleaning.

Over 90 percent of ROUV missions fall in the areas of inspection/determination and monitoring. Some of these missions include the determination of pipeline and/or cable positions after installation, inspection of pipelines and/or cables, location of pipeline "tie-ins," pipeline monitoring for leak detection, and wellhead inspection for offshore oil platforms. Other uses include the structural inspection of offshore platforms and inspection of dams and mineshafts [6].

Surveying is another important function performed by ROUVs. This function can range from visual inspection (via on-board cameras) of the sea bottom to echo-

EXAMPLE 15–5 (*continued*)

sounding for the determination of subsea topography. Martech International reports using an ROUV prior to setting a drilling platform. The ROUV was used for bottom inspection to locate a suitable position for the platform, as well as for inspection of the platform to determine final leveling adjustments [6].

In support of diving operations, ROUVs are used to monitor diver safety and performance, to inform the surface team of diving conditions, or to locate a dive site before any divers are sent down. ROUVs can also provide divers with a powerful, mobile source of light.

Underwater archaeology (search/identification, installation/retrieval) has found ROUVs to be indispensable. The use of ROUVs allows archaeologists to search, and even recover, artifacts from ships sunken in hundreds, or even thousands, of feet of water where diving by humans is impossible. The submersible JASON has allowed the discovery and exploration of such noted shipwrecks as the German World War II battleship Bismarck and the sunken Titanic [7]. In other search/identification roles, ROUVs are used by the US Navy for mine-hunting [8] and for investigation and retrieval of aircraft crashed at sea [6].

In a final role, ROUVs such as SCAMP can be used to automatically clean the hulls of hips while at anchor [6].

Current Submersible Technology

The majority of current submersible technology lies rooted in the methods of classical naval architecture. These methods include traditional propulsion mechanisms, underwater maneuverability systems, and power sources. However, in spite of tradition, there are several new design features currently incorporated into the design of modern ROUVs.

The majority of the advances in ROUV mechanical design have come in the areas of propulsion and underwater maneuverability. Britton [8] describes one such advance that has been incorporated into the Scorpio Scout, an ROUV currently in use by the US Navy. This design is comprised of four six-bladed propellers, each with variable-pitch blades. This design in itself is not exceptional. However, computer control allows individualized pitch control of each blade. In this manner, cross forces can be generated, allowing the Scout to "hover like a helicopter—or roll, pitch, yaw, surge, heave, and sway like an acrobatic seal" [8].

A second advance in propulsion/maneuverability is described by Kimura [3]. This method is known as the thrust vector control jet nozzle. This propulsion system is an improvement on the current "water jet," which uses internal pumps to create a jet of high-speed water for propulsion. Vectoring the jet in different directions accomplishes both forward propulsion and directional control with responses superior to those attained by conventional propellers. This system was first used in 1953 on Jacques Cousteau's Diving Saucer [9] and is currently used for a few ROUVs [6].

Power supply plays a critical role in any submersible design, whether manned or remote-operated. Originally, ROUVs were powered by lead-acid batteries or through a conducting tether connected to a surface support ship. However, new advances in energy storage technology, including zinc-oxide, silver-zinc, and aluminum seawater batteries, have greatly increased the amount of electrical energy that can be carried along with the submersible. Recent progress in "artificial gill" technology has made thermal cycle engines a future possibility [4]. Artificial gills for oxygen extraction from seawater could increase power output by as much as five to eight times.

EXAMPLE 15–5   *(continued)*

Problems and Proposed Direction

It is my belief that ROUV technology is moving in the wrong direction. Each new step in technology, whether in propulsion, energy storage, maneuvering, or observation, seems to force the submersible unit to become larger. This increase in size results in higher hydrodynamic drag and inertial forces, resulting in a substantial increase in power requirements for propulsion and maneuvering. The solution to this dilemma is simple—reverse the current trend in ROUV design by designing smaller submersible units. Through the application of technology not yet considered in ROUV design, this goal can be accomplished.

Research is currently underway at the Massachusetts Institute of Technology in the area of micro-robotics [10]. These micro robots contain "brains" modeled after the insect nervous system and use simple algorithms to perform such rudimentary functions as obstacle detection and avoidance. The smallest of these robots is a mere one cubic inch in size [10]. Obviously, this technology could easily be applied to present ROUV technology.

Visualization by a remote observer is a critical aspect of most ROUVs. For this purpose, large, bulky cameras are often used. However, new advances make this approach unnecessary. SuperCircuits, based in Austin, Texas, has developed a miniature 1.3 ounce real-time videocamera. This camera has excellent resolution in addition to being smaller than a common playing card. The cost: approximately $200 each.

Reducing the size of necessary on-board components can drastically reduce hull size. This means reduced hydrodynamic drag, reduced inertia, and, consequently, reduced powering requirements. The result is a drastically smaller, more cost-effective submersible with no decline in performance. ROUVs designed to perform a mechanical function, such as pipeline repair will, of course, require a larger size. However, those devoted to observation can easily be of reduced size. This reduction in size not only results in a substantial monetary savings, but will also allow the submersible to travel to places where a large submersible physically will not fit.

Conclusion

Remotely operated underwater vehicles have a vast array of applications in a myriad of industries. However, the current trend towards large, bulky submersibles has made this method of submarine endeavor effectively cost-prohibitive for all but the most financially sound corporations. The effective application of current microtechnology will eliminate this barrier, allowing increased ROUV operation in the areas of underwater biology, exploration, archaeology, geology, and observation.

References

[1] Mellin, Torgny A. 1989. Autonomous underwater system for pipeline leak detection and inspection. *Proceedings of the Sixth International Symposium on Unmanned Untethered Submersible Technology.* 15–24. Elliot City, MD: University of New Hampshire Systems Engineering Laboratory.

[2] Yoerger, Dana R. 1991. Robotic undersea technology. *Oceanus* v 34 n 1 p 32–37. Woods Hole, MA: Woods Hole Oceanographic Institution.

[3] Kimura, I., K. Osamura, and T. Takamori. 1990. Thrust vector control jet nozzle for submersible vehicles. *Journal of Fluid Control* v 20 n 3 p 67–81. Kobe, Japan: Kobe University.

EXAMPLE 15–5   (concluded)

[4] Ruggeri, R. T. 1989. Advanced underwater power systems. *Proceedings of the Sixth International Symposium on Unmanned Untethered Submersible Technology.* 153–167. Elliot City, MD: University of New Hampshire Marine Systems Engineering Laboratory.

[5] Vadus, Joseph R. 1976. *International Status and Utilization of Undersea Vehicles 1976.* Washington, DC: US Government Printing Office.

[6] Busby, R. Frank, and Associates. 1979. *Remotely Operated Vehicles.* 10–120. Washington, DC: US Government Printing Office.

[7] Baer, Tony. 1989. Viewing Jason's voyages to the bottom of the sea. *Mechanical Engineering* v 111 n 11 p 36–42.

[8] Britton, Peter. 1989. Submersible acrobat. *Popular Science* v 234 n 4 p 96.

[9] Busby, R. Frank. 1976. *Manned Submersibles.* Washington, DC: US Government Printing Office.

[10] Yeaple, Judith Anne. 1991. Robot insects. *Popular Science* v 238 n 3 p 52–55+.

Source: Used with the permission of the author.

believe the experiments were conducted using sound methodologies. Thus, an important section of a scientific report is the methodology section, which describes the methods of research in enough detail that another researcher can repeat the experiment in order to verify the results. Repetitions are necessary in scientific experiments to show that the results can be trusted.

You can recognize a scientific report by its standard structure:

- An introduction that defines the problem;
- A methods section that describes the methods of research in detail;
- A results section that presents the findings in factual (often quantitative terms); and
- A discussion section that explains the significance of the results.

IMRaD
an organizational structure for a scientific report consisting of an introduction, methods, results, and discussion.

This structure is called the **IMRaD** structure, for "introduction, methods, results, and discussion." Often a literature review follows or is part of the introduction. An abstract is a standard part of a scientific report and usually precedes the introduction (see Chapter 9). The scientific method, and therefore the scientific report, is used in the social sciences and in business as well as in science. Many academic journals consist of scientific reports.

LAB REPORTS

Lab reports are the records of procedures conducted as parts of scientific experiments. They report in detail the experimental procedures, specific results, and discussion of the results. This pattern reflects the pattern of scientific reports, although lab reports are less comprehensive than scientific reports.

The procedures section should be exact about the sequence of steps, the number of trials, and the "conditions" of the trials (that is, how the various trials may have differed from one another). The results section includes factual and specific consequences of the procedures. Tables and graphs may be used to show measurements. The discussion section interprets the results to explain their significance in terms of the goal of the procedure and to explain any surprises.

The writer of a lab report must be meticulous in describing procedures and results. Errors call into question the entire experiment; moreover, they can result in the development of a dangerous product or procedure. A lab technician works with notebook at hand, recording results as they occur rather than trusting memory. The accurate gathering and transmitting of data is one of your legal and ethical responsibilities as a writer of lab reports.

Conventions of Report Writing

Reports can vary not just in their purposes and outcomes but also in their organization, style, and format. Your choices about these report features will both assist and influence readers. Sometimes the conventions you will need to follow will be specified for you. For example, the American National Standards Institute (ANSI) has adopted a 44-page standard that specifies report parts, organization, format, typography, and the use of visual displays. The level of detail specified by ANSI is shown by this example: "The page number appears 3/16 inch below the bottom line of text" (p. 27). The Department of Defense and other units of government have adopted ANSI standards, and all reports written for those organizations must conform to them. Organizations may develop their own style guides, which specify how documents are to be prepared. The following discussion about organization, style, and format assumes that the writer has more options than the ANSI standards allow.

Organization reveals the relationship of ideas to one another and thus influences comprehension. Some types of reports, notably the scientific report, are consistent in the use and order of parts, and variation from this pattern invites readers to question the credibility of the information. Organization also influences reading patterns. Research by James Souther shows that managers do not read reports from the first word to the last but rather pay more attention to certain parts (most notably the executive summary) and may ignore other parts (the appendix). In order to accommodate the managerial reading style, good writers are especially careful in writing the abstract. They may also reverse the conventional beginning-middle-end structure of many written documents and place the end—the conclusions and recommendations—at the beginning so that managers can find "bottom line" information easily. The discussion sections of the report then follow the conclusions, almost like appendixes.

Thomas Pinelli et al., however, found that managers and scientists read NASA reports with close attention to all parts of the report. This research does not support a rearrangement of report parts for managers. When advice like this is contradictory, it is prudent to find out the preferred organization strategies where you work. It is also wise to realize that report writers do have options, and the choice of options will influence readers' responses. We will cover organization more thoroughly in Chapter 16.

Because of the variety of reports and different organizations in which they are written, there is no uniform style for all reports. Some organizations require writers to write in third person, using the pronouns "they" and "it" rather than "I" or "we." This requirement contributes to formality in style. Some organizations encourage writers to use first person ("I"). Either style can be satisfactory, but you should check to see what the preferences in your own organization may be. Before you write your first report, you'll need to review the advice in Chapter 10 on using style effectively.

Some reports are very short, such as handwritten memos, and others are many volumes long. Some reports are letters; some are double-spaced in traditional typescript; others are typeset and bound to look like books. Nevertheless, within organizations and for specific types of reports, such as progress reports, there are conventions for format. Consistency among reports means readers don't have to figure out the conventions for each new report. A sample report of the same type that has been written previously can show you the expectations about format, such as single spacing or double spacing, margins, and page numbers, as well as about the document sections that are expected in your workplace.

In terms of format and pagination, it is not usually necessary to begin each main division on a new page. In fact, it's undesirable to do so. Blank space at the bottom of the page signals *end.* One exception: when a heading for a new division appears at the bottom of the page. Unless a heading is followed by two or more lines of text, you should force a page break to place the new section on a new page where the lines of text can follow immediately.

According to Pinelli et al., readers like to see the illustrations and other visual displays integrated with the text rather than placed at the end, unless there are large numbers of visual displays. Readers look at the visual displays as they read the text, and if they have to flip to the back of the report, they are less likely to read them. Elaborate tables, however, that are useful more for reference than for primary information should go in an appendix.

headings
navigational tools that divide text into sections and subsections

Headings help readers in two ways: they enhance comprehension and they enable selective reading. They announce changes in topics and thus serve as transitions. They also reveal the structure of the information, much as an outline does, which helps readers interpret how topics relate to each other and to the whole report. Headings should at least announce the major sections, such as the introduction, discussion, and conclusion. When the structure of the report is topical, headings for each topic are useful. Headings with a few words are likely to be more informative than headings with a single word (for example, "maintenance costs" tells more than "costs").

Headings provide good signals, but be careful not to use too many. If you use headings more often than every three or four paragraphs, you are probably oversignaling, and your headings aren't working as effective maps for the reader.

Running heads state the title of the report or of a major section of the report, sometimes in a condensed version. Running heads in short reports are valuable as a source of identification if a single page is photocopied. In long reports with major divisions, a running head that names the division helps readers locate parts of the report for selective reading.

running heads
title of a report or major section placed at the top of pages of that section or report

Page numbers are essential for cross-references and to enable selective reading from a table of contents. Preliminary pages, which we describe below, are numbered with lower case Roman numerals (i, ii, iii) so that the text of the report can begin with the Arabic number 1. You don't have to put a number on the title page and contents page, but you do count the pages when you number the others. Pages can be numbered at the top (usually on the right), at the bottom (usually in the center), or on the outside margins (left for an even-numbered page and right for an odd-numbered page) when the report is printed front and back.

REPORT PARTS

Because reports are a genre widely used in the workplace, they have come to have standardized components. In addition to the body of the report, these include the following:

1. The letter of transmittal;
2. Title page;
3. Table of contents;
4. Executive summary;
5. Abstract;
6. Glossary;
7. Appendices; and
8. References.

Not all reports contain all these parts, and long, formal reports generally have more parts than short, informal reports. If the conventions of the organization where you work offer you some choices about what parts to include, choose parts that will be useful to readers and omit parts that readers will not need.

preliminary pages
pages that precede the text of a report, identifying the report and its contents, and giving credit to the people who created it

letter of transmittal
a cover letter that is not part of the actual report but that accompanies it to announce the attached document and project discussed

Preliminary pages precede the text of the report, identifying the report and its contents and giving credit to the people who produced it.

Letter of Transmittal

A **letter of transmittal,** or a cover letter, is not a part of the report itself but often accompanies it. The primary purpose of a letter of transmittal is to announce what the attached document is and what project it relates to. You use it primarily to orient your reader to the report's purpose and topic. Letters of transmittal may accomplish other purposes as well. They can function like prefaces to books, and you can use them to acknowledge assistance of individuals in a more personal way than through citation of sources. You can accomplish a sales purpose by indicating how this report is significant. You can prepare a reader for surprising or disappointing results and start to shape the response to them. We'll cover letters in more detail in Chapter 18.

Title Page

title page
page at the front of a text that identifies the subject of a report as well as its recipient, author, and date

The **title page** identifies the subject of the report as well as its recipient, author, and date. Its obvious purpose is to let readers know what the document is. Other information on the title page depends on what else readers need to know. You can use the title page to provide other identifying information, such as a project number or grant number. If your readers will need to get in touch with you and they do not have your address, the title page should include your and your co-author's phone numbers (including fax numbers if they are available) and addresses (including e-mail addresses). Thus, the title page may not only identify the report but also makes it easy for readers to communicate with the writer. Some title pages include a routing list when a single copy of the report will be distributed among several readers. Some title pages include a descriptive abstract (see Chapter 9). Often copyright material or circulation restrictions will appear on the title page as well.

The arrangement of the information on the title page should correspond with expectations of your readers. Some organizations have standard formats; your sample copy of a report should indicate the standard. If the organization does not prescribe a particular format, follow the convention of placing the title about one-third from the top of the page, followed by the name of the recipient, author, and date. Skip to Chapter 16 for a moment and notice the title page of Example 16–5.

report title
a title that tells the report's topic and the approach the report takes toward it

The main purpose of the **report title** is to inform. A good report title gives two types of information: (1) the topic, and (2) the approach to the topic. The topic defines the subject matter, such as the environmental and efficiency characteristics of electric vans. The topic alone, however, doesn't say why and how the writer has considered the topic. For example, the writer may have analyzed the feasibility of purchasing vans for a company vehicle fleet. Or the writer may describe the vehicles. The title for the report in Example 16–5 indicates both approach and topic:

Purchasing Electric Vans for Texas Tech University: A Feasibility Analysis
(topic)   *(approach)*

This information orients a reader to the report topic and purpose, but it also helps a reader decide whether to read the report at all. Perhaps the company files contain three reports on electric vans, one analyzing the feasibility of purchasing them, one describing the vans themselves, and one offering instructions for conversion. The approach information lets a reader pull the right report from the file.

A report title would be a boring title for a magazine article, but it does the job of informing readers of what the report is about. People assign reports because they need the information in them. Thus, they have an inherent motivation to read the reports. Cleverness may be desirable for titles of magazine articles because it attracts readers who have choices about reading, but it may trivialize a report.

Table of Contents

A **table of contents** (often called simply "contents") lists each main division of the report and the page on which it begins. Its primary function is to let your readers find specific information easily. Its secondary function is to give your readers an overview of the content and structure of your report. Notice the contents for Example 16–5.

The main divisions of long reports may be called chapters, but "chapter" would be a pretentious label for a section of a few pages. Instead, the contents page will probably list the level one headings (and sometimes level two headings). The page number indicates where the section begins.

The first item in the contents list will probably be the abstract. Generally, you won't list preliminary pages unless you have a preface, an unlikely part of any but a very long report. The contents page will list the back matter, however, including the appendixes, references, and glossary. Don't mix a list of tables and figures with the contents page. The visual displays are not sections of the report, so they are out of place in a list of report sections.

List of Tables and Figures

You may list tables and figures (visual displays of data) on a page following the contents page. This list is most useful when there are many visual displays and your report is long. There's less need for such a page for a 10-page report, which your reader can easily skim to locate the visual displays. If you do include the list, format it the same way you formatted the contents

---

table of contents
a list of the main divisions of a report and the page on which they begin; appears at the beginning of the document

page. Include the label and number (for example, "Figure 1"), the title of the visual display, and the page on which it appears.

Abstracts and Executive Summaries

You learned the difference between descriptive and informative abstracts in Chapter 9. A report often contains one or both of these abstracts, but an informative abstract is also called an **executive summary,** particularly in managerial reports. An executive summary appears at the beginning of the report and stands alone. The executive summary follows the conventions of the informative abstract in giving a condensed version of the whole report, including its conclusions and recommendations. Readers then begin the body of the report knowing the outcome, the main arguments, and the key facts. Readers can then go on to read the details in the report with an understanding of their overall significance and to test the conclusions mentally.

executive summary
an abstract that appears at the beginning of the report and gives a condensed version of the entire document

Text of the Report

All reports have a beginning that introduces the issue, a middle that presents the results of investigation, and an ending that applies the results to the problem to show whether and how the problem is solved. The introduction describes the problem or purpose that has occasioned the report. It answers the questions, who, what, when, where, why, and how in order to establish the context for the problem and its significance. It also forecasts the rest of the report by indicating how the report will develop and what its major sections are. A literature review may be included in the introduction as a way of establishing the problem. The ending of the report may include one or more parts: summary, conclusions, recommendations. Each of these parts has a different function and must be labeled accurately.

A **summary** simply reviews the main points of the report. The report may end with a summary if the purpose of the report has been to describe an existing situation or the current state of knowledge. The summary thus wraps up the report and reinforces the main points. The summary can also be used to prepare the readers for your conclusions and recommendations. A review of the main points can be especially useful in a long report that asks readers to deal with a lot of information.

summary
a review of the main points of the document

The conclusions section of a report answers the research question and thus goes a step beyond the summary because answering the question requires evaluation and interpretation.

Recommendations direct action. They advise what should be done based on the conclusions. Managerial reports usually contain recommendations. For an example, see the recommendations section of the report in Example 16–5. The recommendations are listed in a separate, labeled section. The recommendations section does not need to provide much explanation because that has already been covered in the conclusions.

recommendations
section of report that contains advice designed to direct action

Back Matter

After the report's conclusions and recommendations comes the **back matter,** additional pieces of information intended to supplement the main body of the report. Page numbering continues with Arabic numerals.

Appendices include material that is useful for reference but not essential to the argument of the report. A copy of a survey form, detailed tables, or the text of relevant laws would appear in the appendix so that they would not interrupt reading the report. Material that the primary reader of the report does not need but that interests a secondary reader could be placed in an appendix as well. Material in an appendix is much less likely to be read than material in the body of the report, so do not place material there that your primary readers need. Refer to each appendix in the text at the point in the text at which consulting the appendix would provide the reader with useful but not essential information.

If there is more than one appendix, each appendix should have a label and title, much like the label and title of a visual display of data. The label identifies the appendix by a capital letter (Appendix A, Appendix B). The sequence corresponds to the sequence in which the appendixes are mentioned in the text of the report. Each appendix should begin on a separate page and be listed in the table of contents.

As you learned in Chapter 9, a list of **references** (or "Works Cited") includes those works that have been cited in the text. These items are listed and formatted according to the documentation style that has been adopted for the report: MLA, APA, Chicago B, or some other style.

A **glossary** is a mini-dictionary of terms that apply to the topic of the report. Include in it only those terms that will be unfamiliar to your readers. If you believe that most of your readers will need the glossary to read the report effectively, you might move the glossary to the front of your report.

back matter
additional pieces of information that are useful but not essential to the report's argument

appendices
reference material related to the report

references
the *Works Cited* section of a report

glossary
a mini-dictionary of terms that apply to the topic of the report

SUMMARY

A report can be only as good as the work that is reported. Thus, one requirement for writing a good report is to do good work. That means keeping your work on schedule; being thorough in your investigation and analysis; being accurate in your observations and notes; collaborating effectively and efficiently with those whom your work touches; pursuing the goals that have been established for you or your project; and doing whatever else it takes to perform at a high level of quality, competence, and integrity. You'll need many of the techniques of persuasion you learned in Chapter 5.

Good work, however, does not automatically mean that you will write a good report. A good report orients a reader to the project and anticipates and answers the reader's questions. It considers not only the primary reader but also readers to whom the report may be forwarded and future readers. The data in the report must be credible and complete. Furthermore, a good report is well organized and formatted to help your readers find information if they read selectively and to understand the relationship of ideas or topics. A style that emphasizes accomplishment and projects confidence will enhance the report's persuasiveness and increase the chances that the report will influence action and decisions in the way you think will be best.

Successfully bringing together good work and writing for an effective reporting outcome requires planning at the beginning of the project and management throughout it. A lot of the work will be mental: conceptualizing the report in use and the readers' needs for information; planning research to produce that information; selecting strategies of organization and format to enhance comprehension of readers and persuasiveness of the report; and scheduling tasks so that you will have a chance to complete them without cutting corners. Managing these processes involves subdividing the reporting into smaller tasks and completing the tasks in stages. This advice can work in all reporting situations.

This chapter provided you with many good work suggestions for report writing, suggestions that you'll need to remember as you study the next two chapters. Writing Strategies 15-1 will prove useful in planning your reports, and Figure 15-1 will aid you in defining the type of report, the information contained, the readers, and the resulting action for future writing assignments. You will need to refer back often to the conventions of report writing as you plan the style, format, and organization of your reports. Finally, this chapter concluded with a discussion of report parts; use this section as a checklist as you prepare the reports assigned throughout this textbook.

ACTIVITIES AND EXERCISES

1. The case documents in Appendix A (Case Documents 2), on the *Exxon Valdez* oil spill, describe the use of bioremediation to clean up the oil. With a small group of classmates, brainstorm the reports—both managerial reports and research reports—that might have been written in conjunction with this case. Think of who might have written a report, to whom, and for what intended action. You can use Figure 15-1 to help you name reports, but it's fine to define reports according to purpose rather than by a specific name. Present your findings in a memo to your instructor.

2. Analyze the periodic report in Example 15-1 to determine strategies of style. The writer selects words and phrases to encourage a positive interpretation by readers of the data. For example, the first paragraph includes the phrases "on the right foot" and "favorable period." What other words and phrases throughout the report encourage readers to regard the report as good news? Be prepared to share your analysis during class discussion.

3. How does the writer of the report in Example 15-1 convey an attitude of respect for readers? How does that attitude serve the writer's persuasive purpose? How does the writer accommodate different levels of financial sophistication among readers? Be prepared to share your answers during class discussion.

4. Find a report written where you work or in the government documents section of your library. In a memo to your instructor, conduct the following analysis:
   - Try to determine its likely readers and purpose.
   - Analyze the writer's strategies of argument: How is the report arranged (chronologically, topically, or scientifically, or by another method)? How

does the writer establish credibility? What are the features of format, organization, and style? What devices, such as a descriptive abstract and title page, does the writer use to orient readers to the report topic? What devices, such as headings and running headers, does the writer use to make the report easy to use?

■ Finally, evaluate the report: How effective are the writer's choices? How might the writer have made the report easier to understand and use?

5. Write a progress report, addressed to your writing teacher, on your progress on a writing assignment. Report specific tasks accomplished and tasks remaining, and include a completion schedule. Include research and report preparation tasks in your report.

6. Anticipating an assignment for a report for decision making (Chapter 16), conduct a literature review on a topic of interest that you can later incorporate into the report.

7. Assume that you have been assigned to write a literature review on one of the following topics:

a. Use of AZT in slowing down the replication of HIV (see Appendix A, Case Documents 1).

b. Treatment of tuberculosis in people with AIDS (see Appendix A, Case Documents 1).

c. Recent liability cases involving products and accompanying documentation (see Appendix A, Case Documents 4).

Work in the library to find at least eight sources on the topic you select (the references at the end of the appropriate documents in Appendix A may give you a start). If you choose option a or b, create a literature review for your supervisor at the AIDS treatment center of a small, rural hospital in Minnesota. The purpose of your report is to inform your reader on "what is being done," what treatments are currently being used elsewhere, what progress has been made in the treatment of the disease, and what procedures your hospital might want to adopt. If you choose option c, create a literature review for your supervisor in an appropriate company. Assume that the purpose of your report is to inform your reader about "what isn't being done" by industries to protect users or consumers of products who might be injured.

WORKS CITED

American National Standards Institute. *Scientific and Technical Reports—Organization, Preparation, and Production* (ANSI Z39.18-1987). New York, 1987.

Killingsworth, M. Jimmie, and Michael K. Gilbertson. *Signs, Genres, and Communities in Technical Communication.* Amityville, NY: Baywood, 1992.

Miller, Carolyn. "Genre as Social Action." *Quarterly Journal of Speech* 70.2 (1984), pp. 151-67.

Pinelli, Thomas E.; Virginia M. Cordle; Myron Glassman; and Raymond F. Vondran, Jr. "Report Format Preferences of Technical Managers and Nonmanagers." *Technical Communication* 31.2 (1984), pp. 4-8.

——— "Report-Reading Patterns of Technical Managers and Nonmanagers." *Technical Communication* 31.3 (1984), pp. 20–24.

Souther, James. "What to Report." *IEEE Transactions on Professional Communication* PC 28.3 (1985), pp. 5–8.

Reports for Decision Making

Making Decisions about Action: A Procedure ▮ Planning and Organizing the Report for Decision Making ▮ Managing the Report Development Process ▮ Some Difficulties of Making Recommendations

In and beyond the classroom, the approach to decision making shapes the quality of decisions made. Shortsighted decisions, not just in the classroom, result in part from inadequate procedures of inquiry and communication. For example, flaws in decision-making processes and communication have been dramatically revealed by studies of the *Challenger* disaster . . . Short-term and measurable criteria, especially cost, often influence decisions more than long-term impacts on health, safety, the environment, or on political, economic, and social systems, though these social criteria may ultimately determine the effectiveness of the decisions.

Carolyn Rude, "Reports for Decision Making: Genre and Inquiry," *Journal of Business and Technical Communication* (Forthcoming April 1995).

Introduction

I n all organizations, decisions must be made. These decisions result from problems, such as those we discussed in Chapter 2, or from opportunities. Sometimes the decisions are simple: an office manager discovers that paper for the laser printer is running short and decides to order a three-month supply. Sometimes the decisions are far more complex: a researcher who would like a laptop computer to aid in onsite data collection must choose from a bewildering selection of brands and models; an automobile manufacturer, worried about the problem of high production costs, considers whether to transfer its US assembly plant to Mexico.

People base some decisions on hunches or intuition, experience, preferences, or the desire to please others. For example, if your experience in a chemistry class has resulted in good grades, you are more likely to decide to take an additional chemistry class than will a student who has failed chemistry. These methods are legitimate in some situations, but when a decision is complex, these experiential methods may not provide good enough reasons for choice. The decision may require information available only through research. Research reflects awareness that information and judgment can improve the quality of the decision. The automobile manufacturer considering a move to Mexico needs information on the comparative costs of labor; moving, construction, and legal costs of transfer; training of labor; estimated productivity rates; availability of land and community support; environmental and import laws; taxes; impact on present employees; effect on the economy at the present assembly site; and attitudes of customers. A wise decision requires a thorough investigation.

The person who researches the issue will probably recommend an action to a superior in the organization. The person who has the ultimate authority to make the decision may not have the time or even the expertise to investigate all the issues. Thus, the decision maker assigns the investigation to a staff member or group of staff members. In a collaborative investigation, a lawyer may investigate legal aspects, an accountant may investigate financial aspects, and an industrial engineer may investigate production methods and rates. When the investigation is complete, the staff member or group presents the results of the investigation with a recommendation for action in a report to the decision maker. The decision maker bases the decision on the information and recommendations in the report. Even as a new employee, you may have to write or collaborate in writing reports for decision making. Your manager will entrust you with research and analysis and will depend on your recommendations in reaching his or her own decisions. (At this time, you might want to review the discussion of transmitter, decision-maker, advisor, and implementer roles in the section entitled, "What Are Your Audiences' Needs?" in Chapter 3.)

In this chapter, you will learn a procedure for investigation when the expected outcome is a recommendation for action, and you will also learn more about how to arrange your findings in a report. We begin the chapter with a discussion of the kinds of problems that require decisions. We continue with methods of investigation and ways of interpreting findings and offer suggestions for arranging the report and for handling difficult situations in decision making.

The procedure for investigating a problem and recommending an action differs somewhat from other research you may have done, including library research, lab experiments, and mathematical calculations, though all of these methods may provide some of the necessary information for your report. You always need to match your investigation procedure to the type of problem that needs to be solved.

We discussed problem solving in Chapter 2 and recommended that you compare the *actual situation* against the *ideal situation* to find a solution (the *purpose* of your communication). In Chapter 2, we also suggested expressing your goal in a statement, adding the connecting word *but* or *however,* describing a condition that stands in the way of your goal, and then describing some action that would resolve the difficulty (Example 2–2). Finally, in Chapter 2, we contrasted procedural problems (problems of "know how"), informational problems (problems of "know that"), and decision-making problems (problems of "what to do"). In this chapter, we focus again on decision-making problems, problems that are solved by taking action. The procedure for the investigation that leads to a recommendation includes these steps:

Making Decisions about Action: A Procedure

1. Defining the problem.
2. Forecasting a solution.
3. Identifying the research question.
4. Establishing selection criteria.
5. Establishing research methods.
6. Researching and interpreting information.

DEFINING PROBLEMS

As you learned in Chapter 2, decisions about action become necessary when there is a conflict, discrepancy, or inconsistency between an actual or existing situation and the ideal situation (often called a **goal**). A company may desire greater sales on a product than it is achieving. This discrepancy between what the actual situation is and the desired goal represents a problem if the company does not know how to increase sales. Some impediment, such as lack of information, may prevent the achievement of the ideal situation. The solution to such problems is action; the possible actions include doing nothing.

goal
an ideal situation toward which an individual or organization aspires

Problems don't necessarily indicate trouble: the discrepancy between the actual situation and the ideal one can represent opportunity. A company doing well in its present operations may see the possibility for expansion. The discrepancy in that situation can be seen in two ways: between the present rate of return on investments and desired rate, or between the desire to expand and the uncertainty about whether expansion is feasible. Lack of information about what course of action will produce the desired result requires an inquiry.

Problems that require decisions are socially defined; that is, they do not exist in nature. Conflicts or discrepancies exist only if someone identifies ways in which the actual situation or present course of action seems unsatisfactory. Even the shortage of paper for the laser printer is a problem only because people choose to make laser prints and need paper to do so. No biological or physical necessity requires a researcher to use a laptop computer; he or she chooses to use available resources for this purchase rather than for another. A manufacturer must make money to stay in business, but nothing requires the manufacturer to target labor costs for greater efficiency. Because these problems are socially defined, good solutions are also socially defined. To make decisions, the people in authority must agree that the decision is the best one of the possible choices. They must reach a consensus, because before the action occurs and can be evaluated, there is no proof that one decision is best.

Problems are unique in some respects to the individuals or groups that have them. You and your friend may both consider the purchase of a car, but your different interests and preferences, needs, and finances may point to different choices for each of you. Two organizations considering whether to network the personal computers in the office may reach entirely different decisions, even if both own the same type and number of computers. The circumstances, such as the need for staff members to communicate and the physical arrangement of offices, may differ in companies with similar problems and point to different conclusions. An engineering research firm that wants to install a wind tunnel in one of two existing buildings can probably not choose the best option based on what another engineering firm has done.

This is to say that the *problems have meaning in a particular context.* Unlike mathematical problems, problems that require decisions do not have single, correct solutions that can be reached by calculation. A decision maker cannot turn to absolute rules for solving problems, nor can he or she be sure there is a right answer out there waiting to be discovered. The decision that is right in one context may not be right in another context. Complex problems usually offer a great deal of choice in solution.

The social, unique, and ill-defined nature of problems that require decisions indicates the importance of defining the problems carefully, using a systematic procedure for solving them. The first step is defining the problem.

Problems can be defined in an *A but B* structure, with *A* representing the ideal situation or goal, *B* representing the actual situation or the impediment to achieving the goal, and *but* signaling the conflict. Sample problems are summarized in Example 16-1.

When the problems offer simple solutions, such as problem 1, the decision about action (the solution to the problem) is also simple. The only necessary "report" is an oral one from the person who notices the shortage to the staff person in charge of supplies. Then the paper can be ordered. The problem becomes complex only if there is no money for paper, the shortage has been caused by the extravagant use of some staff members, or a particular kind of paper is needed on an emergency basis and the supplier doesn't carry it. In any of those cases, the problem and solution may require analysis and a change in the current way of doing things.

Many purchases, such as the one represented in problem 3, require formal study and sometimes a recommendation report, especially if the cost is great, the products are unfamiliar to readers, and there is more than one basis for making the decision (more, for example, than cost).

Complex problems, such as problems 4 and 5, require investigation and a formal, written report. Too many complicating factors and too many risks exist to change the present course of action arbitrarily. Before the decision can be made, a solution must be forecast and a study of the potential solution must be completed.

FORECASTING A SOLUTION

Once a problem is defined, possible solutions must be forecast to give the investigator some direction in research. As John Hayes observes, "If people can't think up any approaches, then they can't solve the problem" (p. xiv). There may be more than one possible course of action in addition to the present course.

Brainstorming, as we discussed in Chapter 4, yields *possible* solutions. For example, to solve the problem of high costs of production, the US automobile assembly plant may consider not only moving the plant to Mexico but also analyzing assembly procedures or training and supervision with the aim of improving them. It is important to brainstorm all possible solutions because one may be better than another. Usually the simpler and less disruptive action is preferred over a complex action, assuming that the simpler action will achieve the desired goals. Moving a plant to another country would severely disrupt manufacturing; if an easier course of action will solve the problem, it will be preferable.

Decisions regarding purchases often present multiple choices of brands, models, and features. In such a case, the forecasting task is identifying some possible choices. This task may require a needs assessment, which is a detailed analysis of the uses for the purchase. The problem statement may

brainstorming
early stage in the writing process when one tries to create an extensive but unedited list of ideas or solutions

---

EXAMPLE 16–1   Problem Statements, Brief Form

Goal or Ideal Situation	Signal Word	Actual or Existing Situation or the Unknown
1. The office requires paper in order to complete its printing tasks.	But	The supply of paper is almost exhausted.
2. Onsite data collection is necessary.	But	Manual collection is inefficient (error-prone and time-consuming).
3. A laptop computer would enable efficient onsite data collection.	But	Choosing a computer requires knowledge of features, capacities, and brands.
4. To be competitive, the company must produce cars efficiently.	But	Labor costs are 7 percent higher at Plant 1 than at Plant 2.
5. The quality of the students majoring in English must remain high for the programs to remain competitive.	But	As the number of English majors has increased in the past four years, the quality of students, as measured by GPA and SAT scores, has dropped.

---

identify the need for a computer, but a needs assessment will go further to determine whether the computer will be used for word processing only or also for databases, computer-aided design, or desktop publishing. These needs will restrict the number of good choices. Thus, the decision about purchase will be better if the needs are assessed *before* choices are limited. We return to the task of creating a needs assessment later in this chapter.

forecasting
stage in report writing when one tries to select some possible solutions from the list of all possible solutions

Forecasting also requires some selection. From the possible solutions that brainstorming reveals, some can be ruled out as impractical. The investigator chooses the option or options with the greatest promise for further study. If you are investigating the problem of lack of transportation to a job, your brainstorming may yield these possible solutions: using public transportation (bus, subway), carpooling, and purchasing a car. If your job requires you to work later than public transportation is available or you work different hours from your co-workers, you do not need to study the public transportation and carpooling options further. The investigation will test whether and how the purchase of a car will solve the problem. Forecasting has been important, however, because it has let you consider less drastic options before you conclude that you will investigate only the option of purchasing a car.

IDENTIFYING THE RESEARCH QUESTION

Forecasting identifies possible solutions, but it does not provide the information necessary to determine which decision is best. That is the task of research. The first step in research is identifying a research question. The research question focuses the research by stating its objective.

Three basic types of questions focus the research for decision making:

1. *Is the project feasible?* This question points to a study of one solution that forecasting identified. The study will ask whether the individual or company *can and should* follow a particular course of action. The *can* part of the question relates to measurable aspects of choice: money, equipment, and technical know-how. The *should* part of the question relates to values. Your initial research in the purchase of a car may be a feasibility question: Is it feasible to purchase a new car? The *can* questions will relate to cost, loans, and insurance; the *should* questions relate to competing uses for funds and possibly to the environment.

2. *Which is better, option A or option B?* This question points to a comparison of options and is the usual question in the case of a purchase. If you know, for example, that the purchase of the car is feasible, you might compare two or more models to determine which better meets your needs.

3. *Why does this situation occur, and what can be done about it?* This question points to an analytical study, a more detailed examination of the problem. The automobile manufacturer may ask, for example, why the labor costs are 7 percent higher at Plant 1 than at Plant 2 when both plants are building the same model car. When this research question focuses the study, the forecasting of the decision-making procedure will take place after the question is phrased. The researcher will brainstorm some hypothetical answers to the question to forecast the possible solution.

The problem statement and forecast will indicate how the question should be phrased. The same problem might lead to different research questions and different types of study, depending on the need for information. Sometimes forecasting reveals multiple solutions, and the question will be the second one. But if one of the choices seems more attractive at the outset, a study to determine the feasibility of that choice makes sense. If the answer turns out to be no, then you can explore another of the choices.

At the beginning of a study, it is important that you phrase a question rather than make a statement or proposition. The whole idea of research and reporting on the research is to allow for an informed decision to be made. If the best course of action really is unclear, your investigation should be open to various solutions and to the possibility that an unexpected or

research question
question that focuses research by stating its objective; follows forecasting in creating reports for decision making

surprising solution may be the best one. As an investigator, you do not simply choose a solution and defend it in the report to the decision maker. Such a procedure would not only deny the value of information gathering and reasoning in problem solving, but it would also limit the options for solution.

IDENTIFYING SELECTION CRITERIA

criteria
standards for making a good decision in a particular instance; when buying a car, selection criteria might include safety, gas mileage, and dependability

Before the research begins, you should identify the **criteria** for a good decision in the particular circumstance that requires the research and report. If you are buying a car, for example, your selection criteria for choice may include cost, dependability, style, and safety. You may also base your decision on features such as manual or automatic transmission, reputation of the dealer, and ease of maintenance. Different people will have different criteria for their car purchases. For example, some buyers will always use professional mechanics for maintenance, while others will do some maintenance themselves; these plans will determine whether ease of maintenance is a criterion for selection of a car. Even when people with the same problem share criteria, they may rank the criteria differently. Criteria define the information to gather in research and provide the standard for assessing the information.

Because most people want a measure of certainty in decision making, they may think first of criteria that can be measured or that are quantifiable. Or, they may assess only whether an action is possible. For example, if you want a new car but do not have the money for it and cannot get a loan or gift, the purchase is not possible no matter how well it would solve transportation problems. If you would like to open a plastics recycling plant but cannot hire a person with knowledge of the chemical processes necessary in plastics recycling, the project is not possible. If you want to design a car that uses composite material in the bumper, but the bumper built with this material will not meet government crash test specifications, the material is not a feasible choice and that action is not possible.

feasibility
determination that a project is both possible and desirable

Even if a project is technically possible, however, it may not be a good idea if it does not meet other criteria related to project management and the values of the people who will be affected by the project. If a project would jeopardize human safety beyond reasonable limits, it is not feasible even if it is technically possible. For example, if the City Council considers bidding for a state prison in the city to bring in jobs but the citizens are adamantly opposed to a prison, the prison may not be feasible even though there is enough land and labor for the prison. **Feasibility** means that a project is both possible and desirable.

Example 16–2 lists possible criteria for decision making in three categories:

- ■ Technical criteria, which determine whether something can be done.

EXAMPLE 16–2 Criteria for Decision Making

Technical Criteria	Managerial Criteria	Social Criteria
Technology (availability of)	Costs: materials; equipment; renovation; maintenance; financing; transportation; disposal; salaries; licenses	Human safety: manufacturers; operators; bystanders
Physical space; structural requirements		Gender; race; age
Dimensions; capacity	Income: fees; grants; interest	Quality: product; service
Materials: strength; corrosion; weight; availability	Market: demand; need; interest; competition	Environmental impact: long term; short term; manufacture; use; disposal; habitat
Parts: availability; accessibility	Taxes	Human impact: long term; short term; jobs; morale; employment benefits
Material handling requirements	Consistency with organizational goals	
Compatibility with existing systems	Organizational impact: personnel; morale; image; distribution of resources; effect on other projects; changes required in existing operations	Social and cultural issues (e.g., availability of schools, the arts)
Ability to be upgraded		
Adaptability; flexibility		Ethical issues (e.g., conflict of interest)
Ease of manufacture; use		
Maintenance; service; technical support (availability and quality)	Staffing: number; qualifications	Values: personal; organization or community; society
Reliability; longevity; repair record; warranty	Training	Convenience; comfort
Legal restrictions; standards, codes; precedents (e.g., environmental, Americans with Disabilities Act)	Policies (e.g., affirmative action)	Aesthetics
	Time required: building; training; performing procedures	
	Schedule	
	Quality of written instructions	

- Managerial criteria, which relate to day-to-day operations.
- Social criteria, which relate to values and the impact of a decision on people affected by it.

Social criteria may be hard to measure, but they can affect the success of a decision in a powerful way. Criteria for a specific decision may differ from the ones listed in Example 16–2. Furthermore, the categories for the criteria are not fixed. Social criteria could become technical when regulations govern them. For example, in some situations, choices regarding the environment are matters of values, but in other cases government regulations establish limits on activities that could harm the environment. Gender, race, and age could be technical issues if compliance with laws is a question. And, the impact of a new project on existing ones and on the people involved is both a social and a managerial concern. Example 16-2 will help you anticipate the consequences of a decision in a comprehensive way and therefore select appropriate criteria for decision making. Also, Example 16-2 suggests that a good decision may require you to evaluate issues that go beyond the letter of the law and technical specifications.

When you are charged with making recommendations for action to solve a problem, you will identify criteria for decision making early in the process of investigation. A good first step is to consult with the person who assigned the research task to you and anyone else in the organization who knows the problem. You can interview these people to determine the history of the problem, such as solutions that have been tried previously, specific needs, and operating procedures that may affect the outcome of a proposed solution to the problem. You can also use Example 16-2 to help you identify relevant criteria for this project. The criteria may differ from those in Example 16-2, but you can use those criteria in brainstorming. The criteria you select will reflect your ability to anticipate and assess possible future consequences of actions. Establishing criteria means that a decision will be a good one if it results in certain good consequences and avoids other bad consequences.

Identifying and ranking criteria were important tasks, for example, when the library of a medical school outgrew its space in its location on the second floor. Some space on the fifth floor had been planned for a fitness facility but had never been developed. The library administrator wondered if the library would solve its problem of limited space by moving from the second floor to the fifth floor.

The following were some of the criteria in the library case:

- Space: Would the fifth floor provide sufficient room for the library, now and for the foreseeable future?
- Access: Would users be able to get to the fifth floor easily and conveniently? Would existing elevators and stairs accommodate the number of users who would visit the library each day? Were the elevators accessible for users in wheelchairs? Were the elevators convenient to the building entrance?
- Support facilities: Would available restrooms be sufficient for the anticipated number of users? Could a snack bar be installed in a convenient location?

- Building structure: Was the fifth floor built to withstand the weight of the books and equipment in a library?
- Impact of displacement of the fitness facility: How important to employee health, morale, and even recruitment was the planned fitness facility? Was there another suitable location for it?
- Costs: What would be the costs of renovation and moving?

To identify criteria, the investigator must anticipate the completion of the move. The investigator visualizes the fifth floor facility as a library to discover what information a good decision requires. Visualizing the future also helps the investigator and managers anticipate the consequences before they approve the move and start packing books. A decision about a future action cannot be proven right or wrong before the action happens, but thoughtful anticipation of the future provides information that can be used in planning and in preventing mistakes.

The criteria are not equal in importance, and the order in which they emerge in brainstorming does not necessarily reflect their importance. Ranking them will help you determine the order of the investigation and also help you interpret results. In the medical school library example, the two most important criteria are space and building structure. Questions of cost and access are irrelevant if the books will not fit in the space or if the floor would collapse under their weight. These criteria should be researched first.

If the space and structure are adequate, research can proceed on the other questions. If space and structure are not adequate, the research may cease. The research question has been answered: the move is not feasible. The research would proceed only if some arguments for this particular solution are so compelling that they warrant an exploration of modifications to the building to remedy the deficits. Perhaps, for example, the structure could be reinforced, or the space could be expanded. In that case, the direction of research would change to include the new options (though the criteria for decision making would not change).

Cost will, of course, be a **determining criterion.** It is not, however, the first criterion to research. For one thing, the cost analysis cannot be completed until structural and renovation requirements are determined. There's no point in researching costs until the primary criteria have been researched.

Sometimes studies require two or more phases. Through the establishment and ranking of criteria, you plan a preliminary study to address the most important criteria. The report recommends either to continue the study or to stop it. The division into phases may spare the time and cost of unnecessary research. Brainstorm all the possible criteria, however, even if you don't address them. Doing so will help you grasp the entire situation and to consider several criteria in context of the whole problem.

determining criterion
one of the selection criteria that determine if a project is feasible; in many projects cost is a determining criterion

ESTABLISHING RESEARCH METHODS

Criteria establish *what* needs to be researched; the methods indicate *how*. Many projects require several methods. Sometimes the method is a scientific test or experiment, as in a question about materials—for example, whether a storage container would corrode with particular contents. Sometimes the method is library research (as discussed in Chapter 7), although this method is likely to provide only some of the necessary information. Often the methods include surveys, interviews, and site inspections (as discussed in Chapter 8). You will have to pull together information from various sources.

One way to determine methods is to list the criteria you come up with and then to match each one with appropriate methodologies. Example 16–3 shows the planning for the medical school library example we discussed above.

Organize the research itself by methods rather than by criteria. The architect, for example, is a source for three of the criteria. Rather than plan three separate interviews, you can plan one interview to address three separate topics. If the study is a preliminary study, however, and will address only space and structure, you will not need to gather information on expanding the restrooms.

As you proceed with your research, your methods may change somewhat. A person you interview may suggest that you consult a source that you had not considered, or you may discover information that changes your perception of the problem and solution. If you learn from the architect, for example, that the present building structure is inadequate for the library but structural reinforcements could remedy the deficit, you begin to inquire about the nature and cost of the reinforcements. The process of investigation is flexible because it is a learning process. The planning simply gives you a place to begin.

INTERPRETING RESULTS

Decision makers look for good reasons for making a choice. Reasons, to be credible, include both facts and interpretations of the significance of those facts, as we saw in Chapter 9. Claims without the support of facts are not credible. Readers will be convinced that the building structure will not support the fifth floor library only if the investigator cites the architect's information about the number of load-bearing walls, the weight they can support, and the weight of the books. Likewise, facts alone are often ambiguous. If the investigator cites only the figures on load-bearing walls and weights, the readers may respond by saying "so what?" The readers want a plain English response to the question that has been posed: Will the building structure support the weight of the books? A statement that answers the question interprets the facts.

Ironically, an abundance of facts increases the need for interpretive

---

EXAMPLE 16–3   Matching Criteria and Methods

Criterion	Method(s)
Space	1. Interview the librarian to determine linear feet of stacks needed through the next 10 years and the square feet of space needed for offices, study area, and equipment.
	2. Measure the fifth floor area.
Structure	1. Interview the architect who designed the building.
Access	1. Interview the librarian to determine estimated number of users.
	2. In the main library, consult standards for wheelchair access.
	3. Review the building floor plan and walk through the path a user might take from parking lot to fifth floor to judge how a user might perceive ease of access.
Displacement effects	1. Interview a personnel officer to determine the importance of the fitness facility on employee recruitment and retention
Restrooms, snack bar	1. Site inspection: count the number of restrooms and stalls; measure for wheelchair access and grab rails; evaluate the convenience of the location with respect to the library entrance; note possible areas for expanding the restrooms or snack areas (such as adjacent storage areas that could be relocated).
	2. Consult the architect or blueprints about existing plumbing and wiring and ease of expanding restrooms or creating a snack area.
Costs	1. Consult the architect for estimates on renovations.
	2. Consult the building manager for estimates on the cost of moving.

---

statements. Readers may not be able to determine the significance of all the numbers from an elaborate table, for example. The facts do not speak for themselves because there are so many. As investigator, you are closer to the facts and more expert in their meaning; consequently, you have an obligation to report to readers your judgment about the facts' significance. Doing so makes your reasons for your conclusions explicit, and it also gives readers a basis for additional interpretations.

Your report's organization will reflect the method of investigation. Reports for decision making differ from lab reports and reports of scientific investigation because the methods of all of these investigations differ. The report for decision making, like the method of investigation, will emphasize the selection criteria.

Planning and Organizing the Report for Decision Making

As we noted in Chapter 15, reports have a three-part structure: introduction, body, and conclusion. If you followed the suggestions in Writing Strategies 15-1, you will already have a report outline and even some language for your report. You can probably draft the introduction before much serious research begins because the introduction reflects the existing problem. The overall structure of the report will make it coherent, but the report really consists of a number of small segments. You can write these segments as you complete them, not necessarily in the order in which they appear in the report. You may, for example, write the problem statement early. Then you may complete the section on one of the selection criteria before you even investigate the second criterion.

OVERALL STRUCTURE

The report's structure of introduction, body, and conclusion emphasizes the reasoning process, from problem definition to analysis to conclusions. As we discussed in Chapter 15, reports in the workplace are often organized with the conclusions at the beginning and the details following. This order emphasizes the bottom line: the results. It also acknowledges the reading practices of managers, who are likely to read selectively.

Example 16-4 presents two report outlines, the conventional and the managerial. The items in parentheses are optional altogether or optional for that section of the report. For example, you may explain your methods of research in the introduction, or you may explain them where they apply in the body of the report. The problem statement itself may provide sufficient background information, eliminating the need for a separate background section.

conventional structure
standard format for organizing a report that includes an executive summary, introduction, selection criteria, conclusion, recommendations, and appendices, in that order

managerial structure
a standard format for organizing a report that includes an executive summary, introduction, conclusion, recommendations, appendixes, and selection criteria, in that order

Some expectations for organizing reports will probably exist at the place where you work, and you should find out what they are. If managers expect the **conventional structure,** you should provide it. If they expect the **managerial structure,** you organize accordingly. You can ask to see sample reports, or you can ask your supervisor about the preferred structure. You might also choose the managerial structure when readers are likely to agree with your recommendations. On the other hand, the conventional structure might be persuasive when your conclusions and recommendations differ from readers' expectations and preferences. That structure would emphasize the reasoning that led you to your conclusions. Refer to the sample report (Example 16-5, beginning page 524) as you read next how decision making affects the typical structure of a report.

THE INTRODUCTION

As you learned in Chapter 15, the purpose of the introduction is to orient readers to the report. They need an introduction to the topic (the problem the report addresses) and to the report itself (its purpose, scope, and order).

EXAMPLE 16–4 Outlines for Two Report Structures

Conventional Structure	*Managerial Structure*[a]
Executive Summary	Executive Summary
Introduction	Introduction
Problem statement	Problem statement
(Background)	(Background)
(Methods)	(Methods)
Plan of development	Plan of development
(Description of a Mechanism)	Conclusions
(Needs Assessment)	Recommendations
Criterion One	Appendices
Definition, significance	(Description of a Mechanism)
(Methods)	(Needs Assessment)
Findings	Criterion One
Interpretation of findings	Definition, significance
Criterion Two	(Methods)
Definition, significance	Findings
(Methods)	Interpretation of findings
Findings	Criterion Two
Interpretation of findings	Definition, significance
Conclusion	(Methods)
Recommendations	Findings
Appendixes	Interpretation of findings
(Glossary)	(Glossary)
(Supplementary tables)	(Supplementary tables)
(Survey instruments)	(Survey instruments)

[a] James W. Souther recommended this structure as a result of his research on how managers read reports; see "What to Report" and "Identifying the Informational Needs of Readers: A Management Responsibility."

The problem statement indicates the significance as well as the topic of investigation. Perhaps when you have written term papers, you have written about a topic, such as the hole in the ozone layer. But readers in the workplace are not interested in reading term papers about topics. They want to know why the topic is significant to them—how it will help them achieve an efficient, productive organization that meets its goals. The problem statement establishes the conflict that makes a decision necessary.

The sample problem statements in Example 16-1 are summaries or kernels of the problem statement that you will present in your report. The report should provide more details about the problem. It should answer relevant questions, including who, what, where, when, why, and how.

Even if the problem statement repeats what readers already know, you should still present the details of the problem. First, the statement orients readers. Readers may have other projects and problems on their minds. The problem statement efficiently directs their attention to the specific problem that the report addresses. They can scan the problem statement if they know the problem in detail, but they cannot provide missing details if they do not know them.

Second, the problem statement convinces readers that *you* know the problem and its significance and establishes your *ethos.* If readers will be convinced by your recommendations, they must first be convinced that you have been competent in researching and analyzing the problem. You make a good first impression by establishing the key details of the problem clearly and succinctly.

Third, the report will have a future. It will probably be filed, and readers in the future may consult it for a follow-up study. Those future readers are not so well informed about the problem as present readers. For future readers, the report records the history of the problem.

Background Information

Although the problem statement will generally provide sufficient background, in some cases, additional information will help readers interpret the report. Your analysis of readers will guide you in choosing whether to include **background information.** Include only what will help readers understand the information in the report and your interpretation of it. Readers will not appreciate a report that is longer than it has to be, but they will appreciate your ability to anticipate and answer their questions. Ask yourself what readers need to know. Background information can include

background information
additional information to help readers interpret information in a report, including history, description of a mechanism or process, definitions, and needs assessments

- History;
- Description of a mechanism or process;
- Definitions; and
- Needs assessment.

When any of these sections becomes longer than a page or so, the topic should be a separate section (with an appropriate heading) of the body of the report rather than part of the introduction. If only the secondary readers need the information, detailed background information can be placed in an appendix.

Information on the origins of the problem or previous attempts to solve it and the results of those attempts will place the present information in its context or **history.** If this is the second phase of a multiphase study, you should refer to the report from the first phase.

When the report recommends a change in equipment, procedure, or location, you can orient readers to what is unfamiliar by a **description of the mechanism or process** (Chapter 13 will help you with this part). For example, an engineer may study the feasibility of converting to a cogeneration system for the production of heat and electricity for a hospital. The board of directors, who will make the decision, probably includes members who are not familiar with cogeneration. The engineer can help them understand cogeneration by describing the principles of operation and the process of generation and transfer of energy and heat. Visual displays—perhaps a schematic diagram—will help illustrate the process. The board members can understand the results of the report only if they have a concept of the process being studied.

If some unfamiliar but essential terms will be used throughout the report, the introduction is an appropriate place to define them, although you cannot expect readers to memorize a list of unfamiliar terms. Alternatively, **definitions** may be placed in the section in which they are used (Chapter 13 will again help you with this part). The terms will have more meaning in the context in which they are used. When the primary readers are familiar with the terms, definitions may be placed in a glossary at the beginning or at the end of the report.

The problem statement demonstrates that something in your department or company should be changed—perpetuated, modified, or eliminated. Because the **needs assessment** addresses all contingencies that might affect how you might solve this problem, this section sometimes gets a bit long. Thus, you may want to create a separate section for it after the introduction in your report. Typically, the needs assessment addresses issues such as the following:

- We need to solve the problem by December 31, 1996.
- We need to stay within a budget of $10,000.
- We need to keep at least 80 percent of our projects on task while we solve the problem.
- We need to be able to evaluate the success of the solution within 1 percent accuracy.

Methods

At some point in the report, you will need to describe how you got your information. Information is credible only if its source is credible. As with definitions, you may explain your methods in the introduction, or you may explain the methods within the body of the report where they apply. You would favor explaining methods in the introduction if you need only one

---

history
information on the origins of a problem, previous attempts to solve it, and the results of those attempts

description of the mechanism or process
information about the principles of operation or the working of a process

definitions
explanations of terms unfamiliar to the audience

needs assessment
information addressing the contingencies that might affect how a problem might be solved

or two methods that applied to all the topics of investigation. You would favor explaining methods in the body of the report if the methods were unique to each topic. To decide where to describe methods, consider where the descriptions would be most useful to readers. If methods would be repetitive and distracting in the body sections, place them in the introduction. If explaining methods in the introduction would require readers to evaluate methods apart from results, place them in the body sections. The criterion for placement of this information is usefulness to readers.

Plan of Development

Readers appreciate an orientation to the report as well as to its subject. Thus the introduction should include a statement that forecasts the development of the report. The forecasting statement reveals the plan of the report by establishing the topics of the report and the order in which they are discussed. If you are completing one part of a multiphase study, you should also include a scope statement, which details the topics that this report will cover and your reasons for omitting others. You will help readers by telling them what to expect (notice the plan of development in the last paragraph of Example 16-5).

THE DISCUSSION SECTION (BODY)

As you will recall from Chapter 15, you will present the results of research and explain the significance in the body of the report. The most substantial part of the body (or perhaps its entirety) will consist of a discussion of each of the criteria. If some of the options listed previously under "background" are lengthy, they will also be part of the body. A description of a mechanism or needs assessment will precede the discussion of criteria because this is background information that readers need to know before they can evaluate the results of research.

The body of a report for decision making will probably be organized according to the criteria. One exception would be a decision that is based on a scientific test. If your method is the experiment, your sections would consist of methods, results of the test, and discussion of the results.

Each of the criteria, such as space, structure, and access, will require definition and explanation of significance (what the criterion means and why it is important in this context). The subsection on each criterion will also include the specific findings and interpretive statements. It may include an explanation of methods for gathering the information. You can think of each subsection of the body as a mini-report, with an introduction, results, and conclusion. Interpretive statements are important because readers will remember them when they get to the conclusion of the report. Readers cannot be expected to hold numerous facts in their memories, especially if they are not sure of the meaning of the facts. In the library report, the discussion of the criterion of space might conclude with a statement such as the following:

> The library requires 18,000 square feet of space. The available area on the fifth floor includes 21,050 square feet. This space would accommodate the library and allow some flexibility in arrangement of stacks, reading area, and offices. On the criterion of space, the fifth floor area is a feasible location for the library.

Note that although the statement concludes that the space is adequate on the fifth floor, it does not claim that a move should be planned. The other criteria still need to be discussed.

You have some choice about arranging the criteria. Some people arrange their reports according to the order in which they researched the topics (first topic first, and so forth), but the order in which you researched the criteria may not make the most sense to readers or be the most effective. The fundamental guideline is to use arrangement as a tool to persuade and inform readers. Usually this means placing the most important criteria first. This placement helps prioritize the issues and will shape the readers' interpretation of the information. Also, readers will lose interest if you begin with relatively unimportant topics and work up to the most important topics, the order in a suspense novel but not the proper order for a report. Because your readers are reading for information to make decisions, they will be impatient with delays.

Within the overall structure of most important to least important, try to group related criteria. Grouping will aid interpretation by showing relationships. You don't want to make readers leap mentally from technical to social issues and back.

At times you may intentionally violate these arrangement guidelines because of persuasive goals or because of long-term goals. Example 16–5 illustrates a situation when the deciding criterion is placed last. In this report, the investigator explores the conversion of gasoline-powered vans to electric vans to meet the requirements of the Clean Air Act. On the criterion of cost, the electric vans are judged not feasible at the time the report was written. However, the writer is persuaded that the electric vans will meet the other criteria of emissions, range, and capacity. A few years in the future, these criteria will be the most important because the law will require vans that meet certain clean air standards. If the writer placed the cost information first, the readers might dismiss the concept of electric vans because they would have the information they needed for the decision without considering environment and range. The writer sees some wisdom in showing the merits of the vans even though he rejects them for the time being as too expensive. He keeps open the option of reconsidering the vans in the future.

EXAMPLE 16–5   Sample Report

Electric Vans at Texas Tech University:
A Feasibility Analysis

Prepared by Robert Lundberg
for
Joe Garcia, Manager
Vehicle Maintenance Garage
December 10, 19xx

This report analyzes the feasibility of using electric-powered vans to
replace the shuttle busses used at night at Texas Tech. The report
considers environmental issues, range of vehicles, weight and
capacity, recharging stations, and costs.

EXAMPLE 16–5   (*continued*)

Contents

EXAMPLE 16–5   *(continued)*

Executive Summary

In order to comply with the laws that the Texas legislature has mandated, Texas Tech University must purchase vehicles that are able to run on a fuel other than gasoline. Among the possible alternatives are electric-powered vans. These vans are considered for use at Texas Tech to replace the four shuttle vans that circuit the campus each night.

Electric vans meet the environmental requirements because electricity is classified as a zero-emissions fuel. Batteries are expensive and limit the range and speed of vehicles, but the shuttle vans average only 40 miles each night at maximum speed of 20 mph. The vans could be modified to achieve the 12-passenger capacity of the current vans. The recharging stations cost $5,000 to install, and one is needed for each vehicle, but the cost is lower than the cost for other fueling stations, and the vans could be recharged during the day.

The vans are technologically feasible, but the annual cost of $6,300 exceeds the annual cost of gasoline-powered van ($3,100). Until the present vans wear out and research on batteries results in lower prices for electric vans, Texas Tech should not purchase electric vans to replace the shuttles.

ii

EXAMPLE 16–5   (continued)

Introduction

The early sentences answer basic reader questions: WHO, WHAT, WHERE, WHEN, and WHY?

The University Police Department at Texas Tech University uses vans for a shuttle service for students during the evening and very early morning hours. The vans patrol the campus each night, making frequent stops. Although the vans represent an extremely small portion of the vehicles in use at Texas Tech (4 of 138), they are a very important part of the campus fleet. They provide a degree of safety to students who must otherwise walk across the deserted campus at night. However, the vans are not completely beneficial to the campus and to Lubbock. The regular use that these vans get contributes to air pollution. The fact that the vans are constantly running for several hours at a time adds to the emissions that are released into the environment. Using an alternative source of energy could alleviate this problem while still providing security to students.

"However" signals the conflict that creates a problem: the vans provide a necessary service, but they pollute the air.

Following general information about vans in the first paragraph, the writer focuses on the subject of the report (alternative fuels, specifically electricity).

For several years Texas Tech University has been investigating the use of alternative fuels for its vehicles. There have been many reasons for this, but this effort has mainly been part of a larger movement in the United States that is encouraging automobile manufacturers to build low- and zero-emission vehicles. For example, a California law requires that, by 1998, 2% of all vehicles sold in the state must have zero emissions [1,2,3,4]. Texas and other states have similar laws. Automobile manufacturers planning to sell vehicles in such states must develop cars that comply with the laws. The electric-powered vehicle is one of the few alternatives that can comply with the strictest aspects of the law: zero emissions. Thus, the development of electric vehicles is crucial for those companies that wish to be competitive in the area of alternative fuels. The "Big Three" auto makers (Ford, General Motors, and Chrysler) are all currently testing mini-vans and full-size vans [1,2,3]. Each company has focused on vans because the first practical application of electric-powered vehicles is urban delivery and shuttle routes [2,3]. This is an ideal market for the electric van, which can only travel short distances at moderate speeds. The shuttle vans at Texas Tech are an excellent example of vehicles well suited for conversion to electric power.

The information about the "larger movement" in the U.S. offers motivation to readers by establishing the context and significance of the topic.

The motivation is even stronger with this new evidence: Texas Tech has no choice but to comply with the law.

The alternative fuel legislation adopted by the state of Texas applies to government subsidized vehicle fleets. A new law mandates that beginning in 1993, all vehicles purchased by state funded agencies must be able to run on an alternative fuel. By 1998, 90% of all vehicles in use must be converted to alternative fuels. Texas Tech University falls under the jurisdiction of this law. Texas Tech is required to purchase vehicles that can run without contributing

1

EXAMPLE 16–5   *(continued)*

The research question is implicit: Are electric vans feasible for the stated purpose?

This paragraph establishes the purpose of the report and forecasts its development by identifying the criteria to be discussed. The writer reveals the scope of the report when he says that its subject is the shuttle vans, not the entire vehicle fleet.

significantly to the problem of air pollution. Using electric-powered vehicles is one of the alternatives that is available for consideration. Specifically, electric powered vans might be used to replace the shuttle vans on the Texas Tech campus.

This report analyzes the feasibility of using electric vans for this application. It briefly describes the van and then discusses environmental issues, range of the vans, weight and capacity, the recharging stations, and the costs of electric vehicles as they apply to use at Texas Tech University. The analysis assumes the purchase of new vans for the shuttles and the use of existing vans to replace other vans in the vehicle pool that are reaching the ends of their lifespans.

The Electric G-Van

A description of the van answers the basic reader question: What is it?

Southwestern Public Service Company (SPS) has already implemented a fleet of electic vans similar to those proposed for use at Texas Tech. Tim Pillsbury, the distric spokesman for the project, allowed me to ride in one of the "G-Vans" to experience the feel of the electric power. The van accelerates slowly but runs similarly to a gas-powered van. From the outside, the van looks like a standard van with the exception of a large black box that hangs under the vehicle between the wheels (see Figure 1). The box contains the 32 batteries that provide the electricity to power the vehicle. The van is also equipped with a gasoline engine to enable its use for distances that exceed the range of batteries and to serve as a backup to the electric power.

The illustration describes the van more effectively than words.

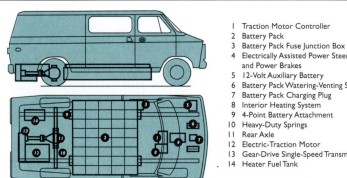

1	Traction Motor Controller
2	Battery Pack
3	Battery Pack Fuse Junction Box
4	Electrically Assisted Power Steering and Power Brakes
5	12-Volt Auxiliary Battery
6	Battery Pack Watering-Venting System
7	Battery Pack Charging Plug
8	Interior Heating System
9	4-Point Battery Attachment
10	Heavy-Duty Springs
11	Rear Axle
12	Electric-Traction Motor
13	Gear-Drive Single-Speed Transmission
14	Heater Fuel Tank

2

EXAMPLE 16–5   (continued)

This paragraph defines the issue of emissions and its significance. This criterion is discussed first because it is the most important. Questions about the environment triggered the study. To be feasible, the vans must meet the environmental criteria established by law.

Environmental Issues (Emissions)

Since the early 1900s, the internal combustion engine has appeared to be a clean, efficient, system to power automobiles with good range, power, and performance. Its major disadvantage, its production of harmful emissions, did not become apparent for many decades. During the last four to five years auto manufacturers and the public have realized that the internal combustion engine will never be able to run as cleanly as other alternative power sources. Therefore, an alternative must be developed to improve the environment. Electricity could be such an alternative.

The writer offers evidence to support the claim that electricity meets the criterion of being a clean fuel. The main evidence derives from citation of authorities.

The main advantage of electricity is that it is an extremely clean source of energy. An electric car produces no environmentally hazardous emissions, and the batteries that are used to store the energy are completely recyclable [1,2,3,4,5,6,7,8]. Researchers in southern California have found that "electric vehicles are 97% less polluting than gasoline powered cars, even when the emissions from the power plants producing electricity for recharging are considered" [5, p. 41]. Therefore, the use of electricity to power automobiles will undoubtedly help clean up the atmosphere.

In addition to providing facts, the writer interprets the meaning of the facts.

Electricity and hydrogen are the only two power sources classified as zero-emissions automobile fuels. Hydrogen is extremely explosive and is therefore not a reasonable alternative to gasoline [1, 6]. Electricity, then, is the only alternative that has proven to be useful in automobiles without directly polluting the atmosphere. Due to this fact, most auto manufacturers are currently developing electric powered vehicles for use in the near future. Texas Tech can take advantage of these new developments to meet the requirements set forth in the legislation. As Lubbock grows larger and becomes more of a metropolitan city, pollution will become more of an issue. If Texas Tech utilizes the technology of electric-powered vehicles along with other local vehicles, this problem will be partially averted. Texas Tech will also get an additional benefit of being recognized as an institution that care about the environment and is willing to take care of it.

The writer looks ahead to see the long-term significance of the information. Note that he does not answer the research question; he draws a conclusion about one issue.

Range of Vehicles (Batteries)

Batteries are used to store the electricity for the vehicle much like a gas tank stores the energy for an automobile today. However, the technology of batteries has advanced little in the last hundred years. Lead-acid batteries, the most common batteries used in electric vehicles, are no more advanced today than they were in 1910 [4]. The limitations are storage capacity, recharge period, and weight.

On the second criterion, the writer also begins by defining the significance of the topic.

3

EXAMPLE 16–5   *(continued)*

The writer interprets the facts for the specific context: the range of vans is short, but it is long enough for the intended use.

Anticipation of better batteries is appropriate because the report will recommend a delay in the purchase of electric vans. In order to keep open the promise of future use, the writer must forecast the reasonable hope of improved batteries.

The conclusion is explicit: on the criterion of battery range and recharge limitations, the electric vans are feasible.

Because batteries have limited storage capacity, the vehicles which they power can only travel short distances. The batteries also have a long recharge period. It takes eight hours to recharge a set of batteries that are discharged in approximately two hours when travelling at maximum speed. The third problem arising from the use of batteries is increased vehicle weight. On a weight-to-power scale, an internal combustion engine currently has an advantage of 300 times the power per unit of weight compared to batteries [5].

Texas Tech uses the shuttle vans on campus only and therefore does not have a great need for an extended range or power. In fact, the vans travel only an average of 40 miles in a single night, which is well within the range of the least technologically advanced battery-electric motor combustion. Thus the short range of the vehicles is not a factor limiting use at Texas Tech.

Much of the current research being done on electric vehicles focuses on finding a lighter, more powerful battery system. Ford Motor Company is using a sodium-sulfur battery that produces three times the amount of energy as lead-acid batteries [1,3,8]. The battery must be kept at 600° F, but this temperature is naturally maintained when the battery is being charged or discharged slowly [1,3]. Otherwise, a fully-charged battery that is not used will last for four days before the chemicals begin to solidify and the battery no longer produces currents [1,3,8]. Texas Tech can utilize this kind of battery because the batteries will always be recharging or slowly discharging—they will rarely be "sitting" idle.

Other companies are trying other types of batteries, including nickel-cadmium, nickel-iron, and several types of fuel cells that use a reaction of a metal with oxygen to produce electricity [1,2,3,4,5,7,9]. Some of these can be quickly recharged, but none of them completely solve the problem of limited range. The battery is probably the key factor that is keeping the electric vehicle from surpassing the internal combustion engine. Its weight and limited range restrict its feasibility to local commuting and delivery services.

The batteries' limitations do not inhibit the feasibility of using electric vans to replace the night shuttle vans at Texas Tech. Based on the campus size and hours of operation (5:00 pm to 3:00 am), the vans currently average 30 to 50 miles per night. This is well within the range of current electric vans. Furthermore, the vans can be recharged during the day, so the long recharge period is not a large problem. Thus, despite the shortcomings of the electric power source (the battery), the use of electric vans for such short range commuting is currently viable for applications such as the Texas

4

EXAMPLE 16-5   *(continued)*

Tech shuttle vans.

As in the discussion of the other criteria, the writer anticipates questions that readers might ask before they make a decision. The question is, How many passengers can the van hold?

Weight and Capacity

According to Mr. Pillsbury of SPS, the 32 lead-acid batteries contained in the black box beneath the electric van weigh a total of 2600 pounds. As a result, the van weighs almost 2000 pounds more than an equivalent gas-powered van. The result of this increase in vehicle weight is a reduced passenger capacity of only five passengers in addition to the driver [10].

The vans that Texas Tech currently uses can carry 12 passengers. Twelve passengers can be accomodated in the electric vans through an adaptation of the suspension to support the additional weight. Thus, on the criterion of capacity, the electric van is a feasible choice for replacing the gasoline-powered shuttle vans assuming the suspension is adapted.

The writer's evidence is the experience of another organization. Claims made on someone else's experience can be risky unless the situations are comparable. In this case, the two organizations would use the same size and type of van for short-range purposes in the same geographic area. Their situations are similar enough to warrant drawing conclusions about one based on the experience of the other.

Recharging Stations

The batteries can provide an ongoing source of electricity only if they are regularly recharged. SPS uses recharging stations that are slightly larger than a typical gasoline pump. The stations inlcude a supply of water to refresh the batteries, decrease maintenance needs, and increase battery life. Each station costs SPS $5,000 to install. After the initial expense, however, maintenance costs were minimal and did not significantly contribute to the cost of using the vans. Thus, fuel costs can be significantly decreased so that the cost of running an electric van is about half that of running a van on gasoline at current prices.

Texas Tech would require one charging station for each electric van. Currently there are four vans being used on the campus. Therefore, four charging stations will be needed if all four vehicles are to be run on electricity. There is plenty of space for these stations since each station can be placed at the edge of the same parking lot where the vans are currently parked during the day. $5,000 per stations appears to be a large investment until the cost of building a fueling station for other alternative fuels is considered. The cost of these other stations is well above the $20,000 cost for the four electric recharging stations. Since Texas Tech must have the capability to use alternative fuels by 1993, the investment in a new refueling station or recharging station must be made. This brings up the final consideration of the feasibility, the cost of the vans themselves.

Costs have meaning in relationship to other costs. $20,000 by itself means less than $20,000 in comparison with another cost. The writer's argument would be stronger if he cited the costs of a refueling station for natural gas or hydrogen or another alternative fuel.

5

EXAMPLE 16–5   *(continued)*

Readers are prepared for this information by the earlier discussion of batteries. The emphasis on high costs explains and prepares for the ultimate recommendation against electric vans on the basis of cost. Note the repetition of words that suggest high cost—"largest," "twice the cost," "elevated."

Costs

The largest cost of running electric vans is the cost of batteries. The lead-acid batteries used by SPS cost $200 each. The $6,400 cost of 32 batteries is twice the cost of the motor that drives the wheels and approximately 17% of the cost of the entire van ($38,000) [3]. The lead-acid batteries are not currently mass-produced in appropriate configurations. This results in an elevated price. If the batteries were mass-produced, the cost of electric vehicles would quickly come down closer to the level of gasoline-powered vehicles [1]. The large amount of intense research now being conducted on improving the batteries suggests that cheaper, more cost-effective batteries will soon be available. At that time, the price of electric vans should fall.

The high cost of electric vans affects the feasibility of using them at Texas Tech. I used the figures in Table 1 to determine the annual cost of using the electric van. The purchase cost includes the adaptation of the suspension that is necessary to increase the capacity to 12 passengers. Figures on maintenance and fuel costs were provided by the manager of the vehicle maintenance garage.

Readers appreciate details on the source of the figures. The explanation increases the credibility of the calculations.

Table 1. Electric van costs (annual)

depreciation	$3,800	38,000 / 10
electricity	201	6,700 kw-hrs / year x 0.03 / kw-hr
batteries	1,600	6,400 / 4
maintenance	700	
TOTAL	$6,301	

Based on these figures, the cost of running the van for its full life (10 years) is $6,301 per year. This includes the original cost of the van and the fuel costs but ignores the cost of the recharging station, a one-time cost that will have to be incurred eventually for alternative fuel vehicles.

These vans can be run on gasoline until 1998 when 90% of the vehicles need to be converted to alternative fuels. Thus, the cost of the fueling stations for natural gas vehicles is not considered at this time. I used the figures in Table 2 for the economic cost analysis for the van, which is capable of conversion to natural gas.

6

EXAMPLE 16–5   *(continued)*

Table 2. Natural gas van costs (annual)

depreciation	$1,800	18,000 / 10
electricity	600	
maintenance	700	
TOTAL	$3,100	

The writer summarizes the conclusions on all the individual criteria. He weighs the individual criteria in the context of the whole problem. In terms of environment, range, and operating costs, the vans are feasible, but the one negative result (purchase cost) cannot be overcome at this time.

Conclusion

The electric van represents the only zero-emissions vehicle available today. However, it can only operate over limited distances and takes a long time to recharge. Thus, its use at Texas Tech would be limited to vans that run short distances for limited hours per day. Shuttle vans are an appropriate use for electric vans.

The costs of running an electric van are relatively low (fueling costs are approximately 50% of the costs of purchasing gasoline). However, the initial cost is so much higher that the annual cost is about double the cost of the gasoline vans. Thus, the electric van is *The writer answers* not economically feasible at this time.

the research   When some of the research on making batteries more efficient
question directly. begins to pay off, the initial cost of purchasing an electric van should come down to a level comparable to the gasoline-powered van. At that *The writer leaves* point, investing in an electric van would be cost effective, and it *open the possibility* would achieve the clean air goals established by law.
that the recommendation
could change in the **Recommendations**
future.
         He returns to 1) The electric van is currently too expensive to be economically
the theme of the   feasible. Texas Tech should not replace the night shuttle vans
initial conflict that   with electric vans at this time.
required the study: 2) Texas Tech should monitor the cost of purchasing electric
the environment.   vans. As the price falls, the van will become economically
            feasible and will eventually become cheaper than
            gasoline-powered vans. The environmental benefits of the van
            are large.
         3) Texas Tech should avoid purchasing new vans with
            conventional gasoline engines to replace vans in its vehicle
            pool. Rather it should maintain the existing vans. Otherwise,
            the university may face expensive conversions by 1998.
         4) As soon as the purchase price is low enough, Texas Tech
            should purchase electric vans to replace the night shuttle
            vans.

7

EXAMPLE 16–5   *(concluded)*

References

1. Shuldiner, H., "New age evs," *Popular Mechanics*, vol. 168.9 pp. 27-29, September 1991.
2. Begley, S., "The power of a voltswagon," *Newsweek*, vol. 117.13, p. 62, April 1, 1991.
3. Frank, L. and D. McCosh, "Electric vehicles only," *Popular Science*, vol. 238.5, pp. 76-81. May 1991.
4. Cogan, R., "Electric Cars," *Motor Trend*, vol. 43.8, pp. 71-77, August 1991.
5. Siuru, B., "Dual-power autos take the wheel," *Mechanical Engineering*, vol. 111.8, p. 236, August 1990.
6. McCosh, D., "Automotive newsfront: Methanol managers," *Popular Science*, vol. 236.1, p. 50, January 1990.
7. Romano, S., "Fuel cells for transportation," *Mechanical Engineering*, vol. 111.8, pp. 74-77, August 1989.
8. Siuru, B., "R&D in the fast lane," *Mechanical Engineering*, vol. 111.10, pp. 62-27, October 1989.
9. Webb, J., "Hydrogen-powered electric cars sets skeptics wondering," *New Scientist*, vol. 130.1775, p. 30, June 29, 1991.

The writer has used the reference note system of citation (common in mechanical engineering). The notes are listed in the order in which they appear in the text. They are identified in the text by their number in brackets.

8

CONCLUSIONS AND RECOMMENDATIONS

In the **conclusions,** you answer the research question. You should be able to claim that the project is (or is not) feasible, option *A* is better than option *B*, or the situation happens because of *X,* which points to the solution of *Y.* To draw these conclusions, you interpret the results from each of the body sections in relationship to one another. The interpretations will explain why you have answered the question as you have.

conclusions
in a report for decision making, conclusions answer the research question

The results of your analysis of each criterion may conflict with one another. If you have three criteria, two of which point to one conclusion and a third that points to another, you will have to decide whether the conclusion based on the third outweighs the conclusion based on the first two. Your ranking of criteria at the outset of your investigation will help. The criteria are of different weights, so you cannot simply decide on the basis of the majority. In the library case, for example, if the space and access are adequate but structure is not, you have to conclude with a negative answer in spite of the fact that on one only of three criteria is the conclusion negative. Be sure to base your conclusions on the stated criteria and not to introduce a new criterion.

Recommendations differ from conclusions. Instead of explaining reasons for answering the research question in a certain way, recommendations direct the particular action that follows from the conclusions. A conclusion statement in the library case would be this: "Even though the fifth floor space will accommodate the library, moving to the fifth floor is not feasible because the building was not designed to support the weight of the books on the fifth floor." Recommendations, however, might include:

recommendations
unlike conclusions that answer a research question, recommendations direct the particular action

1. Investigate the feasibility of reinforcing the building structure to support the weight, considering cost, completion time, and effect on other areas.
2. Investigate the feasibility of relocating the library to the first-floor neurology wing and moving the neurology wing to the space presently occupied by the library. Neurology occupies a space of the size required for the library and does not use all of its space.

Note that the recommendations are phrased as directions and are listed. This phrasing is not necessary, but it is appropriate because the purpose of the recommendations is to tell the decision maker what to do next. Note also that the recommendations do not include explanations. The explanations have already been covered in the conclusions. Careful readers who have read the conclusions, therefore, should almost be able to write the recommendations segment themselves; your ideas should be that clear. However, because some readers might skip the conclusions or because the recommendations segment may be pulled from a report to stand alone, such as at a board meeting, you need to write the recommendations with care. For these reasons, recommendations are usually provided in a list format.

Managing the Report Development Process

Good report writing requires effective investigation and mastery of writing skills, including the ability to conceive of purposes and readers as well as knowledge of various types of reports and of strategies of organization, format, and style. The third requirement for a good report is good management. Reports and the projects to which they relate develop over time. That time has to be managed so that all the tasks can be completed well and on time. Management is also important because many people dread report writing. They would rather do the hands-on work and leave the paperwork to someone else. The intimidation of writing increases when the report will be long. Intimidation invites procrastination, and one cure for procrastination is management. Management requires planning, task analysis, and scheduling. Now is a good time to review planning and scheduling activities as described in Chapter 2 and task analysis as described in Chapter 13. **Task analysis** in report writing means subdividing the overall research and reporting project into the component tasks. These tasks include research tasks, such as conducting interviews and completing lab procedures, and report developments, such as preparing visual displays, writing the executive summary, and editing.

task analysis
in report writing, the subdividing of the overall research and reporting project into component tasks

One way to make any writing task easier is to conceive of it as a series of small projects. Writing that requires a page or two is less intimidating and easier to finish than a long report in its entirety. If you write small sections as you conduct your research or other project, soon you will be able to assemble a lot of the whole from the small parts.

It may be easy to write the introduction right away because it usually consists of a problem statement that explains why investigation took place at all. In other words, the project that results in a report begins with a problem—some inefficiency or lack of information. You know the problem at the outset of your work. Writing it down early not only gives you a head start on the final report but also helps you clarify your thinking about the nature of the problem and the work it requires.

The informal notes you make as your project progresses can become the data for the final report. If the notes are good, it is easy to write the report from them. If you are conducting a lab experiment, you will certainly keep notes on your procedures and results, but even with other types of reports, you can jot down observations and details as you work. Maybe you will find it helpful to keep a small notebook in a pocket just for these notes.

Some sections of reports are consistent from report to report. For example, each progress report during the course of a project will include a description of the project problem and goals. This description may also have been part of the project proposal. Proposals probably include a description of the organization or individual who proposes to do the work. This description will remain relatively consistent from proposal to proposal. When certain sections of reports become standard, they are called boilerplate material. You can store these sections as computer files and then paste them into the new report. This process saves not only writing time but also review time. If the sections have already been approved as accurate by the organization, you don't want to risk introducing inaccuracies by rewriting unnecessarily. If someone else authored your boilerplate materials, acknowledge them according to your company policy.

Some Difficulties of Making Recommendations

So far this chapter has presented an idealized procedure for investigating possible solutions to problems and for writing them in a report. The procedure is sound, and you can trust it to guide you in writing reports for decision making. However, the workplace is never quite so tidy as textbook procedures. You should be aware that even when you follow the procedure, you will sometimes feel uncomfortable with your task and with the recommendations that you believe should be made. The following sections describe some of those situations and suggest ways to increase your comfort with your task.

MAINTAINING AN ATTITUDE OF OPENNESS IN THE WRITTEN REPORT

The early part of this chapter stressed the importance of beginning the investigation with a question rather than with a proposition. The reason for the question is to keep the options for choice open and not close off a good option by a premature decision or feel compelled to defend an inappropriate option. Even though the report will end with a recommendation and is a persuasive document, its purpose is to report an investigation. Instead of establishing a thesis and defending it, the report writer asks a question and explores the answer.

At the time of writing the report, however, you will have reached a decision, and you may feel strongly about it. You have an obviously persuasive purpose as well as an informative one. And, the executive summary will reveal your conclusions and recommendations right up front. The readers— the decision makers—will know the ending before they know the details and the reasoning. Yet they may benefit from approaching the issue with open minds just as you do at the beginning of the investigation. How can you accomplish your persuasive aims while still giving the readers the freedom to make up their own minds?

Sometimes the most persuasive writing is indirect. It persuades because it seems not to, and readers therefore trust it. One way to be indirectly persuasive is to delay specific conclusions until you have presented some evidence. Even though the report conclusions will be available to readers in the executive summary and, in the managerial report structure, before the body of the report, you can maintain a descriptive and inquisitive attitude in the introduction. (This is another reason to write the introduction before you complete the research.) You demonstrate that you began the project with an open mind. You can save your evaluative statements for the body and conclusion, letting these follow the presentation of evidence in each of these sections. It is persuasive to show readers the progress of your reasoning from questioning to evidence to conclusion.

Even though it will be your responsibility to draw conclusions, you should write with respect for the readers' right to draw their own conclusions from your evidence and reasoning.

RECOMMENDING A COURSE OF ACTION TO A SUPERIOR

Especially when you are a new employee, you will probably feel uncomfortable advising a superior in the organization. Nevertheless, recommendations typically move from lower levels in the organizational hierarchy to the upper levels. The subordinate officer recommends; the superior officer directs. The superior delegates because he or she does not have the time and may not have the expertise to conduct the study. Your manager assigns the research task and requests a recommendation assuming that you will be conscientious in your investigation and that you have the right education and experience to interpret what you find.

When you are uncomfortable advising a superior, remember that you are advising, not directing. The decision maker has the ultimate authority to implement the recommendations and has the privilege of rejecting your advice. Your goal, then, is not to reverse roles and tell a superior what to do but rather to provide the facts and the reasons to show why you recommend one action over an alternative.

SAYING NO

It's easier to recommend yes than no. You will probably anticipate what your manager wants to hear, and you will prefer to provide it. Even posing a question of feasibility implies a motivation to pursue that course of action, and you will feel most natural in agreeing. But what if the investigation points to the opposite conclusion?

The purpose of studying the options rather than simply plunging into a course of action is to make a good decision and to avoid one that will later cause regrets. There's nothing wrong with saying no if the reasons for doing so are good. In fact, recommending against an action with high risk of failure will ultimately save the organization time, money, and stress. Even if the decision maker is initially disappointed, the investigation has served its purpose well: to guide the decision.

If you recommend action that conflicts with the expected recommendation, your facts and reasons have to be especially trustworthy and convincing.

DECIDING UNDER PRESSURE

Some emergency situations require quick decisions. Time limits the amount of research that can be completed. Pressures of office politics may also limit the options. You may feel that a superior has already committed to a decision and merely wants your investigation to confirm what is already decided. If your investigation points to a conflicting recommendation, you will feel pulled between the course of action that the study supports and the will of your superior.

The events preceding the explosion of the space shuttle *Challenger* in 1986 reveal the difficulties of making decisions under pressure. Both time and politics pulled against the caution of engineers who questioned whether the O-rings in the solid rocket boosters would seal at the low temperature expected during the launch. Previous experience had raised some questions about the effect of low temperatures on the sealing capacity of the O-rings. But the facts were somewhat ambiguous because no shuttle had been launched at the 36° Fahrenheit temperature that was expected for the launch day. The engineers had serious doubts, but the managers felt that proof of likely failure was insufficient. The shuttle was launched in the cold weather, the O-rings did not seal, and the explosion killed the seven people aboard.

Dorothy Winsor observes that the recommendation of the engineers in the *Challenger* disaster was insufficient to convince the managers for two reasons: the news was bad; and several different organizations were involved (the manufacturer of the boosters, Marshall Space Center, and NASA), increasing the difficulty of communication ("Communication Failures"). Bad news, multiple organizations, a shortage of time for making the decision, and criticism from the press regarding delays all interfered with reasonable decision making. The context in which a decision is made will influence the decision, sometimes for the worse.

To help balance the pressures of time and politics, you can remember that social as well as technical criteria determine what makes a decision good and that judgment as well as calculation represent good thinking. The managers in the *Challenger* case wanted to depend on numbers to predict whether the O-rings would hold. But the numbers were not available because the experience in similar weather conditions was limited. Winsor also observes that the "facts" of the case seem much firmer to us now than they did at the time the decision was being made ("Construction of Knowledge"). The managers wanted to "know" with more certainty than was possible.

Perhaps in the quest for facts, they neglected judgment. The judgment of the experts was more reliable than facts in that case.

SUMMARY

Writing and reasoning are closely related activities. Good writing depends on good reasoning, but it can also be argued that good reasoning, at least for complex problems, requires good writing as well. The process of analyzing a problem, researching solutions, and making decisions is guided by the need to prepare a report that will present and interpret facts and make recommendations. Writing about the problem and arranging the results of your research will help clarify your thinking. Thus, writing well will make you a more valuable employee and more effective in shaping responsible decisions.

When preparing your reports, always define your problem, recognize your readers, pose your research question, identify your selection criteria, and select your research methods. Your criteria include technical, managerial, and social criteria and will be ranked or weighed depending on your situation and research question. This planning will be reflected in how you organize your report. Finally, because researching and writing a report take a great deal of time, you must manage your project closely so you can be confident that your conclusion and recommendations will guide effectively and safely the decisions and actions of your readers. Writing Strategies 16-1 will help you write effective reports for decision making.

WRITING STRATEGIES 16–1   Reports for Decision Making

Problem Definition

What is the problem that requires a decision about action? Express the problem with an *A but B* structure (Example 16–1). Include details that are available (who, what, where, when, why, how). Do not ask the research question yet.

Readers

Who has the authority to make the decision? (Perhaps there will be more than one person at different levels of authority.) What do readers already know about the problem? What are their attitudes to it? Will they be likely to resist or welcome change? What personal values may influence the decision they make? Will some future readers depend on the report to establish the history of a situation?

Reader	Knowledge	Attitudes/Values
a.		
b.		
c.		

WRITING STRATEGIES 16–1 *(continued)*

Research Question

Pose your question as a version of one of these: Is the project feasible? Which is better, option *A* or option *B*? Why should the situation occur?

Selection Criteria

What technical, managerial, and social criteria will determine that a decision is good?

Criterion	Significance
a.	
b.	
c.	

Additional Useful Information

Will readers benefit from a description of a mechanism or process? needs assessment? history and background?

Research Methods

What research methods will give you the data you need for each criterion?

Criterion	Method(s)
a.	
b.	
c.	

Resources

What people, books, or existing reports may be helpful?
What research instruments (survey forms, questionnaires) will you need to develop?

Visual Display of Data

What visual displays may help in the presentation of data?
What type of data do you have? What type of visual display do you need?

Schedule of Tasks

Task	Completion Date
research	
a.	
b.	
c.	
writing, editing	
a.	
b.	
c.	

1. Many of the stories you read in the local or campus newspaper have in their background a report for decision making. For example, a city government considers whether to build an all-purpose arena at a cost of $36 million. The newspaper reports that the issue has been raised, and it presents some of the reasons for building the arena as well as some of the drawbacks. A feasibility study must be completed before the city can formally propose the center to the citizens and ask contractors for bids. Scan the front page of your local and campus newspaper, and identify possible studies in the background of the stories.

   Bring a selection of your findings to class and with your teacher and classmates discuss what a selection of the reports would need to contain.

2. Summarize the problem in the following situations by creating *A but B* statements (see Example 16–1). Be prepared to discuss your answers in class.

   a. A student who desires part-time work to make money for living expenses considers whether to open her own home-based pet care service. Although she has a tight schedule already, she needs the extra money.

   b. The new owners of a feed yard need to organize as a business for legal and tax purposes. The options include a general partnership, S corporation, and C corporation.

   c. A student in mechanical engineering must choose a design project for a senior class. He considers a diver propulsion vehicle that is quite complex and expensive to assemble.

   d. The desuperheater at Acme, Inc., produces steam to heat the seven buildings at the Acme plant. High-temperature, high-pressure steam exits the refrigeration cycle's steam turbine at the plant and enters the desuperheater, where water is added to the steam to drop the temperature and pressure of the steam. The large drop in pressure and temperature produces energy, which is currently wasted.

   e. Livestock graze on 95 million acres of permanent rangeland in Texas. An estimated 88 million of these acres are infested with woody plants that suppress the grass on which the livestock graze.

3. In the following situations (or the ones you identified in Activity 1), identify the research question and brainstorm the criteria for decision making. Work in pairs or small groups for brainstorming. Then join another group to compare the research questions and criteria you have identified.

   a. Moving from a dorm to an off-campus apartment
   b. Preparing for a career in technical communication
   c. Purchasing a laptop computer
   d. Choosing between two brands of pickup trucks
   e. Investigating whether to change from Styrofoam containers to paper wraps in a fast food restaurant
   f. Determining why enrollment has dropped in two high schools and the impact of the drop on staffing and construction in a city school system
   g. Determining whether to pave a section of highway with concrete or asphalt

h.   Determining why the teenage pregnancy rate in a city is higher than the national average in order to decide what to do to lower the rate

4.   Discuss in class why the lists of criteria might differ for two different people investigating the feasibility of moving from the dorm to an off-campus apartment. What might some points of difference be?

5.   People base decisions on hunches, experience, personal preferences, self-interest, and the desire to please others as well as on the procedure(s) of investigation and reasoning that this chapter describes. For each of the claims stated below, determine the basis for the decision. In what circumstances would a decision based on the claim be a sound decision? When might the claim result in a flawed decision? If the claim might result in a flawed decision, what would be a better method of arriving at the decision? Discuss your findings in class.

a.   It worked last time; it will therefore work again.
b.   The choice worked for her, and it will therefore work for me.
c.   You can't solve social problems by throwing money at them; therefore, we will not fund your project.
d.   I will become a lawyer because my father-in-law urges me to do so.
e.   I like rock music, so I will recommend that the Activities Committee schedule a rock concert for fall.

6.   For discussion: What are the differences in meaning of these words: *evaluation, opinion, judgment, interpretation, perception, assumption, bias?* How, in particular, would you distinguish judgment from opinion? Why must some technical communication include judgment and evaluation? How can one convert bias to statements that are acceptable in technical communication? Discuss your answers in class.

7.   Researchers who want specific information about employment in different sectors of the economy will be able to find useful figures in Example 16–6. But someone who wants to know the significance of these figures may be overwhelmed by them. If a report writer turned in this table in place of a written report and said, "Let the facts speak for themselves," readers might be frustrated and avoid studying the table. Its meaning is not obvious. A person would have to study and interpret the numbers to derive meaning from the table.

   In collaboration with one or two classmates, determine an argument that the information in Example 16–6 will support. For example, employment totals and rate of change both support the claim that employment will grow in service-producing sectors while remaining flat in goods-producing sectors. Then help readers understand your argument and make sense of some of the figures in the table in two ways: by drawing one or more graphs to represent some of the numbers (such as a bar or line graph); and by writing interpretive statements. Select data; do not try to include all the information in the table in your graphs and statements. But be sure that your visual displays and written statements make honest claims about the data. Do not claim to represent the whole story of employment by using selected figures.

   When you have prepared your visual displays and statements, exchange your paper with another group. Evaluate the accuracy of the other group's

EXAMPLE 16–6 Employment by Industry, with Projections

No. 658. Employment by Selected Industry, 1970 to 1988, and Projections, 2000

[**In thousands, exept percent.** Figures may differ from those in other tables since these data exclude establishments not elsewhere classified (SIC 99); in addition, agriculture services (SIC 074, 5, 8) are included in agriculture not services. See source for details. N.e.c. means not elsewhere classified. Minus sign (–) indicates decrease]

1972 SIC[1] code	INDUSTRY	EMPLOYMENT				ANNUAL AVERAGE RATE OF CHANGE		
		1970	1980	1988	2000 [2]	1970-1980	1980-1988	1988-2000 [2]
(X)	**Total**	**$1,664**	**102,019**	**118,104**	**136,211**	**2.3**	**1.8**	**1.2**
(X)	Nonfarm wage and salary	70,725	90,043	104,960	122,056	2.4	1.9	1.3
(X)	Goods-producing (excluding agriculture)	32,578	25,659	25,252	25,860	0.8	–0.2	0.1
10-14	Mining	623	1,027	721	705	5.1	–4.3	–0.2
15-17	Construction	3,588	4,346	5,125	5,885	1.9	2.1	1.2
20-39	Manufacturing	19,367	20,286	19,406	19,090	0.5	–0.6	–0.1
24, 25, 32-39	Durable	11,210	12,188	11,436	11,220	0.8	–0.8	–0.2
24	Lumber and wood products	646	691	765	740	0.7	1.3	–0.3
25	Furniture and fixtures	440	465	530	600	0.6	1.6	1.0
32	Stone, clay, and glass products	644	662	600	580	0.6	–1.2	–0.3
33	Primary metal industries	1,260	1,142	772	700	–1.0	–4.8	–0.8
331	Blast furnaces/basic steel products	627	512	277	241	–2.0	–7.4	–1.2
34	Fabricated metal products	1,560	1,613	1,431	1,352	0.3	–1.5	–0.5
35	Machinery, except electrical	1,984	2,494	2,082	2,059	2.3	–2.2	–0.1
3573	Electronic computing equipment	194	354	418	453	6.2	2.1	0.7
36	Electrical and electronic equip[3]	1,871	2,091	2,071	2,014	1.1	–0.1	–0.2
3662	Radio and TV equip	362	378	456	463	0.4	2.4	0.1
3674	Semiconductors/related devices	(NA)	223	262	285	(NA)	2.0	0.7
37	Transportation equipment	1,852	1,900	2,050	2,002	0.3	1.0	–0.2
371	Motor vehicles	799	788	856	786	–0.1	1.0	–0.7
38	Instruments and related products	527	712	749	822	3.1	.06	0.8
39	Miscellaneous manufacturing	426	418	386	350	–0.2	–1.0	–0.8
20-23, 26-31	Nondurable	8,157	8,098	7,970	7,870	–0.1	–0.2	–0.1
20	Food and kindred products	1,786	1,708	1,636	1,563	–0.4	–0.5	–0.4
21	Tobacco manufacturers	83	69	56	40	–1.8	–2.6	–2.8
22	Textile mill products	974	847	729	627	–1.4	–1.9	–1.2
23	Apparel and other textile products	1,364	1,264	1,093	920	–0.8	–1.8	–1.4
26	Paper and allied products	705	693	693	690	–0.2	–	–
27	Printing and publishing	1,104	1,252	1,562	1,751	1.3	2.8	1.0
28	Chemical and allied products	1,049	1,107	1,065	1,084	0.5	–0.5	0.1
29	Petroleum and coal products	192	198	162	140	0.3	-2.5	–1.2
30	Rubber and misc. plastics products	580	727	830	941	2.3	1.7	1.1
31	Leather and leather products	320	233	144	114	–3.1	–5.8	–1.9
(X)	Service-producing	47,147	64,384	79,708	96,376	3.2	2.7	1.6
40-42, 44-49	Transportation and public utilities	4,517	5,146	5,548	6,097	1.3	0.9	0.8
40-42, 44-47	Transportation	2,696	2,962	3,334	3,705	0.9	1.5	0.9
48	Communication	1,130	1,357	1,281	1,344	1.8	–0.7	0.4
49	Public utilities	691	827	933	1,048	1.8	1.5	1.0
50-51	Wholesale trade	3,993	5,275	6,029	6,936	2.8	1.7	1.2
52-59	Retail trade	11,048	15,035	19,110	22,875	3.1	3.0	1.5
58	Eating and drinking places	2,575	4,626	6,282	7,796	6.0	3.9	1.8
60-67	Finance, insurance, and real estate	3,646	5,159	6,677	7,762	3.5	3.3	1.3
70-86, 89	Services	11,390	17,528	24,971	33,717	4.4	4.5	2.5
70	Hotels and other lodging places	(NA)	1,076	1,550	1,960	(NA)	4.7	2.0
72	Personal services	989	901	1,174	1,418	–0.9	3.4	1.6
73	Business services	1,676	3,092	5,570	8,311	6.3	7.6	3.4
734	Services to dwellings and other buildings	295	495	785	1,028	5.3	5.9	2.3
736	Personal supply services	(NA)	563	1,369	2,218	(NA)	11.7	4.1
737	Computer and data processing services	(NA)	304	678	1,200	(NA)	10.5	4.9
7391, 2, 7	Research, management, and consulting services	(NA)	539	811	1,190	(NA)	5.2	3.2
75	Auto repair, services, and garages	391	571	837	1,077	3.9	4.9	2.1
76	Miscellaneous repair services	189	289	347	389	4.3	2.3	1.0
78	Motion pictures	204	217	241	264	0.6	1.3	0.8
79	Amusement and recreation services	468	764	918	1,152	5.0	2.3	1.9
80	Health services	3,053	5,278	7,144	10,139	5.6	3.9	3.0
801-4	Offices of health practitioners	(NA)	1,211	1,850	2,810	(NA)	5.4	3.5
805	Nursing and personal care facilities	(NA)	997	1,319	1,907	(NA)	3.6	3.1
806	Hospitals, private	1,863	2,750	3,300	4,245	4.0	2.3	2.1
807-9	Outpatient facilities and health services, n.e.c.	(NA)	320	675	1,177	(NA)	9.8	4.7
81	Legal services	236	498	852	1,181	7.8	6.9	2.8
82	Educational services	940	1,138	1,557	1,780	1.9	4.0	1.1
83, 4, 6, 9	Social, membership and misc.	(NA)	3,704	4,781	6,046	(NA)	3.2	2.0
(X)	Government	12,553	16,241	17,373	18,989	2.6	0.8	0.7
(X)	Federal government	2,731	2,866	2,971	3,059	0.5	0.5	0.2
(X)	State and local government	9,822	13,375	14,402	15,930	3.1	0.9	0.8
01,2,7, 8,9	Agriculture	3,506	3,426	3,259	3,125	–0.2	–0.6	–0.3
88	Private households	1,794	1,256	1,163	1,103	–3.5	–1.0	–0.4
(X)	Nonfarm self-employed and unpaid family workers	5,639	7,294	8,722	9,927	2.6	2.3	1.1

• Represents or rounds to zero.   NA Not available.   X Not applicable.   [1] 1972 Standard Industrial Classification; see text, section 13.   [2] Based on assumptions of moderate growth; see source.   [3] Includes other industries, not shown separately.

Source: US Bureau of Labor Statistics, *Monthly Labor Review*, November 1989.

presentation and the strategies that the group used to help readers understand the information. Make suggestions for improving the reasoning or presentation.

As a class, discuss the ways in which interpretive statements (including visual displays) may clarify or distort "facts." Also, discuss the role of facts in making interpretive statements credible.

8. Assume that you are a paid staff member of the Minnesota AIDS Project (MAP). Review all the case documents available through this and other organizations in Appendix A, Case Documents 1. Your supervisor has asked you to write a report reviewing what type of information is being sent to the public on HIV and AIDS. You have collected samples, as represented in the case documents. Your supervisor would like you to evaluate the documents that are designed for the public and to recommend where MAP money and efforts should go next. Your supervisor is particularly interested in how effective the documents seem to be in helping the public make wise decisions about changing their lifestyles to include safer sex. In a brief report, complete the evaluation and offer your recommendations. In particular, you must help your supervisor decide how to make MAP money and volunteer efforts be most effective in Minnesota communities.

9. In a brief report to your supervisor, Dean Karen Wartala (see Appendix A, Case Documents 3), recommend what means of future communication should take place between the university and the public (particularly the Coalition of Citizens Concerned with Animal Rights). Some promises and demands were made in the letters between the university and CCCAR, but so far no one has set up a schedule to implement these promises and meet these demands. That's your task. You will need to be as concrete as possible in your recommendations.

As you complete your report to Dean Wartala, keep a list of the criteria that impact your decisions and recommendations. Identify which are social, technical, and managerial criteria. If you see a solution that was not discussed in the series of letters, you can, of course, make that a recommendation. In a memo to your instructor and classmates, attached to your report, explain your criteria and your ranking of those criteria.

10. Assuming that you are a manager at DuPont and using the materials in Appendix A, Case Documents 4, write a brief report to the company's marketing and packaging division recommending changes in the way they label and package dangerous substances such as Vydate. Explain the processes by which you arrived at your recommendations in a memo written to your instructor and your classmates.

WORKS CITED

Hayes, John R. *The Complete Problem Solver.* 2nd ed. Hillsdale, NJ: Lawrence Erlbaum, 1989.

Souther, James W. "Identifying the Informational Needs of Readers: A Management Responsibility." *IEEE Transactions on Professional Communication* PC-28.3 (September 1985), pp. 9–12.

Souther, James W. "What to Report." *IEEE Transactions on Professional Communication* PC-28.3 (September, 1985), pp. 5–8.

Winsor, Dorothy. "The Construction of Knowledge in Organizations: Asking the Right Questions about the *Challenger." Journal of Business and Technical Communication* 4.2 (1990), pp. 7–20.

Winsor, Dorothy. "Communication Failures Contributing to the *Challenger* Accident: An Example for Technical Communicators." *IEEE Transactions on Professional Communication* 31.3 (1988), pp. 101–107.

Proposals

Proposals Solve Problems ▮ The Competitive Nature of
Proposals ▮ Information about Funding Sources ▮ Proposal Review
and Selection ▮ Parts and Structure of Proposals ▮ Proposals Written
Collaboratively ▮ Proposal Management

Proposals serve three purposes. First, they're really important because we get money
to support our programs from them. Second, when we write them, we get clearer
about what it is we're doing. The more we have to write to get someone to buy
into our idea, the more we're forced to articulate clearly our goals, objectives, and
outcomes. And finally, for every proposal we don't get funded, we still raise the
funder's level of knowledge and understanding of the problem we are trying to
solve.

Mary Hartmann, Executive Director, Wayside House, Inc.
Personal interview, September 1993

Introduction

Proposals, like reports, originate with the need to solve problems. A proposal sets forth the solution or the way to meet the need or take advantage of the opportunity; it also requests support for implementing the proposed plan. Proposals, therefore, link organizations or individuals who have problems to solve with the organizations or individuals who can solve them.

Exxon Corporation, for example, had the problem of cleaning up the Alaskan oil spill, and the Environmental Protection Agency (EPA) wanted to test bioremediation as a means of cleaning up spills (see Appendix A, Case Documents 2). Exxon and the EPA had different but related problems. Jointly they invited proposals from biotechnology firms for implementing a plan to use and test bioremediation for cleanup. They reviewed the various proposals they received and selected the ones that offered the best services: expertise of the scientists who would direct the project and testing; a plan that could be trusted to achieve the stated goals; a schedule to ensure completion of the project within stated deadlines; and the budget.

The biotechnology firms that wrote proposals for the Exxon-EPA project make their income by contracting to provide services such as bioremediation and research. They use their expertise to solve problems for other organizations, and they depend on external sources of funding for projects. Proposals bring together two groups—one with a problem and the money to pay for a solution, and one with the expertise to solve the problem—so that both can do their work.

In the Exxon case, the problem and the source of funds are located within the same organization, so the corporation looks outside for solutions. In other cases, especially in social service, education, and the arts, problems originate within the service delivery organizations. They look *outside* for funding and appeal to potential sponsors for support so that they can deliver their services. The Minnesota AIDS Project, for example, notes on its brochure that it receives funding from government grants, the United Way, private foundations, corporations, and individual donors (see Appendix A, Case Documents 1). Only two percent of the Minnesota AIDS Project's earnings comes from its own outreach and education services. It is not a self-supporting or income-producing organization; rather, it sustains itself by writing proposals for support in the form of grants from various organizations. (In this way it is like the biotechnology firms that offered services to Exxon and the EPA.) The variety of sources of its income reflects its different projects and the different interests of its sponsors. One foundation may fund prevention education while a government agency may fund client services.

Of course, many proposals are written locally, that is, within an organizational unit—for example, a proposal to remodel a lab or to purchase new equipment. In this chapter, we focus on how to write persuasive proposals no matter what your situation, but we also discuss sources of funding and the review and selection process for proposals in case you seek support outside your organization.

Proposals are so pervasive in the workplace that it is useful to consider the different purposes that define them. Not all proposals fit into the same mold. In this section, we discuss the research or service delivery proposal and the management proposal.

The Exxon-EPA and Minnesota AIDS Project involved a **research and service delivery proposal:** one that offers research or service delivery in exchange for support, usually including financial support. The service could be design, installation, education, health care, or any other service that someone is willing to buy.

research and service delivery proposal
proposal offering research or service in exchange for financial support

When developing a research or service delivery proposal, you expect to participate in implementing the solution. You say to a potential supporter, in effect, "If you will give us money and time, we will solve the problem that concerns you by completing the tasks described in this document." Grant proposals make this offer to an organization with funds to support research or service delivery. A scientist offers to continue the research to find out if a defective gene causes Alzheimer's disease in exchange for funding for salaries and equipment from the National Science Foundation (NSF). A consulting firm offers to find the best manager for a city to hire in response to the city's request for help with this task. A home health service appeals for support from a government fund created by legislation to provide respite care for caretakers of homebound individuals. An employee offers to research network hardware and software to make the office more efficient in exchange for being freed from other job responsibilities for two weeks. Research and service delivery proposals are usually written to external organizations. Many people make their living by winning contracts to deliver a service or perform research for someone else.

Sometimes people look around them and think that a procedure within their organization could be more efficient, safer, or fairer if it were changed or if different equipment or facilities were available. But they do not have the authority to make the changes themselves. Instead, they must persuade someone with power to make the changes by writing a **management proposal.** It is hard to convince people to change well-established procedures or to purchase new, costly equipment. Even when people agree that a change is needed, they may think that there is no way to accomplish the change. Details of *how* the change might occur are more persuasive than simply telling what might be changed.

management proposal
proposal designed to convince someone in authority to make a change in a procedure or policy

The person who observes a need or an opportunity for improvement usually writes a proposal that offers a solution to be implemented by someone with power and authority. In this type of proposal, you might say, "If you will do these things, then the problems I have pointed out will be solved." A student, for example, who rides her bike to campus is worried about car-bicycle accidents on campus. She proposes to the director of facilities construction that bicycle paths be built so that bikes and cars won't have to share the same street. She doesn't plan to build the bicycle paths herself; rather, she wants to convince the university to invest in the paths. The benefit to her is a safer place to ride her bike.

In order to persuade the facilities director to act on her proposal, she would use her own experiences as a bicyclist to determine the kinds of information and procedures she wants to lay out in this proposal. Then, she might draw on the results of surveys taken from bicyclists and motorists about their gripes and fears about each other. She might also start with some statistics taken from the traffic office, which keeps files on car-bicycle accidents. Or, she might cite some statistics from other universities that had similar problems, installed bicycle paths, and reduced the number of car-bicycle accidents on campus. Finally, she would research the national standards set for width and materials of bicycle paths and plan some routes using a campus map to show how the proposed paths could be implemented.

The Competitive Nature of Proposals

There is always competition for the rewards you seek as a proposal writer. The researcher who wants to study genes and Alzheimer's disease competes with a dozen or more researchers who want to study the same or a similar thing. NSF has limited funds and can fund only the projects with the best chance of achieving the objectives and that identify the most significant problems. The proposal for bicycle paths competes with other campus projects to provide safe and convenient means of transportation—perhaps the proposal for new parking lots will argue more effectively. Even when an employee proposes to study a new system of networking computers for no extra pay, the proposed use of her time competes with other uses.

Of course, all writing is persuasive. In some genres, however, you try to remain objective and delay a commitment to a particular point of view until research is complete. The effort to remain neutral or at least open to a range of possible results is especially important for genres that report research, such as the scientific report or the report for decision making. In other situations, you commit to a particular point of view from the outset. That commitment is most evident in proposals, in which you will *set out to persuade*.

Being persuasive means that you have to show how the interests of both the organization that can provide the service and the one that can provide the funding will be served. That is, you must show that what you propose will serve the sponsor's goals, not just your own.

A proposal persuades in part by spelling out how the change or project will be accomplished, with details of who will do what and when. The **plan of implementation** is the central part of the proposal. This information distinguishes a proposal from a simple suggestion for a change. It is hard to convince people to support your project if they don't understand exactly what you will do. You have to propose a sound and feasible plan for solving a problem. You also have to anticipate and answer reviewers' questions satisfactorily. While doing these things, you also have to convince the potential project supporters that you can provide the service better than anyone else or, in the case of management proposals, that the change is worth the investment of time, retraining, or new equipment.

Competitive proposals reflect awareness of what the potential sponsors are seeking combined with high quality project planning. Persuasion also requires attention to the arrangement, style, and presentation of the proposal.

plan of implementation
central and defining component of a proposal, often suggesting change

Sources of Project Support

Because proposals link organizations with problems to solve with organizations or individuals who can either solve them or who can provide the funds for solving them, the first step in linking the two is to understand the organizations that may support projects and the goals that they have for the support they can give. Successful proposals match their project goals to the goals of the supporting organization.

The discussion in this section applies primarily to research and service delivery projects that are funded externally, that is, by organizations other than the one that provides the service. The support these organizations provide does not have to include money, but typically it does. Foundations, government agencies, professional societies, civic groups, and corporations sponsor most research and service delivery projects.

FOUNDATIONS

Some organizations exist not to make money but to distribute it. Public (local, state, and federal government), corporate, or private foundations are examples of such funding sources. A **foundation** is a nonprofit entity that uses its funds to support certain research, educational, artistic, or charitable projects. Foundations may be funded by donations, taxes (in the case of public foundations), a certain percentage of earnings (in the case of corporate foundations), or permanent endowments. An **endowment** is a sum of money that generates interest or investment income to fund particular kinds of projects. The purpose of foundations is to distribute funds to individuals or groups who can achieve the objectives the foundations have set for the use of their funds.

foundation
a nonprofit entity that uses its funds to support research or educational, artistic, or charitable projects

endowment
a sum of money that generates interest or investment income to fund particular projects

A foundation sets goals but doesn't necessarily have the expertise to achieve the goals with its own staff and facilities. For example, NSF uses tax money to fund scientific research. The problems NSF considers include lack of knowledge about the physical and natural world and very specific problems such as ozone depletion. There would be no way to conduct all the scientific research that NSF sponsors in one place because of the scope and variety of the research, nor would it be desirable to do so. Researchers must maintain a measure of independence from their sponsors and from one another. At the same time, the researcher with the expertise to conduct the investigation doesn't always have the funds that it takes to conduct the research—funds for salaries, equipment, supplies, travel, and **overhead,** such as electricity and office space.

overhead
research costs for such things as space and utilities

Foundations establish their own goals for uses of their funds. Some foundations are especially interested in education, and others are interested in supporting projects in health or agriculture. Some define their interests geographically; that is, they may be willing to fund a range of projects in a certain city or state. Some are willing to pay for equipment and salaries, while others will provide only travel or printing expenses. A private, nonprofit foundation with a substantial amount of money to distribute is probably listed in *The Foundation Directory*. Example 17-1 shows an entry from *The Foundation Directory* that indicates the foundation's geographic and subject matter interests.

GOVERNMENT PROGRAMS

The US government contracts, by means of competitive proposals, for many of the services it needs to have performed. The *Commerce Business Daily* announces projects each day in multiple categories, including research and development, installation of equipment, education and training, transportation, data processing, and social services. Example 17-2 shows an announcement of a need to remove medical waste. Various companies that specialize in medical waste disposal will submit proposals to do the job, and after a review, one will be selected to complete the job and awarded the funds to do it. State and local governments may also seek services or research and award contracts to the organization that presents the best proposal.

Information about Funding Sources

A persuasive proposal not only identifies a workable solution for a significant problem but also targets a potential sponsor's interests. A perfectly good proposal won't get funded if it goes to an organization without an interest in the problem or with restrictions that exclude the type of support being requested. Another way to doom a proposal is to define the problem in a

EXAMPLE 17–1 A Foundation Listed in *The Foundation Directory*

5623
Catto Foundation
110 East Crockett St.
San Antonio, TX 78205   (512) 222-2161

Established in 1967 in TX.
Donor(s): Henry E. Catto, Jr., Jessica Hobby
Catto, A & C Communications
Foundation type: Independent
Financial data (yr. ended 12/31/90):
Assets, $722,509 (M); gifts received,
$604,077; expenditures, $287,041;
qualifying distributions, $286,802, including
$280,633 for 33 grants (high: $60,000; low:
$250)
Purpose and activities: Support
primarily for higher education; grants also
for cultural programs and the environment.
Fields of interest: Higher education,
education, cultural programs, environment.
Types of support: General purposes.
Limitations: Giving primarily in
Washington, DC, and San Antonio, TX. No
grants to individuals.
Application information:

> *Initial approach:* Letter
> *Deadline(s):* None
> *Write:* Jessica H. Catto, Pres.

Officers and Directors:* Jessica
Hobby Catto,* Pres.; Susan R. Farrimond,
Secy.-Treas.; William P. Hobby, Heather C.
Kohout.
EIN: 746089609

Source: Reprinted by permission of the publisher from *The Foundation Directory*, 1993 Edition. Copyright © 1993 by The Foundation Center, 79 Fifth Ave., New York, NY 10003.

way that differs from the way the potential sponsor defines it. One important first step in proposal development, then, is for you to research the potential project sponsor. You should explore printed sources of information including requests for proposals, directories of funding sources, annual reports, and guidelines for proposals available from potential project sponsors. You can

EXAMPLE 17–2   Announcement in the *Commerce Business Daily* of a Contracting Opportunity

F Natural Resources and Conservation Services

Contracting Officer 90C, V.A. Medical Center/Outpatient Clinic, 150 South Huntington Avenue, Boston, MA 02130
F—SERVICE FOR REMOVAL AND DECONTAMINATION OF REGULATED MEDICAL WASTE SOL RFP 523-38-93 DUE 091793 POC Contracting Officer, Edward Camey, (617)232-9500x5536. Contractor shall provide all tools, equipment, and labor to provide for the removal and decontamination of regulated medical waste in accordance with specifications at the V.A. Medical Center/Outpatient Clinics, 150 South Huntington Avenue, Boston, MA 02130. The contract period is from October 1, 1993 through September 30, 1994, with three option years. Contract award is subject to the availability of funds. Solicitation packages will be available on or about August 18, 1993. Requests for solicitation must be in writing. No phone or faxed requests will be accepted. Proposals are due by the close of business on September 17, 1993. SIC 4959 applies to this solicitation, which is a 100 percent small business set-aside. (0228)

Source: *Commerce Business Daily,* August 18, 1993.

make inquiries by phone or in person. This information is essential for you to gauge your audience's interests and expectations.

According to Janis Cavin, "The most common reason for rejection of a proposal is a mismatch between the purposes of the proposal and the goals of the funding source" (7). Researching the funding source's goals is a good investment of your time and energy.

RFPs

request for proposals (RFP)
document identifying a need for a service or for research and inviting proposals for the necessary work

If an organization has a specific problem to solve, it may issue a **Request For Proposals** (RFP). An RFP identifies a need for a service or research and invites proposals for completing the necessary work. Example 17–3 shows

EXAMPLE 17–3   Request for Proposals

REQUEST FOR PROPOSALS
FOR PROFESSIONAL SERVICES
FOR THE LUBBOCK CHAMBER OF COMMERCE
LUBBOCK, TEXAS

I.   General

The Lubbock Chamber of Commerce is seeking professional services for a consultant, architect, or engineer experienced in determining the feasibility of an all-purpose event facility for the Lubbock community.

II.   Scope of Services

The consultant shall determine what size all-purpose event facility the Lubbock community can support and what unique features must be included to make the facility feasible in the Lubbock market. The information provided by the consultant shall include but not be limited to the following:

1. *Market Analysis*
    a. Review historical event statistics of the Civic Center and Auditorium/Coliseum for the past three years (FY 1987–88, 1988–89, 1989–90).
    b. Define market area (primary and secondary geographic area).
    c. Project usage by type of event (i.e., entertainment, sports, conventions, etc.), quantity of events, attendance, and timing.
    d. Demographics by population trends and spending profiles.
    e. Review competitive facilities by location and projected attendance to determine market impact.

2. *Projected Size*
    a. Determine what size facility can be supported by the Lubbock community (number of seats in arena, total square footage).
    b. Determine unique features (arena, exhibit area, dining, meeting rooms, skating arena, etc.) that need to be included to make facility economically feasible.
    c. Determine number of parking spaces required for projected facility
    d. Determine number of acres of land needed for projected facility.

3. *Financial Feasibility*
    a. Prepare estimates of projected revenue by income source (rent, concessions, etc.) for first 10 years of operation of facility.
    b. Prepare estimate of projected operating and maintenance expenses for first 10 years of operation of facility.
    c. Project total cost of construction to include architectural fees, engineering fees, construction costs, furnishings, etc. (excluding the cost of land acquisition).
    d. Provide estimated unit cost per square footage, per arena seat, etc.
    e. Determine economic impact on community (additional revenue generated in hotel/motel tax, sales tax, estimated dollars spent by increased attendance.

> **EXAMPLE 17–3**   *(concluded)*
>
> **III.   Experience and References**
>
> Professional consultants will be selected on the basis of a combination of proven experience, ability to complete the project in a timely manner, and recommendations from recent clients with similar projects. Submit the following information:
>
> 1. A listing of similar projects completed or in progress, including a list of references by name, address and telephone number for each project listed.
> 2. A listing of the individuals, with professional qualifications, who will be assigned to this project.
>
> **IV.   Schedule and Methodology**
>
> The consultant shall be capable of performing the services in the shortest time possible without sacrificing the quality.
>
> 1. Include a projected timetable for the work to be performed in the scope of services outlined in Section II.
> 2. Describe the methodology to be used to complete the project.
>
> **IIV.   Estimated Cost for Services and Basis of Payment**
>
> The Chamber of Commerce desires to obtain the services in Section II at its actual cost not to exceed a maximum fee.
>
> 1. List cost for services to be provided to include professional services and expenses.
> 2. Provide anticipated payment schedule.
>
> **VI.   Contract Negotiations**
>
> Upon selection and approval of a consultant, negotiations will be scheduled to determine final project fees and contract terms.
>
> **VII.   Submission of Proposal**
>
> To be considered, 10 copies of the proposal should be submitted to:
>
> All-Purpose Facility Planning Committee
> c/o Chamber of Commerce
> P.O. Box 561
> Lubbock, Texas 79408
> 14th and Avenue K
> Lubbock, Texas 79401
> Phone (806) 763-4666
>
> Submissions must be received no later than 2 PM Friday, January 11, 1991.
>
> ---
>
> **Source:** Reprinted with the permission of the Lubbock Chamber of Commerce.

a detailed RFP requesting proposals for a feasibility study for an event facility. An RFP may be many pages long if it includes a number of project specifications, as in Example 17–3.

An RFP guides proposal planning and development by defining the problem and goals in specific terms. It provides many clues about objectives, schedule, and budget as well as specifications for equipment or supplies. A consulting firm responding to the RFP in Example 17–3 can use the specific tasks defined in the RFP to establish project objectives. The RFP also specifies that the city wants to see work and payment schedules. This RFP does not tell everything you might want to know; for example, it is vague about money. If the RFP specified an amount, chances are that all the proposals would request that amount or a few dollars less, eliminating the competition that the proposal process aims for. But the RFP is very specific about the outcomes it would like. The RFP serves both parties in the contracting process: the city is more likely to get what it wants if it defines what it wants; the proposing organization does not have to guess at the expectations nor waste time developing proposals for projects that it is not prepared to do.

FOUNDATION DIRECTORIES AND OTHER SOURCES OF INFORMATION

If the problem that the proposal addresses does not originate with the funding organization but rather with an organization that needs funds to provide services, there will be no specific RFP. Instead, proposal developers search for potential sponsors who are interested in the type of project. The Minnesota AIDS project, for example, searched for sponsors with interests in education and health care. A basic source of such information is a directory such as *The Foundation Directory* or the *Directory of Corporate Philanthropy* or a specialized directory such as *Funding in Aging* or *Private Funding for Rural Programs.*

Example 17–1 illustrates how one foundation defines its interests and limitations on the awards it offers. You can use the descriptions in a directory to screen potential sponsors to find one suited to your project. By studying the descriptions, you can easily rule out organizations that are not suitable for the project you are proposing, or you can adjust your project so that it will be attractive to the foundation. This research can make you reasonably confident of a match between your project and the potential sponsor.

Funding organizations may also have detailed guidelines that specify award criteria and methods of selection. Like an RFP, proposal guidelines provide specific information for audience analysis. Annual reports of the potential sponsor also reveal information about its interests by showing the type of projects it has previously funded.

In addition to studying an RFP, funders' guidelines, and annual reports, take the initiative to talk with potential sponsors about the proposed project. These conversations can help you determine whether the project is suitable for a particular organization and also allow you to inquire about the funding schedule and deadlines that must be met. Representatives may be available to talk with you to clarify their expectations, to answer questions, and even to offer suggestions for proposal development. It is in their best interest to receive excellent proposals because they want to spend their money well. They also learn about you in a more personal way than the proposal itself allows. These conversations establish relationships as well as provide an opportunity for information sharing. Such conversations are more likely to occur with private than with government funding sources. Government regulations may require that all applicants receive the same information to ensure a fair award process.

Proposal Review and Selection

You will screen potential projects and sponsors at the beginning of the proposal development process, and the sponsor will screen the proposals you submit. Reviewers read and evaluate the submitted proposals and select the one or ones most likely to accomplish project goals within the limits of time and budgets. The reviewers are people who understand the goals of the funding organization; they may or may not have expertise in the methods of the project being proposed. In order to write for your audience, you need to know whether the reviewers will be specialists or generalists.

peer reviewers
people with expertise in a research area or discipline who judge the relative merits of proposals

Peer reviewers are people with expertise in a field of research or service delivery. If NSF were reviewing a proposal for gene research related to Alzheimer's disease, it would select reviewers who themselves could conduct gene research rather than, say, psychologists or technical communicators. These reviewers can evaluate research methods and determine whether the methods will yield the intended results. Likewise, evaluators of the proposals to dispose of medical waste must be expert in waste disposal methods or in the administration of waste disposal projects. If the proposal will be reviewed by peers, you may assume your audience has some knowledge about the problem and its significance as well as technical knowledge. You will probably want to include technical details of methodology, but you may be able to describe the problem in an abbreviated way.

On the other hand, if the proposal will be read by people with general rather than specific backgrounds in the problem, you need to explain the problem and its significance in terms that a generalist can comprehend. You may describe methodology with less detail about procedures but with more explanation of what the methods will accomplish. For example, a university may have funds to support student research, for which students from throughout the university submit proposals. The reviewers read proposals for projects outside their own disciplines. A biology professor may need to compare a proposal from engineering with a proposal from English. The proposal writers will have to explain what the project is and why it is important in terms that the reviewers can understand.

Proposals are often read and ranked by a team of reviewers, especially when a lot of money is at stake. Different reviewers will see different merits and limitations in the proposals. The team approach to reviewing also neutralizes possible personal bias in an individual reviewer. Some proposals are reviewed "blind"—that is, the name of the proposer does not appear on the proposal that the reviewers see. Thus, they can focus on the merits of the proposal itself rather than letting opinions about an individual intrude in their decision.

The reviewers evaluate according to specific technical, managerial, and social criteria (remember the discussion of some of these criteria in Chapter 16). You may be able to obtain these review criteria from the funding agency if you request them. The criteria would provide you with a good checklist for proposal development. The reviewers will consider the significance of the problem, the responsiveness to the RFP or to the goals and limits set for the use of available funds, the merit of the proposal (including methods), the qualifications of the staff, the adequacy of facilities and other essential resources, and records of past performance.

Figure 17-1 illustrates the process of linking the organization with the problem to the organization that can solve the problem by means of an RFP and competitive proposals. In response to an RFP that describes the problem, need, or opportunity, several organizations develop proposals. After a review, one proposal is selected and the developing organization is funded.

When an organization seeks funds for its own projects rather than responding to an RFP, the RFP stage would be replaced by a search for an appropriate funding source. The proposal would still be one among many proposals submitted to the funding source, only one or some of which would be funded.

UNSOLICITED PROPOSALS

In some cases, you might submit a proposal without an explicit or implicit invitation. Management proposals, in particular, may be unsolicited. The manager or potential project sponsor may not even have recognized a problem, let alone imagined a solution. The proposal may be a surprise (though as a shrewd communicator you will prepare the sponsor before submitting the proposal).

An unsolicited proposal is more difficult to write and less likely to succeed than a proposal that responds to an RFP or an announcement of interests in funding. You must first convince the people who review and decide upon the proposal that there is a problem, that the problem is significant, and that the problem concerns the potential sponsor. In addition, you must convince the reviewers that this problem and solution are more worthy of available funds than other projects to which the sponsor might commit. Convincing a possibly skeptical reviewer that there is a problem is only half your task, because many reviewers who agree there is a problem don't necessarily believe that anything can be done about it. You can help the potential sponsor to imagine the solution by giving complete details of the way the project will unfold and what your results would do to solve the problems.

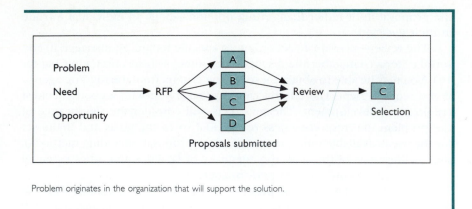

Problem originates in the organization that will support the solution.

FIGURE 17–1
Proposal Request and Selection Process

When you write an unsolicited proposal, you may have only general guidelines about expectations for proposals, such as the ones in this chapter, rather than specific guidelines. Nevertheless, you should make every effort to get some direction from the potential project sponsor. Communication before the proposal is submitted can prepare the audience for the proposal as well as yield ideas about ways to appeal to the reviewer.

Parts and Structure of Proposals

Your proposals may be complete in a one- or two-page letter, or they may require several volumes of text and supplements. Proposals to external organizations are likely to be formal and detailed because you and the proposal reviewer do not share the same knowledge of the problem or even of each other. An internal proposal may be formal and detailed when the change proposed will be substantial or dramatic.

We discuss some standard parts of proposals in this section (you will want to refer to Example 17–5, at the end of this chapter, as you read about these parts). Your proposals may not contain all of these sections or parts, or the parts may carry different names. Before you write a proposal, you should find out the expectations for the proposals: parts required, labels for parts, and arrangements of parts. An RFP and proposal guidelines from the sponsoring organization will list the requirements. One of your first strategies of persuasion is to offer the proposal reviewers the information they seek in the way that they expect to see it. Because you are offering to provide a service for them, you should demonstrate immediately that you are in tune with their wishes by conforming to their requirements for documents. Be careful even with details of wording: if the RFP specifies a part called "Statement of the Problem," don't label your corresponding part "Problem Statement" or "Needs Assessment," and if it asks for an "Overview," don't provide a "Summary."

Each part of your proposal presents significant information that answers questions the reviewers will have. Anticipating and answering the questions satisfactorily is a key to persuasion. Example 17–4 names standard proposal parts and the questions they answer.

The proposal not only wins support for a project but also serves as a planning and management tool for the project. Your details of the plan of implementation, which include dates, tasks, and specific outcomes, provide direction for carrying out the project. Your objectives, schedule, and budget also serve as standards for evaluating progress. When you write a proposal, you make a commitment to carrying out the tasks you propose in the time and for the budget you indicate. Your follow-up progress reports measure what has happened according to what was planned.

COVER LETTER (OR LETTER OF TRANSMITTAL)

A **cover letter** for your proposal may be written by the executive officer of your organization to establish that the organization supports the project. This support assures the funding source that if the grant or contract is awarded, the completion of the project according to the stated plan will be a high priority of the organization and not something that gets lost among other projects. For example, if you were proposing support for a design project from an engineering firm, you might ask your faculty advisor to write a cover letter affirming that the university approves the project. The letter would confirm the availability of facilities and faculty advisement for you and for the proposed project. When there is no executive officer involved, you may write the cover letter, perhaps over your supervisor's signature.

cover letter
like the letter of transmittal, a document written by the executive officer of the organization submitting a proposal that indicates support for the project

SUMMARY OR OVERVIEW

The summary or overview provides a capsule of the whole project, including the problem, objectives of the project, plan, budget, personnel, and facilities. It may highlight special features of the proposal that distinguish this proposal from competing proposals. Such features might include the following:

An unusually well-qualified staff;

A new, efficient method of performing the proposed service; or

An outcome that will exceed the minimum requirements.

Thus, your summary may go beyond answering the main question, "What will this project accomplish and how?" to answer, briefly, "How will this

EXAMPLE 17–4 Proposal Parts and Questions They Answer

Proposal Part	Reviewers' Questions
Cover Letter	Do the executive officer and board of directors of the proposing organization support the proposed project?
Summary or Overview	What is this project about, and what do the writers propose to do? How does this project relate to our interest? How will this project be better than other similar ones we might fund?
Introduction	Who is proposing this project? What are the organization's interests and activities? What kind of track record does it have with similar projects?
Problem Statement or Needs Assessment	What is the problem? What is the need? Is the problem important? Is this the type of problem in which our organization is interested? Does the writer have a grasp of the problem? Does the writer's concept of the problem match our concept?
Objectives	Exactly what will be the outcomes of the proposed work? What will be done? In what order?
Plan	What methods will be used? Are these the right methods to achieve the objectives? When will the project be complete? Can the project be finished in the specified time, or can we predict delays and budget overruns?
Evaluation or Quality Control	What measures will be used to establish that the plan is achieving the objectives? How will we know that the plan is working?
Budget	How much will the project cost? How much will each component cost? Are the costs fair and reasonable for the work performed?
Personnel and Qualifications	Who will do the work? Do they have the qualifications (education, experience, interest) to do the work well? Are they the best qualified of the various people who have applied to do the job? Do they have a record of success in similar projects?
Facilities and Equipment	Do the proposers have the necessary facilities and equipment to do the job they say they will do?
Other Sources of Funding	When our grant ends, will the project die, or is the support broad enough to ensure its continuation? Have others committed to the project?

proposer do a better job than others who have applied?'' All the information appears in more detail in the rest of your proposal, but your summary gives reviewers their first impression of the project and establishes your (and your company's) *ethos*.

INTRODUCTION

Although "Introduction" has different meanings for organizations, an introduction may essentially be a problem statement. If an introduction is requested separate from a problem statement, you should make sure that it focuses on the organization that offers services or requests support rather than on the project itself. Use it to provide a brief history and to state the organization's purposes. You may also review the organization's achievements in activities related to the proposed project. Your introduction is especially important if the proposal reviewers are not familiar with your organization because it provides an opportunity to establish some credibility. Introductions that identify an organization are more common in proposals that request support from foundations for educational and charitable purposes than in proposals that respond to RFPs.

PROBLEM STATEMENT OR NEEDS ASSESSMENT

The problem or need that you have identified is the reason for proposing some corrective action. Therefore, you should use the problem statement to describe a discrepancy between the way things are now and the way they might be and to explain the reason for the proposed project. It is similar to the problem statement for a report (see chapter 16), and it should tell:

- What the problem is;
- When and where the problem takes place;
- Why the problem happens;
- Who is affected by the problem; and
- Why the problem is significant—the short-term and long-term costs (financial and otherwise) of not solving it or the benefits of pursuing a solution.

Some RFPs call for a needs assessment instead of a problem statement; a needs assessment tells what is needed, who needs it, when and where it is needed, and why it is needed.

If you are responding to RFPs, you may see the problem clearly described there and assume that a problem statement is redundant in the proposal itself. One part of your task in developing a problem statement, however, is to convince the reviewers that you understand the problem well enough to plan and carry out a solution. Your own statement of the problem, which draws on the content and even phrases of the problem statement in the

RFP, establishes a sense that you have a firm grasp of the problem. The problem statement may conclude with a description of the proposed solution (though not with the details of implementation).

OBJECTIVES

objectives
a statement in concrete terms of what one plans to accomplish if the proposal is funded

The **objectives** state in concrete terms what you plan to accomplish if you receive the support you request. Objectives answer the question, "What will happen if the proposed project is implemented?" They identify intended results. The next part of the proposal, the plan, will tell how the results will come about. If there are multiple objectives, you should list them, but you may need to explain or elaborate as well.

The following objectives identify project outcomes in specific, even measurable, terms. Note that the verbs *determine* and *characterize* indicate a research proposal (a study), whereas the other verbs suggest service delivery:

To reduce car-bicycle accidents on the central campus from an average of seven per year to zero by 1996.

To determine the optimum seating capacity for an all-purpose events arena in Lubbock.

To reduce oil contamination on eight square miles of Alaska beach by an average of 50 percent.

To determine the degree to which fertilizer amendments enhance the biodegradation of surface and subsurface oil.

To characterize ecological risks associated with fertilizer amendments.

To develop a computerized advisement system for juniors and seniors in mechanical engineering that will be at least 95 percent accurate in its advice.

To recruit, train, and place 100 teenage volunteers in local service agencies.

To raise awareness of AIDS transmission and prevention among teenagers to the point that 75 percent of all high school students in Minneapolis-St. Paul will achieve a score of 85 percent or better on a paper-pencil test covering AIDS transmission and prevention.

PLAN: METHODS AND DATES

Your plan of implementation will make or break the proposal. The reviewers will probably select the proposal with the best plan for solving the problem, assuming the budget is reasonable. Often the biggest barrier to finding workable solutions to problems is knowing *how* to make them happen. *What* will be done has already been established by the problem statement and objectives; the plan specifies how.

The following example illustrates the power of a plan to persuade. A group of engineering students were frustrated by long waits to see their advisor at registration time, and they felt that the advice given was redundant anyway because it was so similar for all students with their major. They thought a computerized advisement system might save students as well as faculty advisors a lot of time. Knowing *what* they wanted, they could have simply suggested the computerized system to the dean. But the dean, who might have responded positively to the idea, nevertheless would probably have defaulted to a standard response to a suggestion for change: it's a good idea, but we don't have the means to implement it. The suggestion would have died a natural death.

Instead, the students developed a plan for creating software and a database that incorporated all the degree requirements as well as prerequisites so that it could tell students what courses they needed in any given semester. They recruited an expert computer student to do the programming, established the contents and structure of the database, and included pilot tests of the program with representative students. By showing how they could achieve the benefits of a computerized system at a reasonable cost, and even offering the people to do the job, they created a solution that would have been hard for the dean to refuse. They created temporary jobs for themselves as well. And, by shifting the focus from what the recipient could do to what they could do for the recipient (shifting from a management proposal to a service delivery proposal), they made it easier for the recipient to agree to the proposal. By focusing on the *how* of their proposal, the students' plan was ultimately persuasive.

The specific methods of proposal implementation vary, depending on the objectives and the nature of your project. The Exxon bioremediation project required scientific tests and statistical sampling of various areas of the beach where the oil spilled. The plan for the proposal to win that job explained the sampling and testing method, including the number of samples to be drawn, dates, and locations, as well as the tests to be performed.

The Minnesota AIDS project included the development, printing, and distribution of the brochure that you see in Appendix A, Case Documents 1. The proposal developers specified in the plan section how many brochures would be printed and how and when the brochures would be distributed. If they had simply written that they would develop and distribute brochures, the reviewers would have legitimately wondered about the locations and means of distribution. Your plan must anticipate such questions and answer in enough detail to convince the reviewers that you have a clear idea of what you will do.

The plan for determining the feasibility of an all-purpose events facility for the city of Lubbock (Example 17–3) included a review of events and uses of existing facilities in the city over the past three years as well as interviews with past and prospective users to determine the city's advantages and disadvantages and estimated use.

The programming, bioremediation, AIDS brochure, and feasibility study examples all reflect research or service delivery proposals. The plan for each of these projects necessarily reveals the chronology of the steps of implementation. The plan for a management proposal is likely to be more descriptive than chronological. That is, it may describe the finished product rather than the steps of implementation inasmuch as the writer does not expect to participate in the implementation. The student who proposes bicycle paths shows their location and provides specifications on width, materials, and traffic control so that the director of facilities will have some guidelines for initiating the project. The employee who proposes networking the computer workstations draws a layout of the wiring, provides specifications on the type of wire, recommends software and hardware, and suggests a schedule for installation.

Whatever your plan for achieving the stated objectives, be specific about what will be done, when, where, by whom, and in what quantities. *Don't leave doubts about whether you know how to get started and complete the task.*

The description of methods lets reviewers envision your plan as it will be implemented. A schedule of tasks and completion dates shows them when the work will be done. A **Gantt chart** offers even more information: in addition to showing completion dates, it shows when the tasks are being accomplished in relation to other tasks. The proposal in Example 17–5 uses such a chart in the Project Schedule section. The tasks listed on the chart have already been described in narrative form, but this chart shows at a glance where the project should be at a given point in time. It also shows that some tasks are simultaneous rather than sequential.

Gantt chart
a schedule of tasks and their relationships and completion dates

EVALUATION AND QUALITY CONTROL

A good proposal includes a section on evaluation and/or quality control. The evaluation methods tell whether your plan has achieved its stated objectives. Alternatively, in this part you promise evidence of quality control and progress according to your established schedule. You have a variety of evaluation methods available; your task is to choose and develop methods to measure what your project aims to do. When your objectives involve education, your evaluation methods often include a test, perhaps a pre- or post test that compares what the students knew before the project began and what they knew after. If scores go up, the project has achieved some success. The objectives may establish how much the scores will rise, and the test will have to be reliable.

If your objective has been to remove a pest or hazard, your evaluation may consist of measurement before and after treatment. The executive summary of the Exxon bioremediation project indicates that samples were taken of the beaches before and after treatment to measure the amount of contamination. The statement that 10 to 70 percent of the oil had been removed contributes to the overall evaluation that "the application of fertilizers is a safe and effective means to enhance" the natural process of biodegradation (Appendix A, Case Documents 2).

If your project is an ongoing construction or service delivery project, your sponsors may request periodic progress reports that indicate whether the project is on schedule. The schedule is determined by the proposal objectives and implementation plan. Your progress reports compare what has happened with what was supposed to happen. If the objectives state, for example, that 500 high school students in Minnesota will participate in AIDS discussion groups by May 15, but 600 students have participated by the target date, you have support for a positive evaluation of the project.

If you find it hard to write the evaluation methods, your objectives may not be specific enough and may need some revision.

BUDGET

You can be sure that proposal reviewers will study your budget carefully, looking at the individual items as well as the bottom line. The funding organization may restrict how you can spend the money you may be awarded; some grants will not provide salaries or equipment, for example. Instead of guessing and using round figures, you should take the time to get specific estimates for materials, printing, travel, and any other expense. Also follow these suggestions:

- Make sure that the plan explains or at least prepares for each item in the budget. Don't introduce surprises at this point. If your budget includes postage, for example, be sure the plan specifies what will be mailed.
- Check your figures to avoid errors in addition.
- Don't inflate the budget, even if you haven't spent all the money that has been offered. Estimate reasonable and fair costs. Don't try to sneak in a new piece of equipment that you want even though the project doesn't really require it. Sponsors don't want to be cheated.
- Display your figures in a table so that the reviewers can easily see the item, the formula for calculating the cost (for example, per piece cost times number of pieces), and the total expense for the item as well as the bottom line.

Don't worry about being the lowest bidder. Funding organizations are looking for quality, safety, timeliness, attention to detail, and sound arguments, as much as or more than low prices.

PERSONNEL AND QUALIFICATIONS

You can attach résumés of the key project staff to a proposal to establish an overall record of qualifications and achievement, but you may be more persuasive and your information may be easier to interpret if you offer a paragraph description that focuses on the project staff's specific qualifications. Your paragraph need not be comprehensive; rather, it should highlight certifications and previous experience with related projects and their outcomes. You might also include testimonials from your clients and customers.

FACILITIES AND EQUIPMENT

You may require specialized facilities, such as a dust-free manufacturing area, or equipment, such as an electron microscope, to complete your proposed project. A description of available facilities and equipment assures the reviewers that the project is feasible in terms of these requirements. If the facilities belong to someone else, you'll need a letter from the owner to confirm their availability for the proposed project.

OTHER SOURCES OF FUNDING AND ONGOING FUNDING

The potential funding source may be interested in other sources of funds you have obtained as a way of determining need and also as a way of assessing the support the project has already gained. If your proposed project will last longer than the support requested, the funding source may want to know your plans for continued funding. Perhaps other agencies will help to support the project if it proves to be successful, or perhaps there will be fundraising activities or client fees. This information helps your funding source assess whether it is investing in a long-term project or one that may fail after the initial support period.

PROPOSAL SUPPLEMENTS

Supplements, often presented in appendices, support the proposal but are not essential to understanding your plan. If you prepare such supplements, they might include the following:

- A map or architectural drawing showing the location of the project.
- Letters of endorsement from experts who think the proposed project has value.
- Résumés of the key people who will implement the project as a way of confirming their qualifications.
- Detailed budget analyses that are only summarized in the text.

- Brochures or other informative literature describing the organization proposing funds.
- Other evidence that supports the proposal but is not central to its arguments.

If your organization is a nonprofit one, you may need to submit evidence of its tax-exempt status, audited financial statements, and the operating budget. You may bind these supplements with the proposal as appendices, or, in the case of a large map, enclose them as separate items. But don't stuff the appendix full of every piece of paper you can find that may be related to the project or your organization. Reviewers are impatient with having to wade through excessive material to find what is really important.

Some organizations require you to complete application forms. The form might be in addition to the proposal narrative, or the proposal itself may have to be typed into the form.

VISUAL DISPLAYS

Visual displays are not a separate part of the proposal, but they are mentioned here as a reminder that visual displays may have a greater impact and be easier to understand than words. Thus, they support the persuasive and informative purposes of your proposals. In addition to presenting the budget in a table, consider drawings or charts to illustrate key concepts or processes and graphs to demonstrate the significance of the problem.

CONCLUSION

Many proposals simply end without a formal conclusion once all the requested information has been provided. Your whole proposal has been a request for support, and to restate it may seem redundant. If you need a formal conclusion, use it to summarize the benefits to the funding source and request support. You will leave the reviewers with a good image of the outcome of the proposed project.

Proposals Written Collaboratively

There are good reasons for collaborating with other writers on proposal writing. The project may require different types of expertise, and the people who are expert in a project requirement should be the ones to write about how the requirement will be met. Involving people who will help to implement the project in developing the proposal gives these people a personal and professional stake in the project that can carry over into their work on the project. Moreover, project deadlines may be so tight that one person could not possibly complete the entire proposal within the allotted time. Note that the announcement in Example 17–2 of the need to remove and decontaminate medical waste was published August 18, but the completed proposals were due September 17. In the intervening month, the contractor also had to request the RFP (here called a solicitation package) in writing. That schedule gave less than a month to develop the proposal, including research, planning, and writing. Collaborative writing is often essential in such instances.

Several years ago the federal government issued an RFP to locate, construct, and manage the supercollider project. That RFP specified requirements for space (a tunnel with a 50-mile circumference was required), details on environmental impact, and even cultural opportunities for the people who would work on the supercollider project. The engineers who could conduct soil tests as part of the environmental impact study were the best people to write the part of the proposal that evaluated environmental impact, but the Chamber of Commerce and civic leaders were better suited to describe the cultural and educational opportunities in the city. Thus, the scope of the project required a division of labor and the contributions of multiple people. Furthermore, the deadlines were tight: proposers had little over a month to gather all the necessary information and to organize and present it. The tasks had to be completed simultaneously rather than sequentially. Collaboration was the only way to complete a credible proposal.

Collaboration will probably require a division of responsibilities, but your whole team will need to review the proposal parts as they develop. Team members should all have a stake in the outcome, not just in their own part. Members responsible for one part can be good reviewers of others' parts because they can spot confusion or unanswered questions that are hard for writers to see in their own work.

A project coordinator keeps the work on schedule and facilitates communication among the different work areas. Each work area may also have a leader to keep its part of proposal development focused and on schedule and to serve as a liaison to the project coordinator. Review meetings provide opportunities to resolve any unanticipated findings or problems, to refine project goals and methods, to make necessary adjustments in the project or in the document, and to edit the sections as they are written.

Proposal Management

Because of the competitive nature of proposals, the likelihood of collaboration, and the likelihood of deadlines that are sooner than you would like, you need to use all your best management strategies as well as your best writing strategies when you develop proposals. The main proposal management tasks include the following:

1. Define the problem or need; describe the proposed solution in broad terms. Study the RFP if one is available. Define the specifics of the problem and the criteria for a solution.

2. Conduct an audience analysis: learn about the potential sponsor's interests, needs, and constraints on awards. Screen the RFP, directories, proposal guidelines, and annual reports. Interview the sponsor for more information on the problem or constraints on the solution. Learn who will review the proposal. Determine that there is a match between what the potential sponsor wants to have done and what you want to do.

3. Plan a solution: get information that will help you devise a sound and feasible plan; refine the details of the plan; develop the required parts of the proposal.

4. Develop a presentation: apply the funding source's guidelines for format and content to this proposal; answer the questions directly and clearly. Create proposal sections using the terms in the proposal guidelines and in the order specified. Use topics and phrases from the RFP as a way of establishing mutual understanding of the problem.

5. Manage the task: organize collaborators; define tasks; set a schedule for completing tasks. If there are collaborators, schedule team meetings throughout the project as well as at the beginning (to plan it) and at the end (to review the results and make last-minute changes). The meetings will keep collaborators on schedule and give you all a chance to assess progress and goals and to work out unanticipated difficulties.

Writing Strategies 17–1 will prompt you for each of these tasks.

SUMMARY

Proposals link organizations that have problems to solve and the organizations that can solve the problems. Many individuals and companies make their living by providing services or research for other companies. They get their work by developing proposals, usually in response to stated needs, for the type of services that they provide. Many charitable and educational organizations, which have limited ability to generate their own income, are supported by grants from foundations and government programs. They, too, develop proposals for the funding they need. Proposals are also the means by which individuals or groups may encourage changes in procedures or facilities to improve efficiency, safety, or fairness. The stakes are high: developing a good proposal requires a significant investment of time and other resources, and there is no sure reward for the investment.

In this chapter, we discussed two kinds of proposals: research and service delivery proposals and management proposals. We discussed several sources of project support, such as foundations and government programs. We also gave you advice on how to get information about funding sources and introduced you to *The Foundation Directory* and the *Commerce Business Daily*. Often proposals are solicited through Requests for Proposals (RFPs). Proposals are then reviewed by peer reviewers or reviewers with general knowledge.

WRITING STRATEGIES 17–1   Proposal Management Worksheet

1. **Define the problem or need.**

   Who

   What

   Where

   When

   Why

   Significance

   Plan summary: State briefly what you plan to do about the problem (or what someone else should do if this is a management proposal).

   Type of proposal: ———research ———service delivery ———management

2. **Analyze the audience.**

   - Who will review the proposal? Who will make the decision about accepting or rejecting the proposal?
   - Are the reviewers specialists, or do they have a general knowledge?
   - What are the interests of the audience in your proposal?
   - How will the proposal benefit the audience?
   - How well does the audience understand the problem?
   - What constraints will influence the audience's response to the proposal? (Consider time, money, competing projects, limits of facilities or equipment, limits in the type of project that the audience will support.)
   - Methods of gaining information about the reviewers and decision maker

     ———study an RFP
     ———locate proposal guidelines from the organization
     ———read descriptions in directories
     ———interview

   - Implications of the responses: What will make this proposal attractive to the audience?

3. **Plan a solution.**

   Objectives: What results will describe an effective solution? What do you want to achieve? What does the audience want to be achieved?

   Plan: What methods or facilities will achieve these results?

   Schedule: On what schedule can the plan be completed?

   Task   Time required

   Evaluation: How will you be able to demonstrate that the objectives have been achieved?

   Budget: How much will the plan cost?

   Staff

   Equipment and supplies

   Postage and printing

WRITING STRATEGIES 17–1   (*continued*)

4. **Presentation: Imagine the document itself, and choose the best strategies of organization, format, style, and visual displays to make it effective.**

   What parts will be required?

   ——cover letter   ——evaluation

   ——summary   ——budget

   ——introduction   ——personnel

   ——problem statement   ——facilities and equipment

   ——objectives   ——ongoing funding

   ——plan   ——supplements

   ——time/task schedule

   What information should be presented in tables, graphs, or other displays in order to increase comprehension or to persuade?

5. **Management: Identify research and proposal preparation tasks, and set a schedule for completing them.**

Research Tasks	Completion Date	Person Responsible
a.		
b.		
c.		
d.		

Writing, Proposal Preparation Tasks	Completion Date	Person Responsible
Writing		
Revision and editing		
Preparation of visual displays		
Preparation of preliminary pages		
Preparation of supplements		
Final proofreading		

To increase the chances that your proposal will succeed, you should look for a match between your goals and capabilities and the goals of the potential funding source. You must develop proposals that meet the guidelines, interests, and expectations of your potential funding source. Your proposal must anticipate and answer questions about the objectives, plan, budget, and evaluation specifically and clearly. Common parts of a proposal include: cover letter, summary or overview, introduction, problem statement or needs assessment, objective, plan, evaluation or quality control, budget, personnel and qualifications, facilities and equipment, and other sources of funding. The planning and reviewing required for proposal development also facilitate good management when the plan is implemented.

1. Study Example 17-5 to determine how it illustrates the principles of proposal development and how it uses the various proposal parts to achieve its goal of persuasion.

    a. Who is the audience for the report? What purpose does the writer hope to accomplish in writing the proposal?
    b. What type of proposal is it (research/service delivery or management)? (The distinction relates to whether the proposers will implement the solution to the problem or whether they are trying to convince managers to implement a solution.)
    c. What are the most persuasive features of the proposal? What is the most detailed section, and why? Use the reviewers' questions in Writing Strategies 17-1 as you consider the ways in which this proposal persuades. How does the proposal answer the reviewers' questions? Do you think the proposal will win the support it seeks (that is, will the project be approved)? What will convince the Iowa State University engineering faculty to approve the project? What will convince Motorola to provide supplies?
    d. How do the visual displays aid with the task of persuasion?
    e. Compare the parts of the proposal with the list of proposal parts in Writing Strategies 17-1. What variations do you notice? What explanations can you suggest for the variations? Is the proposal complete? What parts are omitted, and why?

2. Using *The Foundation Directory* or *Directory of Corporate Philanthropy* in your library, identify two potential sources of funds for the Minnesota AIDS Project, either for client services or for prevention and education. Use the geographic and subject indexes. Briefly explain the criteria you used (such as number and size of awards and interests of the foundation) to select the two sources.

3. Assume that you would like to gain corporate support for a student project involving research, design, or service delivery. (Alternatively, if you happen to volunteer or work for a nonprofit agency, consider seeking support for one of its program objectives.) Use the *Directory of Corporate Philanthropy* or *The Foundation Directory* to identify a potential sponsor for your project. In a brief report, state why you think the sponsor might be interested in your project.

4. A home health care agency wrote a proposal for government funds to provide respite care for caregivers of people with handicaps. They offered to relieve the caregivers, who were often stressed by their 24-hour responsibilities, by providing basic nursing care and living assistance. The RFP required them to describe the project's plan to target services to socially and economically disadvantaged participants, particularly low-income minority individuals. Here was their response:

Services available will be advertised at the following locations:

All senior citizen centers

Adult Day Care Center

Resources United

Coalition on Aging meetings

All hospital senior programs

Civic groups and organizations

Public school counselors

Hospital discharge planners and agency social workers

Physicians, residents, and pharmacists

Caseworkers, MHMR [Mental Health/Mental Retardation] center

Alcohol and drug abuse counselors

Clinics in low-income areas

Assuming you are a reviewer of the proposals submitted, evaluate their response to the question. What unanswered questions might you have? How could the agency be more persuasive in this section of the proposal?

5. According to your instructor's directions, write a proposal for your writing instructor, or an instructor in your major field, proposing a research and report project that you will conduct for course credit. Your broad goal is to convince the instructor that you have identified a significant and feasible project, that your methods will achieve the stated objectives, and that you can complete the project within the time and budget constraints.

6. Assume that the Minneapolis School District has requested proposals from individuals and organizations to head an AIDS task force for education in the high schools. The District has $250,000 to spend on AIDS education over the next two years.

As a member of the Minnesota AIDS Project (MAP; see Appendix A, Case Documents 1), you have been asked by MAP to write a proposal to the District requesting that MAP be granted the funding and authority to head the task force. This goal is slightly different for MAP, as the organization has depended on funds to maintain itself over the years. Now MAP feels it's in a position to take over specific and discrete projects. You have selected a number of MAP staff members and volunteers to work with you (your instructor will place you and others in a collaborative writing team).

You will need to write a detailed proposal on what MAP can offer, how much it will cost (estimate this cost as realistically as possible), what time period it will cover, when you'll be ready to get started, what expertise MAP will bring, how high school students will benefit, why MAP is the best organization to head such an effort, how MAP might draw on other educational and community resources to support the effort, and all the aspects of an effective proposal that were covered in this chapter. In some cases, you can draw inspiration from Case Documents 1 in Appendix A, but in others you may have to do some secondary or primary research to supplement your knowledge about AIDS education.

EXAMPLE 17–5

ISU/MOTOROLA ISDN SENIOR DESIGN PROJECT PROPOSAL

Submitted To:
Dr. David Morgan
Director of Corporate and Computer Motorola Research and Development

Professor Doug Jacobson
ISU Faculty Project Monitor

Submitted By:
S. David Silk
Team Leader

August 20, 1990

TABLE OF CONTENTS

(not shown)

1.   INTRODUCTION

In order for ISDN to take a foothold in tomorrow's telecommunication market, it must accommodate existing non-ISDN equipment as well as current ISDN compatible terminal equipment. Although ISDN standards are still evolving, both the technology

EXAMPLE 17–5   *(continued)*

and the emerging implementation strategy of ISDN are well understood. A team of electrical engineering and computer engineering students (named in Appendix B) wishes to take advantage of this available technology and strategy in order to pursue a senior project. In pursuing this senior project, we seek not only to educate ourselves in this technology, but also to acquire practical application by building a prototype ISDN-compatible terminal adapter (TA).

The purpose of this proposal is to explain in detail the project we wish to pursue demonstrating Motorola's chip set. The following sections discuss project objectives and plan of implementation, design constraints, additional resources, and the project schedule.

2.   OBJECTIVES AND PLAN

2.1   Project Overview

Using the Motorola chip set, we will develop our own board layout implementing IDL to link the chips together to provide a powerful Layer 1/Layer 2 ISDN foundation. Please refer to Appendix C for a preliminary layout. (Note: the layout is subject to change.) Beyond this point, we will specialize by incorporating our own version of Layer 2 and 3, three data ports, a keypad, handset, feature keys, and an LCD with a user interface. All of these features will be incorporated into a prototype that, when finished, will have the basic configuration as illustrated in Figure 1.

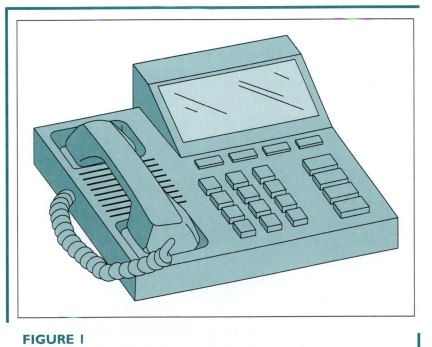

FIGURE 1
Projected Configuration of the Prototype Terminal Adapter

EXAMPLE 17–5   *(continued)*

2.2 Voice Functions

A primary concern in the TA design is the versatility our device will provide the user. Our device will emphasize voice functions and data functions equally. The voice functions listed below will be implemented using either the ISDN technology or individual components.

Auto-dial:	Perform one-touch dialing of stored numbers from a built-in phone number library.
Flash:	Drop caller and await new number to be dialed.
Hold:	Place calls on hold.
Memory:	Store phone numbers to memory for auto-dial.
Mute:	Provide privacy by preventing the caller from hearing your conversation with others in the room.
Program:	Select distinctive ringing pattern to identify incoming caller.
Redial:	Redial the last number dialed manually.
Speakerphone:	Provide two-way speakerphone.

2.3 Data Functions

We will incorporate three data ports into our design to enhance the data features but also to increase the versatility of our device. The first port will provide asynchronous and synchronous data access to the B-Channel with in-band signaling for call control. The second port will provide data access to the D-Channel. This port will support both ASCII communications and generic LAP-B. Our TA will do a LAP-D to LAP-B conversion to allow a linkup to computers with existing X.25 hardware/software. Finally, the third port will provide an out-of-band signaling facility to control the ISDN channels as well as the other two ports. Figure 2 illustrates the three data ports. For simplicity, they will be labeled as they were described above: Port 1, Port 2, and Port 3, respectively.

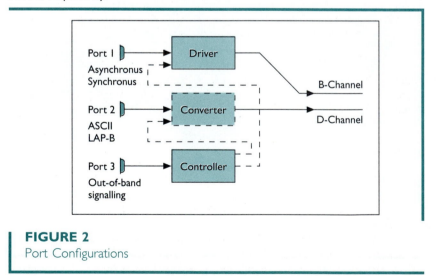

FIGURE 2
Port Configurations

EXAMPLE 17–5 *(continued)*

Here is a brief listing of the port characteristics.

- Port 1:
  1. B–Channel access with in-band signaling for call control.
  2. Support of asynchronous and synchronous data.
- Port 2:
  1. D–Channel access with in-band signaling.
  2. Support of both ASCII communications (contain a PAD-packet assembler and disassembler) and generic LAP-B.
  3. LAP-D to LAP-B conversion to incorporate X.25 hardware/software to provide a linkup with a host. This capability opens up numerous software related features for advanced support of personalized services from the host.
- Port 3:
  1. Support of an out-of-band signaling facility to control the ISDN channels as well as the other two ports.
  2. Contain a EIA-232 interface for a PC/Mac linkup.
     a. Enable a user to control, from the PC, all setup and tear down features.
     b. Enable information to be downloaded from the host by controlling Port 2.
     c. Initiate phone calls by controlling Port 1.
     d. Perform reprogramming of the TA's personalized features.

2.4 Screen Layout and User Interface

Our device will incorporate an LCD capable of displaying several lines of information at once. In addition to the LCD, our device will also incorporate a user interface to facilitate in the operation of the TA. The combination of these two features, an LCD and a user interface, will provide an efficient method for on-screen editing of all functions built into the TA.

2.5 Prototype Testing

Due to the lack of a Central Office in the area that supports ISDN, we will build two prototypes to facilitate the testing of our device. With the two prototypes connected together, we will then be able to test and refine our design fully.

3. DESIGN CONSTRAINTS

3.1 Power Consumption

Due to the ISDN technology involved, our device will not operate on the conventional power levels currently supplied by the telephone companies. Our TA will incorporate a power supply supporting ± 5 volts and ± 12 volts.

As a result of our TA's increased power consumption, it will be designed to be turned off when it is not in use for extended periods of time (e.g., overnight or when a user is on vacation). Also, when it is not in use for short periods of time, it will be

EXAMPLE 17–5   (continued)

designed to "sleep." The ability for our device to be completely turned off will prevent the waste of power. Furthermore, the ability to sleep will allow some nonessential components to be powered down and remain "asleep" until some event "awakens" the device (e.g., the user touches any of the buttons or picks up the handset to make a phone call, or an incoming phone call is received). Since our TA will have its own power supply and be able to be completely turned on and off at the user's will, it will not be dependent on other devices for startup.

3.2   The Ports

Although we have identified and described three data ports, implementing Port 2 is not within the scope of this project. Since the required software used to operate Port 2 is proprietary, we will have to develop complementary software to operate the port ourselves. Accomplishing this task will not be feasible within the 30 weeks allocated for our project. Software for Port 1 and Port 3 will be developed and implemented.

We will, however, complete a well-rounded device that is versatile in design. Section 3.2.1 of this proposal discusses our short-run approach that will actually make a useful long-run solution in the future evolution of our device. Therefore, we intend to design and to incorporate the hardware for Port 2 into our device in the event that our device is recirculated through the university for another session with a new senior design team. A future senior design team will then be able to continue where we left off on Port 2. Guidelines for future implementation of Port 2 are described in Section 3.2.2.

3.2.1   Short-Run Solution for Port 2

As described in Section 3.2, the inability to complete Port 2 to the desired level imposes some limitations to the versatility of our device. Port 2 provides a link to a host via X.25 software/hardware to support an advanced array of personalized services. The major consequence of its absence is the lack of available memory to store the user's personal preferences (i.e., assignment of desired functions to buttons on the TA that can be saved on a per-user basis).

It is important that our device can function as a stand-alone without being dependent on other devices to perform routine operations such as startup. The best solution is to design the TA to read, from internal memory, a generic setup during startup. As it turns out, this short run solution also proves to be a useful feature to have in the long run. In the future when a host is connected to our device, upon device startup the host can simply download the personal preferences superseding the generic setup. This feature will prevent our device from being dependent on the host to start up. Thus, in the event that the host is not functioning properly, the TA will still be able to power-up with its preset generic features and provide immediate service.

In addition to the integration of a generic setup of the features, additional memory will be integrated into our device to enable the user to store alterations to the generic setup. This will eliminate the inconvenience of requiring the user to modify the features each time the device is started up.

3.2.2   Long-Run Solution for Port 2

In order to take full advantage of Port 2, when it is fully operational, the additional features listed below should be incorporated into the device.

EXAMPLE 17–5 *(continued)*

1. The TA should be designed to register with the host with its appropriate user identification information upon startup. Once a user is registered, the user's personal preferences would automatically load on the registered TA. When a user notifies the host of a temporary station change, the host can download the custom phone features to the new TA. Therefore, a user can use any TA in the office complex; the custom TA features follow the user around. This feature would also allow the host to reroute incoming phone calls to the appropriate user if a change in location has occurred. In such a situation, the host can act like a PABX, a Private Address Branch Exchange.

2. The TA should be designed to assist the host computer with a feature called call accounting. The TA can be set up to keep track of all the information pertaining to the calls made on that TA, for instance, date, calling phone number, call starting time, and call length. At regular intervals, the host computer can poll this information to acquire all the necessary statistics on how the telephones are being used and provide detailed information for billing purposes.

3. The TA should be designed to notify the host computer each time it is turned off. Hence, if the TA is off, the host will execute the necessary procedures to take care of incoming phone calls whether the user is unavailable or at another station.

3.3 User Interface

All features of our device, whether it be call setup/teardown or personal phone preferences, will be manipulated via an LCD and a user interface. The integration of an LCD, with the ability to display several lines of information at once, will enable the user to view and/or edit a group of related settings. Also integrated into our device are programmable "soft" keys. The soft keys will be mounted directly under the LCD below their individual captions on the LCD.

As mentioned, there are two modes for the interface, view and edit. When the interface is in the view mode, the user can view listings of the selected features, check a phone number in the call library, or view the standard screen layout to make a phone call. For a depiction of the standard screen layout used to make a phone call, please refer to Figure 3.

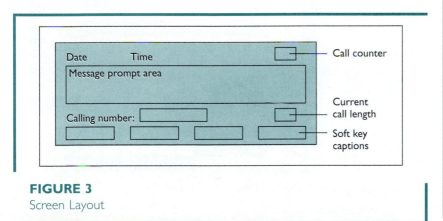

FIGURE 3
Screen Layout

EXAMPLE 17–5 *(continued)*

Conversely, while in the edit mode the user will be able to scroll through the features and modify options, edit the call library, or alter the information displayed in the standard screen layout by adding or deleting features.

3.4 System Software

The core operations of our TA will be controlled by the first three layers of OSI: Layer 1, Layer 2, and Layer 3. As a result of Layer 2 and Layer 3 being classified as proprietary material, we will have to develop and implement complementary versions in order to run our device.

4. FACILITIES AND EQUIPMENT

In addition to the Motorola ISDN chip set, we will need the ADS302 Development System to provide a platform for software development for the MC68302. The university will supply us with the necessary host computer to facilitate in the development of all necessary software. For a detailed listing of what is essential for our project, please refer to Appendix A for a parts listing.

5. PROJECT SCHEDULE

Thirty weeks have been allocated for our project. The schedule in Figure 4 illustrates how we will utilize the thirty weeks. These dates are subject to change.

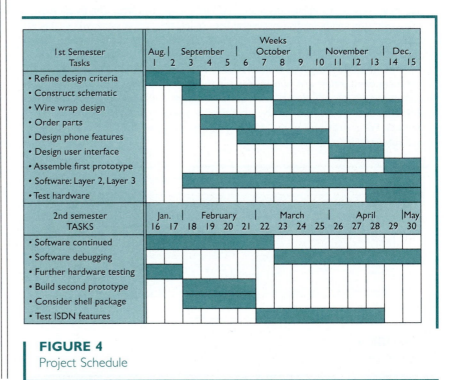

FIGURE 4
Project Schedule

EXAMPLE 17–5   *(continued)*

6. CONCLUSION

We respectfully request the approval of this project by the engineering design faculty and Motorola as well as support in the form of project advice, use of facilities and equipment, and supply of the parts needed to build the TA. This interdisciplinary project will allow electrical and computer engineering students to use the information, concepts, and skills from their backgrounds to study the complexities of ISDN. This project will allow the students to interact as a team and to get firsthand experience at problem solving.

APPENDIX A: Parts

From Motorola:

Quantity	Part Name
3	MC145554 PCM Codec Filter
3	MC145474 ISDN S/T Interface Transceiver
3	MC68302 Integrated Multi-Protocol Processor
1	ADS302 Development System
3	15.36 MHz Crystal
3	16.67 MHz Crystal

From Iowa State University:

Quantity	Part Name	Quantity	Part Name
3	MC1455 Timing Circuit	6	30 pF Capacitors
6	MC1488 Quad Line Drivers	6	25 pF Capacitors
6	MC1489 Quad Line Receivers	6	0.1 μF Capacitors
		3	0.47 μF Capacitors
6	MC6821 Peripheral Interface Adapter	12	Miscellaneous Capacitors
3	MC6850 Asysnc. Comm. Interface Adapter	27	10 kΩ Resistors
		3	10 mΩ Resistor
6	SN74LS04 Hex Inverter	3	700 kΩ Resistor
3	SN74LS32 Quadruple Input Positive OR Gates	3	0.43 kΩ Resistor
		3	29.4 kΩ Resistor
3	1N4001 Diode	18	500 Ω Resistors
33	Miscellaneous Diodes	9	4.7 kΩ Resistors
6	1:1 Ratio Power Transformers	21	Miscellaneous Resistors
6	1 mH Inductor	2	LCS
6	MCM6206 32KX8 Static RAM	2	Power Supplies
6	32x8 EPROM		

EXAMPLE 17–5   *(continued)*

Quantity	Part Name	Quantity	Part Name
6	DB-25 Connectors		
2	Handset		
2	Keypads		

Note: This is a list of parts from the preliminary design of the schematic. As we finalize the design, more parts will be added to this list.

APPENDIX B: Personnel

In order for this project to be successful, it will need a group of approximately eight students: four electrical engineering students and four computer engineering students. In the fall, when the project is officially formed, I will have a commitment from the people I have solicited. Below is a listing of the current members of the design team.

- **Design Team Composition**
    1. Roger Najera, E.E., Communications/Power
    2. John O'Brien, E.E., Analog Electronics/Math Minor
    3. Bill Ryan, Computer E., Software
    4. S. David Silk, E.E., Communications/Telecommunications
    5. Eric Welch, E.E., Communications/Control Systems

    At least three more computer engineers will be added in the fall.
- **Management**
    1. Faculty Project Monitor   Professor Doug W. Jacobson
    2. Technical Writing Advisor   Professor Patricia Goubil-Gambrell
    3. Group Leader   S. David Silk

Source: Reprinted with the permission of the author, S. David Silk.

7. Assume that the Coalition of Citizens Concerned with Animal Rights (CCCAR) has decided that it wants to submit a proposal to the University of the Midwest suggesting alternative ways of reducing the trapping of birds on the experimental crop fields (see Appendix A, Case Documents 3). This proposal will be unsolicited and might be met by some hostility on the part of the University. Therefore, it will require careful and tactful development. Moreover, CCCAR wants the University to bear the cost of funding the proposal so it will have to be very attractive in terms of the solutions it offers.

   Write a detailed proposal to the University that describes the problem, defines what alternative approaches can be undertaken to stop bird trapping on the fields, what their costs are likely to be, and what benefits will be gained from these alternative plans. In addition to looking at the case documents carefully for alternative approaches, you will need to do some work at the library and perhaps make some calls around the community to determine what some cost-effective solutions might be.

Cavin, Janis I. *Understanding the Federal Proposal Review Process.* Washington, DC: American Association of State Colleges and Universities, 1984.   **WORKS CITED**

Developing a Professional Edge

Correspondence

Effective Letters and Memos ▌ Conventions of Workplace Correspondence ▌ Letter Reports ▌ Memo Reports ▌ Letters of Transmittal and Cover Letters ▌ Letters of Recommendation

"Ourselves among others" captures something of today's burgeoning internationalism. But equally important, it can serve as a metaphor for that relationship between the writer of business/technical documents and his or her diversified audience; it is a metaphor of the changing contexts of business/technical writing in which today, more than ever before, writers write to "others" who are "different" from them in some significant way.

Timothy Weiss, "Ourselves among Others: A New Metaphor for Business and Technical Writing," *Technical Communication Quarterly* 1.3 (1992), pp. 23–36.

Introduction

Correspondence is a broad category covering letters, memoranda, messages, and e-mail. This chapter teaches you how to use the conventions and accepted patterns of correspondence to send your audience information that will help them solve problems or persuade them to take action. This chapter will also help you understand how you can modify the standard conventions of workplace correspondence to accommodate the newer communications media, such as fax machines, mail merge software, and e-mail.

Knowing how to write correspondence—and being able to do so quickly and effectively—is one of the most important skills you will need in the workplace. The computer makes it easier and faster to put thoughts into writing and has resulted in more communications being sent both in print and online. Mail merge software, for example, allows us to send hundreds or thousands of "personalized" letters in the time it would have taken us to type only three or four. Fax modems allow us to send our correspondence immediately to anyone with a fax machine or a computer anywhere in the world. Moreover, computer networks, which allow us to send e-mail and documents, have also enlarged and extended our audiences.

This chapter suggests ways to write letters and memos that get the job done without overwhelming either you or your audience. Despite the growing popularity of computer technology and the communication modes it allows, traditional conventions governing workplace correspondence still prevail. In the first section of this chapter, we discuss some general principles about writing effective letters. Next, we provide information about the parts and format conventions of the letter and memo. Finally, we suggest ways to write some of the letters and memos often composed by technical communicators: short reports in letter and memo form, transmittal letters, cover letters, and letters of recommendation.

Effective Letters and Memos

Two of the most common kinds of correspondence you will create are letters and memoranda (memos). All that you've learned about analyzing audience and purpose, arguing with good reasons, and gathering sufficient evidence applies to correspondence. Additionally, workplace correspondence has special conventions governing its format. You should consider these formats flexible **templates** that you can adapt to your audience and situation.

templates
formats for creating documents such as memoranda, transmittal letters, cover letters, and letters of recommendation

In some cases, your audience for correspondence will be quite small, perhaps one or two specific people. In most such cases, you'll know your audience personally. Letters often respond to earlier correspondence or to a telephone call; memos have audiences internal to an organization and generally deal with subject matter known well to both sender and receiver. In other cases, you might be sending a mass mailing, such as a cover letter for a survey. Or, your memo or letter may become part of a paper trail in a legal case or an extended consumer complaint, although you will originally have sent it to only a few readers.

Except for short reports that are sent in the form of letters and memos, correspondence doesn't have a very long life. Correspondence is used to respond to a given issue at a particular time. You may not have much time to prepare correspondence. You can't take a week to write a memo informing the factory that a shipment of a needed part will be delayed. Correspondence must be timely to be effective. It must also be correct; your and the company's reputation can be damaged by errors in content as well in grammar, spelling, or mechanics.

Because you will be very busy at work, the idea of writing a memo or letter may seem time consuming. Before writing correspondence, ask yourself whether this particular matter can be handled by telephone, a quick note, or an e-mail message. Generally speaking, if you think that there will ever be a need for a written record, write the letter or memo. For example, if a complex issue needs clarification, if there are data to be interpreted, if you are providing cost information, or if you need to document the content of a phone call, use correspondence. If all you want to do is remind your colleague that the room for the meeting has been changed, use the phone.

CORRESPONDENCE IN COMMUNICATION LINKS

Because correspondence is a quick way to request or offer information, letters and memos commonly create and sustain many **communication links.** One letter, one memo, leads to another, which in turn leads to another, and each becomes a part of a pattern or paper trail of requests and decisions. Before writing correspondence, you need to be aware of the context into which you are writing and of the documents that have preceded and may follow yours. The three examples that follow are just part of one of those paper trails or communication links.

communication links
the paper trail created by interconnected requests and decisions

Let's say that you have received the following request from a consumer, Ms. Patrick, to replace a computer chip for a personal computer that your company, Systems Tech, produces (Example 18-1). This letter will begin a chain of communication, some oral but much written.

Obviously, the next link in the chain will be your reply to Ms. Patrick. If you cannot meet her entire request, you still want to retain her good will, to explain why you cannot reimburse her for the repair, and to compensate as much as possible for not being able to reimburse her.

Before you write the letter, you need to gather as much information as possible on similar reported malfunctions on this computer model and to check with your own repair shop. You need to determine whether the problem Ms. Patrick reports is a common one, and, if so, you will probably take some responsibility for it. If the problem is an unusual one, you will

EXAMPLE 18–1   Initial Request

July 27, 1994

Systems Tech
1150 Pearson Ave.
Suite 54
St. Paul, MN 55108

Dear Systems Tech,

On January 12, 1993, I purchased a Systems Tech 175 personal computer with hard drive (160 K memory, operating system 6.1) for my place of business. This week the computer malfunctioned because a chip behind the hard drive cracked and began to smoke. I took the computer to a technician located in Denver and was told that the entire mother board would have to be replaced at the cost of $322 (not including labor).

Although the computer is no longer under the year's warrantee available with the Systems Tech 175, I believe that Systems Tech should replace the mother board without cost. The computer is less than two years old and was not damaged by mishandling. Moreover, the computer technician stated that this kind of malfunction is highly unusual. I believe that the computer must have been improperly assembled in the first place or that the chip itself was faulty.

I will not have the computer repaired until I hear from you. I would prefer to have the computer repaired locally, as I need a quick repair, so I would like to be reimbursed for the repair.

Sincerely,

Janet Patrick

Ms. Janet Patrick
Executive Consumer Surveys
135 West 15th St.
Denver, CO 80210

want more information as to how the problem might have happened and how it might be avoided in other machines.

Only by repairing the computer yourself can you control the situation, gather the information you need for your own future manufacturing concerns, and be assured that indeed Ms. Patrick's local computer technician has assessed the problem accurately. Moreover, you need to check on Systems

Tech's previous correspondence with Ms. Patrick—how often has she ordered from you? Has she registered any previous complaints? Has your association with her been positive, neutral, or negative in the past? See Example 18-2 for one way you could respond to Ms. Patrick.

In the meantime, you will send a memo to shipping and the repair shop so that they can anticipate Ms. Patrick's call and the arrival of the computer. You'll also request that your team of computer engineers investigate the problem (Example 18-3). And, of course, in another memo you'll ask the repair shop to alert computer engineering when the computer arrives. Computer engineering will then send you a memo about their findings; the repair shop will confirm when they send the computer back to Ms. Patrick; you will write to her if computer engineering discovers any manufacturing problems with the computer (and if so, you'll write to other owners of the same model to recall or alert them); and so on. You can probably think of other links in the communication chain yourself, most of which would be filled by correspondence. It is this chain or context you must understand as you generate and respond to correspondence.

CORRESPONDENCE STYLE

Correspondence often has a more conversational tone than do other forms of technical communication. For example, you are likely to find the use of contractions, personal pronouns, and first names. Correspondence also makes good use of reader-based principles, addressing the reader in the second person, "you," as demonstrated in Examples 18-1 and 18-3:

- The repair shop will call you. . .
- Although indeed the warrantee on your computer has expired, we want to help you as much as possible.
- I will not have the computer repaired until I hear from you . . . .

As the writer, you can speak about yourself in the first person: "I believe that the computer must have been improperly assembled in the first place . . .". This use of first and second person in correspondence makes for a more conversational exchange.

However, because of this conversational tone, you need to resist any tendency to say more than might be needed. People who need information want it fast. Strange things seem to happen when people begin to write correspondence. They write things they would not ordinarily write or say. For example, some writers make use of archaic phrases and clichés because they think that's what they're supposed to use in correspondence, such as:

Please be advised

Enclosed please find

As per our agreement

EXAMPLE 18–2   One Possible Response

Systems Tech
1150 Pearson Ave.
Suite 54
St. Paul, MN 55108

August 1, 1994

Ms. Janet Patrick
Executive Consumer Surveys
135 West 15th St.
Denver, CO 80210

Dear Ms. Patrick:

We are sorry to hear about your problems with our Systems Tech 175 personal computer. Because each machine is tested by hardware engineers before being shipped to our consumers, we find hardware failure is uncommon in our computers. We are particularly concerned when a problem occurs with one of our customers whom we have served as often as we have served Executive Consumer Surveys.

Although indeed the warrantee on your computer has expired, we want to help you as much as possible. We can do so only if we service the computer ourselves. This procedure would not only allow us to replace any necessary parts with Systems Tech parts but would also help us to determine what exactly caused the problem and to prevent it from happening again. Therefore, we are able to offer you the following:

We will repair the machine at no labor cost.

We will replace the mother board at a parts cost of no more than $250.

We will pay shipping costs to and from Systems Tech.

We will repair the computer within two weeks of its arrival at Systems Tech.

Finally, we will guarantee these repairs and the new mother board for five years from the date of repair.

We value the association we have had with Executive Consumer Surveys and hope that you will find our offer satisfactory. Please call our shipping department at 1-800-555-2233 to arrange for UPS packing and pickup of your computer.

Sincerely,

John Smitzer

John Smitzer
Marketing Supervisor

EXAMPLE 18–3   Memo to Computer Engineers

Systems Tech
1150 Pearson Ave.
Suite 54
St. Paul, MN 55108

MEMORANDUM

DATE: August 2, 1994
   TO: Jane Smalley, Computer Engineering Supervisor
FROM: John Smitzer, Marketing Supervisor   JS
   RE: Possible Problems with 175 Mother Board

Please be aware that a customer has reported a cracked and smoking chip in the Systems Tech 175 somewhere "behind the hard drive," which appears to have damaged the entire mother board. I have checked with the repair shop, and they told me that although the problem is unusual, three such complaints have come in on the 175.

The repair shop will call you when the computer comes in and will copy the other two complaints for you. Please give the computer a quick check to make sure that we do not need to change the 175 design in some way. I promised the customer a fast repair so I hope that we can get someone on this as soon as the machine arrives.

We are in receipt of

Please don't hesitate to call

Feel free to call

Or they lapse into jargon without considering whether the audience can understand their special technical terms and abbreviations:

The RFP for the YETP has been issued by RETC, an agency funded through CETA.

Instead of writing

The Request for Proposal for the Youth Employment and Training Program has been issued by the Regional Employment and Training Consortium, an agency funded through the Comprehensive Employment Training Act.

In the example above, the passive voice adds to the confusion. The active voice—"the Regional Employment and Training Consortium has issued the Request for Proposal"—states the main idea more forcefully, placing the actor and action at the beginning of the sentence.

Writers of correspondence sometimes lapse into imprecise and inaccurate prose in an attempt to sound impressive. Simple, direct language is more effective:

Instead of	*Use*
approximately	about
apprise	tell
initiate	begin
utilize	use
terminate	stop, end

Other symptoms of the bureaucratic prose frequently found in correspondence include abstract language rather than simple, concrete words. Letter writers talk about "equipment" when they mean "a tachometer" or call something a "facility" when they are referring to "an office." Redundancies add more words as well: "a new beginner" instead of simply a beginner, "general consensus of opinion" instead of a consensus, and "round in shape" instead of round.

Some of these same problems exist in all prose, of course, as you've seen in Chapter 10, but in correspondence they are especially inappropriate. Correspondence is, by nature, short, and every word must be necessary. Look at the effect of accumulating all these stylistic problems in Example 18-4.

What do you suppose is the message of this memo? How many times would the state director have to read the memo to get the message? Look at the fourth paragraph. Should memos sound like legislative bills? Some people do think lawyers (solicitor is a fancy name for lawyer) are supposed to talk like this, but is it appropriate in a memo to a busy person who needs a straightforward answer? If state directors issue special-use permits, they know they cannot assign the property rights of the US Government to someone else. And even if the lawyer wanted (just to have it in writing) to remind the state director that this information should be spelled out in the special-use permit, the reminder could be worded in a much clearer and concise way. Consider the revised memo (Example 18-5). Aside from the fact that it's considerably shorter, what other changes can you point out that make this memo more effective and more concise?

CORRESPONDENCE ORGANIZATION

As with other writing tasks, the way you structure your message strongly affects how well your reader responds to it. To achieve the conciseness that your busy reader needs, consider a direct plan in your workplace correspondence, a deductive or top-down approach. This direct plan is especially useful if you have good news to tell your reader, and this plan follows the

MEMORANDUM

DATE: January 12, 1994
TO: State Director
FROM: John Lawbook, Solicitor   JL
SUBJECT: Roland Occupancy Trespass

This responds to your memorandum dated December 21, 1993, requesting that we review and comment concerning the subject Roland trespass on certain lands under reclamation withdrawal.

We appreciate your apprising us of this matter and we certainly concur that appropriate action is in order to protect the interests of the United States.

We readily recognize the difficult problem presented by this situation, and if it can be otherwise satisfactorily resolved, we would prefer to avoid trespass action. If you determine it permissible to legalize the Roland occupancy and hay production by issuance of a special use permit, as suggested in your memorandum, we have no objection to that procedure.

Any such permit should be subject to cancellation when the lands are actively required for reclamation purposes and should provide for the right of the officers, agents, and employees of the United States at all times to have unrestricted access and ingress to, passage over, and egress from all said lands, to make investigations of all kinds, dig test pits and drill test holes, to survey for reclamation and irrigation works, and to perform any and all necessary soil and moisture conservation work.

If we can be of any further assistance in this matter, please advise. We would appreciate being informed of the disposition of this problem.

Source: John O'Hayre, *Gobbledygook Has Gotta Go,* Washington, DC: US Government Printing Office, 1973, pp. 11–12.

MEMORANDUM

DATE: January 21, 1994
TO: Jonathan Leigh, State Director   JL
FROM: John Lawbook, Legal Services
SUBJECT: ROLAND TRESPASS OCCUPANCY

I received your memo on the Roland trespass case. You're right—action is needed. The problem is tough, and we'd like to avoid trespass action if we can. So, if you can settle this case by issuing Roland a special-use permit, go ahead. Please spell out the Government's cancellation rights and right-to-use provisions in the permit as usual.

If we can be of further help, please call.

Source: John O'Hayre, *Gobbledygook Has Gotta Go,* Washington, DC: US Government Printing Office, 1973, pp. 14–15.

typical organization of technical communication—getting to the point in the beginning.

In some cases, you must report bad news or turn down a request, announce a price increase, or deny an application, and you might begin this type of correspondence with some buffers. These buffers will help you keep the good will of your reader—maintain a business relationship, help a reader save face, or ensure some future relationship. You can begin by offering the reader another choice than the one requested, assure the reader that the applications were all outstanding even though the job went to someone else, or offer a discount on the reader's next purchase. Bad news needs to be deemphasized, so placing it more toward the middle of your correspondence will give you a chance to soften the blow your reader might otherwise feel. Our best advice then is to state the good news first, and the bad news early, but only after you have had a chance to prepare the reader.

To test this advice look at Example 18-4 again. The good news is buried. The answer to the state director's questions, ''Is it legal for me to issue Roland a special use permit?'' is yes, although it's a qualified yes. Burying the answer in the middle of the memo forces the reader to search for it, and even though the news is good, the reader might be alienated by that search. Now look at Example 18-5 again. Here the good news is immediately perceived. You probably realize now that it's not just direct language that makes this message clear; it's also the direct plan.

Conventions of Workplace Correspondence

Letters and memos require conventional formats and to deviate from these formats distracts or even irritates readers, causing them to question your knowledge. Make use of what people expect and use your creativity and originality in organizing your arguments and articulating your ideas.

Businesses generally find value in these conventions, although some firms are more flexible than others. Always know the accepted form for correspondence within your company and follow it carefully. If you are not given a corporate style guide, find some example of other work done in the company. Even though you may not be responsible for actually preparing final copies of the letters and memos, your name will be on them so the responsibility for their appropriateness is yours.

In this section we discuss some basic information about the most generally accepted conventions. Some of this discussion is provided *within* the examples in this chapter, so be sure to read the *content* in the figures when looking at the samples.

CONVENTIONS FOR WRITING LETTERS

First impressions are important. Letters should look like a picture in a frame. Study Example 18-6 and notice how the white space surrounding the typed

EXAMPLE 18–6   The Business Letter

October 22, 1994

Ms. Shirley Smith, Director
Office Equipment Supplies Company
P. O. Box 49560
Los Angeles, CA 92571

Dear Ms. Smith:

This letter illustrates how the business letter should be placed on a page so that it appears to be framed by the top, bottom, and side margins. You can make several adjustments in spacing to ensure this "picture-frame" effect.

You can vary spacing at the top of the letter after typing the date. Spacing here can be as few as three spaces and as many as 8 or 12 spaces, depending on the length of the letter.

Another way to adjust the spacing is to change the width of the margins. If your letter is very short, you can use wider margins.

As you learned in Chapter 11, white space invites readers into a document. Placing a letter just slightly higher than the horizontal center of the page also creates an inviting appearance for your letter.

Sincerely,

Richard Lee Evans

Richard Lee Evans
INFORMATION SERVICES, INC.
President

rle/sl

material frames the message. The letter explains how to achieve this picture frame placement, so *be sure to read the letter itself for additional information.*

Most letters have seven main parts: letterhead or heading, inside address, salutation, body, complimentary close, signature, and initials. Example 18–7 illustrates these main parts. The comments in the margin give additional information about the conventions associated with each part as well as its placement.

EXAMPLE 18–7   Main Parts of a Letter

At least 1½″

Letterhead

Coalition of Citizens
Concerned with
Animal Rights
CCCAR

August 2, 1992

3–12 blank lines
Inside address
(single spaced)

Dr. Karen Wartala
Dean, College of Agriculture
University of the Midwest

1 blank line
Salutation
1 blank line

Dear Dr. Wartala:

The Coalition of Citizens Concerned with Animal Rights (CCCAR) was alerted by a concerned person two weeks ago that the University is trapping and killing large numbers of birds in the experimental crop fields on its campus. In our surveillance and investigation we have learned that:

Body (single spaced with double spacing between paragraphs)

- Birds are lured to the traps by bait placed in them, and the number of birds in these large traps has been observed to be at least 50 at times.
- The reported method of killing the birds was suffocation, by placing them, 400–500 at a time, in a bag.
- Approximately 500 birds are killed every day on the campus; 10,000 were killed in a three-month period.

We object to this for the following reasons:

- It is a waste of animal life.
- It is ineffective. Trapping and killing birds will not permanently reduce the bird population in the area. Even temporarily it will have no more than a minimal effect.
- The trapping and suffocation of these birds is cruel in the extreme.

For these reasons, we insist that the University immediately stop the killing of these birds. If you wish to discuss this matter, please contact me at 555-5537 or 555-5856.

EXAMPLE 18–7   *(continued)*

1 blank line	
Complimentary close	Yours very truly,
Signature block	*Ed Younger*
3 blank lines	
	Ed Younger
	Vice President, CCCAR
Initials	EY:sl

Corporations usually have printed **letterhead** for their correspondence, which includes the name of the company, its address, and other information such as telephone numbers and department identification. The date of the letter is usually two spaces below the last line of the letterhead. If there is no letterhead, the **heading** includes the sender's street address, the city, state, and zip code, and the date. Sometimes you will find the sender's address after the sender's name in the signature block.

The **inside address** names the receiver of the letter. It includes the receiver's name, the job title, company name, address, and the city, state, and zip code. An appropriate form of address always precedes a person's name; for example: Ms. Smith; Professor Anderson; Congressional Representative Henderson; or The Reverend Duin. If you do not know the name of the person receiving your correspondence, the first line of the inside address may be the name of the firm, a job title, or a department within the firm. In writing the company name, follow the way the company writes its name. For example, some companies do not abbreviate *company* (Co.).

The **salutation** is the traditional hello used in letter writing. What salutation you use depends on two things: the first line of the inside address and how well you know the receiver. If the first line of a salutation is a company name, you can use Gentlemen, Ladies, or Ladies and Gentlemen, although repeating the company or department name instead will eliminate the need to guess the gender of the readers. The title Ms. eliminates having to decide whether to address a woman as Miss or Mrs., but a better solution is to find out what title the woman you are writing prefers and use it. This strategy avoids the impersonality of "Dear Executive" or "Dear Purchasing Agent."

The **body** of the letter contains the message. Notice in Example 18–7 that the body of a letter is single spaced with double spacing between paragraphs. Bulleted items work in letters as well as reports, and topic headings can improve the readability of letters and memos.

The **complimentary close** is the good-bye in a letter. Its level of formality matches the level of formality in the salutation.

letterhead
stationery used for correspondence that includes the company name, address, and telephone and fax numbers

heading
the sender's name, address, and the date at the top of a letter not printed on stationery

inside address
the recipient's name and address, placed at the top left of the letter, just below the date

salutation
the greeting or hello found at the beginning of letters

body
the message or content section of a letter

complimentary close
the closing or good-bye found at the end of letters

In the **signature block,** if the company name appears two spaces after the complimentary close, the letter writer is legally established as a representative of a company. Otherwise, the company name follows that of the letter writer. **Initials** tell who dictated and who typed the letter.

Professional letters can also have other parts, depending on whether they are needed or not (see Example 18-8). A **special handling line** at the beginning of the letter, usually after the dateline, tells about a letter's transmission. If you are sending a letter by certified mail, registered mail, overnight express mail, or with a return receipt requested, you will include this information. If you fax your letter, you can indicate this mode of transmission by adding the fax number to the inside address. Use an **attention line** only if no name appears on the first line of the inside address. It directs a letter to a particular department or person, but it also directs the letter to someone else if that person is no longer in that department or is on vacation.

The **subject line** saves words in the beginning of a letter by summarizing what the letter is going to discuss, usually in four or five words. You can use it to refer to an earlier letter or to a file reference number. Sometimes this reference is indicated by RE: or Ref:, instead of SUBJECT as in Example 18-8.

If a letter includes an enclosure, mention the enclosure in the body of the letter and after the initials in the **enclosure notation.** You can use the word enclosure or the abbreviations, Enc. or Encl. List the items enclosed if they are important, and if your letter has more than one enclosure, let the reader know how many items to expect.

If you are sending copies of the letter to other people, use the **copy notation.** The notation c: or cc: (standing for carbon copy, a holdover from days before photocopying and computers) is followed by the names of the people to receive copies. Also, pc stands for photocopy and bc for blind copy, marked only on copies that are sent to other readers without the main recipient knowing it.

You can either use the abbreviation P.S. to indicate a postscript or merely add an additional paragraph after the last item in a letter. Handwriting the postscript gives a personal touch to a letter. The postscript is rarely an added afterthought left out of the body of the letter, but instead can emphasize a main point.

If a letter is long, over 250–300 words, continue it on additional pages so as not to crowd the letter on one page. Indicate additional pages with a **continuation page heading** that keeps the letter intact in case pages become separated. You do not use letterhead for these pages, but you should use the same quality and color of paper as in the letterhead. It's a good idea to have at least two lines of a paragraph to continue onto the next page, and never hyphenate a word at the end of one page to carry over to the next page.

EXAMPLE 18–8   Other Letter Parts

July 31, 1994

The ABC Manufacturing Company
2121 Evans Road
San Diego, CA 92021
FAX 619-667-5588

Attention: Mr. H. John Bryan   **Attention line** (considered part of inside address
                                so used on the envelope also)
Dear ABC Manufacturing Company:

SUBJECT: Other Letter Parts   **Subject line**

You will need to use special parts of a business letter when they are appropriate.
Notice the spacing and punctuation used with the attention line.

The subject line can appear even with the left margin, centered two spaces below
the salutation, or centered on the same line as the salutation.

The additional letter parts illustrated in this letter are the enclosure notations,
found below the initials, copy notations, and the postscript. The postscript appears
a double space below the last item on a letter. You can use the abbreviation P.S.
although you don't have to.

Sincerely yours,

Frances Burgus

(Ms.) Frances Burgus

sl

Enclosure 2   **Enclosure notation** (1 or 2 spaces be-
                               low initials)
   1. Bill of materials
   2. Check for $102.01   **Copy notation** (1 or 2 spaces below
cc: Juan Martinez   enclosure notation or initials if no enclo-
                               sure)
The postscript is either indented or even with the margin, determined by how you
arrange the rest of the letter.   **Postscript** (always last item; can be
                               handwritten)

Letters in the workplace follow customary formats usually chosen by the company. Check your company manual or ask your supervisor to see which format your company uses.

CONVENTIONS FOR WRITING MEMOS

You use memos to correspond with people within your company, whether they are sitting in the office next to yours or in a division office across the country. You can use letters for this purpose too, especially if you are writing to a superior in top management, but generally you will use memos for in-house correspondence.

Networking systems that provide e-mail capabilities are popular for in-house correspondence. Although electronic communication may help cut down on the paper load, this off-the-record and ephemeral communication poses problems, both ethical and legal, that its users have just begun to contemplate. Often electronic messages do not become permanent records of correspondence, creating gaps in the information that documents the work history of projects. When people do print out e-mail messages so they have a record of the material covered, the sender sometimes discovers that a hasty, informal, and maybe even ungrammatical and misspelled document is put into the permanent file and identifies him or her as the originator. As a consequence, when you use e-mail for other than run-of-the-mill material, approach your document as an electronic memo and follow the guidelines for effective memos.

Many firms use printed forms for communication within the company. These memos display much more variety than letters in their styles and format conventions. Example 18–9 illustrates a typical memo form. Notice that the heading calls for a date, the writer's name, the receiver's name, and a subject. Job titles usually follow names, and if you are sending the memo to someone outside your department, you would identify both your and the receiver's departments, especially if all who receive copies of the memo would not know this information.

The body of the memo begins two or three spaces after the subject line, and it is usually single spaced with double spacing between paragraphs. Memos do not have closing lines, although you should initial or sign the memo where your name appears in the heading or at the end of the memo. The typist's initials appear below the body, and other notations, such as for enclosures and copies, appear as they do on letters. If your memo is more than one page, you can either use continuation page headings or number the following pages as you would in a report, usually in the upper right corner of the page.

Memo forms can vary from simple, handwritten notes (Example 18–10) to highly structured forms with attached carbon papers and complicated, predetermined routing slips (Example 18–11). Very large businesses may

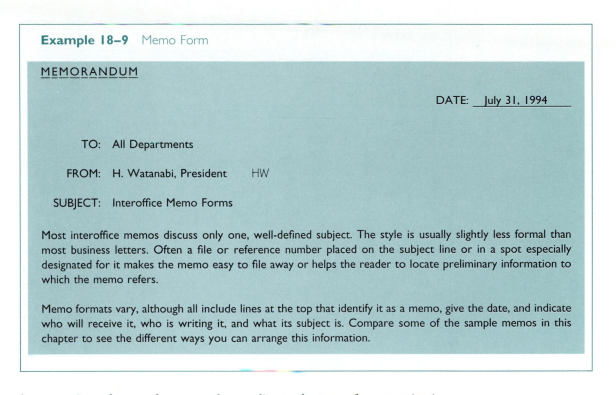

Example 18–9 Memo Form

MEMORANDUM

DATE: July 31, 1994

TO: All Departments

FROM: H. Watanabi, President   HW

SUBJECT: Interoffice Memo Forms

Most interoffice memos discuss only one, well-defined subject. The style is usually slightly less formal than most business letters. Often a file or reference number placed on the subject line or in a spot especially designated for it makes the memo easy to file away or helps the reader to locate preliminary information to which the memo refers.

Memo formats vary, although all include lines at the top that identify it as a memo, give the date, and indicate who will receive it, who is writing it, and what its subject is. Compare some of the sample memos in this chapter to see the different ways you can arrange this information.

have a variety of memo forms, used according to the type of communication being sent.

Although you write memos with a more informal style, what is true in the writing of letters is also true in the writing of memos, and you will want to prepare them carefully. Many times the only identity you have in the workplace is based on your memos.

EXAMPLE 18–10   Informal Memo Form

From the desk of--
Jill Abrams, Executive Director

April 2, 1994

Here's the information I told you about. Get back to me by Friday on this if you can.

Thanks —
Jill

EXAMPLE 18–11   Memo Pack

	Date:
	I.D.#
	REF.#
FROM:	SUBJECT:

MESSAGE

Blue--Originator

Yellow--Purchasing

Green--Accounting

Letter Reports

Often you will use the letter formats discussed in this chapter for your short reports for people outside a company. Short reports are usually defined as those reports under 10 pages. Documents of more than 10 pages need the help of the conventions associated with longer reports (see Chapters 15 and 16).

You can use many of the conventions of effective documentation within the letter formats: headings to chunk your prose, items in a list, and graphics to complement your prose. Example 18-12 shows a sample letter report. (See Appendix A, Case Documents 3 for the background of this letter report.) Notice that after the traditional salutation, a short introductory paragraph begins the letter. This paragraph functions in much the same way that the letter of transmittal or cover letter does. Do not label it as "Introduction," however. This heading sounds out of place in letter reports. The report begins immediately after this paragraph and follows the pattern of organization most

EXAMPLE 18–12   Letter Report

August 10, 1992

Mr. Ed Younger
Vice President, CCCAR
67 East Main
St. Paul, MN 67492

Dear Mr. Younger:

I appreciate receiving your letter stressing your objections to our bird control program. To clarify the program, I am sending you this short report that explains the purpose of the program, details the procedures followed in controlling the birds, and provides information about the trapped birds and what happens to them.

Purpose of the Bird Control Program

The purpose of our bird control program on the campus plots is to protect the experimental plantings that may start new crop varieties. The seed produced on these plots is extremely valuable because it represents the product of genetic crosses carried out over a period of years. With each plant selection, only a few seeds may be produced so it is essential that they be protected. The new plant varieties produced may increase food production worldwide and, thus, may reduce world hunger. We believe the benefits of these plant breeding experiments are important enough that we must protect these experimental plans from being destroyed by pest birds.

Procedures for Controlling Birds

The campus plots occupy approximately 100 acres. The comprehensive bird control program on these plots involves several approaches that we are continually improving as we learn about new techniques. These approaches include the following:

- Plastic strings stretched across the plots vibrate to create a noise that frightens birds.
- Brightly colored, large balloons with metallic designs (scare-eyes) suspended above the plots deter birds.
- A hired, licensed falconer uses trained hawks to frighten birds away from the plots daily.
- We are encouraging sparrow hawks to nest in the plot area to help control the birds naturally.
- Eight traps in the plot area, baited with bread, trap birds. Trapped birds receive food and water until they are removed from the cages.

EXAMPLE 18–12   (continued)

Information on Trapped Birds

The cages trap approximately 150 to 200 birds daily, with the number of birds trapped varying greatly. On unusual days as many as 500 birds may be trapped. Table 1 shows the kind and number of birds trapped from 1960 through 1990.

TABLE 1

Kind and Number of Birds Trapped—1960–1990

| Year | Kind of Bird | | | | Total/Year |
	Sparrow	Blackbird	Starling	Grackle	
1960	6305	581	1086	163	8135
1970	2317	1468	400	131	4316
1980	3843	502	1104	325	5574
1990	3396	395	2160	412	6363
Total/Kind	15861	2946	750	1031	

The program never permanently reduces bird population. Our bird control program attempts to reduce bird population in the plot areas only during the growing season. All songbirds are released. The only pest birds destroyed are blackbirds, starlings, and English sparrows. Destroyed birds are given to the Raptor Center to feed birds housed there. The Raptor Center uses all birds supplied to them.

New Method Used To Euthanize Pest Birds

Beginning August 6, trapped birds will be euthanized with carbon dioxide. Birds collected in cloth bags will be kept in a cool place until they are euthanized with high concentrations of carbon dioxide in a closed chamber. Collecting and transporting the birds in a cloth bag is more humane than using a wire cage, because the birds remain quiet in the dark bag and do not damage themselves by fluttering about, as occurs in a small cage. Euthanizing with carbon dioxide is an accepted, approved method used for small animals that results in loss of consciousness within 45 seconds and respiratory arrest within five minutes. Information from the 1986 Report of the American Veterinary Medical Association Panel on Euthanasia indicates that carbon dioxide is an effective euthanizing agent. Inhaling carbon dioxide causes little distress in birds, suppressing nervous activity and inducing death quickly. Euthanized birds can still be used for food in the Raptor Center.

Alternate Methods for Handling Pest Birds

Transporting and releasing birds is not possible because a suitable release site is not available that would be acceptable to the general public. Releasing significant concentrations of pest birds is not a desirable outcome to those who would be living in a potential release area and would be irresponsible on our part.

EXAMPLE 18–12   *(continued)*

Placing nets over the plots to prevent bird damage is impractical because of the size of the plot area (100 acres). It is also impractical because it would restrict management and evaluation of the plots. Nets would also be subject to damage by storms.

This report provides information about the bird control program on experimental plots at University of the Midwest. If you have further questions about our bird control practices, please call Dr. Pat Lennox, Department Head, Department of Plant Pathology, or Mr. David Obi, Research Plot Coordinator, Department of Plant Pathology.

Yours very truly,

Karen Wartala

Karen Wartala
Dean, College of Agriculture

appropriate to the subject. After the conclusion and recommendations, if there are any, the conventional complimentary close and other letter parts appear.

How is Example 18–12 part of a communication chain? How did the letter report writer establish *ethos* and use *logos?* Do you see any examples of *pathos?*

Memo Reports

Using memo format is common practice for short periodic reports such as progress reports, field reports, accident reports, and trip reports. Such reports are usually under 10 pages and written for an in-house audience. Example 15–2 in Chapter 15 is a report written in the memo format. It begins with typical memo headings. Two or three lines after the subject line, the introductory paragraph explains what is to follow, as in the letter report. At this point, the writer presents the report, following the pattern of organization appropriate for the topic. As in all memos, no signature lines or complimentary closes appear at the end. Topic and audience determine the level of formality of memo and letter reports, but letter and especially memo formats would be inappropriate for a formal technical report.

Letters of Transmittal and Cover Letters

Letters of transmittal transmit, hence their name, reports to recipients. Usually, projects begin as a result of a **letter of authorization,** which authorizes you to work on a project, gives specifications for the project, and tells you

letter of authorization
a letter authorizing someone to begin a project and providing that person with task specifications

(sometimes not specifically enough) exactly what it is you are to do. Reporting becomes the record of the progress of projects, and a final report ends the documentation on most projects. Many companies hire industrial librarians to archive these histories of their projects.

Basically the letter of transmittal tells the reader, Here's the report you asked for, although not as informally as this. At the beginning of the letter or memo, you refer to the authorization and the request for the report along with its topic or project title. Usually the letter of transmittal appears at the beginning of the report, placed either before the title page or after it. If it is after the title page, it directly precedes the table of contents. Example 18-13 is a letter of transmittal.

You may include other information in the letter of transmittal, which acts much like a preface. Remember that the letter of transmittal is usually brief; some of the following items would likely appear in the letter:

- Significant information you wish to stress to the reader, perhaps sections of the report the reader will be particularly interested in or that other people in the organization would find important.
- Very brief summary of the report (more descriptive than informative, especially if an abstract follows within the next page or two).
- Conclusions and recommendations (depending on audience, some writers prefer not to report bad news before the actual report is read).
- Possible action to take as a result of the information in the report.
- Relevant background information not found elsewhere in the report (funding enabling your research; limitations of your results—but be careful not to undermine your report).
- Thanks and acknowledgments for help not mentioned elsewhere in the report (reference librarians, members of your staff, and so forth).

Letters of transmittal frequently end with a statement of appreciation for working on the report (especially if you are reporting to a client on the work your company has done and hopes to continue to do), an indication of what you may have learned from working on the project, an offer to answer questions or provide additional information the reader may need, or any combination of these comments.

Cover letters also may transmit reports, especially ones that are not formally solicited by a letter of authorization. These letters provide the same kind of information as letters of transmittal, except at the beginning you will need to explain more directly why you are sending the report. Cover letters frequently accompany a resume (Chapter 19 provides information on the job application letter), questionnaires for surveys, and any other items you are sending to people who will need an explanation about the information

EXAMPLE 18–13   Letter of Transmittal

September 1, 1994

Ms. Megan Dodd, President
Information Services, Inc.
P. O. Box 4750
Denver, CO 50765

Dear Ms. Dodd:

Here is the report on the results of the survey you asked me to conduct last month on methods being used in the Sales Department to eliminate sexism in our business letters.

As you can see from the report, most of our letter writers favor using Ms. at all times, although several of the supervisors surveyed reported some negative responses to this choice. You will find these comments listed on page 15 of the report.

When I give my report at the meeting next week, I suggest we go over the alternatives listed on page 19 of the report. Although this problem is a sensitive issue to most of the people responding to the questionnaire I sent out, 87 percent indicated they would like to see a policy established for all departments to follow.

I would be happy to answer any questions you might have about the results of the survey.

Sincerely,

Benjamin Evans

Benjamin Evans
Vice President, Sales

Enc.

they are receiving. Depending on the audience, you can use either memo or letter formats.

Cover letters can be extremely brief, especially if someone has requested information informally. Depending on the information requested, your cover letter may merely indicate, "Here's the information you asked for." In many

EXAMPLE 18–14   Self-Adhesive Memos: Routing Request Form and Fax Routing Form

ROUTING - REQUEST

Please

☐ **READ**

☐ **HANDLE**

☐ **APPROVE**

and

☐ **FORWARD**

☐ **RETURN**

☐ **KEEP OR DISCARD**

☐ **REVIEW WITH ME**

To _____

_____

_____

_____

_____

_____

_____

Date _____   **From** _____

Post-It™ brand fax transmittal memo 7671	# of pages ▶	
To	**From**	
Co.	Co.	
Dept.	Phone #	
Fax #	Fax #	

Source: Reprinted with the permission of 3M.

offices, handwritten Post-it™ memos may suffice (Example 18–14). Special self-adhesive paper can also be used as a cover sheet for fax transmissions (Example 18–14). These adhesive notes indicate source and receiver and allow you to send a fax without attaching an additional transmittal sheet.

In some circumstances, your cover letter will need to be extremely persuasive, especially if you are asking your reader to take time to do something for you, such as fill out a questionnaire you are enclosing with your letter. As in any correspondence in which you make requests, tell the readers

EXAMPLE 18–15   Cover Letter

October 22, 1994

Dear Fellow Student:

Finding a parking place on campus has become an increasingly vexing problem. Many of us arrive on campus hours before our classes to ensure we will find a place to park. Lines of waiting cars queue up in parking lots with students offering to take departing students to their cars so they can have their parking slots.

I am exploring solutions to the parking problem on our campus. Would you be willing to take a few minutes of your time to answer the enclosed questionnaire that asks about alternative solutions to the parking problem on campus? The questionnaire asks you to respond to questions about car pooling and other possible solutions to the problem of finding a parking place on campus.

The report of my survey will be sent to the Department of Physical Facilities with recommendations on how to resolve this problem. If you would like to receive the results of this survey, please fill in the separate form attached to the enclosed questionnaire, and I will mail you the data I collect. Your answers will remain confidential.

Your responses will provide valuable information about solving this problem. Please respond by October 31 by mailing your completed questionnaire in the enclosed return envelope.

Sincerely,

Spenser Reid

Spenser Reid

Enc.

the benefits to them in granting your request before you make the request. Example 18–15 shows that you can also offer inducements to the reader. A frequent inducement included in a cover letter is an offer to provide the results of the survey. Assurances that completing the questionnaire will take only a short time may result in more returns.

Letters of Recommendation

People are frequently asked to write letters of recommendation, especially as they move into management positions. Some firms, because of legal considerations, refuse to provide any information beyond a person's employment history with the company. Letters of recommendation, however, are still frequently written either as a result of company policy or from a desire to help someone in the job search. Letters of recommendation are also written for promotions, appointments, and admissions to special programs and schools.

Many letters of recommendation include glowing, but vague, abstract commendations that are of little value to the person you are recommending. Always try to supply concrete, specific detail. Rather than using vague adjectives such as "conscientious," "dependable," "hardworking," "motivated," and "responsible," which have relatively little meaning, try "completed a six-month-long project on time," "initiated a newsletter within the department," or "supervised a crew of seven electricians." In other words, supply good, specific reasons for your argument about the individual's competence.

Begin the letter by indicating your relationship to the person you are recommending: Did you work with this person as a supervisor? As a co-worker? Usually you will mention the length of time you are reporting on, provide a brief record of employment, and give an overall appraisal of your recommendation.

As Example 18–16 illustrates, after the opening paragraph, provide the specific and concrete description of what you know about this person's activities. Follow this description with your concrete and specific evaluation of this person's performance.

The closing paragraph emphasizes how strongly you are recommending this person. It might include a statement of your willingness to answer any questions, with your phone number provided.

Be careful about the language you use to describe the person you are recommending. What might have been intended as a compliment might in fact be seen as inappropriate. Do not make any comments about a person's race, age, sex, national origin, or disability, either negative or positive. For example, don't say, "Mary Smith is a highly skilled technician and is a very attractive presence in the laboratory"; or "Despite the fact that he is 50 [or 20], Jamil Polanski is a diligent and hard worker."

Whether to include negative comments is a tough decision. Some companies fear that if the person does not get the position, negative comments in recommendation letters may be used in lawsuits. For example, in a grievance, the person about whom you made a negative comment might require you (and your company) to prove what to you seemed a matter of professional judgment or interpretation. When you are asked to write a letter of recommendation by a person about whom you cannot write favorably, suggest that the person ask someone else, if you can. In a competitive job market, lukewarm recommendations have extremely negative results. At times you may receive a request for a letter of recommendation when the person has not asked you first if you will provide one. If you have not agreed to supply a letter of recommendation as the result of a personal request, you are under

EXAMPLE 18–16   Letter of Recommendation

June 6, 1994

Mr. Benjamin Evans
Vice President, Sales
Information Services, Inc.
P. O. Box 4750
Denver, CO 50765

Dear Mr. Evans:

I am writing in enthusiastic support of the application of Keith Grey for a position in your department. Mr. Grey worked for me for over a year as a student assistant in the Validation Research Center at San Diego State University. My impressions of his work record are overwhelmingly positive.

Mr. Grey assisted at every stage of the usability tests we performed. He helped arrange for test subjects, maintaining a pool of applicants for this position and corresponding with them. He prepared questionnaire forms, test booklets, sample documentation to be tested, and reports of the tests. He also entered data collected into a series of databases in *Lotus 1-2-3* and *dBaseIII*. He is also proficient on several word processing programs, *WordPerfect, Microsoft Word,* and *Microsoft Word for Windows*. He has designed graphic displays of data for our reports using a number of different graphic programs.

In all he did, he was scrupulously accurate. The only times he redid something were when I changed my mind, and these changes he did cheerfully. He works well without constant supervision, for much of his work involved staffing a one-person office. I often could present a problem I was working on, for example, data collected during a test session that I wanted to display in some way, and he would present to me, very quickly in many cases, several possible solutions. He has an eye for knowing what needs to be done and takes it upon himself to do it.

In addition to these exceptional attributes, he is a congenial, friendly worker. He seems to enjoy his work, is always on time, and has missed not one appointed work time. He works well with the people who call or come into the Validation Research Center, is usually able to answer their questions, and, if not, passes on careful records of their questions to me.

Mr. Grey is obviously an important asset to our Validation Research Center, and I regret to see him leave. I would consider him an asset to any organization and recommend him without hesitation.

Sincerely,

Sherry Burgus Little

Sherry Burgus Little, Director

no obligation to write one, unless you are that person's immediate supervisor and company policy dictates this function as part of your job. It is always risky to give someone's name as a reference without first asking that person's permission, but some people do it.

SUMMARY

In this chapter we have looked at the wide range of correspondence you are going to encounter in the workplace. As with all writing, you will need to consider both the rhetorical dimensions of the project—audience, purpose, and context—as well as the characteristics of the genre. Correspondence often has a smaller audience, often one or two people known to you, and it has a kind of immediacy lacking in other documents you create. You must write correspondence quickly for it to be useful, but you still must follow both the conventions of the genre and the requirements of effective communication.

Correspondence, despite its short length and its transitory nature, is important both in your career and in the life of the corporation for which you work. Correspondence is part of a paper trail or communication link of any number of documents that address a particular issue or problem. Correspondence has to be clear and concise. These documents require special attention to style because they are too short for you to waste space with vague or imprecise sentences or inadequate organization.

Often short reports appear in the form of letters and memos and follow the conventions of correspondence at the same time they solve problems for people or help people make decisions. Letters of transmittal and cover letters introduce longer documents, such as reports and questionnaires. Letters of recommendation, on the other hand, offer professional assessments of people's performance.

Correspondence has clearly defined formats that are well accepted throughout the workplace, and you need to be familiar with those conventions. Make sure you are familiar with the main parts of letters and memos and how to lay these parts out in the document.

ACTIVITIES AND EXERCISES

1. Study the memo in Example 18–4, and without doing any reorganizing, eliminate all unnecessary words. How many words can you eliminate and still convey the message?

2. Pair up with another student. With your resumes available for study, talk about your experiences and educational backgrounds, and then write a letter of recommendation for each other. Write the letter from the perspective of a professional colleague.

3. Collect letters that you receive in the mail. Do you find unnecessary words, redundancies, jargon, and stereotypical phrases that this chapter discusses in your letters? How did the letter writers establish *ethos* and use *logos* in the letters? Are there examples of *pathos?* Does the letter help the reader solve a problem or make a decision? What other features do you find in these collected

letters? As your instructor directs, bring your collection to class for discussion. Working in groups of three, discuss what you find in the collected letters. Report your findings to the class.

4. Example 18–7 is adapted from the case materials provided in Appendix A, Case Documents 3. Find and read the original letter written by Mr. Younger in Appendix A, noting the changes made to Example 18-7. Why were these changes made? What is the effect of these changes?

5. Example 18–12 is based on the short report entitled, "Bird Control on the Experimental Plots University of the Midwest." This letter report was actually never written by Dr. Karen Wartala. Find and read the original report in the case material in Appendix A, Case Documents 3. A number of organizational changes made to this report appear in Example 18–12. Identify these changes. Why were these changes made? Notice that sentences are rewritten as well. Find rewritten sentences and compare them to the original sentences. What is the result of the changes made? What is your reaction to the use of "euthanize" with the description of killing birds with carbon dioxide? Can you identify any ethical problems with the rhetorical decisions reflected in both the original report and the rewritten, fictitious letter report?

6. Explore the ways you can use your word-processing software to assist you in creating effective and consistent correspondence. Create a style guide using the style selection features of your software. Experiment with the mail merge feature of your software and use it to send a letter to your instructor and your classmates. In the letter, discuss how mail merge programs can improve and simplify the correspondence process.

7. If you have access to e-mail (either on a LAN or via Internet) on your campus, consider joining an electronic bulletin board to engage in correspondence with someone who shares your interests. In a memo to your instructor and classmates, describe the nature of this correspondence and how it might differ from more traditional means.

8. In Chapter 8, Exercise 3, you were asked to investigate a technical problem at your university, using an appropriate research tool. If you designed a questionnaire for that project, now write the cover letter that could have transmitted that questionnaire.

9. Look back at the communication link between Systems Tech and Executive Consumer Surveys (Examples 18-1–18-3). Complete the following assignments:

   a. Assume that you did indeed find a faulty chip in the Systems Tech 175 computer. Write a letter to Ms. Patrick explaining your finding and offering what you consider adequate compensation.
   b. Again assuming that you found a faulty chip in the Systems Tech 175 computer, write a letter to the other 175 owners to recall the machine (or to take whatever action you think appropriate).
   c. Assume that when Ms. Patrick's 175 computer arrives and the repair shop checks the machine, they find evidence that a liquid has seeped into the machine. Write a letter to Ms. Patrick explaining your findings and taking whatever action you think appropriate.

10. Select and study one of the sets of Case Documents in Appendix A. With a partner, describe the communication links that have been generated and probably will be generated by the documents you study.

    Next, individually write a letter or memo that might be generated by the documents. For example, you could assume that you work for the Seattle Treatment Education Project and have just written the AZT Step Fact Sheet (Case Documents 1). You need to alert such groups as the Minnesota AIDS project of the availability of the Fact Sheet in a letter that contains a brief summary of the purpose, audience, and content of the Fact Sheet. Or, you might need to alert the university community by memo about the communication link generated by the complaints about bird trapping on experimental fields (Case Documents 2) in case the faculty and staff receive calls or letters about the situation. You would want to summarize the previous correspondence and set a policy as to how to respond to inquiries from the public, press, or special interest groups.

    Finally, exchange letters or memos with your partner. Your partner will now reply to your letter or memo. This reply might take some imagination, but remember that there's usually another link in the communication to be written.

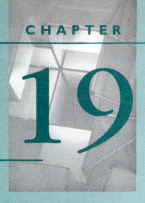

Job Search Materials

Getting Started ▌ Writing the Resume ▌ Writing the Letter of
Application ▌ Preparing for the Job Interview ▌ Succeeding
at the Job Interview

The problem was that prospective employers were leery of a person whose entire
career had been spent with one employer, with one kind of service. He had no
specific training in technical writing, having learned all he knew about it on the job.
He knew his subject matter well enough and was proficient enough to hold his
job, but his rèsumè was flat: no evidence of attempts at self-improvement, no civic
involvement, no volunteer activities, no STC membership. Nothing about specific
interests, hobbies, or participation in sports gave employment managers any hint
of his being able to fill their needs . . . . His problem was that he had really had
only *one* year of experience, repeated 14 times, and there was nothing to show
that he had tried to improve or advance in his trade.

> *Dart G. Peterson, Jr. "The Business of Technical Communication,"*
> *Technical Communication 38.4 (Fourth quarter 1991), pp. 594–97.*

Introduction

Regardless of the career you choose, one of the most challenging and exciting activities you will engage in is searching for a job. You may perform this task a number of times in your professional career, and in every case you will have a good deal of time, energy, and ego invested in a good outcome. This chapter offers suggestions about this important task: how to prepare yourself and find job opportunities; how to analyze what you have to offer prospective employers; and how to contact employers successfully. We discuss the typical documents you write for this task: the letter of application, the resume, and the follow-up letter. We also offer suggestions for carrying out successful job interviews.

As is true for all writing tasks, your written documents will be more successful if you do some planning before you begin to write. Many people slight this important part of the writing process and end up with inadequate job search materials that get them no job offers. You'll be spending most of your life in your chosen career, and your satisfaction is a result of one of the most serious decisions you will make.

Getting Started

What are some things you can do to make sure your decisions are the best ones? Being sure you understand what you will be doing on the job can help ensure a successful job search. Talking to people doing what you think you want to do is probably one of the most effective ways for you to make the right decisions. For example, conducting job information interviews can provide you with valuable insights about how companies function; how their employees are treated; what kind of conditions they work under; what goals and values the company has; and what opportunities exist for advancement, relocation, and further education. Most people enjoy an opportunity to talk about themselves and their jobs. Making plans about what you want to talk about, realizing people are busy so that you don't waste their time, and being courteous before, during, and after the interview will ensure successful encounters.

If you don't know someone you can interview, you can meet people by becoming active in the professional organizations of your field. Most professional organizations have local branches that meet monthly, providing contacts and also professional programs or guest speakers of interest to practitioners. The Institute of Electrical and Electronic Engineers (IEEE), American Society of Mechanical Engineers (ASME), and Society for Technical Communication (STC) are just a few examples of such professional organizations. Most of these professional organizations have student chapters, and becoming a participant in these professional activities is an important step to take well before you begin thinking about your job search. You've probably heard that most job interviews result from personal contacts or referrals, not from answering job advertisements or sending out unsolicited job applications. The **networking** important for such referrals should begin as soon as possible and continue throughout your professional life.

If you live in a place that does not have chapters of national organizations, you can use other resources. Another important source of information about careers is the *Occupational Outlook Handbook,* published periodically by the Department of Labor. Its companion volume, *The Dictionary of Occupational Titles,* identifies job clusters or occupations and gives you job titles within these clusters, as well as short descriptions of occupations. It then provides a decimal number you can use to find more detailed information about job opportunities within the clusters in the *Occupational Outlook Handbook.* You will find information about what kinds of jobs are available, where in the United States such opportunities exist, what background and skills are needed, and what salaries are common.

Of course, your academic advisor or the career counselor at your college or university can also help you explore not only general opportunities for graduates in your major, but also specific companies that hire such graduates. You also will want to talk to the reference librarian at your school and look at the references listed at the end of this chapter before you start making decisions about your job search. The suggestions that follow are ones that you will want to start months and perhaps years before you search for a job.

networking
in the context of a job search, identifying and creating a group of personal contacts who can assist you in identifying jobs and opportunities

ANALYZING YOURSELF

Once you know more about the career you want, you will need to spend some time analyzing what you have to offer a potential employer. It's easy to overlook important skills you have, so we suggest you do some self-analysis by completing Writing Strategies 19–1.

Many books can help you to identify strengths, experience, and skills you will want to highlight in your job search, such as *What Color Is Your Parachute?* by Richard Bolles, *Who's Hiring Who?* by Richard Lathrop, and *Guerrilla Tactics in the New Job Market* by Tom Jackson. Many people skip this analysis step, thinking it might be too time-consuming. But, completing a critical self-analysis can frequently make the difference between getting the job you want or not. Doing a careful analysis at this stage generates information for your letter of application and resume, as well as for your job interview. Although you will not necessarily use all the information you generate in this exercise, you can select those pieces of information that will best show what you have to offer prospective employees.

IDENTIFYING JOBS

Many people look for job advertisements in newspapers, trade or professional journals, placement services, or employment agencies. Again, the most successful source of jobs is referrals by people you know, such as contacts

WRITING STRATEGIES 19–1   Self-Analysis Worksheet

Personal

Name:

Address:

Permanent address (if still in school):

Home telephone:

Business or residence hall telephone:

Social Security number:

Education

(Fill out this information for each institution. Do not include high school.)

Name of college:

City, state:

Degree(s) earned:

Certificates:

Major(s):

Minor(s):

Month and years attended:

Relevant courses:

Honors and awards:

GPA:

Extracurricular activities:

Courses liked best:

Courses liked least:

Courses with best grades:

Achievements and accomplishments:

Special Training: (Workshops, seminars, training schools)

Name:

Date:

Certificates/licenses earned:

Skills/capabilities:

Personal Interests

Hobbies/personal interests:

Membership in organizations:

   Name of organization:

   Involvement:

   Accomplishments/achievements:

WRITING STRATEGIES 19–1   *(continued)*

Work History

(Paid or unpaid. Complete information for each position, whether with the same company or not.)

> Company name:
> Job title:
> City, state:
> Dates of employment: from   to   (include months)
> Nature of job—duties, responsibilities:
> Awards/achievements/accomplishments:
> Significant contributions made to employer:

Skills

(Complete for each identified skill)

> Related skill:
> Level of competency:
> How demonstrated:

Other Awards/Honors, Information

References

(Complete for each reference)

> Name:
> Address:
> Telephone:
> Relationship or title:

you've made before looking for a job, your professors, and your professional colleagues. Take advantage of internships offered by your school, summer job experiences, and volunteer work to find contacts and information about available jobs. Participating in job or career fairs sponsored by school placement services or professional organizations can also help you identify job opportunities. You can also use electronic bulletin boards that post job opportunities.

Once you have targeted jobs for which to apply, start gathering as much information as you can about the companies you've identified. Read information about the company in newspapers, annual reports, brochures, and advertisements. Trade journals and business publications such as *The Wall Street Journal* and *Forbes* publish information about the activities and accomplishments of companies. *F & S Index of Corporations and Industries* will direct you to publications about companies. Often, your library will have information files on local companies including recent news releases to help you tailor your job search to the needs, interests, and philosophy of your targeted companies. In each state, the Office of the Secretary of State, Corporate Division, will have information about the status of corporations, the names of the corporate officers, and the addresses of the corporations. The *CPC Annual,* published by the College Placement Council and available in school placement centers and most libraries, provides names, addresses, and general descriptions of companies as well as suggestions for conducting successful job searches. *Thomas' Register of Manufacturers* and Dun and Bradstreet's *Reference Book of Manufacturers,* available in most libraries, are two more sources of job information.

Again, school placement centers and university career counselors provide information about all parts of the job search. Most offer workshops on writing letters of application and resumes. Some provide help with preparing for job interviews and videotapes of practice interview sessions to suggest ways to handle typical interview questions. Some provide an electronic resume review that allows employers to review resumes by computer. State offices of economic development offer similar help, if you do not have a school placement center available. Employment agencies can also help in your job search; some charge you for such services, although others charge the company that hires you. *Be sure to understand exactly how payment will be handled* before signing any agreements, because some agencies can charge as much as your first month's salary.

These sources can make your job application stand out from all the rest. Being thorough is time well spent.

Writing the Resume

Once you have completed the tasks described in the preceding section, you're ready to prepare your resume. You should always have an up-to-date resume on file. Even though you may not be considering changing jobs or looking for one, resumes are frequently part of proposals and reports. The resume will be your prospective employer's first glimpse of your experience and accomplishments. You have already completed much of the work for the resume by preparing the self-analysis worksheet in Writing Strategies 18-1, for resumes *always* include the following information:

- Name, address, and phone numbers
- Education
- Experience

Resumes *usually* include the following information:

- Career objective
- Skills, licenses, certificates

- Honors and awards
- Professional activities
- Information on how to obtain references

You include in your resume those things you want to advertise or highlight, based on your careful self-analysis and the position for which you are applying. Your resume headings emphasize what you have to offer or have accomplished, such as "Computer Programming Languages" or "Community Service."

Although resumes alone seldom get jobs, they can determine whether you have a chance at a position. Because a resume sells you to potential employees, its main purpose is to get you an interview. Competition for jobs is fierce; often the hiring process, in its earliest stage, is not one of choosing the best applicant but is instead screening out those that don't promise the best fit for the job. Hundreds of resumes may result from a single job advertisement; personnel directors, with only a general knowledge of what is needed, may do the initial screening of applicants, spending only seconds on each resume. Managers will then receive the resumes that have made it past the personnel directors and will decide whether to schedule an interview.

What are some of the things that can help ensure that your resume will go in the "yes" pile? Generally, you will want to think about the overall appearance of your resume, content, persuasive strategies, and format.

APPEARANCE

A resume should invite your reader to read it. It should be placed attractively on the page, with careful use of white space (wide margins and space between headings), and easy-to-read type. It should be carefully proofread and edited; some personnel directors and prospective employers discard resumes with typos and misspelled words. They might infer that if something as important as a resume is carelessly or hastily done, the quality of work this person will produce will be poor. Use all your knowledge about document design when you plan your resume.

It's usually not a good idea to have a fancy resume (that is, one that deviates too much from the more traditional formats) or one that has been done by a professional resume service, as people who look at many resumes recognize the distinctive styles of most resume services. For professional positions you should design your resume and then have it prepared on a word processor to look typeset. Print your resume on a laser printer and make photocopies on quality white or cream-colored paper.

Remember that your resume must be easy to skim, and the important details of your education, experience, and accomplishments must stand out. Use headings, columns, lists, and boldface or italics to organize your resume.

CONTENT

Most resumes are one page, although they increase in length with the level of your education and the length of your work experience. Your resume is not a complete history of everything you have done, but rather a careful selection of your qualifications and skills. Leave out information such as your marital status, age, health, and other characteristics that might invite discrimination. Include personal interests and hobbies, especially if this information relates to your qualifications for a position. Perhaps you are an excellent speaker, work with disadvantaged children, or have organized a literacy campaign in the community. These personal activities can highlight your talent and motivation.

References are not always listed on resumes. You may choose to indicate that references will be furnished upon request, although it is safe to assume prospective employers will know this, and you may wish to save space on your resume for other information or provide different references for different job applications. You should prepare a list of references separately to give to prospective employers before or during the job interview. This list must include names, titles (showing their relationship to you), addresses, phone numbers, and even fax or e-mail addresses (always make it easy for the prospective employer to reach your references). Sometimes letters of reference can be part of a portfolio you bring to the interview, although confidential recommendations are more convincing.

PERSUASIVE STRATEGIES

In resumes for technical positions especially, you need to be sure your readers can understand what you are talking about and that they can grasp the information quickly and easily. Avoid acronyms, abbreviations, and technical jargon. Both the technical supervisor and the nontechnical reader must understand your resume, although you should use the terminology of your career field. Keep paragraphs and sentences short, and use phrases when possible. Leave the subject *I* out of sentences so that sentences begin with dynamic action verbs. Again, listing allows the reader to scan through your qualifications quickly without missing any of the important details.

Pay particular attention to your writing style as you describe what you have to offer prospective employers. The following guideline will help you select words and phrases you need to advertise your skills and persuade your readers:

1. Use action words—powerful, dynamic verbs:

advised	created	designed	presented
improved	operated	trained	organized

established	evaluated	supervised	supported
analyzed	cut	invented	edited
conducted	produced	prepared	adapted
maintained	assessed	oversaw	applied
directed	coordinated	developed	studied
planned	identified	wrote	inspected
managed	installed	sold	programmed

Try to stay away from phrases like "Responsibilities included" or "Duties were."

2. Emphasize accomplishments and achievements instead of describing what you did. Try to make these quantifiable, such as "Saved $20,000," "Improved sales by 20 percent," or "Supervised 10 people with a budget of $450,000." In choosing what to describe, always remember you are emphasizing what you have to offer your employer.

3. Keep information simple. Don't try to pack too much information into each phrase or you will dilute your key selling points.

4. If possible, include all experience and skills you can offer, whether they are from education, work experience, volunteer experience, or social activities.

5. Stress important information in the beginning of your resume.

Finally, if you do not have much relevant experience, you will probably want to highlight another aspect, such as your education. However, don't list all the relevant courses you have taken, but instead concentrate on the skills you have acquired as a result of your education that are of value to your prospective employer. For example, you might refer to your ability to program in several different computer languages or your document design skills.

Because the format of your resume will help you emphasize your most impressive qualifications, we discuss the three most common formats below.

THE CHRONOLOGICAL RESUME

The **chronological resume** is the most common in the workplace. As in all resumes, your name, address, and phone number should be displayed in the beginning (Example 19-1). In this resume, the most recent information is considered the most relevant and important. Experience and education are listed in reverse chronology, the most *recent* listed first. Dates (both month and year) are usually listed first, followed by the name of the organization, the job title, and the job duties. You may wish to highlight the work

chronological resume
a resume in which experience and education are listed in reverse chronology, with the most recent information listed first

EXAMPLE 19–1   Chronological Resume Using Paragraphs

RESUME OF KAHUKU V. OADES

46480 Madrid Court
San Diego, California 92124
(619) 888-2319

Education

Sept. 1988–present — San Diego State University, San Diego, California. Bachelor of Science degree in Biology in May 1993.

Experience

Jan. 1992–present — Undergraduate researcher, Department of Biochemistry, San Diego State University. Developed and performed smooth muscle cell isolation protocols, conducted spectrophotometric enzyme assays, performed protein purification, and attended local, national, and international professional meetings.

Sept. 1990–Jan.1992 — Research Assistant I, Center for Research on Aging, Department of Medicine, University of California at San Diego. Maintained cell lines in primary culture, performed assays including ELIZAS, cytotoxicity, RIA protein assays, and radioisotope assays, cultured bacterial cells including thermophilic bacteria, partially purified proteins, maintained a cryostorage facility.

June 1990–Sept. 1990 — Lab Helper, Clinical Immunology Department, Scripps Clinic and Research Foundation. Prepared buffers and reagents, ordered laboratory supplies, maintained cryostorage tanks.

Honors

1992 — Received National Institute of Health: Minority Biomedical Research Support (MBRS) Grant.

1988 — Received Samuel Gompers Secondary School, Math/Science/Computer Science Magnet Program Certificate.

Publications

Muscle Membrane Plasma MgATPase is an Ecto-MgATPase. K. Oades, P. J. Yazaki, H. B. Cunningham, and A. S. Dahms. Abstract, *MARC/MBRS Symposium Proceedings,* October 1992.

The Skeletal Muscle Transverse Tubual MG-ATPase Identity with Mg-ATPases of Smooth Muscle and Brain. H. Brad Cunningham, Paul J. Yazaki, Kahuku V. Oades, Ron Domingo, Rodger A. Sabbadini, and A. Stephen Dahms (submitted for publication).

References

Available upon request.

you have done instead of the length of time you did it. In that case, place dates in parentheses after the job title and organization name. You can also list these actions with bullets rather than use paragraph form (Example 19–2). Bullets also make the resume easier to scan, but don't use so many bullets that they become distracting.

Again, because most people have experience with the chronological style of resume, it provides them with a familiar guide for conducting the job interview. Because this style follows a chronological order, you'll find it easy to prepare. You simply structure it by providing the dates, company names, and job titles, and the other information about your education you have already listed in your self-analysis worksheet. This format allows you to highlight a steady employment and education record.

Unfortunately, this format also draws attention to any time gaps that might appear, gaps that might be interpreted negatively. If you are a student or are seeking a career change, this format also emphasizes your recent jobs that may not be directly related to the position for which you are applying. Putting yourself through college waiting tables will show your determination but doesn't relate directly to a position as computer programmer, for example.

THE FUNCTIONAL RESUME

The **functional resume** highlights your qualifications without emphasizing specific dates or work history (Example 19–3). As you can see, this style de-emphasizes a work history with gaps and downplays unrelated work experience and education, which might best be handled in the job interview. Instead, the functional resume emphasizes professional growth and accomplishments that relate directly to the position for which you are applying. The tasks or functions you can perform receive primary attention.

On the other hand, this format does not provide the conventional kinds of information employers normally expect on a resume, and some potential employers may be uncomfortable with what they consider omissions.

The functional resume usually begins with the job objective, which also emphasizes what the job applicant has to offer—a persuasive strategy in itself. Making the objective general rather than listing a job title avoids limiting positions for which you may be considered.

Finally, you can use capital letters, italics, bold, and different-sized type to emphasize information. Achievements can then be highlighted by concrete accomplishments and quantifiable details.

THE COMBINATION RESUME

The **combination resume** allows job applicants to take advantage of both the chronological and the functional styles. It is similar to the functional

functional resume
a resume that highlights qualifications without emphasizing specific dates or work history

combination resume
a combination of chronological and functional resumes, listing qualifications and providing company names and dates in a separate section to minimize gaps in work history or temporary jobs unrelated to a career goal

EXAMPLE 19–2   Chronological Resume Using a Bulleted List

RESUME OF KAHUKU V. OADES

46480 Madrid Court
San Diego, California 92124
(619) 888-2319

Education
September 1988–present

San Diego State University, San Diego, California. Bachelor of Science degree in Biology in May 1993.

Experience
January 1992–present

Undergraduate researcher, Department of Biochemistry, SDSU.

- Developed and performed smooth muscle cell isolation protocols
- Conducted spectrophotometric enzyme assays
- Performed protein purification
- Attended local, national, and international professional meetings

September 1990–January 1992

Research Assistant I, Center for Research on Aging, Department of Medicine, University of California at San Diego.

- Maintained cell lines in primary culture
- Performed assays including ELIZAS, cytotoxicity, RIA protein assays, and radioisotope assays
- Cultured bacterial cells including thermophilic bacteria
- Partially purified proteins; maintained a cryostorage facility

June 1990–September 1990

Lab Helper, Clinical Immunology Department, Scripps Clinic and Research Foundation.

- Prepared buffers and reagents
- Ordered laboratory supplies
- Maintained cryostorage tanks

Honors
1992

Received National Institute of Health: Minority Biomedical Research Support (MBRS) Grant.

EXAMPLE 19–2   *(continued)*

1988   Received   Samuel   Gompers   Secondary
                             School, Math/Science/Computer   Science
                             Magnet Program Certificate.

Publications

Muscle Membrane Plasma MgATPase is an Ecto-MgATPase. K. Oades, P. J. Yazaki,
H. B. Cunningham, and A. S. Dahms. Abstract, *MARC/MBRS Symposium Proceedings,*
October 1992.

The Skeletal Muscle Transverse Tubual Mg-ATPase Identity with Mg-ATPases of
Smooth Muscle and Brain. H. Brad Cunningham, Paul J. Yazaki, Kahuku V. Oades,
Ron Domingo, Rodger A. Sabbadini, and A. Stephen Dahms (submitted for publi-
cation).

References available upon request.

resume, but it includes company names and dates in a separate section. It
allows you to emphasize the skills and capabilities that are relevant, while
deemphasizing gaps in work history or temporary jobs that are unrelated to
your career goal. Both the combination and functional resume are harder to
write than the chronological resume, demanding careful and critical analysis
of the information you have compiled as a result of your self-analysis, the
job position for which you are applying, and your assessment of the organization to which you are applying.

Study Examples 19-4, 19-5, and 19-6. These resumes illustrate a writer's
search for an acceptable format to allow him to discuss his long work history,
which has given him excellent qualifications. This writer believed that providing a detailed history, one that spans 12 years of different positions, would
prevent his highlighting skills and qualifications, making it hard for the reader
to assess what he has to offer. Listing all these positions and providing a
detailed description of each job would take two pages and mean repeating
activities performed on different jobs, burying significant information about
his performing increasingly more sophisticated tasks. Thus, the chronological
resume format (Example 19-4) prevents readers from easily understanding
what he has to offer. How could this resume (Example 19-4) have been
improved by following guidelines about persuasive strategies and appearance?

In another attempt, the writer identified those skills he wished to advertise in a combination resume (Example 19-5). However, these skills are
listed with no relation to each other so that the reader is still left trying to
make sense of the long list.

In Example 19-6, however, the writer has been able to classify his
skills into easily grasped categories, while still providing a work history and

EXAMPLE 19–3 The Functional Resume

CANDIS L. CONDO

4566 Polk Avenue, San Diego, California 92105   714-283-9668

OBJECTIVE

TECHNICAL WRITER in an organization that needs precise, clearly written documents.

WRITING EXPERIENCE

As *Job Analyst* at San Diego Trust and Savings Bank, developed precise job descriptions based on information obtained through questionnaires and interviews. Have prepared more job descriptions in the first quarter of this year than were prepared in the entire previous year.

As Technical Writer at the same bank, wrote and revised entire Personnel Policy Manual, finishing project four months ahead of schedule.

As Technical Writer for RAIR, Inc. a computer timesharing company, wrote custom software documentation for custom programs.

As College Instructor, taught basic composition, technical writing, and business writing courses at Ohio State University, San Diego State University, American Institute of Banking, and San Diego City College.

RESEARCH AND ANALYTICAL SKILLS

Analyze both exempt and nonexempt jobs and develop written descriptions containing sufficient detail for recruiting, selecting, placing, training, and evaluating bank staff.

Research, prepare findings, and make recommendations on sensitive legal issues, such as sexual harassment, solicitation/nonsolicitation, and age discrimination.

Conduct salary surveys, prepare results, and make recommendations on the appropriate "slotting" of both exempt and nonexempt jobs within bank.

VERBAL COMMUNICATION SKILLS

Schedule and conduct interviews at all levels up through senior management to obtain information necessary to describe accurately job content.

Solicit information from employers and survey services by letter, telephone, and personal visits.

PROFESSIONAL AFFILIATION

Member of the Society for Technical Communication

PERSONAL DATA

Avid scuba diver and jogger, enjoy classical music, successful public speaker.

EXAMPLE 19–4   One Case: A Chronological Resume

This resume buries important information.

BRENT MOSBARGER
4266 RIVER DRIVE
SAN DIEGO, CA 92224
789-7234

EMPLOYMENT HISTORY

Date: 3/91 to 1/93 J. S. King Electric, San Diego, California
Project: South Elementary School
Remarks: I supervised this .5 million dollar electrical project, which consisted of 5 new buildings and the remodel of 3 existing buildings.

Date: 2/90 to 3/91 DDD, Escondido, California
Project: North County Hospital
Remarks: Journeyman, I worked on roughing the shell and almost all of the mechanical and backup generator controls.

Date: 12/87 to 1/90 J. S. King Electric, San Diego, California
Project: Midway Junior High School
Remarks: Lead journeyman, I supervised all underground work, consisting of large concrete encased duct banks and setting all handholes and transformer pads to finished grade. I also installed all of mechanical controls and most of the interior pendent mounted lighting.

Project: Blue Onion Restaurant
Remarks: Lead journeyman, I laid out and installed the dance floor lighting system, consisting of mounting the unistrut grid, line and control circuits, mounting of trusses and fixtures. I also installed the class A F.A. system and ended up taking over the project for the 4 weeks to completion.

Project: GE Building 99
Remarks: I supervised this project which consisted of service change from 1200 amps to 1600 amps. Several sub-panels, 225 kva x-former, and miles of G-4000, 1500, and 2600 wiremold. I had 11 electricians working for me.

Project: Software, Inc.
Remarks: I supervised this project. It was the addition of 4 class 10 clean rooms. Consisted of remote ballast panels, variable speed drives for 100 hp 3 phase fan motors, a complete pneumatic control system, toxic gas sensors with master and slave monitors. This was a very complicated control job.

EXAMPLE 19–4   (*continued*)

Date:	5/87 to 12/87 Jones Electric, Denver, Colorado
Project:	State Prison Complex
Remarks:	Journeyman/foreman. I did the auto-body building and worked in the other 12 buildings on their contract.

Date:	3/85 to 5/87 New Mexico Electric, Albuquerque, New Mexico
Project:	Federal Building
Remarks:	Journeyman, computer facility for GSA payroll, consisting of a large UPS system, 400 Hz motor generator sets, and some 15 kv splicing. Most of the conduit was G.R.C.

Project:	Department of Energy
Remarks:	Journeyman, installed wiring systems, UPS systems, and computers designed to measure and correct air quality. All of the conduit used on this job was G.R.C. up to 4 inch.

Project:	Lees Southwest
Remarks:	Journeyman, 150,000 sq. ft department store. A commercial job with a great deal of lighting and an elaborate energy management system.

Date:	9/84 to 2/85 Fairbanks Electric, Fairbanks, Colorado
Project:	Tunnel Lighting
Remarks:	Journeyman, installed HPS fixtures in 2 highway tunnels between Denver and Central City, Colorado.

My apprenticeship was spent working for Gleen Electric and McLeod Industries in Denver, Colorado, and Spring Electric in Omaha, Nebraska.

Education: San Diego State University, Electrical Engineering
Graduation: May 1995

Experience Highlights: My experience with most of the projects was from ground breaking to the final inspection. I spent 8 months estimating for J. S. King Electric Company. I hold a C-10 license through the state of California. I am comfortable with prints and specs and have a good knowledge of code and construction.

educational information. When this writer interviews for a job, he can provide a complete list of all the jobs with dates, titles, and job descriptions if the interviewer wants to see it.

Obviously, your audience and your content will dictate the resume format you choose. You may want to have more than one resume on hand, using different styles depending on the ways you choose to present yourself. As

EXAMPLE 19–5 One Case: Combination Resume

This resume leaves too much interpretation up to reader.

RESUME OF BRENT MOSBARGER
4266 River Drive
San Diego, California 92224
789-7234

EDUCATION
Candidate for Bachelor of Science degree in Electrical Engineering from San Diego State University in May 1995.

WORK ACHIEVEMENTS
Administered change order.

Supervised and motivated up to 15 people.

Managed electrical projects with budgets of over one million dollars.

Possess a California State Electrical Contractors license.

Ensured that all systems installed were properly operating.

Possess a working knowledge of the National Electrical Code Book.

Qualified as industrial and commercial journeyman electrician.

Resolved conflicts and discrepancies in plans and specifications.

Responsible for ensuring that projects meet plans, specifications, and deadlines.

Participated in estimating and submitting bids on large electrical projects.

Coordinated time schedules between different trades and activities of my trade.

WORK EXPERIENCE

12/87–1/93	J. S. King Electric, San Diego, California. Foreman/Journeyman electrician.
5/87–12/87	Jones Electric, Denver, Colorado. Journeyman electrician.
3/85–5/87	New Mexico Electric, Albuquerque, New Mexico. Journeyman electrician.

REFERENCES
References available upon request.

EXAMPLE 19–6 One Case: Revised Combination Resume

This resume is the most persuasive choice.

RESUME OF BRENT MOSBARGER
4266 River Drive
San Diego, California 92224
789-7234

EDUCATION
Candidate for Bachelor of Science degree in Electrical Engineering from San Diego State University in May 1995.

TECHNICAL
Possess California State Electrical Contractor's license.

Ensure that all systems installed are properly operating.

Possess a working knowledge of the National Electrical Code Book.

Qualified as industrial and commercial journeyman electrician.

MANAGEMENT
Administered change orders.

Supervised and motivated up to 15 people.

Managed electrical projects with budgets over one million dollars.

Resolved conflicts and discrepancies in plans and specifications.

Responsible for ensuring that projects meet plans, specifications, and deadlines.

Participated in estimating and submitting bids on large electrical projects.

Coordinated time schedules between different trades and activities of my trade.

WORK EXPERIENCE
J. S. King Electric, San Diego, California. Foreman/Journeyman electrician.	12/87–1/93
Jones Electric, Denver, Colorado Journeyman electrician.	5/87–12/87
New Mexico Electric, Albuquerque, New Mexico Journeyman electrician	3/85–5/87

REFERENCES
References available upon request.

you can see, the information you generate on the self-analysis worksheet helps you select the skills and accomplishments you choose to highlight. You can vary this emphasis quite easily once you've completed the worksheet.

The **letter of application,** sometimes called a cover letter, accompanies your resume. It is best written after you have written your resume. The letter of application, like the letter of transmittal (see Chapter 18), serves as a tool to transmit the resume to the prospective employers and to emphasize to readers significant details on the resume—those skills and accomplishments that relate to the specific position and company addressed. The letter of application is asking for something—in this case, a job interview—so you want to be persuasive. You need to establish an appropriate *ethos* and use good reasons as to why the reader would benefit from hiring you (try some *because-clauses*). And as in the letter of recommendation (see Chapter 18), you are evaluating—yourself this time—and you want to be concrete and specific. The content and organization of your application are designed to do the following:

Writing the Letter of Application

letter of application
a cover letter accompanying the resume that summarizes significant details of the resume

- Create a successful and immediate link to reader.
- Emphasize the skills, experiences, and accomplishments you have to offer the prospective employer.
- Request action—a job interview.

CREATING SUCCESSFUL LINKS

To create an immediate connection with your reader, address the letter to a particular person. You may need to call the company and talk to a receptionist or operator to identify who will be receiving the letter. Be sure to ask about pronunciation, spelling, and preferred title of your reader while you are at it. Often, the job ad itself will include a contact name. Use it in the address and the salutation and, of course, spell it correctly. If the ad lists an S. Baker as the contact and you do not know which title to use in the salutation, use the full name or call the company to ask what the S. stands for.

In the first paragraph, often in the first sentence, mention how you learned about the position. If someone has told you to apply, mention this referral. If you are answering an ad, name the source of the ad (remember: names of newspapers and trade journals are in italics or underlined, months should be spelled out, and dates require special punctuation). You want to make clear in this beginning that you are applying for a specific position. The following example gets the job done directly and simply:

> I read your ad for the position of a nutritionist in the May 3, 1993, issue of the *San Diego Tribune.* I would like to be considered an applicant for this position.

However, this opening sentence is typical, and the person receiving letters of application will probably see dozens if not hundreds of similar opening sentences. Try writing a more interesting beginning, one that still does the job but grabs the attention of the reader, setting your letter and you, of course, off from all the others, as in the following:

> Your company's advertisement for a nutritionist in the May 3, 1993, issue of the *San Diego Tribune* calls for someone with a combination of scientific and interpersonal skills. My B.S. degree with a double major in nutrition and biology and my experience working with children and older clients provide me with the qualifications you are seeking.

Notice also that this opening leads the reader smoothly into the next section of the letter, a demonstration that the education and experience do indeed match those desired. This paragraph also demonstrates the "you" approach, addressing the reader's needs rather than your own. The last example especially emphasizes the "Here's what I can offer you" rather than what the first example suggests: "Here's what I want."

If you are not responding to an ad or able to mention a referral in the beginning of your letter, you may consider mentioning a fact about the company you have learned as a result of preparing for the job search. Did you read a recent report about a product or process the company is working on? Did you read about a new service that requires the experience, interest, or education that you can offer the company? Mention the information and where you read it. Most readers will be impressed with your thoroughness, especially if you can demonstrate sincerely and enthusiastically how your background and this information are linked, as in the following:

> A recent article in *HNS News* detailed the advancements Hughes Network Systems is making in digital cellular communications. I have a career interest in this technology and believe that both my education and work experience qualify me for a position as engineering technician in your department.

Such openings encourage the reader to pay more attention to what you say in the rest of your letter.

EMPHASIZING WHAT YOU HAVE TO OFFER

The "you" approach is especially critical in the letter of application. Here you must discuss your skills, talents, experience, and education so that your reader will invite you to an interview. It is difficult to write about yourself with a confident tone without appearing to be bragging, but still you must discuss what sets you apart from all the rest and how your skills and interests will benefit your potential employers. Remember that you need proof—

concrete, detailed, good reasons—as to why you should be hired (or interviewed). For example, rather than saying that other people love to work with you, you can state: "On our last seven international proposal projects, I was specifically added to six of the groups at the committee leaders' requests."

In two or three short paragraphs, you must select concrete and specific details that illustrate the skills, experience, and education you have to offer. In Example 19-7, the writer lists his qualifications but not much more than what is already on his resume. What do you think will make the letter more persuasive?

The revised draft of the same letter of application (Example 19-8) is slightly more persuasive. Although it still suffers from some choppiness, the first paragraph now provides a good link to the detail of the middle paragraphs. The last sentence also now reflects the "you" approach, stressing what the job applicant has to offer, rather than what the job applicant wants. How do you think the letter could be improved further?

Notice in the final version of the letter of application (Example 19-9) that the writer has managed to provide more persuasive detail, emphasizing his qualifications match the job mentioned. Early in the letter, the writer directs the reader's attention to the resume and the additional details offered in the resume. Rather than simply stating that the resume is enclosed, the writer points to good reasons in the resume for considering him. He is experienced in "C" programming both at school and on the job, skills required for the Computer Sciences Corporation position. Also, the writer carefully establishes his *ethos.* He demonstrates his interest and professionalism by researching the position and analyzing his own qualifications.

Rather than repeating in your cover letter what is on the resume, in essence you want to "teach" your reader how to understand the information provided in the resume. For example, you might add up the years of experience from the jobs you have listed in your resume and call the reader's attention to that impressive length of time in your letter. In doing so, you are persuading the reader that your qualifications match the reader's needs.

ASKING FOR ACTION

The last paragraph of your cover letter should be a call for action, a request for the interview in which you can demonstrate that you are as impressive in person as you are on paper. Some writers may feel comfortable being assertive here and saying, "I will call you next week to set up an appointment for an interview." Be aware that just one word can change the tone of the ending. Notice the difference in these two sentences:

- *If* we arrange an interview, I can give you more information about my qualifications. [This leaves the action up to the reader.]

EXAMPLE 19–7   First Draft of Letter of Application

This letter is ineffective in highlighting resume and persuading reader.

4810 Campanile Drive
San Diego, CA 92155
August 9, 1994

Ms. Cheryl Herbers
Computer Sciences Corporation
Applied Technology Division
4045 Hancock Street
San Diego, CA 92110

Dear Ms. Herbers:

I saw your job advertisement for UNIX system programmers in Sunday's issue of the *Union-Tribune*. I would like to work for Computer Sciences Corporation as a programmer.

In December, I will graduate from San Diego State University with a Bachelor of Science degree in Electrical Engineering. I have taken classes in computer organization and in "C" programming. I also enjoy programming "C" in my leisure.

Since December of 1990, I have held a job requiring the use of IBM personal computers. I also have experience working in the UNIX and Cyber environments. This work experience and my coursework have well suited me for a programming position.

I have included with this letter my resume. There you will find more information about my coursework, skills, and experience.

I will be available mornings and afternoons until August 30. Would it be possible to schedule an interview?

Sincerely,

Brent W. Warren

Brent W. Warren

Enclosure

EXAMPLE 19–8   Second Draft of Letter of Application

This letter is more persuasive but could still be improved.

4810 Campanile Drive
San Diego, CA 92155
August 9, 1994

Ms. Cheryl Herbers
Computer Sciences Corporation
Applied Technology Division
4045 Hancock Street
San Diego, CA 92110

Dear Ms. Herbers:

In Sunday's issue of the *Union-Tribune,* I found your job advertisement for UNIX system programmers. My experience in "C" programming and with the UNIX operating system has well suited me for a programming position.

I gained this experience at SDSU, where I will be graduating in December 1994 with a Bachelor of Science degree in Electrical Engineering. I have taken several classes requiring "C" programming skills. One of these classes also required using the UNIX operating system.

I also have experience using IBM-compatible personal computers. Since December 1990, I have held a job requiring the use of these types of computers. I have also had a personal computer at my home for several years. On this computer, I program in Quick C as a leisure activity.

I have included my resume with this letter. On it you will find more information about my coursework, skills, and experience.

I will be available mornings and afternoons until August 30. Would it be possible to schedule an interview? Please call me at 619-333-4848.

Sincerely,

Brent W. Warren

Brent W. Warren

Enclosure

EXAMPLE 19–9   Third Draft of Letter of Application

This letter is the most persuasive version.

4810 Campanile Drive
San Diego, CA 92115
August 9, 1994

Ms. Cheryl Herbers
Computer Sciences Corporation
Applied Technology Division
4045 Hancock Street
San Diego, CA 92110

Dear Ms. Herbers:

In Sunday's issue of the *Union-Tribune,* I found your job advertisement for UNIX system programmers. My experience in "C" programming and with the UNIX operating system has prepared me for the position at Computer Sciences Corporation.

As you can see from the enclosed resume, I gained this experience at SDSU. I will be graduating in December 1994 with a Bachelor of Science degree in Electrical Engineering. During my studies, I have taken seven classes requiring both "C" programming and the UNIX operating system. In one class, I wrote a program that broke down enrollment figures of various courses into categories. The program then calculated grade point averages and unit totals in each category and printed the results to the screen in an attractive format using graphics routines.

In addition to "C" programming experience, I have used IBM compatible personal computers both on the job and at home. Since December 1990, I have run income tax programs and entered data into an income tax database. For several years, I have also used a personal computer at home. On this computer, I program in Quick C as a leisure activity.

I believe my interest and experience qualify me for a position as an UNIX system programmer. May I have an interview to discuss my qualifications with you in more detail? Please call me at 619-555-4848. I am available mornings and afternoons until August 30; after that I can most easily be reached in the early mornings.

Sincerely,

Brent W. Warren

Brent W. Warren

Enclosure: Resume

- *When* we arrange an interview, I can give you more information about my qualifications. [This assumes the action will take place.]

Whatever your approach, end by making it easy for the reader to contact you. Even though your telephone number is on the resume, provide it in the letter. Let the reader know a good time to reach you. If you are applying for a position out of town, you may want to let the reader know when you will be in the vicinity. Some companies may not want to pay for you to come in for an interview, but if you will be in their location at a particular time, they may be willing to see you. It's better not to thank the reader in advance for anything, and beware of using stereotyped expressions found often in letters like "at your convenience," "please do not hesitate to call," or "feel free." Finally, as the letter in Example 19–9 illustrates, the enclosure notation after the signature block indicates your resume.

Preparing for the Job Interview

You've already done a good deal of work to prepare for a job interview. In completing your self-analysis, you have gathered some of the facts you will need in your interview: names, addresses and dates for your work history and references; dates and details about education and training (including paperwork supporting this information such as copies of certificates and required licenses); skills and capabilities you want to emphasize, as shown on your resume. You have also gathered information about the companies to which you have applied.

Now it's time to do further research for the interview itself. You want to learn as much as you can about the firm interviewing you (and remember— it's also the firm you are interviewing): know well what kinds of jobs you can do for your prospective employer and what products or services the firm offers. Outline specifically what you have to offer, especially those points you want to be sure to cover in the interview. Develop a list of appropriate questions to ask during the interview, but avoid salary or fringe benefits questions at this point. You can take your list of questions to the interview or even prepare a briefing sheet for your interviewer to focus your discussion.

Anticipate questions you may be asked and how you will respond. Some likely questions include the following:

- Tell me something about yourself (usually at the beginning of the interview to set you at ease). However, remember to avoid supplying irrelevant information in response to this question.
- Why do you want to work for us? (A great opportunity to show you have done your homework.)
- What are your strengths?

- What are your weaknesses? (A tough question to handle. Do you have a weakness that is really a strength? For example, are you a perfectionist about your work, wanting to do the best job possible at all times? When discussing a perceived weakness, it's always a good idea to discuss as well what you are doing to remedy the situation or how a weakness may somehow work in your favor.)
- What are the favorite courses you've taken? In which ones did you make your best grades?
- What work experiences have you had that are particularly meaningful?
- What interesting projects did you complete? What skills have you learned?
- Where do you see yourself in five years? Ten years?

Most of these typical questions don't have right answers. In most cases, the interviewer just wants to see what you will do with the questions. Usually, the resume acts as an outline for the interview. If your resume discusses a particular job you have had, anticipate such questions as, "I see you worked for four years with General Dynamics, Electronic Division, as a technician. What were some of your major activities? What accomplishments are you most proud of? Why did you leave?"

Remember that you have rights as an interviewee. Companies are not allowed to discriminate against candidates. Decisions about whether or not to hire you must be made on the basis of your qualifications rather than on personal items. Therefore, you should be wary of any company that asks questions about your marital status, your plans to have children, your age, or your race, religion, or ethnicity. If you have special concerns about a spouse being able to find work, for example, you can go ahead and ask those questions. Once you have brought up this subject (or any related one), the company is then free to pursue that line of questioning to a certain extent. Companies often try to be sensitive to issues of family life and disability, so if you have an interview you might be asked if you require any special accommodations such as a wheelchair accessible room or the like. Don't be afraid to ask for what you need.

If you have any questions about the most current equal opportunity and affirmative action guidelines in your state, ask the reference librarian for information on the subject as you plan for your interview. The following list offers some questions that are lawful and unlawful for interviewers to ask, topics that interviewers may not raise, and documents that interviewers may not require legally:

Unlawful Questions
Are you naturalized or native-born?

When did you acquire citizenship?

Are your parents or spouse naturalized or native-born?

When did your parents or spouse acquire citizenship?

What is your first language?

How did you acquire your ability to speak, write, or read a foreign language?

What is your first language?

Do you wish to be addressed as Mr.? Mrs.? Miss? Ms.?

Are you married? Single? Divorced? Separated?

Where does your spouse work?

What are the ages of your children, if any?

How old are you?

What is the date of your birth?

Do you have a disability?

Have you ever been treated for any of the following diseases?

Have you ever been arrested?

Unlawful Documents
Proof of age, birth certificate, or naturalization record.

Unlawful Topics
Birthplace of applicant and relatives.

Religious denomination, affiliation, church, parish, pastor, or religious holidays observed.

Lineage, ancestry, national origin, descent, parentage, or nationality.

Nationality of applicant's parents or spouse.

Name, addresses, ages, number, or other information about spouse, children, or other relatives *not employed by the company.*

Applicant's *general* military experience.

Clubs, societies, or lodges of applicant (except for professional societies).

Lawful Questions
Are you a citizen of the United States?

Do you intend to become a citizen of the United States?

If you are not a citizen, have you the legal right to remain in the United States?

Do you intend to remain permanently in the United States?

Are you between 18 and 65 years of age? If not, state your age.

Do you have any impairments—physical, mental, or medical—which would *interfere* with your ability to perform the job for which you are applying?

Have you ever been convicted of a crime? (give details)

Lawful Topics

Languages applicant speaks and writes fluently.

Applicant's academic, vocational, or professional education and the public and private schools attended.

Applicant's work experience.

Name, addresses, ages, number, or other information about spouse, children, or other relatives *already employed by the company.*

Applicant's US military experience or experience in the state militia.

Applicant's service in particular branch of the armed forces.

Applicant's membership in any organization which the applicant considers relevant to ability to perform on the job.

If you are asked an unlawful question during an interview, seek the advice of your career counselor or major professor.

Plan what you will wear at your interview and dress appropriately. Avoid something too flashy or too casual. If possible, visit the job site and observe the dress of employees; then dress accordingly.

Prepare a briefcase or a file folder with materials you plan to take on the interview. Include extra resumes; names, addresses, and telephone numbers of references and employers; and paper and pen. Although you probably will not take notes during the interview, you'll want to jot down as much as you can remember about the interview immediately afterward: names of people you met; topics discussed; and details you learned about the job, company, and interviewer. You may want to make up a portfolio to take with you to the interview. Include in it work samples, writing samples, letters of recommendation, copies of licenses or certificates—anything you think will help you show employers what you have done and can do. Arrange all your papers so you can find them easily.

Practice the interview. If your placement center has workshops for the job interview, videotape yourself in a mock interview to identify mannerisms you may not be aware of. Do you fidget while you talk, for example, rubbing your nose or stroking your chin? Do you jingle change in your pocket, tap your finger or pen on desktops, fiddle with your hair? Practicing an interview in front of a mirror can also be helpful. Interview for jobs every chance you have, but don't waste your time and others' by interviewing for jobs you know you would not take. The more experience you have, the more relaxed you will be. Talk to others who have been interviewed to find out about their experience.

Have an idea of what to expect by becoming familiar with interview routines. Some interviews are brief, lasting 20 to 30 minutes, especially first ones. Others may consume an entire day or two, in which you're being introduced to a number of different people. Even if your travel to the interview has been paid for by the company, don't make long distance telephone calls at the company's expense or abuse the snack and drink refrigerator in your hotel room.

Succeeding at the Job Interview

With your preparation done, you will find the interview far less stressful than you might imagine. The most important rule for an interview is to be yourself. Acting a role is not going to fool an experienced interviewer. Of course, you will want to act professionally: you will want to arrive promptly, perhaps five or ten minutes early, dress appropriately, be friendly but not too casual, and have your thoughts and questions carefully organized. Because you have done your homework, try to relax and enjoy the opportunity to tell the interviewer about yourself and what you can do. Many say the first 20 seconds of an interview are crucial for making a positive first impression. Take cues from your interviewer: wait until you are asked to sit down, for example. It is generally best to take the initiative when you are shaking hands, especially if you are a woman. Some male interviewers may think it impolite for them to initiate a handshake. Don't allow awkwardness to intrude into these first few seconds of the interview process. When you shake hands, do so firmly. Greet the interviewer by name if you are sure of the pronunciation. Smile. Look your interviewer in the eye.

When the questions begin, answer questions fully and honestly. If you don't know an answer, say so. Try not to answer questions with either a yes or no, but also don't talk too much. Listen attentively and show your interest and enthusiasm. You'll want to make lots of eye contact with the interviewer, sit up straight, appear poised and at ease. Don't chew gum or smoke, even if invited to. If you are asked a question such as, "Tell me something about yourself," bring the conversation around to those things you know you want to talk about. Ask in return, "Would you like to hear how my educational background provides important skills for this position?" By countering open-ended questions like this with a question, you can often be sure the interview stays on your target: tell the interviewer what you have to offer. Some interviewers are not terribly skilled at interviewing. They sometimes talk excessively about themselves, their jobs, or the company. By preparing for the interview, you know exactly what facts you want to be sure to cover. Use every opportunity to draw the conversation into these areas.

Try to be positive and courteous at all times. Don't whine or knock other employers. Keep away from personal matters: don't discuss your marriage,

EXAMPLE 19–10   The Follow-Up Letter

143 South Camino del Rio
San Diego, CA 94623
September 21, 1994

Ms. Julie Wesson
Southwest Research Institute
P. O. Drawer 28510
San Antonio, TX 78228-0510

Dear Ms. Wesson:

Thank you for the time you spent last Wednesday discussing the instructional designer position with me. I appreciate the detailed description you gave me of the position and its responsibilities, especially because my interest and educational experience so closely parallel it.

As you asked during the interview, I am enclosing the instructional video that I scripted, storyboarded, and produced, transporting pictures, sound, and animation from one medium to another. I believe this video reflects the experience I have had in analyzing, designing, developing, and evaluating instructional design projects for the past two years.

Once again, thank you for your interest in me and my qualifications and for taking the time to talk to me. If you would like any further information, please call me. I sincerely hope to hear from you soon.

Very truly yours,

Jesse Wilson

Jesse Wilson

Enclosure: Video

---

financial condition, or your politics. Be alert to hints that the interview is over and before leaving thank the interviewer for the interview.

Sometimes you will be interviewed by a panel of three or four interviewers. For management positions especially, some interviewers may try to test you in a stress interview, being antagonistic or deliberately placing you in a stressful situation to see how you will react. Sometimes panels assign members to assume antagonistic roles with other members becoming rescuers when the applicant gets into trouble. Such tactics are not usual, but should you face situations that you find unnaturally stressful, realize you are being tested. Try to remain calm and in control of yourself.

Immediately after the interview, find an opportunity to write notes about the interview. Jot down important details, like names, titles, and job descriptions you will want to remember.

After an interview, especially one that went well, you will want to write a **follow-up letter** (Example 19–10). A follow-up letter, usually not more than three paragraphs, thanks the interviewer for the interview, makes a specific reference to something that you and the interviewer might have discussed, and emphasizes again your interest in the position and what you can offer the company.

follow up letter
a letter sent after a job interview that emphasizes the applicant's continued interest in the position

Sometimes during the interview additional information is requested. Such a situation provides a perfect opportunity, of course, for writing a follow-up letter. Enclosing the information with the letter, you can make the letter a natural development of the interview. You will want to create the same kind of connection for situations that have not developed from the interview. Perhaps you will want to call to the interviewer's attention pertinent facts that you want to emphasize once more. You may have discovered during the interview a mutual interest in sports or hobbies. Mentioning such one-of-a-kind information creates in the interviewer's mind a personal picture of you that will make you stand out from all the competition. Establish courteous but personal *ethos*.

SUMMARY

In this chapter, we looked at the job search process and the materials you will have to prepare to carry it out successfully. A successful job search begins with careful preparation on your part. Some of the preparation involves analyzing your own strengths and weaknesses as well as your experience and background. Another part of the preparation is finding the jobs to apply for.

After you have identified a possible job lead, you will need to make successful contact with a potential employer by creating an effective resume and letter of application. You have several resume formats to choose from and are now more aware of the advantages and disadvantages of those choices. Always keep a current resume. Your letter of application should be persuasive and reader-oriented. The letter of application teaches the reader how to read your resume in order to highlight your most impressive and relevant qualifications.

Once you have been offered an interview, you will have to prepare for it by doing more research on the company, rehearsing your answers, and gathering your materials. After the interview, you will need to write an appropriate follow-up letter.

Many job applicants need to remember that finding the right fit is a two-way street. Not only is the company trying to find the right applicant, but applicants are also trying to find the right job for them. You will receive negative responses. You may even find yourself in the position of realizing that a job may not be the right one for you. The most important thing is not to become discouraged. Be aware that you have much to offer a prospective employer, and applying and interviewing for a position means that you are interviewing the company as well. It's always wise to keep the door open for future positions even though your application for a particular position is not successful.

ACTIVITIES AND EXERCISES

1. Complete the self-analysis in Writing Strategies 19–1.

2. Prepare an interview schedule (a list of questions you would ask) for a job information interview. Ask others about questions you would want to ask about positions that interest you.

3. Study the combination resume in Example 19–6. What suggestions could you make for making it even more effective? Look especially at the verbs used to identify the skills the candidate offers. Can the verbs be made even more dynamic and descriptive?

4. Locate an advertisement for a job for which you would like to apply. Prepare a resume and write a letter of application for the job as advertised.

5. With another student in class, practice the job interview for the position advertised in Exercise 4. These mock interviews could be practiced in front of the class with open class discussion following the mock interview about alternative ways to handle questions.

6. Interview people who have the job you hope to have. Ask them about their resumes, letters of applications, and job interviews. What advice can they offer you about the job search?

7. Assume that you are the public relations director for the university involved in the controversy depicted in Appendix A, Case Documents 3. You are relocating and applying for a similar job in another university. Write an appropriate cover letter to explain your role in this controversy in the best possible light. Decide how to turn this experience into a way of describing your skills and abilities.

   After you have prepared your letter, assume that you have been invited for a two-day job interview. Because you have brought up the controversy in your letter, you will undoubtedly be asked about it. With a classmate, do a mock interview in which you handle questions about this experience and event.

WORKS CITED

Bolles, Richard Nelson. *What Color Is Your Parachute? A Practical Manual for Job-Hunters & Career Changers*. Berkeley, CA: Ten Speed Press, 1988.

Jackson, Thomas. *Guerrilla Tactics in the New Job Market*. 2nd Ed. NY: Bantam, 1991.

Lathrop, Richard. *Who's Hiring Who: How to Find the Best Job Fast*. 12th Ed. Berkeley, CA: Ten Speed Press, 1989.

Oral Presentations

Preparing Oral Presentations ▌ Delivering Oral Presentations

As much time as college-educated workers spend writing, they generally spend more in oral communication.
Paul Anderson, based on a meta-analysis of survey research conducted in workplaces in "What Survey Research Tells Us about Writing at Work." In *Writing in Nonaca-demic Settings,* ed. Lee Odell and Dixie Goswani (New York: The Guilford Press, 1985), pp. 3–83.

Don't tell me the details of how you got the data, just tell me what the data means.
A request from a project manager at Hewlett-Packard Labs upon listening to an extemporaneous presentation, as reported by Frederick Gilbert, "The Technical Presentation," Technical Communication 39.2 (1992), pp. 200–201.

Introduction

Like many people, you may dread the times when you have to stand up before your colleagues or strangers and make presentations. By giving a lot of attention to audience analysis, preparation, practice, and your own *ethos* as a speaker, you can do a great deal to reduce your fears and improve the quality of your oral presentations.

Although the majority of this book has provided instruction on how to design written documents, this chapter helps you communicate technical information through oral presentations suitable to your listeners' needs.

Keep in mind that oral presentations encompass an enormous variety of experiences, including the following:

- A two-minute pitch to sell yourself to a potential employer;
- A technical presentation to a local society concerning the complexities of composting;
- A ten-minute slide show to upper-level managers about a proposed product;
- A scripted talk detailing research results at a professional conference; or
- An impromptu explanation of networking systems to interested colleagues.

To prepare for these varied experiences, you apply the same rhetorical principles for developing written documents; that is, you analyze your purpose and audience, and then you develop your presentation. You need to consider your own knowledge, attitudes, and needs or purpose as well as those of your listeners. Once you complete that analysis, you then use the information you gathered to construct and deliver your presentation.

Preparing Oral Presentations

We begin with a guide for preparing oral presentations (Writing Strategies 20-1). As you might guess, most people focus immediately on the "Text and Delivery" column, determining the content they will include and how they will present this content to their listeners. However, focusing first on the delivery is equivalent to drafting a document without taking time to plan it. Therefore, we recommend that you begin your development process first by considering:

- Knowledge—what you and your listeners know about the subject;
- Attitudes—how you and your listeners feel about the subject; and
- Needs—why you need to communicate this information orally or your purpose in doing so.

Responding to these questions helps you to determine what content you should include and how you will present this information. When you give an oral presentation, you are often persuading; responding to these questions

WRITING STRATEGIES 20–1   Oral Presentation Development Grid

	Speaker	**Listeners**	**Text and Delivery**
Knowledge	What do I know about this subject? Where does my knowledge come from?	What do my listeners know about this subject? What limitations are there to their knowledge?	What content should I include? In what order? How much detail?
Attitudes	How do I feel about this subject? How do I feel about addressing this particular audience?	How do my listeners feel about this subject? Will their reaction be mixed?	How should I present this information (formally or informally)? What kind of visual aid should I use?
Needs	Why do I need to communicate this orally? Am I informing, teaching, or persuading?	Why do my listeners need to hear this? How will they put the information to use? What kinds of approaches would appeal to them?	How will I meet or exceed the listeners' needs and expectations for this presentation?

will also help you present good reasons (*logos*) for your argument, establish your *ethos,* and use *pathos* when appropriate.

Completing this grid also helps you determine exactly what it is that you want to accomplish in your presentation. That is, it helps you to clarify your purpose, your listeners' attitudes and expectations, and the type of oral presentation you wish to deliver. As an example, one speaker responded to these questions as a way to brainstorm how she would design and deliver a presentation on AIDS to a local community group (see Example 20-1). In this case, after brainstorming responses to the questions in the grid, she highlighted specific responses and then used these to develop an outline for her presentation.

CHOOSING A TYPE OF ORAL PRESENTATION

You can construct formal or informal presentations to meet your listeners' needs. Usually, speakers consider four types of oral presentations: the impromptu presentation, the extemporaneous presentation, the scripted pre-

---

EXAMPLE 20–1   A Speaker's Completed Oral Presentation Development Grid

	Speaker	Listeners	Text and Delivery
Knowledge	My own knowledge comes from: Experience Magazine articles Newspapers Pamphlets Medical reports Financial reports	They may know about AIDS in general but I doubt they are aware of the emotional scarring and financial debts the disease leaves behind.	I should include excerpts from my sources as well as my own experience. I should use facts about the rising medical costs that do not increase the quality of care. I should tell how painful it is to watch someone die and not be able to do anything for them, not be able to comfort them in any way.
Attitudes	I feel there should be more financial support of AIDS patients as well as free or low-cost counseling available to family members. I want to reach this audience so I will have to make a personal appeal to them.	Many people feel that AIDS is brought about by carelessness, although many times it is not. Such people feel that the patient is at fault and cannot relate to my point of view. Other people may support my point of view. Regardless of the method of exposure, the outcome is always tragic.	Although I might begin by writing a script, I want to practice so that I won't read the presentation. Especially during my personal stories, I want to stop and just talk to the people. For facts, I want to display actual statistics and graphs (using an overhead projector).
Needs	I feel I need to communicate this because otherwise people will continue their	I think my listeners will take notice because these facts are just too strong to ignore.	I think I will meet the expectations if I can find the statistics to back me up. I want to put some of these

EXAMPLE 20–1   *(continued)*

	Speaker	**Listeners**	**Text and Delivery**
Needs	biased way of thinking and nothing will get done to help the people living with AIDS and their families. The number of people living with AIDS is increasing steadily and in turn so are the families affected by this.	The fact is, anyone can experience AIDS in one way or another. Most can relate to the death of a loved one. Most can also relate to being in debt. Probably not as much as I am referring to, but in debt nonetheless.	statistics on a handout and leave this with people, so that they will remember this information and know how easily they could be affected by AIDS. I want the listeners to share this information with others and just be made aware of the consequences of AIDS.

sentation, and the memorized presentation. Each type has specific advantages that make it the most appropriate for a specific situation. In this section, we define each type of presentation, list some scenarios when it might be used, and note advantages and disadvantages of which you should be aware.

The **impromptu presentation** is one that is delivered at a moment's notice, unprepared and unrehearsed. Impromptu speeches are, by their nature, unplanned. You have no notes and give little thought to organization.

You already make impromptu presentations daily. For example, perhaps a group stops by and asks you to explain why you have chosen a particular computer system. Or a project team decides to brainstorm on a product related to one that you are currently developing and asks you to offer advice.

You will generally use an impromptu presentation when your listeners need immediate feedback from you or when you need immediate feedback from your listeners. Thus, you mainly use it to respond to questions. With an impromptu presentation, you need to "think on your feet" and quickly provide a snapshot of what your listeners need to know or what you need to find out. Listeners want responses to questions, although keep in mind that they may not be asking a precise question because of their lack of knowledge on the subject. Respond with shorter speeches (under two minutes) and follow up with questions to determine your listeners' understanding of the information. Let listeners ask questions to determine the direction of your responses.

Do not use impromptu presentations to deliver complex information, even if the subject is familiar and comfortable for you. Use it to respond to

impromptu presentation
an unrehearsed and unprepared presentation delivered at a moment's notice

questions and brainstorm new ideas. Think of it as a dialogue with your listeners. When you are called upon to give an impromptu presentation, take a few seconds to glance around the room and decide what you really want to say. Be brief and be gracious.

Impromptu presentations obviously take little preparation time. However, you prepare so little that you risk treating your subject in a disorganized or incomplete manner, so monitor your response carefully and be careful not to touch on complex information that would require a more thoughtful, organized, presentation. Generally speaking, though, your audience will understand the circumstances of your presentation and will want to hear information in your own words that can quickly address their questions or confusion.

Many people confuse extemporaneous and impromptu presentations, but they are actually very different. An **extemporaneous presentation** is one that you plan carefully in advance, prepare thoroughly, and then deliver in a relaxed, conversational manner. The best lecturers you heard in college probably used this method of presentation.

extemporaneous presentation
a presentation planned thoroughly in advance and delivered in a conversational manner

The extemporaneous presentation is most preferred by speakers and is the type of oral presentation you are most often called upon to deliver. It allows you the best opportunity to shine as a speaker because it gives you the time to prepare thoroughly and, through careful practice, deliver a relaxed, conversational presentation. When you give this kind of presentation, you'll do so from a set of notes containing your exact ideas, the sequence to be used when presenting the ideas, and the supportive material that you will need. You do not, however, prepare the exact language of the presentation. Some people call the extemporaneous presentation an outlined talk.

You are most likely to design an extemporaneous presentation when you want to have a solid control of your information but also want to keep things on a comfortable, face-to-face level with the audience. Imagine the following scenarios:

- A group of technicians asks you to explain differences in asbestos removal processes.
- A local club asks you to present the current status of drug control efforts in your company.
- Your board of directors asks you to analyze and present current marketing efforts.

In such situations, your listeners want well prepared information, but they want to hear your voice rather than a stilted, read speech. When planning an extemporaneous presentation, respond by using Writing Strategies 20–1 for a thorough knowledge of the subject and your listeners' attitudes and needs.

We suggest that you prepare an extemporaneous presentation in three steps:

1. In an outline or list, write down all of your ideas in the order of their introduction.
2. In a set of brief notes, write down no more than a dozen key words that will act as reminders during your presentation.
3. Practice to work out your general treatment for each part of your talk, decide how to emphasize your main points, and develop transitions that are clear and concise. In addition, use visual aids to guide both you and your listeners.

Extemporaneous talks can be created rather quickly and delivered with a natural speaking voice. Based upon listeners' reactions, you can speed up or slow down, eliminate unnecessary material, or add something that you discover is needed. However, you can easily run over your time limit, leave out crucial information, or encounter difficulty in finding the appropriate phrasing that will explain what you mean accurately and clearly. If you get tongue-tied or tend to forget your message, you may want to opt instead for developing a scripted presentation. Keep a watch handy to make sure you stay within the time allotted.

A **scripted presentation** is a verbatim copy of what you will present. Although initial preparation is similar to that of an extemporaneous speech (that is, you would use Writing Strategies 20–1), the similarity ends there. For a scripted presentation, you determine every word and many gestures in advance, and when you present it, you simply read to your listeners. Even though you will be reading your paper, it's essential that you remember hearing a written paper is very different from reading one, and you need to take listeners' needs into consideration by providing a clear road map of your points.

scripted presentation
a verbatim copy of what you will read orally

You need to design a scripted presentation when you must deliver technical information clearly and accurately. Use it for situations when you must be precise and when small slips in phrasing could be embarrassing or damaging, such as in the following scenarios:

- A professional association accepts your paper on desktop video-conferencing for presentation at a national meeting.
- An NCAA committee asks you to present basic nutritional concepts regarding the use of training tables before collegiate games.
- The Alaska Department of Environmental Conservation (ADEC) asks you to present your firm's research findings on the effectiveness and safety of bioremediation.
- A government commission asks you to testify regarding occurrences of sexual harrassment in the workplace.

Such a presentation demands immense skill so that the manuscript doesn't "sound" like an essay.

Listeners want well prepared, accurate, and clear information on which to take notes and use in the future. Again, respond by using the grid in

Writing Strategies 20–1 for a thorough knowledge of the subject and your listeners' needs. Remember that listeners need more help following an oral argument than readers need following one in print. Prepare a scripted presentation by writing the text much as you would any technical document; however, avoid multisyllabic terms and remember to define new information. Use examples and visual displays whenever possible.

If you expect to be nervous during your presentation, or if you are speaking to government or corporate officials, a scripted presentation ensures that you will communicate clearly and concisely. However, it takes a long time to prepare, and when you are delivering it, you cannot easily adjust it in light of the reactions you get from your listeners.

memorized presentation
a fully scripted presentation that is memorized and delivered as if it were impromptu

A **memorized presentation** is probably the most difficult presentation you will be called upon to give, and fortunately you will not be asked to do so very often. The memorized presentation, like the fully scripted presentation, requires that you prepare fully and carefully much in advance of your delivery. Then, you memorize your entire presentation and practice delivering it as if it were impromptu. The invention of a variety of teleprompting devices means that you will be called upon to deliver memorized presentation rarely, but, there are still occasions when you may be called upon to give short memorized presentations, such as in the following scenarios:

- You are asked to give a eulogy for a colleague who has died.
- The screening committee for the Board of Regents of a college considers you a candidate. They send you the questions they will ask in advance and expect you to address them at the interview.
- Your company is putting together a new public relations video and has asked you to make a short presentation in it.

In preparing a memorized presentation, you need to create a comfortable script that allows you to say things in your own words. You don't want to write a talk that sounds when you deliver it as if you are at your first rehearsal of the school play. Tape record yourself as you practice and make sure you don't sound stilted. Listeners want you to be well prepared but to sound conversational. They are used to watching professionals deliver their lines on television and have relatively high expectations. However, they want to see you as a human being speaking to them, not stiffly reciting lines.

Prepare a memorized presentation by writing the text as you would for a scripted presentation. Use examples and visual aids whenever appropriate. Keep it as short as possible.

The advantage of a memorized presentation is that you can get things exactly right, while still appearing to be speaking casually. Unfortunately, these presentations take a long time to prepare, and they do not allow you flexibility if you are asked a question and have to leave your script and then resume. You can also forget a word or sentence and easily become flustered.

Keep in mind that you can combine any of these presentation types into a speech that works for you. For example, while you might begin by reading from a manuscript, you might allow yourself flexibility when describing an example or relating an anecdote.

FOLLOWING FOUR ESSENTIAL GUIDELINES

Whenever you need to prepare an oral presentation, the following guidelines will help you be more effective and comfortable:

1.   Make it *short*.

- It takes twice as long to listen as to read.
- Stick to a few main points.
- Time your presentation a bit short so that you *can talk slowly*.

Let's say that you need to give a presentation to your supervisors at the Environmental Protection Agency (EPA) about bioremediation in the cleanup efforts in Prince William Sound after the *Exxon Valdez* oil spill. You have 30 minutes, so obviously you will have to choose the most important points to present. Because your audience is knowledgeable about environmental issues and probably somewhat familiar with bioremediation, you can probably focus on their main concern—the safety and effectiveness of the process in this case. Your audience is also concerned about whether it should promote bioremediation in any other situations. In your 30-minute slot, you decide you need to speak for 20 minutes and then leave 10 minutes for questions. When you practice your presentation several times, you find you generally speak for about 18 minutes—a time you are comfortable with because you can then slow down, emphasize, or add any information or definitions if your audience appears confused at any point in the presentation. Running a few seconds short on your speech is seldom a problem; running over time often irritates your audience and damages your *ethos*.

2.   Make the organization obvious.

- Acknowledge your introduction (if you've been introduced).
- Tell your audience what you're going to tell them—capture their attention, define any essential terms, and give necessary background.
- Then tell them—inform them, persuade them, teach them (entertain them too, if appropriate).
- Then tell them what you told them—summarize, conclude, recommend.

In your presentation on bioremediation to the EPA, you need to remind your audience of the definition of bioremediation, preview the points you will cover, and prepare the audience for your conclusions and recommendations. You also want to capture your audience's attention and motivate them to listen to you, a task that should not be too difficult with this knowledgeable and interested audience. Although you are informing them about bioremediation, you want to persuade them to accept your recommendations. Thus, you might begin your presentation as follows:

> Today, I want to address the following question: Do we have an effective and safe method for cleaning up shore-line oil spills? Is it a method that we can recommend or implement with confidence?
>
> I believe that with bioremediation we can say yes to the first question. Bioremediation is effective. Bioremediation is safe. But our yes must remain a cautious one. Bioremediation is safe and effective only in certain circumstances. And, bioremediation must be backed up by other clean-up procedures. Any EPA endorsement of bioremediation must include those other procedures.
>
> I will quickly review the bioremediation process for you, using the cleanup efforts in Prince William Sound following the *Exxon Valdez* oil spill as a test case. First, I will describe the environment of the Sound and the extent of pollution in that environment. Second, I will discuss the safety of bioremediation from two points of view: the precautions necessary to protect the workers involved in the bioremediation process and the environmental consequences of bioremediation. Third, I will assess the effectiveness of bioremediation in removing oil hydrocarbons from the spill site. Finally, I will detail when and to what extent the EPA can feel comfortable recommending and endorsing the bioremediation process.
>
> In reviewing the bioremediation process, let me remind you that bioremediation is basically biodegradation, a natural process. . . .

Your audience will immediately understand your approach to the topic of bioremediation and be looking for the first three parts of your presentation. Also the audience will be expecting you to end with a cautious endorsement of bioremediation and will listen for the benefits you highlight and the qualifications to those benefits. In your introduction, then, you have prepared your audience to *hear* you.

3. Make the ideas simple and vivid.

- Add mini-previews and summaries throughout the presentation to help your audience keep track of where you are and remember what you have said.
- Use obvious and clear transitions. Design your presentation much like a road map with all the turns, curves, and changes in direction and speed obviously marked.
- Repeat your main terms often. Don't rely on pronouns—your audience might lose track of what your "it," "their," "they," and so forth refer to.
- Try to show your audience how things *look*.
- Explain ideas in plain language before using mathematics or formulas.
- Explain the purpose of any procedure or idea.
- Incorporate rhetorical questions to keep your listeners' attention (e.g., So I ask, what are the real needs of our clients?)

As you precede with your presentation, at times you will be teaching your audience about the bioremediation process in Prince William Sound. You will need to describe vividly the environment and the process, such as:

> Typically about *three million oil-eating micro-organisms* exist in every ounce of sediment in Prince William Sound. After bioremediation, *thirty million such organisms* were present in each ounce of sediment. And, the oil-eating ability of *those thirty million micro-organisms* increased *ten fold*.

Changes in your tone or voice level and repeating phrases or using similar structures help emphasize the essential words in your presentation (such as those phrases italicized above).

Transitions, mini-summaries, and mini-previews again help you announce changes in direction within your presentation. For example, after you finish explaining the bioremediation process, you might conclude that part and announce the next part of your presentation:

> Thus, we see that bioremediation in Prince William Sound relied on a greatly increased number of microorganisms to break down oil hydrocarbons into carbon dioxide, water, and microbial biomass. But how safe was this process to the environment and to the workers?
> First, the workers—what application procedures and protective clothing were designed to ensure worker safety . . . ?

Transitions such as *thus, but,* and *first* in the part of your presentation above and your preview of the next part of your presentation help your audience follow your presentation. And, the brief summary of your description of bioremediation will help your audience remember your presentation.

4. Summarize and be ready for questions.

- Repeat your main points in your conclusion.
- Repeat each question for the benefit of the audience.
- Reword clumsy questions.
- Wait until after the speech to give handouts to listeners.

In your presentation on bioremediation, you would conclude with a summary of the three main parts of your presentation and your recommendations as to when bioremediation will be effective, perhaps as follows:

> I have reviewed the process of bioremediation, as used in the Prince William Sound cleanup, as a natural biodegradation process which relies on microorganisms to break down oil hydrocarbons. As we saw, the plant, bird, and animal life, the overall environment of the Sound, was conducive to this process. Worker safety was ensured by protective clothing. Environmental safety was ensured by birds and mammals' instinctive avoidance of unnatural substances. Finally, I have detailed this most effective aspect of bioremediation—the speed with which the oil disappears and the environment returns. The EPA can then be confident in endorsing bioremediation with the five precautions I detailed earlier:

1. Bioremediation must be backed up by hot water spray and use of absorbents.
2. Bioremediation works best after bulk oil is removed by manual and natural processes. . . .

Following these four guidelines—make it short, make the organization obvious, make the ideas simple and vivid, and summarize and be ready for questions—will help you prepare and deliver more effective oral presentations.

CREATING VISUAL AIDS

visual aids
anything provided by the speaker for the audience to look at during a presentation, including, but not limited, to handouts, charts, videotapes, posters, overheads, and slides

Oral presentations nearly always benefit from **visual aids** (see Chapter 12 for specific suggestions on how to develop visual displays of data). A visual aid is anything that you give your listeners to look at during your presentation. It might be something they can touch and hold, such as a handout, or it might be a table, graph, drawing, illustration, map, or photograph presented via slides, overheads, blackboards, posters, videotape, or computer programs.

Remember that there's always the risk of overdoing visual aids. Frederick Gilbert, president of a technical communication consulting firm, describes an all too familiar scene: "A technical expert stands next to the overhead projector with a huge pile of transparencies, reading the hard-to-read text to a half-sleeping audience. This has been called 'the talking overhead projector'" (201). Gilbert's solution is not to eliminate visual aids but rather to hold your listeners' attention by:

- Using fewer rather than more visual aids;
- Using color graphs and charts rather than words;
- Never beginning or ending with visuals; and
- Keeping visuals big and bold. (201)

Never overcrowd your visual aids, a common problem on overhead transparencies, and label clearly all the lines, bars, and other parts of any visual display of data.

Despite the possibility of overdoing it, you run a greater risk by not using visual aids. Raised on MTV, Nintendo™, and computers, audiences increasingly demand a visually oriented presentation. Visual aids can also increase the effectiveness of your presentations by:

- Helping you attract and hold your listeners' attention (you are giving people a place to gaze that is directly related to your message).
- Helping your listeners follow the organization of your presentation and keep track of your main points (you can design a poster that displays the main points of your talk).

- Helping you explain your material (drawings, graphs, charts, and other visual aids communicate material with an economy and effect that you cannot achieve with words alone).
- Helping you remember what you want to say (especially in an extemporaneous presentation, visual aids remind you what you intend to say next and help you not to forget relevant material).

Increasingly, people recommend using a **storyboard** to integrate visual aids into your presentation. You can also use a storyboard to integrate gestures and supporting materials into your presentation. A storyboard is simply a two-column tool developed by people who write movie or video scripts. In the left column, you write a general outline or list (for an extemporaneous presentation) or a script (for a manuscript presentation) of your talk. In the right column, you indicate in words or sketches the visual aids your listeners will see as those words are spoken. Although Example 20–2 shows a rather elaborate storyboard for a manuscript presentation on collaboration via desktop videoconferencing, a storyboard doesn't have to be this elaborate to be effective. You can simply write a few notes in the margins to help you integrate your visual aids into your presentation.

As in any writing or speaking situation, consider your listeners' expectations. If they expect you to present a slick, packaged slide show and you draw lines on an overhead, assume that no matter how excellent your content may be, you are likely to be judged otherwise. Or if you have a set amount of time to present, your listeners may become disturbed if you take time to draw diagrams on a blackboard or overhead. In the information age in which we live, listeners expect you to come prepared (except, of course, in the case of impromptu presentations) and to use innovative technologies to develop and present your speeches (see the section later in this chapter on Employing New Technologies).

storyboard
a two-column script used to prepare oral presentations (as well as video presentations) in which the left column is used for the text and the right column for the visual aids that will accompany it

Delivering Oral Presentations

When you deliver an oral presentation, you are a bit like a stage performer, with some words to deliver and some visual aids to manage. For your delivery to go smoothly, you need to prepare your stage and listeners, recognize special needs of your listeners, and cope with your own anxiety.

When delivering oral presentations, consider the following:

1. Set the *stage* and the *audience*.

   - Get there early and get ready—try out all audio-visual equipment.
   - Remove distractions if you can.
   - Cool off the audience (open windows if you need to).
   - Choose a room that is a little too small.
   - Choose a room with the entrance in the rear.

EXAMPLE 20–2   An Example Storyboard for a Manuscript Presentation

Oral Presentation	*Visual Aids*
Collaboration via Desktop Videoconferencing: Implications for Technical Communication	
Are we satisfied with the ways in which we use technology to collaborate? Are we satisfied that we have enough access to information in our world? Could we link ourselves with other companies?	
I do not think that we should be satisfied; in fact, one purpose for this conference is to examine the past and design the future of technical communication. The purpose of my presentation is to promote innovative ways of using technology for collaborating, to describe one means for linking collaborators and providing access to numerous companies. . . .	Show poster entitled "Collaboration via Desktop Videoconferencing." List purposes for the presentation.
We have taken a configuration designed for business applications—Cameo™ (for audio and video) over an Integrated Services Digital Network (ISDN™ or a broadband services phone line)— and have applied it to our company's needs. Figure 1 shows the basic configuration of the multimedia system [discuss this extemporaneously], and the following video shows how collaborators exchange and view similar texts, all the while seeing each other and verbally discussing their responses. . . .	Show illustration of the desktop videoconferencing system. Show video of two collaborators exchanging information and collaborating.
[the storyboard continues]	

2. Carry an ''Insurance Policy'' (for extemporaneous presentations).

   ■ Put notes on cards and number them clearly in case you drop them.
   ■ List your main points.
   ■ Write on one side only.
   ■ Talk from visual aids and other materials.
   ■ Have a plan in mind if the overhead projector breaks or the bulb in the slide projector burns out.

3. Use *visual aids*.

   ■ Make print on overheads or any visual aids large and easy for the audience to read (practice the document design techniques in Chapter 11 and the visual display principles in Chapter 12).

- Label all overheads and slides and mark your script with when to display and remove them.
- Practice with your visual aids prior to the presentation so that you don't have to adjust to them during your presentation.
- If you are in a big room, ask someone to turn your overheads if necessary so you can stay at the lectern.

4. Talk *enthusiastically* and *slowly.*

- Keep in mind that you are talking to people, not reciting words.
- Use your voice to clarify your message.
- Show enthusiasm and interest.
- Keep good posture.
- Keep your eyes on your listeners' noses or foreheads.
- Place your hands at your sides except when highlighting something on a visual aid or making gestures. If you're nervous, write your gestures on your script (and practice them so they don't look wooden).
- Move on transitions and make purposeful gestures.
- Do not go over your allotted time!

PREPARING YOUR STAGE AND LISTENERS

Your presentation will be more effective if you think deliberately about the arrangement of your stage and the needs of your listeners. If you are going to speak in an unfamiliar location, remember to visit the room beforehand and arrange your presentation area. If you are speaking in a room that is larger than you had expected, make sure that your visual aids will be visible to all listeners and that everyone in your audience can hear you. If you are using slides as visual aids, be sure that you will have a source of light for reading your notes when the room is dark. Most importantly, make sure that all of your props or materials are available and easy to access. If possible, place them on a table in front of you in the order that you will be using them.

One of the best ways to prepare your listeners is simply to look at them before you start to speak. This helps you to create a personal bond with them, gets their attention, and helps establish your *ethos* as a sincere and interested speaker. If you have difficulty looking straight at your listeners' eyes, look at their foreheads or noses. Your peripheral vision will still allow you to recognize their reactions and needs. Throughout your presentation you can continue to look at your listeners (even with a manuscript presentation) in order to see if you need to change your rate of speaking, speak more loudly, or give your listeners more time to study a visual aid. Avoid any disruptive behaviors you might have—fiddling with rings or paper clips, for example.

One way to help your audience is to anticipate the traps they might fall into when listening. Sometimes listeners prejudge ideas and decide to tune out even though they do not yet know the content of your message. Listeners also misplace their attention by concentrating on remembering all the facts rather than important ideas, taking too many notes, or pretending to be listening. Or listeners let trigger words elicit an emotional reaction to your content rather than continuing to listen to your view on the subject.

To help listeners avoid these traps, you can provide handouts that include your major facts, stating that you will distribute these after your presentation. You can also keep listeners' attention by being truly enthusiastic about your content. In addition, if you limit your major points to no more than seven (plus or minus two), you will be keeping within the limits of the human mind to store and process incoming information. When distractions occur— a baby starts to cry, someone mows the lawn outside your window, air conditioning breaks—acknowledge that these conditions are affecting the audience and enlist their good will in your endeavor.

RESPONDING TO QUESTIONS

You also help listeners by responding to their questions either during your presentation (if you have stated that you will allow questions any time during your presentation) or after your presentation is finished. Handling questions can be tricky, so remember to do the following:

- Greet each question politely and listen carefully.
- Repeat the question for the other listeners (this allows you to re-phrase the question to see if you have heard it correctly, and it gives you additional time to develop your response).
- Always be accurate with your answer.
- Answer each question completely while being as brief and specific as possible.
- When you don't understand a question, say so and ask the person to repeat or rephrase it.
- If you don't know the answer, say so.
- When a question seems truly absurd, don't say so; clarify it and continue.
- Don't counter one question with another question.
- Never answer by simply saying yes or no; this quick response of-fends listeners.
- When questions come in the middle of your presentation, use them as transitions if possible. (Some of the preceding advice is from Connolly 34–35.)

Answering questions completely and honestly will maintain the *ethos* you have established while delivering your oral presentation.

DESIGNING A TECHNICAL PRESENTATION ENVIRONMENT

At some point you may be asked to help design a technical presentation environment. Gerald McVey has applied ergonomic and architectural design principles to the design of the technical presentation environment and offers the following advice (24–27).

Choose a room that accommodates the scheduled number of listeners, your designed seating arrangement, and the activities programmed for the room. For example, if you want:

- A classroom style (with chairs), allow 15 square feet per person.
- A classroom style (with tables), allow 25 square feet per person.
- A U-shaped arrangement, allow 35 square feet per person.
- A circular arrangement, allow 42 square feet per person.

For computer terminals, allow an extra 5 square feet to the guidelines above.

The configuration of a room contributes to the effectiveness of your presentation and to the listeners' comfort. The basic dimensions of a room should be 1.2 to 1 (length to width), with seating beginning at a distance of two times the projected image size. Rectangular rooms provide the greatest number of good viewing locations. Trapezoidal rooms work well for special applications. Avoid circular rooms; they have serious acoustical problems. To determine the right height of a room, divide the room length by six and add 4½ feet to that dimension. Then you will be able to use a projection screen at least 4 feet in height.

Chairs should have a height adjustment for a range from 14 to 19 inches. A chair's seat should be 18 to 20 inches wide, 17 inches deep, and sloped and padded. Chairs with short armrests do not interfere with computers and provide additional comfort.

You will have good **acoustics** in rooms with acoustical ceiling tile and floor carpeting. Hard surfaces tend to make sounds bounce around the room. Place lighting on major and supplementary task areas such as chalk or dry-mark boards, desks, and other work surfaces. Listeners give more attention to well-lit displays.

acoustics
the quality of sound in a room

If you want to use color, confine it to surfaces outside the line of sight of the audience (e.g., the back and side walls). Remember that bold colors are stressful; therefore, use off-white or light, chalk green or gray for end walls planned as backgrounds for visual displays. However, be aware that some members of your audience might be color-impaired.

Keep the air temperature within a range of 68° to 74°F and air velocity should be 15 to 25 feet per minute for low-activity rooms, and 25 to 50 feet per minute for rooms with greater activity.

display systems

equipment used to display visual aids, including television monitors, slide projectors, overhead transparency projectors, and plasma displays attached to computers

Display systems include a television monitor, a slide projector, an overhead transparency projector, or plasma displays hooked up to computers. Effective display systems have high legibility, are comfortable when viewed from any angle, have no apparent flicker, and are adaptable for the inclusion of new presentation devices.

RECOGNIZING LISTENERS' SPECIAL NEEDS

hearing impaired people

audience members whose inability to hear may require special development and placement of audio aids and the support of a signer

visually challenged people

audience members who may require special placement and presentation of visual aids as well as alternative media

Although **hearing impaired people** will be able to see your visual aids and view the signing of your presentation (provided that you have planned for a signing professional to be available), speakers sometimes assume that listeners who are visually challenged, because they can hear, will have little trouble understanding your presentation. Quite to the contrary, your visual aids present unique difficulties for **visually challenged people.** To illustrate this fact, turn back to back with a partner, and have one person describe the flow chart of the human information processing system (Example 20–3) while the other person draws it. How much difficulty did you and your partner have?

Parkin and Aldrich studied home readers or people who tape record materials for the visually challenged. They found that the most frequent problem for such readers was communicating information given in maps, diagrams, and other visual displays. In fact, they often found that home readers simply gave up trying to describe visual aids, hoping that the text alone gave sufficient information. To illustrate how difficult it is to describe a visual display of data, Example 20–4 presents a complex graph and its oral description.

Although there are few guidelines for developing oral presentations to be delivered to the visually challenged, the National Braille Association's instructions for reading a table are as follows (this same manual includes suggestions for presenting illustrations, graphs, and diagrams to the visually challenged):

1. Read the title, source, captions, and any explanatory keys.
2. Describe the physical structure of the table. Include the number of columns, the headings of each column and any associated subcolumns, reading from left to right. If column heads have footnotes, read them following each heading.
3. Explain whether the table will be read by rows (horizontally) or by columns (vertically). The horizontal reading better conveys the content. On rare occasions it is necessary to read a table both ways.
4. Repeat the column headings with the figures under them for the first two rows. If the table is long, repeat the headings every fifth row. Always repeat them during the reading of the last row.

EXAMPLE 20–3   Flow Chart of the Human Information Processing System

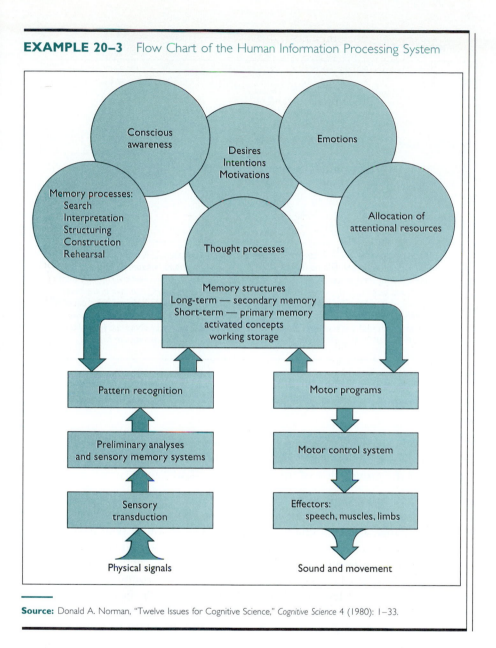

Source: Donald A. Norman, "Twelve Issues for Cognitive Science," *Cognitive Science* 4 (1980): 1–33.

EXAMPLE 20–4   A Complex Graph and Its Oral Description

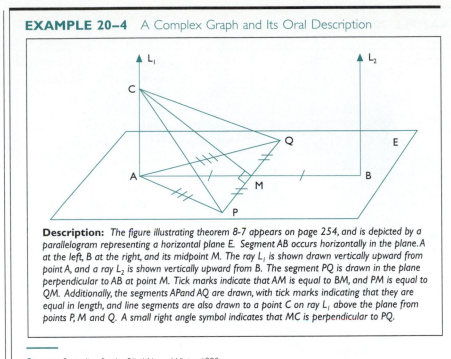

Description: *The figure illustrating theorem 8-7 appears on page 254, and is depicted by a parallelogram representing a horizontal plane E. Segment AB occurs horizontally in the plane. A at the left, B at the right, and its midpoint M. The ray L_1 is shown drawn vertically upward from point A, and a ray L_2 is shown vertically upward from B. The segment PQ is drawn in the plane perpendicular to AB at point M. Tick marks indicate that AM is equal to BM, and PM is equal to QM. Additionally, the segments AP and AQ are drawn, with tick marks indicating that they are equal in length, and line segments are also drawn to a point C on ray L_1 above the plane from points P, M and Q. A small right angle symbol indicates that MC is perpendicular to PQ.*

Source: *Recording for the Blind News,* Winter 1990.

5. Indicate the last row by saying "and finally . . ." or "last row . . .".

6. At the completion of the reading say, "End Table 3."

As in any oral presentation, listeners have varied skills. Those who have become visually challenged later in life will have conceptions of what figures, tables, diagrams, and flow charts look like; however, those who have been visually challenged from birth will have greater difficulty with visual arrangements and the visual imagery often used to describe them (Hartley). Of course, you'll also need to consider how cultural background and ability to read and understand English might make a difference in how you deliver your oral presentation. Therefore, be aware of these listeners' needs and plan for *all* members of your audience, and your attention to these needs will be appreciated.

public speaking anxiety (PSA)
a common response to having to present material orally in front of others

COPING WITH ANXIETY

Communication specialists have long been interested in how to help speakers cope with **public speaking anxiety,** often referred to as PSA. Scholars have developed three very different approaches to help speakers cope with anxiety:

- Behavioral—This approach provides speakers with additional skills or training in preparing for and delivering oral presentations.
- Cognitive—This approach focuses on changing negative or irrational thoughts about speaking to rational or positive ones.
- Affective—This approach focuses on helping speakers associate positive feelings with public speaking.

Research indicates that all three approaches work (Allen, Hunter, and Donohue). This research also shows that using two approaches is superior to using only one. Thus, to cope with anxiety, we would recommend that you use the grid presented in Writing Strategies 20–1 to help you prepare for your presentation (a behavioral approach), that you write down your negative or irrational thoughts about speaking in one column and work with others to list comparative rational or positive thoughts in the right column (a cognitive approach), and that you begin to view the impromptu presentations that you make daily as a positive form of public speaking (an affective approach).

Remember: Don't allow irrational thoughts to conquer your confidence; rather, prepare your presentation and be fully satisfied if your public "performance" comes within 80 percent of your best practiced performance.

Employing New Technologies

New technologies are constantly being created to aid face-to-face presentations and presentations across distances. Keep in mind that the minute this textbook hits the press, new technologies may well have replaced the ones described below. However, our description will give you an idea of how you might use these technologies in oral presentations. Even if your school does not have these technologies yet, your workplace most likely will.

Numerous software programs can help you present your information publicly. **Presentation support software** often includes graphics programs and other desktop publishing software. In some cases, you can print out your presentation and then make overheads for public display.

If you have a computer projection system available for your presentation, you can design a computerized presentation (PowerPoint™ is such a tool). Hypercard™ can also be used to develop a series of "cards" that can be linked and cross-referenced very easily. You could, in effect, make each notecard in your presentation a Hypercard card. When delivering your presentation, you can access these cards and project only those cards needed for your presentation. Or, if you remember a specific card discussed at one meeting and recorded your meeting on a card, you could search all the cards

presentation support software
graphics programs and other desktop publishing software used to develop visual aids for oral presentations

and locate your idea and place it in your final presentation. The value of this computer technology is that the electronic text you have developed can now be used in your oral presentation.

Other emerging technologies that are now available include the following:

- Cameraman™ by Parkervision. Cameraman hooks up to a camcorder or video camera and enables you to make a video recording of your talk without anyone behind the camera. You can hook up Cameraman for a practice session and then sit back and evaluate your presentation on video.
- Pointmaker™ by Boeckeler Instruments. Pointmaker allows you to draw arrows, circles, lines, and freehand drawings on video presentations.
- AirMouse™ by Selectech Technologies. AirMouse is a hand-held, remote-control mouse that needs no pad or hard surface to operate. As a presenter, you can point and choose from displayed menus remotely rather than running back and forth to a computer to continue your presentation.
- Laser Pointer-2000™ by Laserex. Laser pointers allow you to point out information on display from a small device in the palm of your hand.

videoconferences
use of television cameras and monitors and digital transmission devices to provide interactive video and audio for meetings at distant sites

Videoconferences provide speakers and listeners with simultaneous audio and video across distances; that is, you see and hear each other simultaneously from different locations. According to Doug Robertson and Rowan Carroll,

> Videoconferencing, once priced out of reach of most users, is now within the grasp of a broader audience. With an installed base of almost 15,000 units, and the industry posting 70 percent growth rates for the last three years, 2,000 companies have turned to videoconferencing to overcome competitive pressures and to improve global communication. Videoconferencing is hot and the temperature is rising. (32)

Videoconferencing systems combine television monitors and cameras, special coder/decoder systems called codecs, and digital transmission services from long distance carriers and local telephone companies to provide full interactive video and audio meetings for speakers and listeners. Thus, as a speaker you might be located in Minneapolis while your listeners are located in Orlando. With the price of these systems dropping, advice for delivering presentations will change in the future as speakers view listeners on television monitors.

Videoconferencing systems have also been developed for use from your desktop computer. Desktop videoconferencing units such as FlexCam™ by VideoLabs or Cameo™ by Compression Labs integrate camera and microphone systems that run on computers. This means that you can deliver your presentation live from your computer to anyone else who has a computer equipped with videoconferencing equipment and connected to digital transmission services. The University of Minnesota, for example, has used the Cameo system and the Integrated Services Digital Network (ISDN™) phone service to allow university students to mentor high school students across distances (see Duin et al.). Essentially, the university students have acted as speakers in highly interactive impromptu presentations with the high school students.

If you are thinking that you will never use these technologies, think again. Technologies will continue to be designed to bring speakers and listeners closer together across distances, and as we increase our use of desktop videoconferencing, the demand for clear impromptu presentations and accurate extemporaneous presentations will also increase. Therefore, remain aware of new technologies and plan for using them throughout your professional career.

SUMMARY

This chapter provides you with a basic oral presentation grid (Writing Strategies 20–1) to analyze your and your listeners' knowledge, attitudes, and needs before designing the content of your presentation and the delivery of that content.

Oral presentations include impromptu, extemporaneous, scripted, and memorized presentations. The impromptu presentation is one that is delivered at a moment's notice, unprepared and unrehearsed. The extemporaneous presentation is delivered from a set of notes and any supporting materials you need. For a scripted presentation, you determine every word and gesture in advance, and when you present it, you simply read to your listeners. Memorized presentations require that you prepare your speech and then memorize it. Speakers most prefer the extemporaneous presentation because it allows them to respond to their listeners and speak in a conversational tone of voice.

When preparing your presentation, remember to keep it short, make the organization obvious, make the ideas simple and vivid, and summarize and be ready for questions. Because oral presentations nearly always benefit from visual aids, use visual aids but don't rely totally on them. Also, keep your listeners' special needs in mind, and if listeners are visually challenged, plan for complete oral descriptions of any visual aids used in your presentation. To coordinate your visual aids with your presentation, you can create a storyboard that designates when each visual aid should be used.

When delivering your presentation, remember to set the stage, carry notes on cards (for extemporaneous presentations), use your visual aids, and talk slowly yet with enthusiasm. Think deliberately about the arrangement of the room and the auditory and visual needs of your listeners. When responding to questions, repeat each question and answer it completely while being as brief and specific as possible.

Finally, be on the lookout for new technologies that will aid both the preparation and delivery of your presentations.

1. Look back on the examples from the EPA presentation on the bioremediation cleanup process in Prince William Sound and review Appendix A, Case Documents 2. Using notecards for an extemporaneous oral presentation, prepare the complete presentation and deliver it to your classmates and instructor.

2. You can use the oral presentation development grid in Writing Strategies 20-1 to begin brainstorming for a presentation that you need to design and deliver. If you don't have a specific subject in mind, use any of the case studies in this textbook. For example, you could brainstorm for:

   ■ A presentation to the Coalition of Citizens Concerned with Animal Rights (CCCAR) in which you describe the University of the Midwest's efforts to control birds in the experimental crop fields on its campus (or change the listeners to a group of professors whose raw data comes from the experimental crop fields).
   ■ A presentation to colleagues in your workplace about the impact of product liability suits on the development of documents accompanying your products.
   ■ A presentation to colleagues in your workplace about the need to develop a policy on sexual harassment.

   Again, after you have determined your and your listeners' understanding of and attitude toward the subject as well as your needs, you can then concentrate on determining what content to include and how to deliver the presentation.

3. Locate any visual aid (graph, table, flow chart) and, following the guidelines from the National Braille Association, write out how you would orally present this visual aid to the visually challenged. Bring your description to class and compare with other studies.

4. To gain an understanding of how you might mix and match oral presentation types, work with a partner and determine the type of presentation that you would deliver in each of the scenarios in Example 20-5.

5. Working in small groups, use the advice in this chapter to redesign your classroom as an ideal technical presentation room. Draw the layout and compare with other groups.

6. Note the layout for a computer laboratory in Figure 20-1. Imagine that you are to deliver an oral presentation in this room. Describe the strategies you would use to arrange your "stage" and hold your listeners' attention.

7. The feedback form shown in Example 20-6 was created as a way to have students give each other feedback on their presentations. Critique this feedback form, and, based on your critique and the suggestions in this chapter, design a new feedback form.

8. Locate the Exxon Case Study document entitled, "A Report Detailing the Effectiveness and Safety of Bioremediation in Prince William Sound" in Appendix A, Case Documents 2. Example 20-7 lists all the major headings in the body of this document. Assume that you will have only five minutes to present this information to a local group of Green Peace volunteers. What are your listeners' attitudes and needs? What information/sections will you present orally and why? What

EXAMPLE 20–5 Scenarios for Exercise 4

Context for Speaking	Listeners' Need (1) to learn (2) to do (3) to learn to do something	Type of Presentation (1) impromptu (2) extemporaneous (3) scripted (4) memorized (5) combination
Instructing clients to fill out a purchase agreement form		
Teaching a local community group the best ways to avoid exposure to HIV		
Describing your own rights in a legal contract with a landlord		
Encouraging customers to purchase exercising equipment		
Describing product liability laws to corporate officials		
Encouraging local residents to vote against the building of a nuclear power plant in your county		
Explaining how to solve a telecommunications problem to a staff member		

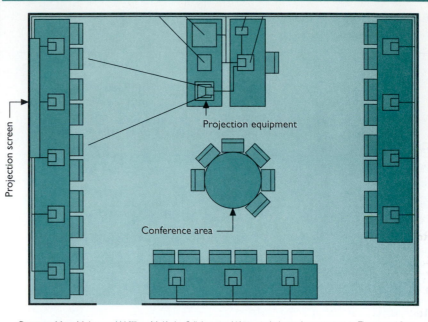

Projection screen

Projection equipment

Conference area

Source: Mary M. Lay and William M. Karis, *Collaborative Writing in Industry: Investigations in Theory and Practice.* Amityville, NY: Baywood Press, 1991. Reprinted with permission.

FIGURE 20–1

Room Layout for a Computer Laboratory

type of presentation will you deliver? How might you structure and deliver this speech? Be innovative.

9. Note the document entitled, "What Are Universal Precautions?" from the case study materials on AIDS in Appendix A, Case Documents 1. Read through this document and prepare *three* different types of presentations:

- A one-minute impromptu speech in response to the question, "What are universal precautions one might take to avoid exposure to HIV?".
- A three-minute extemporaneous speech.
- A five-minute scripted speech.

What are the differences in your preparation and delivery processes in these three types of presentations? Which type of presentation was most effective? Easiest? Most difficult? Best for a classroom audience?

EXAMPLE 20–6   Oral Presentation Feedback Form

Speaker _____   Person giving feedback _____

	Poor	Satisfactory	Good	Excellent

Does the speaker appear:

1. Well prepared and knowledgeable.   ___ ___ ___ ___
2. At ease, yet professional.   ___ ___ ___ ___

Has the speaker:

3. Determined how the *audience* and *purpose* affect the *content* of the presentation.   ___ ___ ___ ___
4. Developed an effective *strategy* for the presentation.   ___ ___ ___ ___
5. Designed effective *visuals/handouts*.   ___ ___ ___ ___

Rate the speaker's:

6. Volume, pitch, clarity, and speed.   ___ ___ ___ ___
7. Verbal fillers (uh, um, okay).   ___ ___ ___ ___
8. Gestures.   ___ ___ ___ ___

Additional Comments:

EXAMPLE 20–7   Headings in Exxon Case Study

How Tiny Organisms Helped Clean Prince William Sound	
First Large-Scale Use	
Accomplishments in Alaska	
Questions and Answers:	
The Process Safe Substances The Role of Bioremediation Outlook for 1991	
Scientific Summary—Test Findings	
Indigenous Organisms of Alaska	
Special Fertilizers	
Project Monitors	

WORKS CITED

Allen, M.; J. Hunter; and W. A. Donohue. "Meta-Analysis of Self-Report Data on the Effectiveness of Public Speaking Anxiety Treatment Techniques." *Communication Education* 38 (1989): 54–76.

Connolly, James E. *Making More Effective Technical Presentations.* Minneapolis, MN: University of Minnesota, 1982.

Duin, Ann Hill; Linda A. Jorn; Craig J. Hansen; Elizabeth Lammers; Lisa Mason; and Sandra Becker. *Mentoring via Telecommunications/Multimedia.* Narrative report submitted as part of an EDUCOM award application, 1993.

Hartley, James. "Presenting Visual Information Orally." *Information Design Journal* 6.3 (1991): 211–220.

McVey, Gerald F. "The Planning and Ergonomic Design of Technical Presentation Rooms." *Technical Communication* 35.1 (1988): 23–28.

Parkin, A. J., and F.K. Aldrich. "How Can Studying from Tape Be Made Easier? *New Beacon* 71 (1987): 340–341.

Robertson, Doug, and Rowan Carroll. "Videoconferencing: Matching Systems and Peripherals with Applications." *Presentation* 27.5 (1993): 32, 35–36.

Tape Recording Manual. 3rd ed. Washington D.C.: National Braille Association. National Library Services for the Blind and Physically Handicapped, Library of Congress, 1979.

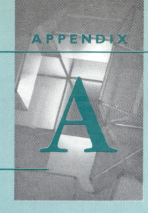
Case Documents 1

Case Documents 1: MAP brochure: used with permission of the Minnesota AIDS Project; Taking Care: used with permission of the author, James Rothenberger, School of Public Health, University of Minnesota; Monthly Surveillance Report: Used with permission of the Minnesota Department of Health AIDS Epidemiology Unit; Tuberculosis and AIDS Project; AZT Fact Sheet: used with permission of the Seattle Treatment Education Project.
 Case Documents 2: all documents used with permission of the Exxon Company, USA.
 Case Documents 4: all documents reprinted with permission from 642 R.2d 339 and 280 S.E. 2d 510,
Copyright © 1981 by West Publishing Company.

What is the Minnesota AIDS Project?

Care, Prevention and Advocacy

The Minnesota AIDS Project (MAP) has long been a leader in the fight against AIDS. We are a statewide, private non-profit agency combining the strength of dedicated volunteers and trained professional staff in this fight. Dedicated to serving those most affected by HIV/AIDS, MAP provides direct service and practical support to partners, families and care-givers. MAP works to stem the spread of HIV infection throughout the state by targeting prevention education and assistance to those most at risk. We provide advocacy on both an individual and policy-making level.

For All Minnesotans

The Minnesota AIDS Project has evolved into a statewide network of volunteers and staff in order to represent and serve all in need. Founded in 1983 by concerned members of the gay community, MAP has grown significantly in response to the AIDS epidemic. Today with a core of 500 volunteers and staff of over 40, MAP has offices located in the Twin Cities, St. Cloud, Duluth, Rochester, and Marshall. MAP represents the rich diversity of Minnesota's many communities.

What does the Minnesota AIDS Project do?

Prevention Education

We inform people about HIV/AIDS, how it is transmitted and how it can be prevented through:

- the AIDSLine, a toll-free information and counseling hotline
- targeted street education
- community education and seminars
- advertising campaigns
- brochures, pamphlets and posters
- practical information about HIV/AIDS

Client Services

We support people living with AIDS and HIV infection, as well as those who care about them through:

- case management
- housing
- legal services
- support groups for people newly diagnosed with HIV
- emergency assistance
- life enhancement activities
- practical support such as transportation and home helpers
- buddy program

What can you do about AIDS?

Protect Yourself and Those You Love

Make a commitment to educate yourself and tell your loved ones about HIV/AIDS. Practice safer sex and do not share needles. Call the AIDSLine with your questions.

Join the Fight

Help confront AIDS in Minnesota by contributing:

- energy, skills, concern and volunteer time
- funds - donations of all sizes are needed in the struggle against AIDS. Individual contributions provide critical support for many programs.

To find out about the services available to your community, as well as how you can help, contact the office of the Minnesota AIDS Project nearest you.

The Minnesota AIDS Project receives 64% of its funding from government grants and the United Way for specific program purposes. Private foundations, corporations, and individual donors contribute 34%. Earnings from outreach and education services total 2%.

A United Way Agency

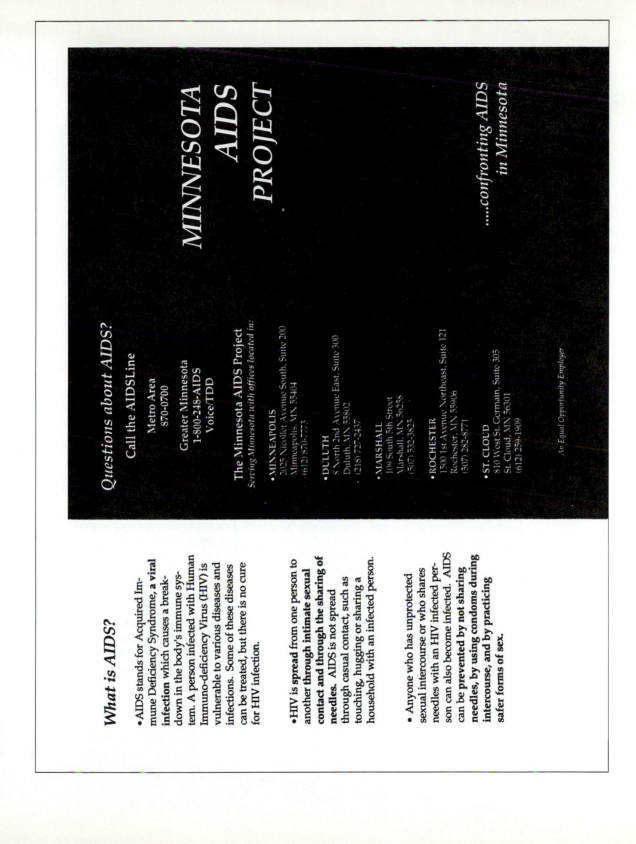

MINNESOTA AIDS PROJECT

.....confronting AIDS in Minnesota

Questions about AIDS?

Call the AIDSLine

Metro Area
870-0700

Greater Minnesota
1-800-248-AIDS
Voice/TDD

The Minnesota AIDS Project
Serving Minnesota with offices located in:

•MINNEAPOLIS
2025 Nicollet Avenue South, Suite 200
Minneapolis, MN 55404
(612) 870-7773

•DULUTH
8 North 2nd Avenue East, Suite 300
Duluth, MN 55802
(218) 727-2437

•MARSHALL
109 South 5th Street
Marshall, MN 56258
(507) 532-3825

•ROCHESTER
1500 1st Avenue Northeast, Suite 121
Rochester, MN 55906
(507) 282-8771

•ST. CLOUD
810 West St. Germain, Suite 305
St. Cloud, MN 56301
(612) 259-1909

An Equal Opportunity Employer

What is AIDS?

•AIDS stands for Acquired Immune Deficiency Syndrome, a viral infection which causes a breakdown in the body's immune system. A person infected with Human Immuno-deficiency Virus (HIV) is vulnerable to various diseases and infections. Some of these diseases can be treated, but there is no cure for HIV infection.

•HIV is spread from one person to another through intimate sexual contact and through the sharing of needles. AIDS is not spread through casual contact, such as touching, hugging or sharing a household with an infected person.

• Anyone who has unprotected sexual intercourse or who shares needles with an HIV infected person can also become infected. AIDS can be prevented by not sharing needles, by using condoms during intercourse, and by practicing safer forms of sex.

Taking Care
A Manual for
AIDS Caregivers

In the next decade, AIDS will affect all Americans. Many of us will be caring for a patient or a loved one who has AIDS.

There are no rules for taking care of a person with AIDS, but there is information in this manual that can help. It was produced by people with years of experience in health care, AIDS education and counseling.

1989 Midwest AIDS Training and Education Center (MATEC)
School of Public Health, University of Minnesota. Funded by the U.S. Public Health Service, Health Resources and Services Administration, Education and Training Center Program, Grant # HHS BRT 000033-01-0.

MATEC Coordinator:
James Rothenberger
Publication Coordinator:
Patricia Ohmans

First, the facts

With AIDS,
it's what you do
that matters.

- AIDS stands for Acquired Immune Deficiency Syndrome. This serious infection, caused by the Human Immune Deficiency Virus (HIV) destroys the body's natural ability to resist infections. People with AIDS become seriously ill and often die of diseases that most people's bodies easily resist. Some of these illnesses can be treated, but there is no cure for HIV infection itself. (This brochure uses "AIDS" when referring to a condition, such as "AIDS patient," and "HIV" when referring to the virus that causes AIDS.)

- Infection with HIV can, by itself, be a serious health condition, but not everyone infected with HIV has AIDS. Some people seem unaffected by the virus. Others develop only limited symptoms. However, anyone infected with HIV can pass the virus on to others.

- People infected with HIV can be male or female. That is why this manual refers to a person with AIDS alternately as "he" or "she." HIV infects white people and people of color, heterosexuals, and homosexuals. With AIDS, it's not who you are, it's what you do that matters.

- HIV is spread through close sexual contact, in which blood, semen or vaginal secretions are exchanged. It is also spread through the shared use of intravenous needles, and perinatally, from mother to child around the time of birth. Some cases of HIV transmission through non-intact skin in contact with HIV-infected blood and through transplantation of infected organs and tissues have also been documented.

- People risk becoming infected with HIV when they have unprotected anal or vaginal intercourse with an infected person, or when they share needles with someone who is infected. A pregnant woman infected with HIV can pass the virus on to her unborn child through shared blood, or to an infant through breastmilk. Before 1985, many people received blood transfusions contaminated with HIV. Now the blood supply is screened and it is safe to receive blood.

What are universal precautions?

They are the best way to avoid exposure to HIV.

What are universal precautions? They are the simplest way to ensure that you will not be infected with HIV when you care for a person with AIDS.

• **Wear protective clothing.**
All health care workers should routinely use appropriate barrier precautions when in contact with blood or other body fluids.

Wear gloves for touching blood and body fluids, mucous membranes, or non-intact skin of all patients, for handling items or surfaces soiled with blood or body fluids, and for performing venipuncture and other vascular access procedures. Change gloves after contact with each patient.

Wear masks and protective eyewear or face shields during procedures that are likely to generate droplets of blood or other body fluids to prevent exposure of mucous membranes of the mouth, nose and eyes. Wear gowns or aprons during procedures that are likely to generate splashes of blood or other body fluids.

• **Wash your hands.**
Wash your hands and other skin surfaces immediately and thoroughly if they are contaminated with blood or other body fluids. Wash your hands immediately after gloves are removed.

• **Be careful with needles and sharps.**
Take precautions to prevent injuries caused by needles, scalpels and other sharp instruments or devices during procedures; when cleaning instruments; during disposal of used needles; and when handling sharp instruments after procedures.

To prevent needlestick injuries, do not recap, bend or break needles. Do not remove them from disposable syringes or otherwise manipulate needles by hand. After they are used, disposable needles and syringes, scalpel blades and other sharp items should be placed in puncture-resistant containers for disposal; the puncture-resistant containers should be located as close as practical to the use area. Large-bore reuseable needles should be placed in a puncture-resistant container.

- **Keep resuscitation devices handy.**
  Although saliva has never been implicated in the spread of HIV, try to minimize the need for direct mouth-to-mouth resuscitation. Keep mouthpieces, resuscitation bags or other ventilation devices available in areas in which the need for resuscitation is predictable.

- **Protect chapped or irritated skin.**
  If you have oozing lesions or dermatitis, wear gloves or avoid direct patient care and patient care equipment until the condition resolves.

- **Be careful if you're pregnant.**
  Pregnant health care workers are not known to be at greater risk of contracting HIV infection than health care workers who are not pregnant. However, if you develop HIV infection during pregnancy, the infant is at risk of infection from perinatal transmission. Because of this risk, pregnant health care workers should strictly adhere to precautions to minimize the risk of HIV transmission.

Wear gloves.
Wash your hands.
Be careful with needles.

Minnesota Department of Health AIDS Epidemiology Unit Pediatric/Adolescent HIV/AIDS Monthly Surveillance Report November 1, 1992

The following five tables summarize information on transmission category, race, and vital status for cases of HIV infection and AIDS in children and adolescents which have been reported to the Minnesota Department of Health. In order to relate the age of the person as closely as possible to age at time of HIV infection, the information reflects age at time of first positive test for HIV infection or time of diagnosis of AIDS. In accordance with the U.S. Centers for Disease Control criteria for pediatric AIDS surveillance, we have defined "pediatric" to include those less than 13 years of age at time of diagnosis or test, and "adolescent" to include those 13 to 19 years of age at the time of diagnosis or test.

TABLE 1   Pediatric and Adolescent (0–19 years of age) AIDS Cases by Exposure Category

Exposure Category	Male Number (%)	Female Number (%)	Total Number (%)
Men who have sex with men	0 (0)	0 (0)	0 (0)
Injecting drug use (IDU)	0 (0)	0 (0)	0 (0)
Men who have sex with men and IDU	0 (0)	0 (0)	0 (0)
Hemophilia/Coagulation disorder	2 (33)	0 (0)	2 (14)
Heterosexual	0 (0)	1 (13)	1 (7)
Transfusion, blood/components	1 (17)	1 (13)	2 (14)
Mother with/at risk for HIV infect.	3 (50)	5 (63)	8 (57)
Other/Undetermined	0 (0)	1 (13)	1 (7)
TOTAL	6 (100)	8 (100)	14 (100)

TABLE 2   Pediatric (<13 years of age) and Adolescent (13–19 years of age) Cases of HIV Infection (Including AIDS) by Race/Ethnicity

Race/Ethnicity	<13 Years Number (%)	13–19 Years Number (%)	Total Number (%)
White (not Hispanic)	18 (56)	37 (59)	55 (59)
Black (not Hispanic)	11 (34)	21 (34)	32 (34)
Hispanic	2 (6)	2 (3)	4 (4)
Asian/Pacific Islander	1 (3)	0 (0)	1 (1)
American Indian/Alaskan Native	0 (0)	2 (3)	2 (2)
TOTAL	32 (100)	62 (100)	94 (100)

TABLE 3   Pediatric (<13 years of age) Cases of HIV Infection (Including AIDS) by Exposure Category

Exposure Category	Male Number (%)	Female Number (%)	Total Number (%)
Hemophilia/Coagulation disorder	7 (37)	0 (0)	7 (22)
Transfusion, blood/components	3 (16)	1 (8)	4 (13)
Mother with/at risk for HIV infect.	9 (47)	11 (85)	20 (63)
Other/Undetermined	0 (0)	1 (8)	1 (3)
Total	19 (100)	13 (100)	32 (100)

TABLE 4   Adolescent (13–19 years of age) Cases of HIV Infection (Including AIDS) by Exposure Category

	Male Number (%)	Female Number (%)	Total Number (%)
Exposure Category*			
Men who have sex with men	28 (60)	0 (0)	28 (48)
Injecting drug use (IDU)	2 (4)	1 (9)	3 (5)
Men who have sex with men and IDU	3 (6)	0 (0)	3 (5)
Hemophilia/Coagulation disorder	13 (28)	0 (0)	13 (22)
Heterosexual	0 (0)	5 (45)	5 (9)
Transfusion, blood/components	0 (0)	1 (9)	1 (2)
Other/Undetermined	1 (2)	4 (36)	5 (9)
TOTAL	47 (100)	11 (100)	58 (100)

* Exposure category data are pending for 3 males and 3 females.

TABLE 5   Pediatric (<13 years of age) and Adolescent (13–19 years of age) Cases of HIV Infection (Including AIDS) by Vital Status

	<13 Years Number (%)	13–19 Years Number (%)	Total Number (%)
Vital status			
Alive	27 (84)	58 (94)	85 (90)
Dead	5 (16)	4 (6)	9 (10)
Total	32 (100)	62 (100)	94 (100)

AIDS PUBLIC POLICY IN MINNESOTA

A series of public policy position papers from the Minnesota AIDS Project

Number 5   August 1992

TUBERCULOSIS AND AIDS

In 1990 the Centers for Disease Control (CDC) received the first report of clusters of drug resistant TB cases among hospitalized AIDS patients.[1] Tuberculosis is an old disease staging a comeback from its recent obscurity. It poses an immediate threat to those infected with HIV and a longer-run threat to the entire public. TB is preventable and largely still curable, but aggressive public health programs must be fielded immediately.

BACKGROUND

Tuberculosis (TB) caused by Mycobacterium tuberculosis, is a contagious, airborne disease of the lungs. Sometimes TB affects other parts of the body. It is a major cause of death in many parts of the world today and was the leading cause of death in the early years of this century in the United States. As recently as 1950 Minnesota reported 5,000 new cases. TB has been called a "social disease with medical aspects" because it can be found wherever poverty, poor nutrition, homelessness, and substance abuse occur.

Robert Koch identified the cause of TB in 1882 in Germany. Control programs have existed for over a century. Clinics, case finding programs, isolation in TB sanitariums, and most important—improvements in the standard of living—helped bring tuberculosis under control. Antibiotics introduced in the 1940s and '50s (streptomycin and isoniazid) were very effective in curing the disease. By the 1960s TB had become relatively uncommon in the United States.

TB causes fatigue, weight loss, coughing, and hemorrhage. These and other symptoms become more severe as the disease progresses. Detection of TB can be done by mass screenings using the Mantoux skin test or chest X-rays. Sputum cultures provide definitive diagnosis but may take weeks to complete. Polymerase chain reaction (PCR) tests provide much faster results but are not ready for general clinic use.[2]

CURRENT STATUS OF TUBERCULOSIS CONTROL EFFORTS IN MINNESOTA

In 1990 there were 25,701 cases of TB in the United states, 9.4% more than in 1989.[3] While AIDS may be poised to make a comeback in many parts of the country, it is not as serious a threat in Minnesota at the present time.

	Number of New Tuberculosis Cases	Rate per 100,000 in 1991
United States	26,283	10.4
Minnesota	102	2.3
Metro Area	45	
Greater Minnesota	47	

In Minnesota, surveillance and monitoring of TB cases is a state responsibility, while TB treatment services are provided by county and city clinics located in the Twin Cities metro area and Rochester. Public Health nurses and physicians provide TB treatment services in most state counties. Federal grants currently supply $86,020 for 2.5 FTE Minnesota Department of Health TB surveillance staff (1992-1993 budget). $30,000 in state funds are available to supply free TB medicine to patients (1991-1992 budget).[4] Current federal spending on TB programs is $20,000,000. Prevention efforts are cost effective, repaying the money spent 4 to 1.[5]

HIV & THE REEMERGENCE OF TB

If tuberculosis has been brought under control in the United States, it has not been eliminated. Some metropolitan centers are seeing shocking numbers of new TB cases. Cities such as New York which combine high rates of IV drug use, alcoholism, homelessness, and poverty offer ideal conditions for TB to infect large numbers of people. These are the same cities that have the highest rates of HIV.

Persons living in substandard, crowded conditions are

more likely to be exposed to the airborne bacteria which causes TB and, owing to poor health, are more likely to develop active disease. The larger the pool of active TB cases, the more likely is transmission.

Infection by Mycobacterium tuberculosis does not automatically cause disease. TB often remains inactive, or latent, causing no symptoms. Experts believe that the lifetime chances of healthy people developing TB after exposure are about 10%. Tuberculosis tends to be much more aggressive in persons whose immune systems have been damaged by HIV or cancer treatments. HIV infection can activate latent TB and, in combination, TB and AIDS take more severe forms. Since Active TB is contagious, cities with large numbers of HIV cases see the most dramatic growth in TB.

While TB can generally be cured, the disease is becoming resistant to drugs which are inexpensive, easy to administer, and have few side effects. One reason for this is that patients most likely to be infected (chemically dependent homeless persons, for example) are the least likely to complete the lengthy course of treatment. Antibiotics help these patients feel better so they stop taking their medicine—long before all the TB germs have been destroyed. When they become ill again they return for more medicine. The result is hardier, more drug-resistant strains of TB. Multi-drug-resistant TB is difficult to treat, no matter how responsible the patient is.

Of the 102 TB cases in Minnesota, 3 were resistant to 1 drug and 1 case was resistant to two drugs. 2 of the 102 cases were HIV+ and neither were drug resistant.[6]

Effective hospital infection control procedures are a critical component of good medical care. In a recent study of a cluster of TB cases, it was found highly probable that TB had been acquired in hospital. The airborne TB germs were transmitted through ventilation ducts during hospitalization. In the same study it was reported that 18% of the health care workers in the hospital tested positive for TB.[7]

AIDS Public Policy in Minnesota
is produced by the Minnesota AIDS Project
2025 Nicollet Avenue South
Minneapolis, MN 55404
612/870-7773

Michael Jefferis, Series Editor
Tom Flynn, Director of Education
Lorraine Teel, Executive Director

Desktop Publishing: Brian Cockayne

Reviewed by: Mandy Carver, Susan Lentz, Scott Mayer, Mary Frances Skala

Supported through a grant from the
MINNESOTA AIDS FUNDING CONSORTIUM

The necessary lab work (culturing colonies of M. tuberculosis) and then testing various drugs on the colonies to determine drug resistance and find an effective treatment can take from a month and a half to 16 weeks. During this period of time ineffective treatments may be in use which further increase TB drug resistance.[8]

THE LAW

Persons with active cases of tuberculosis may become involved in legal action if their behavior can be shown to be a public health threat—for example, TB patients who are infectious and are failing to carry out prescribed treatment. MN Statutes sections 144.4171 - 144.4186 establishes a process beginning with a report to the Commissioner of Health and ending with remedial steps which the Commissioner can pursue in court, ranging from education to involuntary commitment for treatment.

The Commissioner may seek a court-ordered health directive regarding an individual who poses a health threat to others—that is, someone who appears either willing to transmit a communicable disease to another person, or unable to avoid transmission owing to mental incompetence. MN Statute 144.4182 permits emergency proceedings whereby a person posing a health threat can be taken into custody and held for up to 72 hours. The hold can be continued for up to 10 days.

RECOMMENDATIONS

TB is a dangerous disease, and in a mobile society local outbreaks of infection can become national epidemics in a short period of time. Such was the case with HIV. We are thus making recommendations for both local and national action.

In Minnesota:

- Meeting the housing, nutritional, and medical needs of poor and homeless people, chronic alcoholics, drug abusers and HIV infected persons is the first line of defense against the spread of tuberculosis, since these unmet needs will foster the continued spread of TB.

- Health education programs must be implemented to provide information about TB to the general public.

- Persons infected with HIV must receive education about their risk for TB

- The same data privacy standards protecting all Minnesotans should be applied to tuberculosis patients' records. The goal of insuring compliance with medication protocols can best be met by assuring the patient's confidentiality.

- HIV education and AIDS service organizations and TB surveillance and treatment programs should coordinate their efforts to prevent HIV and TB transmission.

RECOMMENDATIONS *cont.*

Nationally

• Tuberculosis surveillance activity should be increased rationally so that early treatment can be commenced before isolated cases develop into clusters and then into general outbreaks of infection. This includes making TB testing available at homeless shelters, community health clinics, etc.

• Immediate, effective, and complete treatment for tuberculosis must be made available through in-patient and/or out-patient treatment programs

• A continuum of intervention should be available for those with active TB who do not comply with treatment. This may include incentives—meals, transportation or pocket money. Court-ordered detention must be reserved as a last resort and only when the patient has adequate legal counsel.

• Infection control procedures which protect patients and health care workers alike must be followed in hospital and clinics. (For example, isolation in rooms with negative air flow; room air exhausted outside; UV radiation; barriers, etc.)

• Medical schools must re-establish tuberculosis as an active specialty area. There are few young researchers entering the field to replace those who have reached retirement age. New drugs, diagnostic tests, and vaccines need to be developed. Research into the genetic mechanisms by which Mycobacterium tuberculosis becomes resistant to drugs is needed.

REFERENCES

[1] B. R. Edlin, et al, New England Journal of Medicine, June 4, 1992, p. 1514; Vol. 326, No. 23.

[2] Lawrence K. Altman, New York Times, February 11, 1992, p. B6.

[3] AIDS Action Briefing: Tuberculosis and HIV: Challenges in Policy and Practice, May 1992.

[4] Charlotte Hagenmiller, Minnesota Department of Health.

[5] op. cit. AIDS Action.

[6] op. cit. Hagenmiller.

[7] op. cit., Edlin, et al.

[8] op. cit., Edlin et al p 1520.

RESOURCES

MAP
Minnesota AIDS Project

AZT

What is it?

AZT (also known as zidovudine, azidothymidine, and Retrovir) is an antiviral that works to slow down the replication of HIV. This compound was developed in 1964 as an anti-cancer drug. AZT did not prove to be very effective as an anti-cancer agent, but the manufacturers continued to make small amounts for research purposes. In 1985, *in vitro* (test-tube) studies showed that AZT was a potent inhibitor of HIV. Human trials confirmed AZT's ability to inhibit HIV and it was soon approved for use in individuals with CD4 counts (T cell counts) less than 200 who had previously had an AIDS-defining opportunistic infection (such as *Pneumocystis carinii* pneumonia) at a dosage of 1,500 mg per day. Because AZT was only being used in people whose immune system was already severely compromised and because the dosage was so high, many individuals experienced severe adverse reactions to AZT, such as bone marrow toxicity. Further studies of AZT showed that lower doses of 500 to 600 mg were just as effective as the old dosage of 1,500 mg and the incidence of side effects was greatly reduced.

It has also been shown that individuals who begin AZT therapy early, with CD4 counts above 200, tolerate therapy much better with toxic side effects appearing in less than 3% of individuals.

How Does it Work?

AZT was the first antiviral approved for use in people with HIV. It does not kill the virus, but rather works by interfering with the virus's ability to integrate into the cell's genetic material (i.e. DNA). This has the effect of decreasing the rate of infection to other cells in the body. Under normal conditions, when HIV enters a cell, it is able to integrate itself into the cell's DNA. Once the virus accomplishes this, it is able to reproduce , making many HIV virions which in turn, can infect other cells. When an individual takes AZT, this process is interrupted. AZT, like ddI and ddC, is known as a nucleoside analog. It works by acting as a decoy for a newly forming chain of DNA. However, this decoy causes the premature termination of the DNA chain, leading to inhibition of reverse transcription. AZT has been shown to block HIV replication in CD4 cells (T cells), monocytes, and macroph-

ages -- all of which are cells that can be infected with HIV. In addition, AZT does cross the blood-brain barrier well which is essential in helping to prevent and fight HIV-related neurological complications.

Who Should Use AZT and How Much?

AZT is officially recommended for HIV infected individuals who have a CD4 count of less than 500, with or without symptoms. Many studies have confirmed added benefits for individuals who begin AZT before symptoms develop in addition to a decreased incidence of side effects in these individuals.
The current dosages recommended are either 500 mg a day or 600 mg a day. Many physicians recommend 500 mg for individuals who are asymptomatic and 600 mg for individuals experiencing HIV-related symptoms. There has been research which suggests a dosage of 300 mg is also effective. However, physicians are often reluctant to prescribe this dosage as it is the very minimum required for effectiveness. An individual who is only taking 300 mg a day absolutely could not miss a single dose or they would not derive any benefit from treatment that day.

What Does the Research Show?

Much research has been conducted on AZT in the past seven years. This section covers some of the larger more recent studies completed. The results presented represent statistically significant findings. For additional study information, you may call the STEP office.

The Multicenter AIDS Cohort Study

In April, 1992, results of the Multicenter AIDS Cohort Study (MACS) were published. This study assessed the survival of over 2,500 men in relation to AZT use and prophylaxis for *Pneumocystis carinii* pneumonia (PCP). The investigators found that the use of AZT before the development of AIDS, regardless of PCP prophylaxis, significantly reduced death in all follow-up periods. According to a press release from the U.S. National Institute of Allergy and Infectious Diseases, "Early treatment with AZT reduced the risk of death at six months by 57 percent, whether or

Seattle Treatment Education Project, 127 Broadway E, Suite 200, Seattle, WA 98102 (206) 329-4857 not it was combined with a drug that prevented PCP. Moreover, investigators found that, after 2 years, early treatment reduced the risk of death by 33 percent, when compared to those who did not take AZT before a diagnosis of AIDS."

European/Australian AZT Study

In February, 1992 the discontinuation of a large European/ Australian study was announced[4]. The study was designed to determine the efficacy of AZT in people with HIV who are asymptomatic. The trial was stopped early after preliminary analysis showed that the incidence of disease progression in individuals receiving AZT was half that of the placebo group. In this study of nearly 1,000 individuals, disease progression was defined as a decrease in CD4 count below 350 mm^3 or the development of clinical symptoms. For those individuals with CD4 counts between 500 and 750, the probability of disease progression was 18% for the placebo group, versus 9% for the treatment group. Disease progression in individuals with CD4 counts between 400 and 500, was 38% for the placebo group versus 20% for those treated with AZT.

Italian AZT Study

In March, 1992, results from an Italian study of AZT were published[1]. This study examined the influence of long-term AZT therapy on survival in individuals with AIDS. 271 people with AIDS were enrolled, 159 of whom received AZT, 112 of whom did not. The overall median survival time for those receiving AZT was 22.1 months, and for those patients who were not treated, the median survival time was 10.6 months. The estimated 1 year survival rate for the AZT treated group was 85%, whereas the estimated 1 year survival rate for the untreated group was 46%. The estimated 2 year survival rate for the AZT treated group was more than twice that of the untreated group.

Australian AZT Study

In a multicenter study of AZT in Australia, the survival of 308 men with AIDS using AZT was compared with the survival of 482 historical controls who did not use AZT. The median survival time after diagnosis with AIDS in the group who received AZT was nearly 3 times higher than the historical control group.

AIDS Clinical Trials Group Studies

The ACTG 016 compared AZT with placebo in 711 mildly symptomatic individuals. 11 out of 260 individuals with CD4

counts between 200 and 500 mm^3 progressed to advanced ARC or AIDS versus 34 out of 253 individuals who received placebo.

The ACTG 019 study compared AZT (either 1,500 mg a day or 500 mg a day) with placebo in 1,338 asymptomatic individuals. In people with CD4 counts below 500 mm^3, 8% of the placebo group progressed to AIDS compared with only 2% of the group receiving 500 mg of AZT, and 3% of the group receiving 1,500 mg AZT.

Concorde Trial

The Concorde trail enrolling 1749 people from 73 sites in England, Ireland, and France, has generated an immense amount of controversy, deliberation, and confusion. The mean length of follow up in this large study was three to four years. Its purpose was to determine the optimal time for initiating antiretroviral therapy with AZT in asymptomatic individuals. That is, is it better to give AZT to people when their CD4 counts fall below 500 (as is the current recommendation in the USA) or is it better to wait until the individual develops symptoms? The dose of AZT used in this study was 1200 mg/day, at least twice the current recommended dosage in the USA. The trail lasted from 10/88 to 10/91 and is the largest and longest trail of AZT use in asymptomatic individuals. Subjects were randomized to receive AZT (immediate group) or placebo (deferred group). The original intent was that individuals in the placebo arm would receive AZT if symptoms developed. However, in the fall of 1989 when the initial results of a portion of the ACTG 019 trail were released, a change was made in the protocol of Concorde. They now gave those in the placebo arm the option of receiving AZT when their CD4 count fell below 500. Of the 872 individuals in the placebo group, 282 (32%) chose to receive AZT at that time. However, due to the "intent-to-treat" analysis used by the Concorde researchers, these 282 individuals who began AZT were considered as belonging to the placebo group when the analysis was done. The interpretation of the data by the Concorde investigators was that there was no difference in the disease progression or survival between the immediate and deferred groups. When CD4 counts instead of clinical end points were used to evaluate the results, the immediate group showed an advantage over the deferred group. The investigators go on to say that CD4 cell counts as marker of clinical disease progression in asymptomatic individuals on AZT monotherapy is unreliable. Their caveat on CD4 counts is that they are probably valuable in natural history studies, but not so reliable in evaluating drug treatment trails, where markers of viral load would be more helpful. The investigators also stated that the results of their trail do not indicate that "early intervention" is bad or harmful, but suggest that they are not sure there is a benefit either. They go on to say that some individuals will benefit from early intervention and others will not. The difficult task is deciding which course any single individual will follow with regards to progression. There has been a large amount of discussion generated from this study. Although additional data was presented in Berlin, there are immunologic and virologic analyses still being performed. The entire study data will appear in publication at some future point. Most major HIV/AIDS treatment newsletters throughout the country have written on this topic including *Treatment Issues, AIDS Treatment News*, and others. Therefore the reader is invited to search out other commentaries on this issue. It was not the purpose of the Concorde to ascertain the efficacy of AZT in symptomatic individuals. The benefit of AZT therapy in this group has clearly been previously established. Also, it was not the purview of Concorde to ascertain the efficacy of AZT in combination with other antivirals in any patient population with regard to disease progression or survival.

Some concerns about this study are as follows. The "intent-to treat" analysis skews the data in favor of no difference between the two groups : any potential benefit which the 282 individuals who received AZT were considered in the placebo group. The dosage of AZT used in this trail could possibly have cytotoxic effect, thereby decreasing a perceived benefit. This point is further made in the following report on ACTG 019 when the effects of 500 mg/day of AZT is compared with 1500 mg/day. It is generally acknowledged that the benefit of AZT monotherapy is limited and decreases over time. Therefore, any benefit from treatment with AZT is going to be seen early on. As the length of time on treatment in creases, either a change in monotherapy or the institution of combination therapy has suggested benefit and has been advocated by some for many years. Seen in this way, the results of Concorde are not all that surprising. Keeping individuals on single nucleoside antiretroviral will no doubt lead to decreased effectiveness of that drug (no matter what it is) over a prolonged period of time such as three to four years. However, the conclusion drawn from the study investigators are in contrast to the five other studies of early AZT use in asymptomatic persons. Studies from Italy, EACG 020 (Europe/Australia), the MACS study in the USA, ACTG 019, and from John Hopkins have concluded that early treatment with AZT is advantageous in survival and/or disease progression. Recent accounts if the "non-latency" of the virus even during the asymptomatic stage, offer the rationale that intervention with drugs which inhibit the infection of new cells (like nucleoside analogues do) are going to be more effective at earlier stages of disease in contrast to later ones. The clinicians I speak with or have heard from , are not making any changes in their recommendations regarding treatment strategies based on Concorde.

In an on going follow up of disease progression in individuals previously enrolled in ACTG 019 with less than 500 CD4 cells, the duration of AZT's efficacy was studied. These are people who initially were randomized to one of three treatment arms: placebo, 500 mg/day AZT, 1500 mg/day AZT. The trail for this subset of participants lasted almost two years until it was stopped in 8/89 and those in the placebo group were offered AZT. Looking at disease progression for a subsequent follow up period of 2.6 years, 232 of 1556 people progressed or died. When considering all 1556 individuals, decreased disease progression is seen in those initially randomized to receiving 500 mg/day of AZT compared to 150 mg/day AZT or placebo. This benefit has not been shown for survival. When these 1556 individuals are broken down into two groups with either more or less 300 CD4 cells at entry, in those with less than 300 CD4 cells, no difference in progression among the three groups is discernible. However, in those with greater than 300 CD4 cells at entry, decreased progression was seen in the 500 mg/day group compared with the 1500 mg/day or placebo groups. These results argue favorably for instituting antiretroviral therapy at early stages of infection rather than later. Dr. Paul Volberding, protocol chair and presenter of this study put forth two hypotheses: in patients with less than 300 CD4 cells, the benefit of AZT may be around 18 months; in patients with 300-500 CD4 cells, the benefit of AZT use could be greater than two years.

Conclusions Drawn From Studies

Conclusions that can be drawn from the above mentioned studies as well as countless others not mentioned are:

*AZT prolongs survival in symptomatic individuals.

*AZT decreases frequency and severity of opportunistic infections.

*AZT delays progression to AIDS.

*AZT delays progression to symptomatic disease.

*AZT improves cognitive or neurologic function.

*AZT improves weight gain.

*AZT increases CD4 cell numbers.

*AZT increase CD8 cell numbers.

----Seattle Treatment Education Project, 127 Broadway E, Suite 200, Seattle, WA 98102 (206) 329-4857----

*AZT decreases serum and cerebrospinal fluid (CSF) p24 antigen levels.

*AZT increase skin-test reactivity.

*AZT increases platelet counts.

*AZT decreases the incidence of AIDS Dementia Complex

*AZT increases quality of life.

Many researchers have proposed that AZT's antiviral actions only comprise about one third of AZT's total benefits. This is why many physicians will not discontinue AZT therapy in individuals who no longer appear to be deriving antiviral benefits from the drug. These physicians will add another antiviral, such as ddI, to the treatment regimen, yet still keep the person on AZT to derive other benefits. For example, AZT improves the response to the pneumococcal vaccine in individuals with HIV infection. *In vitro* studies have also shown that AZT can inhibit some bacteria such as *Salmonella, Escherichia coli, Enterobacter, Shigella*, and others. It has also exhibited inhibitory activity against the Epstein-Barr virus *in vitro*.

What About Side Effects?

Although AZT has been shown to have a lot of benefits for the HIV infected person, it is by no means the perfect drug. AZT's most serious side effects are anemia (decreased red blood cell counts) and neutropenia (decreased number of neutrophils, one type of white blood cell) due to suppression of the bone marrow. In early studies of seriously ill individuals and extremely large doses of AZT (1,500 mg), anemia and/or neutropenia was seen in nearly half of the study participants. Even with the high doses, these side effects were seen much less frequently in individuals without an AIDS diagnosis. Now that individuals are beginning AZT therapy before they are seriously ill and using a much lower dose, the incidence of anemia and neutropenia have drastically decreased to around 3%. For the small percentage of individuals who do develop bone marrow suppression, there are other options available. Many individuals choose to continue taking AZT and take a drug called EPO (erythropoietin) to combat the anemia. Other individuals choose to try decreasing the dosage of AZT. Some individuals will decide to discontinue AZT therapy and try another antiviral approved for individuals who can not tolerate AZT called ddI.

AZT also has other, less serious (but still bothersome) side effects. Many individuals will experience headaches, nausea, hypertension, and a general sense of not feeling well during the first weeks of therapy. These effects are probably a combination of the body adjusting to the drug and the person's anxiety over beginning therapy. However, these effects will gradually subside in nearly all individuals.

After long-term therapy with AZT, a very small percentage of individuals may develop muscle weakness or pain exacerbated by exercise (myopathy or myalgia). The incidence of these side effects has also drastically decreased with the reduction in dose from 1,500 mg to 500 or 600 mg. If myopathy or myalgia does occur it usually goes away once AZT therapy is discontinued or the dosage is reduced. Additionally, myopathies and myalgias are not always caused by AZT and can be due to HIV itself. The underlying cause should first be determined by your health care provider before changes in your dosage are initiated.

A rumor that long-term AZT use causes lymphoma has caused undo worry in some individuals considering AZT therapy. The study which sparked this rumor showed that 8 out of 55 individuals (14.5%) who had used AZT for a median of two years developed lymphoma. Further analysis of this study showed that of the eight individuals with lymphoma, the median CD4 count when they *started* AZT therapy was 26. Their median CD4 count when they developed lymphoma was 6. Even before AIDS, it was a well known fact that immune suppression due to other causes increased the risk of lymphoma. The vast majority of researchers agree that the severe, prolonged immune suppression of the individuals who developed lymphoma in this study is what caused the lymphoma, not long-term AZT use.

Interestingly, in the ACTG 019 study mentioned in the research section, the placebo group reported just as many side effects as the AZT treatment group. This brings up the psychological factor. If a person is reluctant to begin AZT therapy because they believe it is a dangerous drug with horrendous side effects, that person is likely to develop side effects. Likewise, if an individual believes that a placebo is actually a wonder drug, it is likely that person will experience short-term benefits, even increases in CD4 counts. If an individual is considering AZT therapy, try to keep an open mind. Don't anticipate side effects or you will be more likely to experience them.

What about Viral Resistance?

The matter of viral resistance has raised concern in some individuals as to when to start AZT. Laboratory studies have shown that HIV can mutate and new strains of HIV can evolve that are resistant to AZT. Based on these findings, some individuals have argued that AZT therapy should not be instigated early, but rather saved for more advanced disease when the person needs it most. When looked at closely, this view really doesn't hold up and may actually increase the chances of developing resistant strains. First of all, AZT is not the only drug available. If resistance does occur, an individual can always consider using ddI or ddC. Studies have shown that HIV is not "cross-resistant" to these antivirals, meaning that if the HIV strain is resistant to AZT, it will not be resistant to ddI and vice versa. Also, studies have shown that the mutation of HIV which causes AZT resistance is directly proportional to the rate of viral activity. Therefore, if AZT is initiated earlier in the course of HIV, when the viral load is less, the virus may have a more difficult time mutating. Studies have been done to corroborate this theory, reporting that individuals who begin AZT when they are asymptomatic, develop far fewer AZT-resistant strains of HIV than individuals who initiated therapy after symptoms began.

AZT in Combination

For further information on combination therapies, one may call the STEP office and request the combination therapy fact sheet.

*Acyclovir: The combination of AZT and high-dose acyclovir has been studied by a number of different researchers. Unfortunately, the majority of the studies did not show any increased antiviral benefits from the addition of acyclovir. However, if individuals suffer from recurrent herpes outbreaks, acyclovir should be considered to decrease the incidence of outbreaks, thereby decreasing additional immune suppression caused by the herpes viruses.

*Alpha Interferon: Some researchers believe the combination of AZT and alpha interferon may increase antiviral effects. Studies have been done, but there is no definitive answer yet. It appears the added antiviral benefits are minimal, if any, and the combination appears to have synergistic toxicity. Side effects included liver dysfunction, decreased neutrophils and platelets, and fatigue.

*ddC: The combination of AZT and ddC looks very promising. The two drugs appear to work synergistically. This means that the benefits of the drugs used together are greater than simply the sum of their effects. Sort of like adding two plus two and coming up with six instead of four. Because the drugs do not have overlapping side effects, the combination is well tolerated and does not result in additional toxicity. The FDA advisory board has recommended that the combination of AZT and ddC be approved.

*ddI: The combination of AZT and ddI works similar to the combination of AZT and ddC. AZT and ddI also do not have overlapping side effects. ddI is already approved for use in individuals who have failed AZT therapy. Although combination therapy is not one of the official uses for ddI, many physicians are prescribing ddI for use in combination with AZT and insurance companies have been paying for it in many cases.

References:
1. Vella S, et al. Survival of zidovudine-treated patients with AIDS compared with that of contemporary untreated patients. JAMA 1992;267:1232-1236.
2. Swanson CE, Cooper DA. Factors influencing outcome of treatment with zidovudine of patients with AIDS in Australia. AIDS 1990;4(8):749-756.
3. Graham N, et al. The effects on survival of early treatment of human immunodeficiency virus infection. N Engl J Med 1992;326:1037-1041.
4. Burroughs Wellcome News Release, February 3, 1992.

July 1993

Case Documents 2

How Tiny Organisms Helped Clean Prince William Sound

Alaska's Prince William Sound today is essentially clean. There are several reasons for the rapid cleanup of the area, but among the least known and most important factors in the recovery process were microscopic petroleum-eating organisms that were assisted by a process known as bioremediation.

Prince William Sound is home to naturally-occurring single-celled creatures that continuously feed on hydrocarbons from natural petroleum seepages and evergreen droppings. Immediately after the 1989 *Valdez* accident, these micro-organisms increased their numbers to respond to the oil from the spill. However, their ability to "eat" this oil was limited by the available amounts of nitrogens and phosphorus—the basic nutrition the organisms need to multiply and biodegrade oil.

To help the microbes, and enhance their natural cleaning action, scientists from the US Environmental Protection Agency and Exxon applied special fertilizers to oiled shorelines in Prince William Sound. These fertilizers provided the microbes with more nitrogen and phosphorus, allowing them to reproduce and consume oil at a quicker pace. This process of stimulating the microbes is known as bioremediation.

Under typical conditions there are about three million oil-eating micro-organisms in each ounce of sediment in Prince William Sound. After bioremediation the number of microbes rose to as high as 30 million and the oil-eating ability of the organisms increased approximately 10-fold. Repeated tests have shown that in Prince William Sound bioremediation was both effective and safe.

Never before had such a large-scale use been made of the bioremediation process in an oil spill. And it was undertaken in Alaska with extreme care because of concern over potential side effects from the fertilizers, such as growth of algae.

First Large-Scale Use

The Alaska project began in 1989 when a joint scientific panel formed by the US Environmental Protection Agency (EPA) and Exxon recommended that fertilizers containing nitrogen and phosphorus be applied to test areas to determine the feasibility of large-scale bioremediation work. Ordinary garden fertilizers could not be used because tidal action would wash the nitrogen and phosphorus away too quickly, rendering them ineffective. To meet this challenge, the scientists applied two patented products—Inipol EAP22 and Customblen—developed especially for the slow release of nutrients. These compounds are designed to stick to oil and remain there until consumed by microbes.

In 1989, tests on 74 miles of shoreline were so encouraging that bioremediation was used again in 1990 on some 400 patches of oil on shores in Prince William Sound and the Gulf of Alaska. Scientists found that two summers of bioremediation achieved a level of biodegradation that would otherwise have taken many more years.

The Exxon-EPA project provided scientists with new insights and knowledge about bioremediation:

Accomplishments in Alaska

- First, there is a better understanding about the processes through which microbes convert oil and other hydrocarbons into substances such as carbon dioxide, water, and microbial biomass. Scientific data showed that the pace of natural degradation can be accelerated, within limits, by controlling the amount of nutrients.
- Second, laboratory and field tests showed conclusively that bioremediation can be used safely, and that the microbes reduced their number as the oil is eaten.
- Third, the operational techniques pioneered in Alaska have significantly advanced knowledge about how to use bioremediation safely and effectively to treat future oil spills. The Alaska project demonstrated for the first time on an actual spill that accelerated microbial activity can be a major factor in cleansing oil from beaches. In future oil-spill cleanup operations, government and corporate officials can now consider bioremediation as a proven tool to supplement more traditional processes, such as spraying with hot water or the use of sorbents.
- Finally, the Alaska project, which took bioremediation from the experimental stage to proven application for the right conditions,

demonstrated how a large corporation can join with Federal and State agencies to find an effective solution to a complex problem. The US Environmental Protection Agency, Exxon and the Alaska Department of Environmental Conservation managed to launch the Alaska bioremediation project quickly and efficiently.

Questions and Answers

THE PROCESS

What Is Bioremediation?

To understand bioremediation one must first understand biodegradation. Biodegradation is a natural process in which oil hydrocarbons are broken-down, over time, by micro-organisms that then produce simple substances such as carbon dioxide, water, and microbial biomass. Bioremediation is a human effort that quickens this natural process. In some instances, bioremediation means transporting oil-eating microbes to a spill site. In Alaska, however, bioremediation consisted only of feeding nitrogen and phosphorus to the environment's own oil-hungry micro-organisms.

Where Do the Microbes Come from and How Do They Work?

The Gulf of Alaska region, including Prince William Sound, is rich in native oil-eating organisms, which have evolved over millions of years to degrade hydrocarbons produced by evergreen forests and natural petroleum seepage. These organisms are always at work reproducing and converting hydrocarbons into carbon dioxide, water, and microbial biomass.

SAFE SUBSTANCES

Is Bioremediation Safe?

Yes. Extensive studies indicate that bioremediation in Alaska posed no adverse environmental consequences. Although birds and mammals instinctively avoid unnatural substances, scientists made doubly sure that they remained away from newly treated beaches by placing brightly colored balloons and flagged ropes in bioremediation areas. In addition, because application procedures were carefully designed and workers wore protective clothing, there were no health problems for workers involved in the bioremediation process. As expected, field monitoring results show that once the oil is eaten, microbe populations return to pre-spill levels and there are no long-term effects from bioremediation.

What Are the Substances Employed in the Alaska Bioremediation?

The fertilizers used in Alaska are essentially the same as two of the common nutrients used by the home gardener, nitrogen and phosphorus. However, these nutrients were especially prepared to ensure that there would be no environmental damage to the adjacent waters. Inipol EAP22, a sticky liquid fertilizer with the viscosity of honey, contains a water soluble form of nitrogen with an oleophilic form of phosphorus in a biodegradable base. Inipol EAP22 was applied to oiled rock and gravel surfaces. Customblen, a slow-release formulation of soluble nitrients encased in a polymerized vegetable oil with a nitrogen to phosphorus ratio of 28:8, was applied in some areas to reach subsurface deposits through rain and tidal action.

If a Liquid Substance Was Applied, How Do We Know It Wasn't Washing the Oil into the Water and onto the Ground?

Extensive laboratory tests proved conclusively that the liquid fertilizers were sticking to the oil and accelerating the biodegradation process, not washing oil off the rocks and other shoreline surfaces.

THE ROLE OF BIOREMEDIATION

If Bioremediation Is So Safe and Effective, Why Use Anything Else?

Bioremediation works best after bulk oil has been removed by manual and natural processes. Traditional removal methods are more effective for the heavy concentrations of crude oil that are encountered immediately after a spill. Thereafter, bioremediation is effective in removing oil both on the surface and in subsurface sediments.

OUTLOOK FOR 1991

Will Bioremediation Be Used in Prince William Sound This Year?

This spring, Prince William Sound will be examined by a team of scientists representing the US Coast Guard, the Alaska Department of Environmental Conservation, and Exxon to determine shoreline conditions. If the team locates segments that can benefit from bioremediation, nutrients may be applied to those areas. Even without bioremediation, the biodegradation process will continue as a normal and natural part of the ecology of Prince William Sound.

TEST FINDINGS

Scientific Summary

A comprehensive US Environmental Protection Agency/Alaska Department of Environmental Conservation/Exxon monitoring program produced the following conclusions in December 1990:

Bioremediation proved an effective way to accelerate biodegradation of oil both on shoreline surfaces and in those limited areas where oil penetrated beneath the surface. The process was effective in removing residual oil that had been weathering for almost one and one-half years.

Initial bioremediation accelerated the rate of biodegradation by three of four times for both surface oil and subsurface sediments sampled to a depth of 12 inches. Second applications replenished nutrients and accelerated oil removal five to ten fold. Fertilzier reapplication was desirable every 3–5 weeks to sustain and enhance microbial action.

Bioremediation produced no adverse environmental consequences. Even immediately after application, nutrients applied in recommended strength pose no risk.

Even without bioremediation, biodegradation would have proved an effective means of removing oil. But the cleansing process would have taken much longer.

Indigenous Organisms of Alaska

In Prince William Sound, an active community of microorganisms (or heterotrophs) is present at all times. This community has made a significant contribution to the ecology of the region over millions of years.

These organisms are active in cold climates and in both fresh and sea water. Before the spill, scientists estimate that hydrocarbon-degrading microbes constituted one tenth of one percent of the total heterotroph population in Prince William Sound. The introduction of the oil and the fertilizers increased their number so rapidly that they soon represented more than 10 percent of the microbe population.

Special Fertilizers

Two products especially useful for beach areas were used in the Alaska bioremediation program:

Inipol EAP22, manufactured by Elf Aquitaine of France, is classed as an oleophilic or oil-soluble compound which sticks to oil and remains there until consumed by biodegradation, thus preventing nutrients from entering the water. Inipol EAP22 is a microemulsion which contains a water soluble form of nitrogen with an oleophilic form of phosphorus in a biodegradable base. It was developed following the 1978 grounding of the tanker *Amoco Cadiz* on the coast of France and is especially effective in the removal of surface oil.

Customblen, produced by Sierra Chemicals of Milpitas, California, is a granular slow-release formula which encases the nutrients in polymerized vegetable oil. It is similar to fertilizers sold to home gardeners as Osmocote and used on many golf courses. Customblen's nutrients are released gradually during storms and tidal action, making it especially effective feeding oil-eating microbes in subsurface deposits.

Project Monitors

The 1989 bioremediation study was launched by the US Environmental Protection Agency's Office of Research and Development and Exxon under the Federal Technology Transfer Act of 1986. This act encourages collaborative efforts between public agencies and private companies for the economic, social and environmental benefit of the nation.

The EPA and Exxon tested the capacity of bioremediation to treat oiled sections of shoreline. EPA was responsible for oversight and management of the study, while Exxon funded logistical, laboratory, and field activities. In addition, the Alaska Department of Environmental Conservation participated in the planning, operation, and monitoring of the project. Independent organizations were employed to measure the results of bioremediation and ascertain whether there were any adverse environmental consequences.

A REPORT DETAILING THE EFFECTIVENESS AND SAFETY OF BIOREMEDIATION IN PRINCE WILLIAM SOUND

In December, 1990, The Alaska Department of Environmental Conservation (ADEC), The US Environmental Protection Agency (USEPA) and Exxon jointly submitted to the US Coast Guard a report detailing the effectiveness and safety of bioremediation in Prince William Sound. A reprint of the Executive Summary of this report follows. For further information about this report or Exxon's bioremediation work in Alaska, contact R. C. Prince, Ph.D. or R. R. Chianelli, Ph.D., Exxon Research and Engineering Company (201) 730-0100.

Executive Summary*   The joint ADEC/USEPA/Exxon biodegradation monitoring team successfully organized and implemented a comprehensive program for assessing the utility of fertilizer amendments for enhancing the biodegradation of surface and subsurface oil, and for characterizing the associated ecological risks. Interim reports, with analyses of the data then available, were presented in July and September. The Final Report presents all the monitoring data, and reinforces the earlier conclusion that biodegradation is an important mechanism in the removal of oil from shorelines in Prince William Sound, and that the application of fertilizers is a safe and effective means to enhance this natural process.

- Chemical analyses of the remaining oil on the beaches indicates that biodegradation has already removed 10–70% of the oil in individual samples.
- By employing ratios of degradable and undegradable fractions of the oil components, we have derived an estimate of the baseline oil degradation rate. This is approximately
    2.2 g oil/Kg sediment/year on the surface
    1.1 g oil/Kg sediment/year in the subsurface
- The activity of oil-degrading bacteria in surface sediments and subsurface sediments sampled at a depth of 30 cm was enhanced three to four fold relative to the unfertilized sediments. This en-

---

* Reprinted as originally released.

hancement was sustained for 32 days after the initial fertilizer application. A second application replenished nutrients and stimulated relative microbial activities five to ten fold.

- The microbial populations on the fertilized areas have consistently higher numbers of hydrocarbon degrading bacteria than the corresponding unfertilized areas. Adding fertilizer resulted in a five to ten-fold increase in hydrocarbon degraders.
- Fertilizer application resulted in no adverse ecological effects.

This report provides evidence that fertilizer application is an effective means to enhance the activity of oil-degrading bacteria in surface and subsurface sediments with minimal environmental impact. The elevated activity of hydrocarbon-degrading bacteria and the measured increase in their populations together provided convincing evidence that fertilizer application indeed enhances oil degradation. Reapplication of fertilizers is warranted every 3–5 weeks.

BIOREMEDIATION TECHNOLOGY DEVELOPMENT AND APPLICATION TO THE ALASKAN SPILL

R. R. Chianelli, T. Aczel, R. E. Bare, G. N. George,
M. W. Genowitz, M. J. Grossman, C. E. Haith, F. J. Kaiser, R. R. Lessard,
R. Liotta, R. L. Mastracchio, V. Minak-Bernero, R. C. Prince,
W. K. Robbins, E. I. Stiefel, J. B. Wilkinson, and S. M. Hinton
(Exxon Research and Engineering Company)

J. R. Bragg and S. J. McMillen
(Exxon Production Research Company)

R. M. Atlas
(University of Louisville, Louisville, Kentucky)

Introduction: Shortly after the *Exxon Valdez* ran aground on Bligh Reef (March, 1989), a task force was assembled at Exxon Research and Engineering Company to identify novel technologies for oil spill cleanup. One technique identified by this task force for potential application to the cleanup of oiled beaches in Prince William Sound (PWS) and the Gulf of Alaska (GOA) was enhanced bioremediation. Bioremediation is the process in which microorganisms are used to biodegrade oil, and enhanced bioremediation employs application of nitrogen and phosphorous or other nutrients that may be limiting the degradation process.

Biodegradation is a natural process in which bacteria consume petroleum and break it down to biomass and carbon dioxide. The bacteria that consume the petroleum are "hydrocarbon-oxidizers" that destroy petroleum molecules by adding oxygen to them. The newly formed oxygenated molecules are further consumed until only the final products, biomass and carbon dioxide, remain. The process of converting the oil to biomass and carbon dioxide is called mineralization.

Biodegradation, together with physical processes such as evaporation, is a major process by which petroleum is removed from the environment following an oil spill. The role of these processes for a large spill such as that of the *Amoco Cadiz* was described by Gundlach et al. [Gundlach, E. R., P. D. Boehm, M. Marchand, R. M. Atlas, D. M. Ward, D. A.

Wolfe, 1983]. This study showed that biodegradation was acting as rapidly as evaporation even during the first days following the spill. It is thought that the high rate of biodegradation along the coast of Brittany was due to runoff of agricultural fertilizers from farms near the coast. The addition of nitrogen and phosphorus from this agricultural runoff enhanced the biodegradation on the beaches [Atlas, R. M., 1990] presumably by removing the nitrogen and phosphorous limitation resulting from low naturally occurring levels of these elements in sea water.

Although nutrient-enhanced bioremediation had been used extensively for land farming and ground water applications, its use in marine spills had been limited prior to the *Exxon Valdez* oil spill. However, since the large *Amoco Cadiz* oil spill in 1978, efforts have been made to develop enhanced bio-remediation techniques. In 1985, an oleophilic, or "oil-loving", fertilizer INIPOL EAP-22 was applied to a small spill of diesel oil in Ny-Alesund, Spitsbergen, Norway. The results of this application were inconclusive due to the high rate of natural cleaning [Sveum, P., A. Ladousse, 1989]. Nevertheless, extensive laboratory studies suggested that enhanced bioremediation might be applicable to stranded oil on the beaches of PWS and the GOA. This paper reports on laboratory studies of nutrient enhanced bioremediation and the resulting application of nutrients to over seventy miles of beaches in PWS and the GOA in 1989, the largest shoreline bioremediation project ever attempted. Questions that arose in Summer of 1989 were answered in laboratory studies during the Winter of 1989–90, which contributed to the approval of nutrient reapplication in Spring 1990.

EXXON COMPANY, U.S.A.
POST OFFICE BOX 2180⅍HOUSTON, TEXAS 77252-2180

THE PRINCE WILLIAM SOUND
BIOREMEDIATION STORY

The Challenge: Effective cleanup following the 1989 *Valdez* spill.

The Response: After the spill, the **US Environmental Protection Agency (EPA)** approached **Exxon** to discuss ways to clean up the oil. A joint effort between the **EPA, Exxon,** and the **Alaska Department of Environmental Conservation** was organized to test the effectiveness and safety of bioremediation—a process in which nitrogen and phosphorus fertilizers are fed to naturally-occurring, oil-hungry microbes, allowing them to multiply and consume oil faster.

The Application: **Inipol EAP22,** a liquid based oleophilic ("oil-loving") fertilizer was sprayed on selected beach sites in Prince William Sound and the western Gulf of Alaska. Inipol EAP22 sticks to rocks and encourages surface cleaning.

Customblen, a slow-released fertilizer, was applied to encourage subsurface cleaning.

These fertilizers were applied to 74 miles of shoreline in 1989, and on some 400 patches of oil in Prince William Sound and the Gulf of Alaska in 1990. Laboratory and field tests showed conclusively that bioremediation is safe for the environment and effective for oil removal.

The Results: Bioremediation achieved a level of cleanup in Prince William Sound during a 24 month period that would have otherwise taken many more years to accomplish.

Bioremediation

for Shoreline Cleanup Following
the 1989 Alaskan Oil Spill

James R. Bragg
Roger C. Prince
John B. Wilkinson
Ronald M. Atlas

Background on Bioremediation

The Bioremediation Process

In recent years many biological products and application techniques have been proposed for oil spill cleanup and site remediation. Proposals have included a wide range of nutrient products, natural or "bioengineered" microorganisms, bioreactors, or *in-situ* treating conditions. The following terms define the processes as applied in Prince William Sound and discussed in this report:

- *Biodegradation*: the natural process by which microbes (bacteria and fungi) consume hydrocarbons and produce carbon dioxide, water, biomass, and partially oxidized, biologically inert by-products. The process is basically one of oxidation, wherein the bacterial enzymes catalyze the insertion of oxygen into the hydrocarbon so that the

molecule can subsequently be consumed by cellular metabolism. Some molecules are completely degraded to carbon dioxide and water, while others are altered and then incorporated into biomass. The process releases energy used by the microorganism for sustenance and for growth of additional microbial cells.

- *Bioremediation*: the process of accelerating the rate of natural biodegradation of hydrocarbons by adding fertilizer to provide nitrogen and phosphorus, which, following a spill, are usually limited in availability relative to the amount of hydrocarbons (see Figure 1.3).

Bioremediation as used in Prince William Sound and the Gulf of Alaska involved simply adding fertilizers to the beaches to supplement naturally available nitrogen and phosphorus. No trace metals or other micronutrients were included in the fertilizers, and no microorganisms were added. The bacteria involved were strictly native bacteria already present on the Alaskan shorelines.

Microbial Metabolism of Hydrocarbons

Hydrocarbon compounds are abundant in the environment, occurring both naturally and from anthropogenic (man-made) sources, and microorganisms have become adept at utilizing them for food. For example, as shown in Figure 1.4, total annual input of hydrocarbons to the oceans of the world is about 180 million metric tons. Most of this is biogenic (made by living organisms), such as alkanes produced by phytoplankton. The annual contribution to this total from all sources of petroleum hydrocarbons is estimated at 3.2 million tons (National Research Council 1985). Yet, almost regardless of their source, microbial biodegradation naturally consumes these hydrocarbons, maintaining an ecological balance (Atlas 1972, 1981, 1985; National Research Council 1975; Colwell and Walker 1977; Jordan and Payne 1980; Bossert and Bartha 1984; Floodgate 1984).

Hydrocarbon-degrading microbes have been studied extensively, and they have been found in all marine environments sampled (Robertson et al. 1973; Mulkins-Phillips and Stewart 1974; Bunch and Harland 1976; Atlas 1978; Roubal and Atlas 1978;

FIGURE 1.3   Bioremediation is the acceleration of natural biodegradation by addition of fertilizer to bring amounts of nitrogen and phosphorus into balance with the supply of carbon.

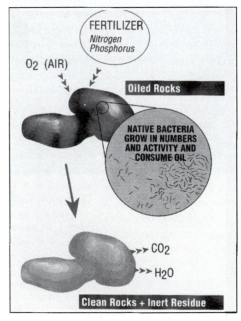

Gunkel et al. 1980). In general, researchers have found that in marine areas far from any source of petroleum hydrocarbons the numbers of oil-degrading organisms are lower than near sites of petroleum discharge. Hydrocarbon-degraders make up less than 0.1% of the total microbial population in areas removed from unusual hydrocarbon sources such as natural submarine seeps or municipal/industrial discharge sites (Atlas 1981; Azoulay et al. 1983), but increase their fraction of the total microbe population to 10% or greater near sources of oil discharge (Mulkins-Phillips and Stewart 1974).

Hundreds of different microbial species have been identified that are capable of metabolizing hydrocarbons, including bacteria, filamentous fungi, yeasts, cyanobacteria, and microalgae. However, each species has specific enzymes that give it the ability to degrade

only a few of the many types of molecules in complex hydrocarbon mixtures, and this has important implications regarding how completely a given mixture of hydrocarbons can be consumed by a given microbial community.

The precise chemistry involved in the metabolic pathways by which microbes cleave the hydrocarbon molecules is varied, complex, and dependent on the specific chemical bonds to be broken and enzymes available to the microbes (Cerniglia 1984; Gibson 1971, 1977; McKenna and Kallio 1965; Perry 1977, 1979, 1984; Pirnik 1977; Ratledge 1978; Singer and Finnerty 1984; Van Der Linden and Thijsse 1965). However, all result in oxidation of all or part of the original hydrocarbon molecule. The original molecule may be broken into a smaller molecule containing one or more oxidized carbon atoms, be incorporated into cell structure, or be mineralized to carbon dioxide and water. Often the smaller molecule is an intermediate species that will be degraded further, but perhaps at a different rate from that of the original molecule.

One of the questions addressed in the Alaskan project was whether the more complex hydrocarbons such as polycyclic aromatic hydrocarbons (PAHs) could be broken down by the indigenous microorganisms. As described later in this report, these microbes demonstrated the ability to break down most types of hydrocarbon molecules present in North Slope crude oil. In large part, the breadth of hydrocarbon degradation realized following the *Exxon Valdez* spill was likely the result of the large diversity of microbial species found in the intertidal zones.

Even when conditions are ideal for microbial growth, not all hydrocarbon components of crude oil will be completely degraded to carbon dioxide and water. The very high molecular weight asphaltenes, resins, and complex aromatics biodegrade slowly (Wong et al. 1984). However, these materials are relatively harmless to the ecosystem because of their low solubility in water and their low biological activity (Hughes and Staffort 1983).

Role of Nitrogen and Phosphorus

Where does the need for nitrogen and phosphorus originate? These elements are vital for all living

FIGURE 1.4 *Most hydrocarbons that enter the marine environment are synthesized by living organisms such as phytoplankton.*

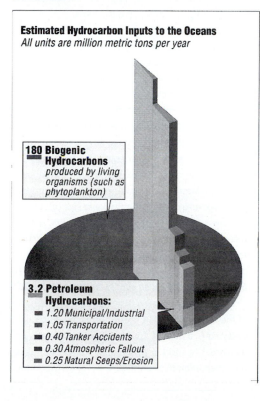

Estimated Hydrocarbon Inputs to the Oceans
All units are million metric tons per year

180 Biogenic Hydrocarbons
produced by living organisms (such as phytoplankton)

3.2 Petroleum Hydrocarbons:
- *1.20 Municipal/Industrial*
- *1.05 Transportation*
- *0.40 Tanker Accidents*
- *0.30 Atmospheric Fallout*
- *0.25 Natural Seeps/Erosion*

things. Nitrogen is an essential constituent of proteins and nucleic acids. Some bacteria can use atmospheric gaseous nitrogen, but most require the nitrogen in forms that are highly water soluble. Ammonia, nitrate, urea, and amino acids are among the most readily assimilable forms for most microorganisms. Phosphorus is an essential component of nucleic acids and the lipids of cell membranes and, as phosphate, plays a central role in biological energy transfer processes. The most readily assimilated form of phosphorus is phosphate, the form used in most fertilizers.

Normally, the beach environment is in relative equilibrium with respect to the amount of hydrocarbon to be degraded, the number of degrading microorganisms, and the amounts of nitrogen and phosphorus needed to balance microbial metabolism. However, after a spill, the large amounts of hydrocarbons introduced as food far outweigh the amounts of nitrogen that are available for balanced cell production. It is not clear whether the availability of phosphorus is also limiting, but it is prudent to assume phosphorus is also limited in supply. Unless more of these nutrients are added, microbial cell growth and the rate of biodegradation of oil are constrained. In the bioremediation process, more nitrogen and phosphorus are added, microbial growth and metabolism increase in response to the added food supply, and the rate of oil degradation increases. In effect, bioremediation is accelerating a natural process.

Under ideal conditions, the added nutrients should be applied so that they are in proximity to the target oil residue and remain there until needed by the microbes. However, in a marine beach environment, many physical processes such as tidal flushing and waves can dilute and diffuse added nutrients, reducing their concentrations below effective levels. The application strategy and fertilizer composition then become important elements for achieving success in bioremediation. For example, a slow-release solid fertilizer makes nutrients available to the microbes over a period of days and weeks rather than all being immediately water soluble and subject to rapid washout. An oleophilic liquid fertilizer adheres to oiled surfaces, providing proximity to the oil and slowly releasing nutrients.

Mass transport processes that control the rate of nutrient dissolution and dispersion are complex in the intertidal environment. Still, the applied nutrients must be able to remain in place and persist over time to be effective.

Role of Oxygen

The fundamental process of hydrocarbon biodegradation is oxidation. Unless adequate oxygen transport is maintained, oxygen availability can become the limiting component, even if sufficient nitrogen and phosphorus are available. In a beach environment, oxygen transport is affected by the permeability and porosity of the sediment and by the frequency and extent of flushing of shoreline sediment by seawater in tides and waves. During rising and falling tides, subsurface sediment receives oxygen dissolved in the seawater, providing the sediment has sufficient permeability to permit significant movement of seawater. During low tides, moist films of water and oil in the sediment are in contact with air, and oxygen diffuses into the films.

In the event that the beach sediment has low permeability, such as low-energy marsh mud flats, oxygen and nutrient transport into the subsurface may be low or negligible (Johnson 1970; Gibbs and Davis 1976). However, in such cases, penetration of oil may also have been insignificant.

For each potential bioremediation project, the availability of oxygen needs to be considered in advance of treatment. As described later, oxygen transport was investigated for the bioremediation project in Alaska and found not to be a rate limiting factor because of the generally large size of the sediment, its high permeability to seawater, and the ample content of dissolved oxygen in seawater flushed through the sediment during each of the two daily tide cycles.

Shoreline Bioremediation prior to the Exxon Valdez Spill

Prior to the *Exxon Valdez* accident, bioremediation of marine oil spills had been studied in laboratory tests and in a limited number of field trials (see, for instance, Lee and Levy 1986; 1989). Interest in using bioremediation to clean marine oil spills was stimulated following the *Amoco Cadiz* spill in 1978, when it became clear that biodegradation of hydrocarbons had played a major role in natural oil removal (Gundlach et al. 1983). There was evidence that fertilizers in water from agricultural runoff had accelerated biodegradation.

Even prior to the *Amoco Cadiz* spill, a pilot bioremediation project was initiated in 1976 at Spitsbergen, Norway, where small amounts of Forcados crude oil were spilled and microbial response monitored (Sendstad et al. 1984). After two years, agricultural fertilizer was added, and the rate of degradation with fertilizer was estimated to be increased about tenfold.

An experiment was conducted on Baffin Island (Sendstad et al. 1982) to determine the effect of mineral fertilizers on the biodegradation of Lago Medio crude oil. Oil was buried under gravel and sand. A rate enhancement of about fivefold was interpreted for fertilized crude samples versus those from untreated areas.

Recognizing the need for a liquid fertilizer that would allow slow release of nitrogen and phosphorus and that would tend to remain with the oil to be degraded, Elf Aquitaine developed Inipol EAP22 (Ladousse and Tramier 1991), which is an oil-external microemulsion that contains nitrogen and phosphorus nutrients. Because it contains an external oil phase, it readily sticks to oil residues and is therefore called an "oleophilic" fertilizer.

This fertilizer was first tested against a water-soluble agricultural fertilizer in an experimental spill in Norway (Sendstad et al. 1984; Halmo 1985). A threefold increase in biodegradation of alkanes was observed for Statfjord oil fertilized with either Inipol or agricultural fertilizer relative to oil from unfertilized areas, but with no statistical difference in rate between the two fertilizers.

Bioremediation was used following an actual spill in November 1985 of about 88,000 liters of marine gas oil on the shoreline in Ny Alesund, Spitsbergen (Sveum 1987; Sveum and Ladousse 1989). Inipol EAP22 was applied the next summer, and extensive monitoring of hydrocarbon levels was conducted. The increase in biodegradation rate of treated oil was about six to nine times higher than for untreated oil.

In summary, available experience suggested that addition of fertilizer might have significant potential to help eliminate oil stranded in intertidal zones. However, because prior tests were limited and results might be site specific, success of bioremediation following the *Exxon Valdez* spill could not be assured without field testing. ❖

Case Documents 3

```
┌─────────────────────────┐
│   Coalition of Citizens  │
│     Concerned with       │
│     Animal Rights        │
│         CCCAR            │
└─────────────────────────┘
```

August 2, 1992

Dr. Karen Wartala
Dean, College of Agriculture
University of the Midwest

Dear M. Wartala:

The Coalition of Citizens Concerned with Animal Rights (CCCAR) was alerted by a concerned person two weeks ago that the University is trapping and killing large numbers of birds in the experimental crop fields on its campus. In our surveillance and investigation we have learned that:

- Birds are lured to the traps by bait placed in them, and the number of birds in these large traps has been observed to be at least 50 at times.
- The reported method of killing the birds was suffocation, by placing them, 400–500 at a time, in a bag.
- Approximately 500 birds are killed every day on the campus; 10,000 were killed in a three month period.

We object to this for the following reasons:

- It is a waste of animal life.
- It is ineffective. Trapping and killing birds will not permanently reduce the bird population in the area. Even *temporarily* it will have no more than a minimal effect.
- The trapping and suffocation of these birds is cruel in the extreme.

For these reasons, **we insist that the University immediately stop the killing of these birds.**

If you wish to discuss this matter, please contact me at 555-5537 or 555-5856.

Ed Younger

Ed Younger
Vice President, CCCAR

cc:
Dr. Pat Lennox, Head
Plant Pathology Dept.
University of the Midwest

President Hans Petersen
University of the Midwest

Lois Waseca, Head
Agronomy and Plant Genetics
University of the Midwest

F. Franklin Jefferson, Director
Agriculture Experiment Station
University of the Midwest

KTTP-TV
WDCO-TV
KNSP-TV
KURE-TV
WDCO-Radio
Minnesota Public Radio
KFCI-Radio
Associated Press
Minnesota Daily
Saint Paul Pioneer Press
Star Tribune
Twin Cities Reader

BIRD CONTROL ON THE EXPERIMENTAL PLOTS UNIVERSITY OF THE MIDWEST

The purpose of our bird control program on the campus plots is to protect the experimental plantings that have been initiated for development of new crop varieties. The seed produced on these plots is extremely valuable because it represents the product of genetic crosses that have been carried out over a period of years. With each plant selection there may be only a few seeds produced so it is essential that they be protected. The new plant varieties that are produced will be used to increase food production on a worldwide scale and, thus, may have a dramatic impact on reducing world hunger. We believe the benefits of these plant breeding experiments are important enough that we must protect these experimental plants from destruction by pest birds.

The campus plots occupy approximately 100 acres. The bird control program on these plots is a comprehensive one that involves several approaches that are being continually improved as we learn about new techniques. They include the following:

Plastic strings are used that when stretched across the plots vibrate to create a noise that frightens birds.

Brightly colored, large balloons with metallic designs (scare-eyes) are suspended above the plots which are known to have a deterrent effect.

A licensed falconer has been engaged who uses trained hawks to frighten birds away from the plots daily.

A program has been initiated to encourage the nesting of sparrow hawks in the plot area as a further means of biological control of birds.

Birds are trapped in large cages baited with bread. The trapped birds are supplied with water and food until they are removed from the cages. There are eight traps lcoated in the plot area. Approximately 150–200 birds in total will be trapped daily on the average. The number of birds per trap will vary a great deal. On unusual days as many as 500 birds may be trapped. The average number of birds trapped per year is usually about 5,000 with a range from year to year of 3,000–10,000. Bird populations are never permanently reduced as a result of the program. Our bird control program attempts to reduce bird populations in the plot areas only during the growing season. All songbirds are released. Only pest birds like blackbirds, starlings and English sparrows are destroyed.

The birds cannot be transported to another site and released because a suitable release site is not available that would be acceptable to the general public. The release of significant concentrations of pest birds is obviously not a desirable outcome to those who would be living in a potential release area and would be irresponsible on our part.

The birds that are destroyed are given to the Raptor Center for feed for birds that are housed there. The Raptor Center uses all the birds that are supplied to them.

Beginning the week of August 6, 1992, trapped pest birds will be euthanatizing with carbon dioxide. Birds will be collected in a cloth bag and kept in a cool place until they will be euthanatized with high concentrations of carbon dioxide in a closed chamber. Collecting and transporting the birds in a cloth bag is more humane than using a small wire cage because the birds remain quiet in the dark bag and do not damage themselves by fluttering about as occurs in a small cage. Euthanatization with carbon dioxide is an accepted, approved method that has been used extensively for small animals and results in loss of consciousness within 45 seconds and respiratory arrest within five minutes. Information obtained from the 1986 Report of the American Veterinary Medical Association Panel on Euthanasia indicates that carbon dioxide is an effective euthanatizing agent. Inhalation of carbon dioxide causes little distress in birds, suppressing nervous activity and inducing death quickly. A further advantage in using carbon dioxide is that the euthanized birds can still be used for food in the Raptor Center.

Placing nets over the plots to prevent bird damage has been suggested. This approach is impractical because of the size of the plot area (100 acres). It would also be impractical because it would restrict management and evaluation of the plots and the nets would be subject to damage by storms.

If there are further questions about our bird control practices call:

Dr. Pat P. Lennox, Department Head
   Department of Plant Pathology   555-8200

Mr. Dave Obi, Research Plot Coordinator
   Department of Plant Pathology   555-3779

```
┌─────────────────────────────┐
│ ┌─────────────────────────┐ │
│ │  Coalition of Citizens  │ │
│ │     Concerned with      │ │
│ │     Animal Rights       │ │
│ │        CCCAR            │ │
│ └─────────────────────────┘ │
└─────────────────────────────┘
```

August 13, 1992

Dr. Karen Wartala
Dean, College of Agriculture
University of the Midwest

Dear Dr. Wartala,

Thank you for providing the Coalition of Citizens Concerned with Animal Rights (CCCAR) representatives with the opportunity to meet with you and other University members regarding the trapping and killing of birds in the experimental crop fields. We feel that a great deal was accomplished in clarifying the issues and look forward to a resolution that will satisfy all concerned.

I would like to review the following conclusions of that meeting. It was agreed that:

- The Department of Plan Pathology commits itself to researching bird deterrent methods with the ultimate goal of eliminating the current method of trapping and killing by next season of late June to mid-August, 1993.
- The Department of Plant Pathology will consult with the attached names and organizations, as well as any other resources at the deparatment's disposal, for information and advice regarding alternatives. Please provide CCCAR with progress reports on or by December 1, March 1, and June 1.
- Several diverse alternative methods will be correctly applied and given an adequate length of time, and monitored and documented to assess their effectiveness.
- CCCAR will be provided with documentation of previous bird statistics from 1955 to present, and will be provided with new statistics at the end of the 1993 season.
- The Department of Plant Pathology will work with CCCAR member Jim Hamar during the remainder of the 1992 season to determine a means by which birds trapped can be released.

Please direct all questions, comments and correspondence to me. Again, please accept our sincere thanks for your cooperation in this matter.

Respectfully,

Malka Rothstein-Abrams

Malka Rothstein-Abrams
CCCAR
(655) 555-0304

cc.—President Hans Petersen, Vice President F. Franklin Jefferson, Dr. R. Lois Waseca, Dr. Pat Lennox, Dave Obi, KTTP-TV, WDCO-TV, KNSP-TV, KURE-TV, WDCO-RADiO, Minnesota Public Radio, KFCI-Radio, Associated Press, *Minnesota Daily, St. Paul Pioneer Press, Star Tribune, Twin Cities Reader*

Bird Control Resources

The Humane Society of the United States   (202) 452-1100
Guy Hodge, Naturalist
2100 L Street NW
Washington, DC 20037

International Alliance for Sustainable Agriculture
Terry Gips, Executive Director   (612) 331-1099
Newman Center, University of Minnesota
1701 University Avenue SE
Minneapolis, MN 55414

Professor Ronald Johnson   (402) 472-6823
202 Natural Resources Hall
University of Nebraska
Lincoln, Nebraska 68583

David Tressemer (Bird Control Consultant)   (303) 449-0486
Boulder Colorado
(has citations on USDA Department of Agricultural Ornithology, c. 1880–1930)

August 22, 1992

Ms. Malka Rothstein-Abrams
Coalition of Citizens Concerned with Animal Rights

Dear Ms. Rothstein-Abrams:

I am pleased that Dave Obi, Richard Jones, Pat Lennox, Donna Harrigan, and I were able to meet with you and the others from the Coalition of Citizens Concerned with Animal Rights (CCCAR) on Friday, August 10 to discuss the bird control program in the crops research plots on the campus. I thought this was a productive meeting in that additional information was provided by both CCCAR and the University, that all agreed on the importance of the crops research and the necessity to control the damage done by the birds, and that some agreements on future actions were made.

I have read your letter of August 13 to me, and want to respond so there will be no misunderstanding of what these agreements were.

- The CCCAR and the University agreed that the crops research is important, that it must be continued, that birds do damage to the research plots, and that this damage must be controlled.
- The University stated that the bird control measures currently being used would continue throughout the remainder of this season, but that between the end of this season and the start of the control period in 1993 the University will explore additional and alternative methods for controlling the bird damage. This does not mean that the Unviersity agrees to change its current program. It means that we will carefully explore valid, reliable, effective, feasible, and humane alternatives, and if one or more can be found that will meet our requirements, they will be adopted.
- The University is willing to work with CCCAR during the remainder of the current season to transport the trapped birds to a release site provided that an acceptable site can be found that meets all community and state regulations.
- The University will make available to CCCAR the bird trapping statistics from the 1950s to the present.

While we did not agree to provide CCCAR with periodic progress reports, we will be happy to keep you informed on the alternative approaches that are being explored, and will inform you prior to the beginning of the 1993 control season of the methods that will be used.

Sincerely,

Karen Wartala

Karen Wartala
Acting Dean

Case Documents 4

**AETNA CASUALTY AND SURETY
COMPANY, a Connecticut Corporation, et
al.,
Plaintiffs-Appellees,
v.
JEPPESEN & COMPANY, a Colorado
Corporation, Defendant-Appellant.**

No. 79-3075.

United States Court of Appeals,
Ninth Circuit.

Argued and Submitted Nov. 10, 1980.

Decided April 20, 1981.

Airline insurer sued publisher of instrument approach charts seeking indemnity for money paid in settlement of wrongful death actions filed by representatives of passengers killed in a plane crash. The United States District Court for the District of Nevada, Peirson M. Hall, Senior District Judge, 463 F.Supp. 94, apportioned damages between the airline and publisher on the basis of a finding that the publisher was 80% at fault and the airline 20% at fault. Publisher appealed. The Court of Appeals, Merrill, Circuit Judge, held that: (1) the district court did not abuse its discretion in denying the publisher's motion for jury trial; (2) the district court's finding that the instrument approach chart was defective was not clearly erroneous; (3) the district court was clearly erroneous in finding the members of the airplane crew to be free from negligence; (4) the district court did not err in determining that, if called upon, the Nevada Supreme Court would take the position of the California Supreme Court and apply principles of comparative fault; and (5) California law apportions indemnity according to the extent that each party's fault contributed to the original accident, even if the original basis for liability of each defendant differed.

Vacated and remanded.

[1] JURY ⇐ 25(4)
230k25(4)
Formerly 170Ak2046

District court did not abuse its discretion in refusing to grant relief from waiver of jury trial where mere inadvertence was excuse offered by counsel. Fed.Rules Civ.Proc. Rule 38(b), 28 U.S.C.A.

[2] JURY ⇐ 25(6)
230k25(6)
Formerly 170Ak2037
District judge's expression of belief that nature of case was such that bench trial would be preferable to jury trial did not amount to statement that case raised only equitable issues, which would not entitle party to jury trial, so as to relieve party from its failure to make timely demand for jury trial. Fed.Rules Civ.Proc. Rule 38(b), 28 U.S.C.A.

[3] AVIATION ⇐ 237
48Bk237
In action against publisher of airport instrument approach chart for damages arising from deaths of passengers killed in airplane crash, district court was not clearly erroneous in finding that chart was defective because graphic depiction of "profile" view of approach, i. e., the presentation of a side view of approach with descending line depicting minimum allowable altitudes as approach progresses, which covered distance of three miles from airport, appeared to be drawn to same scale as graphic depiction of "plan," i. e., a depiction as if one were looking down on the approach segment of flight from directly above, which covered distance of 15 miles.

[4] PRODUCTS LIABILITY ⇐ 11
313Ak11
Under Nevada law, plaintiff can recover for injuries caused by use of product with defective design which makes it unsafe for its intended use, so long as plaintiff is unaware of defect at time of use.

[5] AVIATION ⇐ 237
48Bk237
In action against publisher of airport instrument approach chart for damages arising from deaths of passengers killed in airplane crash, district court was clearly erroneous in finding members of airplane crew to be free from negligence in allowing

themselves to be misled by variance in scale in chart between "plan" and "profile" depictions of approach to airport.

[6] PRODUCTS LIABILITY ⇔ 3
313Ak3

If called upon, Nevada Supreme Court would apply California law and apply principles of comparative fault to products liability suit.

[7] INDEMNITY ⇔ 13.2(2)
208k13.2(2)

California law apportions indemnity according to extent that each party's fault contributed to original accident, even if original basis for liability of each defendant differed.

*340 Philip R. McCowan, San Jose, Cal., argued, for defendant-appellant; Bruce C. Janke, San Jose, Cal., on brief.

Rex A. Jemison, Beckley, Singleton, Delanoy, Las Vegas, Nev., for plaintiffs-appellees.

Appeal from the United States District Court for the District of Nevada.

Before MERRILL and KENNEDY, Circuit Judges, and HARRIS,[FN*] District Judge.

FN* Honorable Oren Harris, Senior United States District Judge of the Eastern District of Arkansas, sitting by designation.

*341 MERRILL, Circuit Judge:

This appeal is taken from judgment granting Aetna Casualty and Surety Company indemnity from Jeppesen & Company for money paid by Aetna in settlement of wrongful death actions filed by representatives of passengers killed in a plane crash. We reverse.

On November 15, 1964, a Bonanza Airlines plane crashed in its approach to Las Vegas, Nevada, on a flight from Phoenix, Arizona. All on board were killed. Wrongful death claims filed on behalf of the passengers were settled by Bonanza, with Aetna as Bonanza's

insurer paying to the extent of Bonanza's coverage.

Jeppesen publishes instrument approach charts to aid pilots in making instrument approaches to airports. Aetna contends that the chart for the Las Vegas Airport was defective, and that product defect was the cause of the crash. Asserting product liability on the part of Jeppesen, it brought this action in the District Court for the District of Nevada as Bonanza's subrogee, seeking to recover from Jeppesen the sums it has paid in settlement of the wrongful death claims.[FN1] Following bench trial, the court found that the chart was defective; that the defect was the proximate cause of the crash; that Bonanza was negligent in failing to discover the defect and alert its pilots; and that the crew members were not negligent in relying on the defective chart. The court apportioned damages between Bonanza and Jeppesen on the basis of its findings of comparative fault: 80 percent to Jeppesen and 20 percent to Bonanza. It is from that judgment that Jeppesen has taken this appeal.

FN1. The same theory of product defect had been the basis of an action brought against Jeppesen by representatives of the deceased crew members. The case was bifurcated and the issue of liability was sent to the jury, which found for plaintiffs. Before the issue of damages was tried, the case was settled. The parties stipulated that the court should vacate the jury verdict as if motion for new trial had been granted. This was done, and the case then was dismissed with prejudice.

1. Jury Trial

[1] Jeppesen first contends that the court abused its discretion in denying motion for jury trial.

Jeppesen was very late in requesting a jury (five years after commencement of the action), giving as explanation that counsel had misunderstood the federal rules. Fed.R.Civ.P. 38(b) requires that a demand for trial by jury on any issue triable as of right be made any time after the commencement of the action but not later than 10 days after service of the last pleading directed to such issue. (By Jeppesen's calculations, it was 50 days late in requesting a jury, counting from the filing of its third amended complaint. By Aetna's calculations, Jeppesen was 22 months late, since the third complaint did not add any new issues which would have reopened the time for making a jury demand.)

Although the trial judge has discretion under Fed.R.Civ.P. 39(b) to grant relief from waiver, this court has held that it is not an abuse of discretion to refuse such relief where mere inadvertence is the excuse offered by tardy counsel. Mardesich v. Marciel, 538 F.2d 848 (9th Cir. 1976). We find no abuse of discretion here.

[2] Jeppesen contends further that denial of the motion was abuse of discretion, since it was predicated on a mistake of law: the judge's belief that the case raised only equitable issues which would not entitle appellant to a jury trial. We do not read the record as indicating that the court believed it was without discretionary power to grant a jury trial, but rather that the nature of the case was such that a bench trial would be preferable to a jury trial. We find no abuse of discretion.

2. Finding Respecting Product Defect

[3] Jeppesen contends that the record does not support the court's finding that the instrument approach chart was defective.

Jeppesen approach charts depict graphically the instrument approach procedure for *342 the particular airport as that procedure has been promulgated by the Federal Aviation Administration (FAA) after testing and administrative approval. The procedure includes all pertinent aspects of the approach such as directional heading, distances, minimum altitudes, turns, radio frequencies and procedures to be followed if an approach is missed. The specifications prescribed are set forth by the FAA in tabular form. Jeppesen acquires this FAA form and portrays the

information therein on a graphic approach chart. This is Jeppesen's "product." The parties do not dispute that the information thus contained in Jeppesen's Las Vegas approach chart is in all respects accurate. The defect, if any, is in the graphic presentation of that information.

Each chart portrays graphically two views of the proper approach. The top portion is the "plan" view, depicted as if one were looking down on the approach segment of the flight from directly above. The bottom portion depicts the "profile" view, presented as a side view of the approach with a descending line depicting the minimum allowable altitudes as the approach progresses. The plan view is regarded as a superior method of presenting course and course changes; the profile view as a superior method of presenting altitude and altitude changes. Each chart thus conveys information in two ways: by words and numbers, and by graphics.

The plan view correctly shows the minimum altitude at a distance of 15 miles from the Las Vegas Airport as 6,000 feet. The profile view does not extend beyond three miles from the airport. Both plan and profile views correctly show the minimum altitude at a distance of three miles from the airport as 3,100 feet. The "defect" in the chart consists of the fact that the graphic depiction of the profile, which covers a distance of three miles from the airport, appears to be drawn to the same scale as the graphic depiction of the plan, which covers a distance of 15 miles. In fact, although the views are the same size, the scale of the plan is five times that of the profile.

Aetna produced as witness an aviation psychologist, who testified that most Jeppesen approach charts have the same or roughly the same scale for both plan and profile views; that a pilot and navigator would come to take this for granted, and, when faced with the Las Vegas chart would assume that the altitude shown on the profile as proper for three miles distant would, reading it as drawn to the same scale as the plan, be proper for 15 miles distant. The theory of Aetna was that the crash was due to pilot reliance on this faulty assumption, invited by the difference in scale. It contends that this difference in scale created a conflict between the information conveyed by the graphics of the chart and that conveyed in words and numbers, and that this conflict rendered the chart defective.

Jeppesen disputed Aetna's claim that it was the custom of the chartmakers to draw the profile and plan views to the same scale. In addition, Jeppesen produced experienced pilots as witnesses who testified that they had never made assumptions such as those attributed to the Bonanza flight crew, and had never heard of any pilots who had. Jeppesen contends that Aetna has failed completely to make out a case of product defect. We cannot agree.

While the information conveyed in words and figures on the Las Vegas approach chart was completely correct, the purpose of the chart was to translate this information into an instantly understandable graphic representation. This was what gave the chart its usefulness this is what the chart contributed to the mere data amassed and promulgated by the FAA. It was reliance on this graphic portrayal that Jeppesen invited.

The trial judge found that the Las Vegas chart "radically departed" from the usual presentation of graphics in the other Jeppesen charts; that the conflict between the information conveyed by words and numbers and the information conveyed by graphics rendered the chart unreasonably dangerous and a defective product.

[4] Under Nevada law a plaintiff can recover for injuries caused by use of a product with a defective design which makes it *343 unsafe for its intended use, so long as the plaintiff is unaware of the defect at the time of use. Worrell v. Barnes, 87 Nev. 204, 207, 484 P.2d 573, 575-76 (1971). See Restatement (Second) of Torts s 402A. On these facts, we conclude that the court's finding that the product was defective is not clearly erroneous.

3. Finding Respecting Crew Negligence

[5] The district court found that the

members of the crew had relied on the graphic portrayal contained in the chart and were misled into assuming that it was safe to fly at an altitude of 3,100 feet 15 miles from the field; that they had acted upon that assumption; and that in making that assumption, which resulted in the fatal crash, they were free from negligence. It is here where we part company with the district court.

To find that Jeppesen's product defect was a proximate cause of the crash we must hypothesize pilot reliance on the graphics of the chart and complete disregard of the words and figures accompanying them. We reject outright a standard of care that would consider such conduct as reasonable attention to duty by a pilot of a passenger plane. The only testimony as to standard of care that of expert pilots (including Aetna's aviation psychologist) is flatly to the contrary. We hold that the district court was clearly erroneous in finding the members of the crew to be free from negligence in allowing themselves to be misled by variance in scale between the plan and the profile. For purposes of apportioning damages, the extent to which that negligence contributed to the crash remains to be decided.

4. Choice of Law

[6] Jeppesen contends that the district court failed to apply the law of Nevada with respect to apportionment of damages. The district court held that there was no applicable Nevada law, and that if presented with the question the Nevada Supreme Court would adopt the position of the California Supreme Court and apply the principles of comparative fault. Jeppesen contends that the case of Reid v. Royal Insurance Co., 80 Nev. 137, 390 P.2d 45 (1964), rejects the principles of comparative negligence and denies indemnity to a party who is himself at fault, however slight that fault may be. We disagree.

Reid purports to deny indemnity only in the circumstances of that case, where the parties were equally at fault and had equal knowledge of the danger and opportunity to guard against it. It does not attempt to state a general indemnity rule nor address the issue whether indemnity should be denied to a party whose fault was a lesser cause of the injury. This court has already noted that Nevada decisions on implied indemnity (including Reid), are less than clear. See Santisteven v. Dow Chemical Co., 506 F.2d 1216, 1218 (9th Cir. 1977). We find no error in the court's choice of law.

5. Apportionment of Damages

[7] The district court apportioned damages on the basis of the comparative fault doctrine adopted in California in Li v. Yellow Cab Co. of California, 13 Cal.3d 804, 532 P.2d 1226, 119 Cal.Rptr. 858 (1975), noting that this doctrine has since been enlarged to apply to product liability cases. Daly v. General Motors Corp., 20 Cal.3d 725, 575 P.2d 1162, 144 Cal.Rptr. 380 (1978). Under this system, a defendant remains strictly liable for injuries caused by a defective product, but plaintiff's recovery is reduced to the extent that its lack of reasonable care contributed to the injury. Daly, supra, 20 Cal.3d at 736-37, at 575 P.2d at 1168, 144 Cal.Rptr. at 386.

In computing damages, the district court applied the comparative fault doctrine by looking to "the consequence of fault on the part of each of the parties and the possible damages which might arise from such fault," thus taking into account the potential for harm that could flow from the conduct of each of the parties. The court regarded the extent of Bonanza's fault in failing to alert its crew members to the variance in scale as limited to this flight alone. Jeppesen, however, was charged with the potential loss of life that might *344 occur on any flight using its charts. Jeppesen asserts that this method of apportioning damages was in error, and we agree.[FN2]

> FN2. Even accepting the court's standard, Bonanza's duty was not limited to the passengers on this flight. It is hardly fair to take potential future harm into consideration in computing Jeppesen's liability but not Bonanza's.

California law apportions indemnity according to the extent that each party's fault contributed to the original accident; this is true even if the original basis for liability of each defendant differed. Safeway Stores, Inc. v. Nest-Kart, 21 Cal.3d 322, 579 P.2d 441, 146 Cal.Rptr. 550 (1978) (negligent defendant found 80 percent at fault while strictly liable defendant found 20 percent at fault; indemnity apportioned accordingly); see also Pan Alaska Fisheries, Inc. v. Marine Constr. & Design Co., 565 F.2d 1129, 1139 (9th Cir. 1977). Inasmuch as the district court properly held that Nevada would follow California law, we believe that Nest-Kart states the rule applicable here. Accordingly, we reverse and remand with instructions to reapportion indemnity under the standards enunciated in Nest-Kart, with the negligence of the plane's crew appropriately considered.

6. Other Issues

We find no support in the record for appellant's claim that the trial judge denied it due process by prejudging the merits of this case, nor do we find that the court gave collateral estoppel effect here to the action brought against Jeppesen by representatives of the deceased crew members. See footnote 1, supra.

Judgment is vacated and the matter remanded for a reapportionment of damages. No costs are awarded.

END OF DOCUMENT

Robert ZIGLAR, Administrator of the
Estate of Lizzie Irene Ziglar, deceased
Plaintiff,

v.

E. I. DU PONT DE NEMOURS AND
COMPANY; Stoney Venable; Midkiff and
Carson
Hardware Company; Basil G. Gordon; and
A. B. Carson, Defendants.

No. 8017SC730.

Court of Appeals of North Carolina.

July 21, 1981.

Administrator of estate of farm laborer who died after ingesting insecticide/nematicide brought products liability suit against retailer and manufacturer of the highly toxic chemical, which was a clear liquid packaged in a translucent one-gallon container similar to a plastic milk jug. The Superior Court, Surry County, W. Douglas Albright, J., entered summary judgment in favor of defendants, and administrator appealed. The Court of Appeals, Vaughn, J., held that: (1) retailer could not be held liable; (2) summary judgment in favor of manufacturer was precluded by existence of genuine issues of material fact concerning adequacy of packaging, adequacy of warnings on container's label, and adequacy of first aid instructions on label; and (3) evidence presented jury question as to whether laborer was contributorily negligent in drinking the liquid, even though she first had to break seal of container and drink what had a distinct odor like rotten eggs.

Affirmed in part and reversed in part.

[1] NEGLIGENCE ⊛ 136(14)
272k136(14)
Usually, it is jury's prerogative to apply standard of reasonable care in a negligence action, and summary judgment is therefore appropriate only in exceptional cases where movant shows that one or more of the essential elements of a claim do not appear in pleadings or proof at the discovery stage of the proceedings.

[1] JUDGMENT ⊛ 180
228k180
Usually, it is jury's prerogative to apply standard of reasonable care in a negligence action, and summary judgment is therefore appropriate only in exceptional cases where movant shows that one or more of the essential elements of a claim do not appear in pleadings or proof at the discovery stage of the proceedings.

[2] PRODUCTS LIABILITY ⊛ 1
313Ak1
When defendants moved for summary judgment in products liability case, they assumed task of demonstrating that plaintiff would be unable to prove at trial sufficient facts to establish following essential elements of products liability action sounding in tort: evidence of standard of care owed by reasonably prudent person in similar circumstances, breach of that standard of care, injury caused directly or proximately by the breach, and loss because of the injury.

[2] JUDGMENT ⊛ 185(2)
228k185(2)
When defendants moved for summary judgment in products liability case, they assumed task of demonstrating that plaintiff would be unable to prove at trial sufficient facts to establish following essential elements of products liability action sounding in tort: evidence of standard of care owed by reasonably prudent person in similar circumstances, breach of that standard of care, injury caused directly or proximately by the breach, and loss because of the injury.

[3] POISONS ⊛ 6
304k6
Retail seller of insecticide/nematicide could not be held liable for failing to give warning to buyer about dangers inherent in using poison which appeared like water and was packaged in a translucent one-gallon container similar to plastic milk jug, in view of plaintiff's failure to show that retailer had reason to know or duty to discover by exercise of reasonable care product-connected danger complained of or that retailer should have known that buyer would not appreciate

possible harm involved in using toxic pesticide which was packaged in clear, plastic container and looked like water.

[4] PRODUCTS LIABILITY ⇐ 14
313Ak14
A retailer must exercise reasonable care in the sale of a dangerous product, and performance of due care necessarily requires him to warn purchaser of any hazard attendant to product's use.

[4] PRODUCTS LIABILITY ⇐ 25
313Ak25
A retailer must exercise reasonable care in the sale of a dangerous product, and performance of due care necessarily requires him to warn purchaser of any hazard attendant to product's use.

[5] PRODUCTS LIABILITY ⇐ 14
313Ak14
Supplier's duty to admonish, with respect to products manufactured by another, arises only if two circumstances simultaneously exist: supplier has actual or constructive knowledge of a particular threatening characteristic of the product and the supplier has reason to know that the purchaser will not realize the product's menacing propensities for himself.

[6] PRODUCTS LIABILITY ⇐ 13
313Ak13
A seller of a product made by a reputable manufacturer, where he acts as a mere conduit, is under no affirmative duty to inspect or test for a latent defect and cannot be held liable for failing to inspect or test in order to discover a defect and give warning concerning it.

[6] PRODUCTS LIABILITY ⇐ 14
313Ak14
A seller of a product made by a reputable manufacturer, where he acts as a mere conduit, is under no affirmative duty to inspect or test for a latent defect and cannot be held liable for failing to inspect or test in order to discover a defect and give warning concerning it.

[7] PRODUCTS LIABILITY ⇐ 13

313Ak13
Restatement of Torts provision that burden on seller of requiring him to inspect chattels which he reasonably believed to be free from hidden danger outweighs magnitude of the risk that a particular chattel may be dangerously defective is particularly sound where product is sold by supplier in its original sealed container.

[8] JUDGMENT ⇐ 181(33)
228k181(33)
In products liability suit brought against manufacturer of insecticide/nematicide, a clear liquid packaged in a translucent one-gallon container similar to a plastic milk jug, summary judgment in favor of manufacturer was precluded by existence of genuine issues of material fact concerning whether manufacturer was negligent in manufacturing inherently dangerous toxic substance without taking reasonable precautions to decrease risk of its lethal confusion with ordinary, harmless drinking water, whether warnings on container were adequate, and whether first aid instructions on container were adequate.

[9] PRODUCTS LIABILITY ⇐ 24
313Ak24
A manufacturer must execute the highest or utmost caution, commensurate with risks of serious harm involved, in the production of a dangerous instrumentality or substance.

[10] PRODUCTS LIABILITY ⇐ 24
313Ak24
Even though a negligence standard is applied in products liability case, manufacturer must be more careful if in the manufacture of dangerous articles for his conduct to be deemed reasonable than would otherwise be necessary in the manufacture of products with less dangerous propensities.

[11] PRODUCTS LIABILITY ⇐ 24
313Ak24
The law requires a manufacturer to eliminate dangerous character of goods to the extent that the exercise of reasonable care, considering all of the circumstances, enables him to do so.

[12] PRODUCTS LIABILITY ☞ 14
313Ak14

A product is defective if it is not accompanied by adequate warnings of the dangers associated with its use, and these warnings must be sufficiently intelligible and prominent to reach and protect all those who may reasonably be expected to come into contact with it.

[13] PRODUCTS LIABILITY ☞ 14
313Ak14

Manufacturer's duty to warn unquestionably requires him to be particularly careful in labeling poisons so they may be properly identified and used.

[14] POISONS ☞ 6
304k6

In products liability suit brought against manufacturer of insecticide/nematicide, a clear liquid packaged in a translucent one-gallon container similar to a plastic milk jug, evidence presented jury question as to whether farm laborer was contributorily negligent in drinking some of the poison after breaking the seal to open the container, even though the liquid had a distinct odor like rotten eggs.

[15] APPEAL AND ERROR ☞ 1073(1)
30k1073(1)

Trial court's refusal to permit introduction of two affidavits on day of hearing on defendants' motions for summary judgment in products liability suit was not prejudicial error, particularly in view of fact that plaintiff suffered no prejudice from admission of material in the record. Rules of Civil Procedure, Rule 56(c), G.S. § 1A-1.

**511 *148 Plaintiff, as administrator, filed a negligence claim against defendants for the **512 wrongful death of Mrs. Lizzie Irene Ziglar, who died shortly after she drank some poisonous insecticide. The court entered summary judgment for two of the defendants, the manufacturer and the retail seller of the insecticide.

The essential facts, as gleaned from the pleadings, interrogatories, depositions and affidavits, are these. In May 1974, Stoney Venable, a tobacco farmer, was using an experimental pesticide, "Vydate L Oxamyl Insecticide/Nematicide," in his tobacco plant beds. Du Pont manufactured the highly toxic chemical, which was, at that time, a clear liquid, with "a terrific odor-like rotten eggs," packaged in a translucent one-gallon container, similar to a plastic milk jug. The warning "Danger-Poison" was printed in red bold-face letters that were approximately 3/17 of an inch in height, on the front panel of the label on the Vydate L container. This warning included the symbol of a red skull and crossbones which was about 4/17 of an inch in height and 4/17 of an inch in width. The back panel of the label provided antidote and first-aid information and again emphasized the words "Danger" and "Poison" in red, bold-face type and included two additional red skulls and crossbones.

On 7 May 1974, Venable purchased another sealed container of Vydate L from a clerk at Midkiff and Carson Hardware Store. The employee who sold the product to Venable did not caution him in any way about the use of the product around humans. Venable placed the chemical in the back of his pickup truck under some old coveralls to keep it from falling over. Later that same *149 day, he put a blue mason jar of iced water and paper cups in the back of his truck, for the refreshment of his field hands and drove to where they were pulling tobacco plants.

Upon arriving there, Venable offered the water to the workers and poured himself a cup from the mason jar. A short while later, Mrs. Ziglar, a laborer, walked over to the truck to get a drink of water. She did not, however, pour from the mason jar; instead, she opened the container of Vydate L, which was located on the same side of the truck as the carton of paper cups, and drank some of the poison. She immediately commented, "(t)his tastes bitter," whereupon Venable jumped up and told her not to drink any more. After some discussion with Mrs. Ziglar, Venable drove her to his house where he gave her some warm salt water, in accordance with the instructions on the Vydate L label. A local doctor administered injections of the antidote to her, but she died shortly after she was admitted to a hospital. Her death was caused by a combination of the ingestion of the poisonous chemical and sickle cell disease in crisis.

Plaintiff now appeals from the entry of summary judgment on his negligence claims against Du Pont and the hardware store.

David B. Hough, Winston Salem, for plaintiff-appellant.

Smith, Moore, Smith, Schell & Hunter by J. Donald Cowan, Jr., Greensboro, for defendant-appellee, E. I. Du Pont De Nemours and Co.

William G. Reid, Pilot Mountain, for defendant-appellees, Midkiff and Carson Hardware Store, Basil G. Gordon and A. B. Carson.

VAUGHN, Judge.

Though this appeal is interlocutory because the judgment entered did not adjudicate all of the claims in the case or dispose of the cause as to all of the parties, Bailey v. Gooding, 301 N.C. 205, 270 S.E.2d 431 (1980), we have elected, in our discretion, to treat the "appeal" as a petition for a writ of certiorari and shall proceed to address the merits of the case. G.S. 7A-32(c); App.R. 21(a).

*150 The sole issue is whether the manufacturer and retail seller of an inherently dangerous toxic substance were entitled to summary judgment on plaintiff's products liability claims.[FN1]

FN1. We note at the outset that this litigation was pending prior to 1 October 1979, the effective date of the new products liability act in Chapter 99B of the General Statutes. The statute does not, therefore, apply to the instant case, and we express no opinion as to whether its provisions might require a different analysis or result than that rendered herein. See generally Blanchard and Abrams, North Carolina's New Products Liability Act: A Critical Analysis, 16

Wake Forest L.Rev. 171 (1980).

**513 [1][2] It is elemental that it is usually the jury's prerogative to apply the standard of reasonable care in a negligence action, and summary judgment is, therefore, appropriate only in exceptional cases where the movant shows that one or more of the essential elements of the claim do not appear in the pleadings or proof at the discovery stage of the proceedings. Ragland v. Moore, 299 N.C. 360, 261 S.E.2d 666 (1980). See, e. g., Moore v. Fieldcrest Mills, Inc., 296 N.C. 467, 251 S.E.2d 419 (1979); Strickland v. Dri-Spray Division Development, 51 N.C.App. 57, 275 S.E.2d 503 (1981). Consequently, when defendants moved for summary judgment in the instant case, they assumed the task of demonstrating that plaintiff would be unable to prove at trial sufficient facts to establish the following essential elements of a products liability action sounding in tort: "(1) evidence of a standard of care owed by the reasonably prudent person in similar circumstances; (2) breach of that standard of care; (3) injury caused directly or proximately by the breach, and; (4) loss because of the injury." City of Thomasville v. Lease-Afex, Inc., 300 N.C. 651, 656, 268 S.E.2d 190, 194 (1980) (citing Prosser, Handbook of the Law of Torts s 30 (4th ed. 1971)). We hold that the defendant retail seller of the insecticide, Midkiff and Carson Hardware Store, has met this burden with respect to plaintiff's negligence claim against it and affirm the summary judgment entered in its favor. Nonetheless, we reverse the order of summary judgment for the defendant manufacturer because plaintiff did establish, by a forecast of his own evidence, the necessary elements of a products liability claim against Du Pont on several theories.

[3] The sum and substance of plaintiff's claim against the defendant Hardware is that it was negligent due to its "abject failure to give any warning whatsoever" to the purchaser, farmer *151 Venable, about "the dangers inherent in using a poison which appears like water in a plastic beverage jug." [FN2] We disagree.

FN2. Plaintiff also alleged in the

complaint that the Hardware was negligent because it sold the poison: (a) in a clear liquid form when it should have known that it was available in an amber color and (b) in a container which had no safety devices to prevent ingestion by humans. In his brief, however, plaintiff has relied on a single basis to show defendant's negligence: its failure to warn Venable verbally about the dangerous possibility that someone might mistakenly drink Vydate L as water. Accordingly, that is the sole issue we address in determining whether the trial court properly entered summary judgment for the hardware store.

[4][5] It is indeed true that a retail seller must exercise reasonable care in the sale of a dangerous product and that the performance of due care necessarily requires him to warn the purchaser of any hazard attendant to the product's use. Restatement (Second) of Torts s 401 (1965). See Wyatt v. Equipment Co., 253 N.C. 355, 117 S.E.2d 21 (1960). The supplier's duty to admonish, with respect to products manufactured by another, however, only arises if two circumstances simultaneously exist: (1) the supplier has actual or constructive knowledge of a particular threatening characteristic of the product and (2) the supplier has reason to know that the purchaser will not realize the product's menacing propensities for himself. Stegall v. Oil Co., 260 N.C. 459, 133 S.E.2d 138 (1963); Restatement, supra, s 388. See generally Annot., "Manufacturer's or seller's duty to give warning regarding product as affecting his liability for product-caused injury," 76 A.L.R.2d 9 (1961). Neither circumstance appears on this record for the following reasons.

[6][7] First, plaintiff did not present any specific facts, as opposed to mere general allegations, in response to defendant's motion for summary judgment, which tended to show that the Hardware knew or should have known that the manufacturer's written **514 warnings on the product's label were inadequate to warn others, who could be expected to come into contact with the Vydate L, of its poisonous character. For example, plaintiff might have asserted the Hardware's actual or constructive knowledge about the defective nature of the manufacturer's warnings by showing that other customers had complained about Vydate L's dangerous propensity for being confused with water, that it had received special instructions from the manufacturer regarding this danger, or that the manufacturer had notified it that Vydate L was now available in a safer form, with amber coloration. See, e. g., Wilson *152 v. Chemical Co., 281 N.C. 506, 189 S.E.2d 221 (1972). In the absence of facts similar to these, we must conclude that any insufficiency in the manufacturer's warnings, in light of the poison's colorless form and packaging in a translucent container, constituted a hidden defect which the Hardware had no duty to detect or remedy. [FN3] For, it is well-established that a seller of a product made by a reputable manufacturer, where he acts as a "mere conduit," [FN4] "is under no affirmative duty to inspect or test for a latent defect, and therefore, liability cannot be based on a failure to inspect or test in order to discover such defect and warn against it." 2 Frumer and Friedman, Products Liability s 18.03(1)(a) (1979); Cockerham v. Ward, 44 N.C.App. 615, 262 S.E.2d 651, review denied, 300 N.C. 195, 269 S.E.2d 622 (1980) (affirming the entry of summary judgment for the seller in a products liability case). See Restatement (Second) of Torts s 402, Comment d (1965), which explains that "(t)he burden on the seller of requiring him to inspect chattels which he reasonably believes to be free from hidden danger outweighs the magnitude of the risk that a particular chattel may be dangerously defective." This rule is particularly sound where, as here, the product is sold by the supplier in its original, sealed container. See Davis v. Siloo, Inc., 47 N.C.App. 237, 267 S.E.2d 354, review denied, 301 N.C. 234, 273 S.E.2d 444 (1980) (affirming the dismissal of negligence claims against the distributors of a toxic substance); 63 Am.Jur.2d, Products Liability s 40, at 51 (1972). See also G.S. 99B-2(a). Thus, plaintiff has failed to show the first prerequisite to a retail seller's duty to warn: that the Hardware had reason to know about, or a legal duty to discover by the exercise of

reasonable care, the product-connected danger complained of.

> FN3. The manufacturer's compliance with the minimum statutory labeling requirements for toxic pesticides under federal and state law further supports the conclusion, that in these circumstances, any deficiency in the written warnings on the Vydate L label was not reasonably discoverable by the retail seller. See 7 U.S.C. s 136(q)(2)(D) and G.S. 143-443(a)(3) (mandating the use of skull and crossbones, the display of the word "poison" prominently in red on a contrasting background, and the inclusion of antidote information).

> FN4. There is no evidence in this record that the defendant Hardware did anything more than simply serve as a "middleman" between the manufacturer and a willing purchaser in an ordinary commercial sale.

Second, plaintiff has also not demonstrated that the Hardware should have known that the purchaser, Venable, would not *153 appreciate the possible harm involved in using a toxic pesticide which was packaged in a clear, plastic container and looked like water. In the instant case, Venable testified that he frequently administered toxic chemicals in his professional pursuit of farming "to make it pay off." At a minimum then, he should have been generally aware of the dangers involved in using pesticides and the special need to store such substances carefully. In addition, however, Venable also had reason to know of the peculiar dangers associated with Vydate L, for he said that when he purchased the second container of the poison on 7 May 1974, he had previously read the manufacturer's label and warnings, had also read about the product in an agricultural bulletin and understood it was an experimental pesticide. These facts persuade us that any tendency of Vydate L to be mistaken for harmless water should have been plainly observable to Venable, a professional user of toxic substances, which thereby obviated **515 any obligation of the Hardware to warn him

further. It is manifest that a retail seller has no duty to warn of an obvious hazardous condition which a "mere casual looking over will disclose." Restatement (Second) of Torts s 388, Comment k (1965); Annot., supra, 76 A.L.R.2d 9, 28 (1961). Moreover, there is simply no compelling reason to require a seller "to warn a person who in his occupation or profession regularly uses the product against any risk that should be known to such a regular user." 63 Am.Jur.2d, Products Liability s 51, at 61 (1972).

We thus hold that no legal duty of the retail seller to warn the purchaser was triggered in this case as a matter of law. In sum, the Hardware did not violate any standard of reasonable care by failing to give verbal warnings, about the myriad circumstances in which Vydate L might be confused with drinking water, in addition to the general warnings provided by Du Pont on the sealed container's label. In this situation then, the Hardware has plainly shown that an essential element of plaintiff's negligence claim is missing defendant's breach of due care, the absence of which properly authorized the judge to enter summary judgment in its favor.

[8] On the other hand, however, the defendant manufacturer, Du Pont, did not successfully negate the existence of an essential element of plaintiff's negligence claim against it. Viewing all the evidence in this record in the light most favorable to plaintiff *154 with the benefit of every reasonable inference arising therefrom, Page v. Sloan, 281 N.C. 697, 706, 190 S.E.2d 189, 194 (1972), we hold that genuine issues of material fact, concerning the reasonableness of Du Pont's conduct, were raised on three bases.

[9][10] The first basis of plaintiff's products liability claim is that Du Pont did not exercise the required degree of due care in its general manufacture and packaging of Vydate L. A manufacturer must execute the "highest" or "utmost" caution, commensurate with the risks of serious harm involved, in the production of a dangerous instrumentality or substance.[FN5] See Davis v. Siloo, Inc., 47

N.C.App. 237, 267 S.E.2d 354, review denied, 301 N.C. 234, 273 S.E.2d 444 (1980); see also Luttrell v. Mineral Co., 220 N.C. 782, 18 S.E.2d 412 (1942). Here, a jury question was raised about whether Du Pont was sufficiently cautious in the production of Vydate L by plaintiff's prima facie showing that: (a) Du Pont knew that the insecticide was a dangerous substance; (b) Du Pont manufactured the highly toxic chemical as a colorless liquid and packaged it in clear plastic jugs; and (c) Vydate L, in this form, could be easily mistaken for water in its appearance. See Restatement (Second) of Torts s 388 (1965). In this regard, the relevant facts are as follows.

> FN5. This standard of care is not to be confused with strict liability which is not recognized in this State in products liability cases. Smith v. Fiber Controls Corp., 300 N.C. 669, 268 S.E.2d 504 (1980). Simply put, even though a negligence standard is applied, a manufacturer must be more careful in the manufacture of dangerous articles for his conduct to be deemed reasonable than would otherwise be necessary in the manufacture of products with less dangerous propensities. See also Cockerham v. Ward, 44 N.C.App. 615, 619, 262 S.E.2d 651, 654 (1980), where the Court explained: "a manufacturer is not an insurer of the safety of products designed and manufactured by him, but is under an obligation to those who use his product to exercise that degree of care in its design and manufacture which a reasonable prudent man would use in similar circumstances."

Du Pont noted, in its own information bulletin about the product, that Vydate formulations were highly toxic and that its exposure to humans should be avoided. Du Pont also admitted, in its answers to plaintiff's interrogatories, that the insecticide was a clear liquid in a translucent container during its experimental marketing in 1973 and 1974, and Venable said he purchased the product in this form in May 1974. Dr. Modesto Scharyj, an expert witness in pathology,

testified that Vydate L was a "completely colorless chemical; and for that reason it was easy for (him) to *155 understand how (decedent) could possibly confuse that chemical with water," at least while the container was sealed. The affidavits of decedent's **516 fellow laborers, as well as the deposition of her employer, moreover, all tend to support the conclusion that she drank the poisonous chemical intending to refresh her thirst, with some water, as she had previously been advised.

[11] The law requires a manufacturer to eliminate the dangerous character of goods to the extent that the exercise of reasonable care, considering all of the circumstances, enables him to do so. See Cashwell v. Bottling Works, 174 N.C. 324, 93 S.E. 901 (1917). It is not without significance, therefore, that Du Pont began bottling Vydate L in gray, opaque containers, on 24 May 1974, shortly after this tragic accident occurred, as requested by the State of North Carolina, and that it added amber coloration to the colorless poison in January 1975. Thus, on this record, a critical factual issue, and one not susceptible to disposition by summary judgment, was whether Du Pont was negligent in manufacturing an inherently dangerous toxic substance without taking reasonable precautions to decrease the risk of its lethal confusion with ordinary, harmless drinking water.

[12][13] Another basis of plaintiff's claim is that Du Pont did not provide the kinds of warnings on the product's label which were reasonably necessary to notify persons of Vydate L's poisonous character, especially in light of the chemical's marked resemblance to water. It is well-established that a product is defective if it is not accompanied by adequate warnings of the dangers associated with its use and that these warnings must be sufficiently intelligible and prominent to reach and protect all those who may reasonably be expected to come into contact with it. Prosser, Handbook of the Law of Torts s 96, 99 (4th ed. 1971); see Corprew v. Chemical Corp., 271 N.C. 485, 157 S.E.2d 98 (1967). The manufacturer's duty to warn

unquestionably requires him to be particularly careful in labeling poisons so they may be properly identified and used. See Fowler v. General Electric Co., 40 N.C.App. 301, 307, 252 S.E.2d 862, 866 (1979); Epstein, Products Liability: The Search for the Middle Ground, 56 N.C.L.Rev. 643, 653 (1978). While it is true that Du Pont fulfilled the applicable statutory requirements for the labeling of a poisonous insecticide (see note 3, supra) and that Vydate L had a "slight *156 sulfurous" odor,[FN6] we cannot say that such warnings were entirely adequate as a matter of law. Rather, such facts were but a part of the total circumstances to be weighed and considered. Indeed, the following facts tended to show that Du Pont had not taken every reasonable precaution to admonish against the risk of confusion of the chemical with a drinkable beverage: (1) again, the insecticide was distinctly similar to water in appearance; (2) there was no evidence that decedent could read or write; and (3) the skull and crossbones symbols on the label were small only 4/17 of an inch in height and 4/17 of an inch in width. Further, we would note that it should not have been unforeseeable to Du Pont that Vydate L would be used in close proximity to farm laborers, who might be illiterate, since it intended the insecticide "to be used mainly as a spray or transplant water treatment" on tobacco, which is generally known to be a labor-intensive crop.

FN6. We agree with plaintiff that at least a question of fact is raised as "to whether or not an odor (of the insecticide) in an open tobacco field on a hot and muggy day suffices as a warning of a dangerous poison to a farm worker who has labored and sweated under the hot sun for several hours."

Two decisions of this Court are particularly instructive here: Davis v. Siloo, Inc., 47 N.C.App. 237, 267 S.E.2d 354, review denied, 301 N.C. 234, 273 S.E.2d 444 (1980), and Whitley v. Cubberly, 24 N.C.App. 204, 210 S.E.2d 289 (1974). In Davis, the decedent was killed by aplastic anemia resulting from exposure to Petisol 202 in the course of his employment. The plaintiff administratrix filed a negligence claim against the manufacturer, alleging, among other things, that the label on the product's container inadequately admonished the user to avoid prolonged skin contact with the chemical. The Court affirmed the denial of **517 defendant's motion to dismiss the negligence claim and held, in pertinent part, that:

"the manufacturer of the dangerous substance will be subject to liability under a negligence theory for damages which proximately result from the failure to provide adequate warnings as to the product's dangerous propensities which are known or which by the exercise of care commensurate with the danger should be known by the manufacturer, or from the failure to provide adequate directions for the foreseeable user as to how the dangerous product should or should not be used with respect to foreseeable uses."

*157 47 N.C.App. at 245-46, 267 S.E.2d at 359. In Whitley, the plaintiff's intestate also died from aplastic anemia after the administration of a certain drug duly prescribed by her physician. Two of the bases of the administrator's claim against the drug manufacturer were that it improperly marketed and over-promoted the drug and failed to provide sufficient warnings about the drug's dangerous tendencies to the medical profession, as well as consumers thereof. This Court held that such allegations raised genuine factual issues and therefore reversed the trial judge's order of summary judgment for the defendant manufacturer. Significantly, the Court further stated:

"That Parke, Davis may have fully complied with all applicable Federal laws in its marketing and labeling Chloromycetin would not in itself free it of liability for harm caused by use of the drug if it were shown that such use and resulting harm was caused by the company's negligent acts in overpromoting the drug, the dangerous properties of which it was aware or in the exercise of due care should have been aware."

24 N.C.App. at 207, 210 S.E.2d at 292. These two cases provide authoritative support for our holding that plaintiff's allegations regarding the inadequacy of Du Pont's warnings raised factual issues, legally sufficient to withstand a

motion for summary judgment, because defendant did not come forward with "uncontradicted evidentiary material to show that it was not negligent" in this respect. Whitley, supra. We likewise do not believe that Du Pont's compliance with all statutory labeling requirements would necessarily exonerate it from liability for its possibly negligent acts in failing to take greater steps, i. e., using more prominent written warnings and symbols on the product's label, to prevent this toxic colorless liquid from being mistaken for water.

Plaintiff presented ample evidence to support its negligence claim on another competent ground: that the product's label was further deficient because the first-aid instructions listed thereon were not clear or complete. Those instructions provided, in part, as follows: "If swallowed, give a tablespoon of salt in a glass of warm water and repeat until vomit fluid is clear." Stoney Venable testified that, almost immediately after decedent drank some of the poison, he read the label on the jug and noted the foregoing *158 advice. He then got decedent to accompany him in his truck to go to his house to get some warm salt water. This took approximately eight minutes. Decedent became unconscious a short while later. Plaintiff contends that Du Pont's emphasis on the use of warm salt water was misleading because it incorrectly focused attention "on obtaining salt water rather than on inducing regurgitation" and that this improper focus on the actual remedy to be pursued "precipitated a time-consuming rush in search of salt water and caused the quick induction of vomiting to be overlooked altogether." This contention is well-supported by the following facts. First, Du Pont admitted, in its own answers to interrogatories, that it was "desirable to induce vomiting by whatever means is immediately available in order to remove as much of the substance as possible from the body." In addition, Du Pont clearly stated in its information about Vydate L which was sent to poison control centers across the country that the first-aid treatment for ingestion was to "(i)nduce emesis or perform gastric lavage." Moreover, Du Pont advised, in a 1977 publication of information **518 for physicians regarding the symptomatology and treatment of cases involving Vydate L, that an appropriate first-aid step to be taken before the arrival of a physician was to "drink 1 or 2 glasses of water and induce vomiting by touching back of throat with finger or blunt object." We believe that such facts raised a substantial question as to whether Du Pont was negligent in not instructing more plainly, on the product's label itself, that, in cases of accidental ingestion, vomiting should be immediately induced by whatever means available.[FN7]

> FN7. Many situations could arise where, as here, a glass of warm salt water is not instantly obtainable as compared with the almost constant availability of a finger or two.

Plaintiff's evidence challenging the completeness of the label's antidote information was as follows. The label advised: "Atropine sulfate should be used for treatment. Administer repeated doses, 1.2 to 2.0 mg intravenously every 10 to 30 minutes until full atropinization is achieved." In its 1977 information bulletin to physicians about Vydate L poisoning, however, Du Pont distinguished between the procedures to be followed in cases of mild or severe intoxication. The directions in the bulletin for mild intoxication were the same as those printed on the product's label in 1974, but a different treatment was advised for *159 severe intoxication: an initial intravenous dosage of 2 to 4 mg to be repeated every three to ten minutes. Du Pont offered no evidence showing why it did not provide the additional antidote information for cases of severe poisoning, or why it was not negligence for it to fail to do so, on the product's label during its experimental marketing. Du Pont thus failed to negate plaintiff's claim that it was negligent for only providing adequate antidote instructions for cases of mild poisoning alone. Moreover, the following facts clearly demonstrate that plaintiff's claim was well-substantiated. Venable testified that, after attempting the warm salt water treatment at his house, he drove decedent to a physician's office, and when they arrived there, she was already unconscious and "slumped over" in the back of his truck. Venable took the Vydate L jug into the office and explained what had happened. The doctor then went out to the truck and gave her a shot. The rescue squad was called, and she was taken to the hospital, which was some twenty to twenty-five minutes away. From these facts, it would be reasonable to infer: (1) the doctor read the antidote information on the jug's label and administered one injection of atropine in the amount suggested, which was only sufficient for mild intoxication; (2) that decedent was suffering from severe intoxication of the chemical when the antidote was given since she was unconscious at the time; (3) that if the physician had read about the different treatment for severe poisoning on the label, he would have administered greater quantities of atropine and instructed the rescue squad team to give her additional shots every three to ten minutes during the twenty-five minute trip to the hospital; and (4) that decedent's death might have been prevented if she had received larger amounts of atropine.

Plaintiff also alleged that Du Pont was negligent because it did not exercise reasonable care in disseminating medical information about Vydate L to poison control centers from the outset of the product's experimental marketing in 1973 and 1974. The record indicates that an index card, including treatment information for Vydate L poisoning, was not filed with the National Clearinghouse for Poison Control Centers until August 1974, three months after this fatal accident occurred. These facts, standing alone, would not, however, support plaintiff's products liability claim because, even if Du Pont did not exercise due care in disseminating this information earlier, there is no evidence *160 anywhere in this record tending to establish a causal connection between such an omission of care and the resulting injury.[FN8]

FN8. Plaintiff might have established the essential causal link of such a claim by showing, for example, that either the initial treating physician, or the hospital physicians, had attempted, in their efforts to revive decedent, to locate this medical information by contacting the nearest poison control center.

**519 [14] In sum, we hold that plaintiff substantiated a products liability claim against the defendant manufacturer on three grounds: (1) its negligent manufacture and packaging of Vydate L; (2) its failure to provide adequate warnings on the product's label to notify others of its toxicity; and (3) its negligent provision of ambiguous and incomplete first-aid instructions on the label.[FN9] We further hold that the defense of contributory negligence was not established in this case as a matter of law. "(P)roximate cause is ordinarily a question of fact for the jury, to be solved by the exercise of good common sense in the consideration of the evidence of each particular case." Prosser, Handbook of the Law of Torts s 45, at 290 (4th ed. 1971); see Williams v. Power & Light Co., 296 N.C. 400, 403, 250 S.E.2d 255, 258 (1979). Though it may be true, as Du Pont contends, that decedent should have known that this poisonous liquid was not water in a beverage jug because she first had to break the seal of the container, and the liquid had a distinct odor like rotten eggs, such factual occurrences were merely a part of the total circumstances to be considered by a jury in deciding whether decedent's death was caused by her omission of due care for her own safety or by the manufacturer's failure to exercise reasonable care in the production of an inherently dangerous substance. See Hale v. Power Co., 40 N.C.App. 202, 252 S.E.2d 265, review denied, 297 N.C. 452, 256 S.E.2d 805 (1979).

FN9. Though the statute does not apply to this litigation, see note 1, supra, we would comment that our holding is consistent with the tenor of the new products liability act which recognizes claims for personal injury or death resulting from "the manufacture, construction, design, formulation, development of standards, preparation, processing, assembly, testing, listing, certifying, warning, instructing, marketing, selling, advertising, packaging or labeling of any product." G.S. 99B-1(3).

[15] Plaintiff also assigned as error the trial court's refusal to permit the introduction of two affidavits on the day of the hearing on defendants' motions for summary judgment. Rule 56 of the Rules of Civil Procedure plainly provides that, on a motion for summary **161** judgment, the adverse party may serve opposing affidavits "prior to the day of hearing." G.S. 1A-1, Rule 56(c). Thus, it would seem that the judge did not abuse his discretion in denying plaintiff's request to introduce the affidavits; however, for purposes of this case, it suffices to say that plaintiff has suffered no prejudice from the omission of this material from the record for two reasons. First, neither affidavit contained facts which would further support the existence of a duty to warn by the retail seller, and, thus, even if such evidence should have been admitted and considered, it would not alter our affirmance of the judgment entered in the defendant Hardware's favor. Second, while the evidence in these affidavits did tend to enhance plaintiff's negligence claim against the defendant manufacturer, any consideration of the propriety of the judge's exclusion thereof, is rendered unnecessary by our decision reversing the judgment entered for Du Pont on the record as it now stands. Plaintiff will presumably have, therefore, the opportunity to present and develop this evidence, as he wishes, at the full trial of the matter.

The order of summary judgment entered for Midkiff and Carson Hardware Store is affirmed.

The order of summary judgment entered for E. I. Du Pont De Nemours and Company is reversed.

Affirmed in part; reversed in part.

WELLS and BECTON, JJ., concur.

END OF DOCUMENT

B

Diagnostic Test

PART ONE: GRAMMAR

Parts of Speech

In the space to the left of each of the sentences below, write one of the following numbers to identify the part of speech of the underlined word:

1. noun	4. adjective	7. conjunctive adverb
2. pronoun	5. adverb	8. conjunction
3. verb	6. preposition	9. article

_____ 1. Do technical communicators have <u>any</u> ways to improve their writing?

_____ 2. Mary arrived at the meeting early <u>and</u> greeted the later arrivals with enthusiasm.

_____ 3. Hail <u>appears</u> to be a part of every April in the Midwest.

_____ 4. <u>That</u> driver is a menace to civilization.

_____ 5. The baseball season <u>opener</u> was marred by a ticket mixup.

_____ 6. Putting up a fence along the <u>outside</u> of the property required a building permit.

_____ 7. To create <u>the</u> most effective documents, you have to plan carefully.

_____ 8. Since he was bitten by that cat, he hasn't been feeling <u>well</u>.

_____ 9. <u>Those</u> people are the best neighbors one could have.

_____ 10. I had plenty of notice about the meeting change; <u>therefore</u>, I will be able to attend.

Fragments, Comma Splices, and Run-On Sentences

Determine whether the examples below contain a fragment, a comma splice, a run-on sentence, or whether they are correct. In the space on the left, write one of the following letters to identify the error. If there is no error write **C** for correct. Correct the error above the example.

1. F (Fragment)
2. CS (Comma Splice)
3. RO (Run-on)
4. C (Correct)

_____ 11. Even though I always have the best intentions.

_____ 12. In preparation for the opening day of fishing season, they cleaned and sorted all of their gear, examining every lure for the smallest flaw or spot of rust.

_____ 13. To be an excellent communicator, one must plan documents carefully and then one must follow through on all details of writing, editing, and proofreading.

_____ 14. One candy bar is often good, two candy bars is often better.

_____ 15. Some days everything goes wrong from the early morning on. Including not being able to find a place to park and being out of coffee.

Dangling Modifiers

One sentence in each pair contains a dangling modifier. Underline the error and, in the space at the left, write the letter (a or b) that identifies the **correct** sentence.

_____ 16. a. After accepting the job offer, the boss congratulated her new employee.
      b. After accepting the job offer, the new employee was congratulated by the boss.

_____ 17. a. Deciding to adopt a child, the parents enthusiastically selected a little girl.
      b. Deciding to adopt a child, the little girl was enthusiastically selected by the parents.

_____ 18. a. To meet government standards, some specifications were drawn up.
      b. To meet government standards, we drew up some specifications.

_____ 19. a. Although nearly 400 years old, we didn't think that the ruins were in bad shape.
      b. Although nearly 400 years old, the ruins didn't seem in bad shape.

_____ 20. a. Opening the window to let out a huge bumblebee, the car accidentally swerved into an oncoming car.

b. Opening the widow to let out a huge bumblebee, the driver accidentally swerved into an oncoming car.

_____ 21. a. Taking our seats, the lecture began.

b. The lecture began as we took our seats.

_____ 22. a. Eating popcorn and drinking beer, we enjoyed the evening.

b. Eating popcorn and drinking beer, the evening passed pleasantly.

_____ 23. a. To write well, a writer must take care with grammar.

b. To write effectively, good grammar is essential.

_____ 24. a. Having been sworn in, the lawyer questioned the witness.

b. The witness, having been sworn in, was questioned by the lawyer.

_____ 25. a. Ready to set up camp, we were hit by a storm.

b. Ready to set up camp, the storm hit.

Pronouns

Whenever you find an incorrect form of a pronoun, underline it and write the correct form in the space at the left. If there are no errors, write **C**.

_____ 26. These clothes don't belong to you or I. How did they get here?

_____ 27. She doesn't deserve the promotion any more than me.

_____ 28. Theirs only one way to do this job, he said vehemently.

_____ 29. Who do I love? Let me consider the possibilities.

_____ 30. The chef at the Peking Garden, who we think makes the best dim sum in the Twin Cities, is considering moving from the area.

_____ 31. Will you be using some of your's or will you rent dishes and tables for the party?

_____ 32. Its a matter of principle whether or not the dog sleeps in it's bed or on our bed.

_____ 33. Whom did you select as your delegate to the meeting?

_____ 34. Who's rules will we follow for the card game—Hoyle's or the house's?

_____ 35. Most of the professors think the scholarship should go to some young student whom everyone agrees is seriously committed to a career in technical communication.

Pronoun Agreement

In the space to the left of each sentence, copy the correct form from within the parentheses.

_____ 36. The credit union sends a statement to (its, their) members every other month.

_____ 37. The man who grafted the new apple on the dwarf stock really knew what (he was, they were) doing.

_____ 38. Everyone in the women's caucus cast (her, their) vote for Mary.

_____ 39. Each of the men going through the rehabilitation program had to learn to respect (himself, theirself, themselves).

_____ 40. Neither Patty nor Peggy wanted to bring (her, their) children to work that day.

_____ 41. The rabbis who gathered for the hearing began (its, their) deliberations immediately.

_____ 42. Professor Walzer tells anyone interested in technical communication what (he or she, you, they) must do to succeed.

_____ 43. Nobody in (his or her, their) right mind should ever travel abroad without the appropriate immunizations.

_____ 44. Kathy always blames someone else for her errors; she never accepts responsibility for (it, them).

_____ 45. Not one of the students was willing to agree that (he or she, they) was responsible for the group's failure.

Subject-Verb Agreement

In the space at the left of each sentence, copy the correct form from within the parentheses.

_____ 46. The game warden made sure each of the fish (was, were) legal sized.

_____ 47. Low on the list of priorities (is, are) the creation of more Ph.D. programs.

_____ 48. There (is, are) many indicators of economic health in the country.

_____ 49. The reason for his unusual actions (is, are) a variety of childhood afflictions.

_____ 50. None of the secretaries in any of the College's ten departments (have, has) been promoted in the last four years.

_____ 51. The doctor or the nurse practitioner (remains, remain) with the patients until they are transferred out of Emergency.

PART TWO: USAGE

Usage

In the space to the left of the sentence, write the number of the expression from within the parentheses that is preferable according to careful, conservative language users.

_____ 52., 53. Of all the (1. medium, 2. media) used in business, television is the (1. medium, 2. media) that has been underused.

_____ 54., 55. When I (1. proceeded, 2. preceded) with my plan to enroll in medical school, I thought of all the women who had (1. proceeded, 2. preceded) me.

_____ 56. His kindness (1. complements, 2. compliments) her financial acuity.

_____ 57. The longer we talked, the (1. farther, 2. further) we were from reaching a consensus.

_____ 58. Shoppers with (1. fewer, 2. less) than 10 items in their carts can go through the express checkout.

_____ 59. Gee, I (1. could have, 2. could of) had a V-8™!

_____ 60. When she moved to this country, she quickly got tired of the (1. continual, 2. continuous) solitication calls on the phone as soon as she sat down to dinner.

_____ 61. The shirt had so much starch that its (1. course, 2. coarse) texture rubbed the skin off his neck.

Usage

From each pair of words, underline the word that matches the synonyms listed next to it.

62. Imply/Infer—to draw a conclusion about something.
63. Cite/Sited/Sighted—to acknowledge an authority or a source.
64. Disinterested/Uninterested)—being impartial or objective.
65. Accept/Except—excluding or omitting.
66. Eminent/Imminent—outstanding or distinguished.

PART THREE: PUNCTUATION

Commas

Insert commas where they are needed in the following sentences. Then, before each sentence, write one of the following numbers to indicate the rule that governs the punctuation.

1. non-restrictive clause or phrase   2. an appositive
3. a noun in direct address   4. a parenthetical element
5. a speaker in dialogue   6. an absolute phrase
                7. correct: no comma needed

_____ 67. Dr. Gross who is an authority on the rhetoric of science is the coordinator of the program in scientific and technical communication.

_____ 68. A small chunk of its hide having been torn off by the barbed wire the dog lay calmly in the sun for a day until it began to heal.

_____ 69. My bank account I must admit could use a serious infusion of cash.

_____ 70. All of the students who attended the white water canoeing weekend and practiced what they had learned ended up in the water at least once nevertheless.

_____ 71. The part that failed, the O-ring was worth very little money in itself.

_____ 72. The hook having been removed from the steelhead the large fish turned slowly up the stream and moved with dignity up the river.

_____ 73. His golf game not to exaggerate at the least presents a danger to anyone within half a mile.

_____ 74. After seventeen hours on the plane and another two in the car we arrived at our final destination, Tokyo.

_____ 75. My tennis racket having outlasted my knees looks barely used.

Apostrophes, Colons, Commas, Dashes, End Marks and Quotation Marks

In each of the spaces at the left, write **C** if the punctuation is correct or **W** if it is incorrect. Within the incorrect sentences, correct the faulty punctuation by adding, rewriting, or changing marks. *Make no unnecessary changes.*

    76. Because Cindy was the only professor in the department who won the teaching award she deserved the large cash award she received.

    77. The twins room was so tidy that visitors were astonished.

    78. We stayed within our budget all year, however at Chanukah we spent more than we planned when we bought gifts for the grandchildren.

    79. His approach to life was marked by three beliefs; work hard, treat people fairly, and take time to have some fun.

    80. The raptor tried it's wings once and lifted itself into the stiff breeze off the lake.

    81. We did everything we could to get rid of crabgrass but short of uprooting the entire lawn our yard will never look like the neighbors.

    82. Their idea of decorations consisted of stringing vines around the room and then covering them with little twinkling lights.

    83. The new parents bought the child a plush tan and white teddy bear with bright eyes.

    84. Our plan leaves promptly on June 20 1994 at five oclock.

    85. I've always loved cross-country skiing moreover, the snow in Upper Michigan is so deep that even falling generally isn't too painful.

    86. The teachers rule was clearly stated on her syllabus ''Late papers are not accepted.''

    87. When I got older and had a family of my own I realized how much it cost to feed six children.

    88. For the party Pam brought raspberry cheesecake a chocolate cake and fresh fruit because none of us would have dared to compete with her on making deserts..

    89. Rather than giving up, the child kept trying to ride his bike without training wheels.

    90. Being a technical communicator is an excellent profession for one gets paid well and is confident that he or she is making life easier and safer for others.

Subordinating, Coordinating, and Emphasizing Information

Rewriting the following sentences so that the ideas are cordinated, subordinated, or given proper emphasis. Create a **single,** grammatically correct sentence. You might have to eliminate a vague pronoun reference or use subordinating or coordinating conjunctions to accomplish your goal.

1. HyperCard is another application to use in designing hypertext. It is a product of Apple Corporation. (Emphasize the company whose product it is.)

   _____

   _____

2. There are many things to consider when visiting Russia. You have to bring a raincoat, comfortable shoes, and clothes that you can wash by hand. You have to get appropriate shots. You need to bring sufficient money in small bills that are easy to exchange.

   _____

   _____

3. We were fishing on Lake Superior, five miles out and near the Huron Islands, when a sudden and violent storm came in from the West. (Emphasize the storm.)

   _____

   _____

4. A new way to develop mechanical interfaces such as toggles and knobs was discovered. The company made a lot of money from this process. The new engineer invented it. (Emphasize the engineer.)

   _____

   _____

5. There are national and international standards for creating safe and effective warning labels. (Emphasize the standards.)

   _____

   _____

Subject-Verb Agreement

_____ 1. A so-called labor saving device in the kitchen is often one of those technological inventions that always (seem, seems) to take more time to clean up after than to use.

_____ 2. Denny, and my other brothers, (refuses, refuse) to spend time helping with the family business.

_____ 3. (Has, Have) one or the other of you thought of running for public office?

_____ 4. There (appear, appears) to be agreement that bioremediation and mechanical cleaning devices will both be needed to reduce the effects of the oil spill.

_____ 5. The national effort to create a digital information service (is, are) hampered by the failure of many in government to understand its potential uses.

Dangling Modifiers

One sentence in each pair contains a dangling modifier. Underline the error and, in the space at the left, write the letter (a or b) that identifies the **correct** sentence.

_____ 1. a. Eating our popcorn, the movie began.
      b. Eating our popcorn, we watched the movie begin.

_____ 2. a. After driving all day, the motel was a welcome sight.
      b. After driving all day, we found the motel a welcome sight.

_____ 3. a. Although only a first-year student, Joe was already attended advanced physics classes.
      b. Although only a first-year student, advanced physics courses were already attended by Joe.

_____ 4. a. Having sworn to uphold the law, the police officer arrested her next door neighbor.
      b. Having sworn to uphold the law, the neighbor was arrested by the police officer next door.

_____ 5. a. Ready to move into the new house, we found the roof leaking.
      b. Ready to move into the new house, the roof was leaking

Usage

Underline the correct word from the choices offered in parentheses.

_____ 1. Are you feeling (1. all right? 2. alright?)

_____ 2. When I want (1. advice 2. advise), I'll pay an expert to give it to me.

_____ 3. She never was willing to (1. agree to 2. agree with) my plan.

_____ 4. (1. Irregardless, 2. Regardless) of your feelings on the matter, you must do the legally correct thing

_____ 5. All of the other kids could throw the ball (1. farther, 2. further) than our youngest son.

Commas

In the following sentences, place commas where they belong and write an explanation for your decisions.

_____ 1. The weather this spring day I believe is the most pleasant so far this year.

_____ 2. When you read Julie falls asleep immediately.

_____ 3. Mike who holds down two jobs and finds himself perpetually short on sleep has decided to return to college.

_____ 4. The pike was fully two-feet long give or take an inch or two.

_____ 5. Stella's farmhouse which is located in Baraga on 100 acres of trees is filled with her paintings and homemade quilts.

Punctuating Clauses and Phrases

Each of the following sentences contains one adjectival clause or a participial phrase. Underline the clause or phrase. Insert commas where they are needed. In the space at the left of each sentence, write one of the following:

R (if the clause or phrase is restrictive)
N (if the clause or phrase is non-restrictive)

_____ 1. Minnesota's legislature which has been dominated by the Democrat/Farm/Labor party is in conflict often with its Independent Republican governor.

_____ 2. Mary who is an executive director of a social service agency often takes Friday afternoon off to volunteer at Shalom Home.

_____ 3. Mr. McMillan only used military time in the computer lab which had the stripped down look of an army barracks.

_____ 4. Jim is a man who needs no introduction.

_____ 5. Those of you who think that writing grammar examples is easy should try it.

_____ 6. The toggle that is used to adjust the cabin lights often stuck in the on position.

ADDITIONAL SOURCES OF INFORMATION

Gibaldi, Joseph, and Walter S. Achtert. *MLA Handbook for Writers of Research Papers.* 3rd ed. New York: MLA, 1988.

Hacker, Diana. *A Writer's Reference.* 2nd ed. Boston: Bedford Books of St. Martin's Press, 1992.

Lunsford, Andrea, and Robert Conners. *The St. Martin's Handbook.* 2nd ed. New York: St. Martin's Press, 1992.

Rude, Carolyn D. *Technical Editing.* Belmont, CA: Wadsworth, 1991.

Index